Reservoir Fisheries of India

by
V.V. Sugunan
Central Inland Capture Fisheries Research Institute
Barrackpore, India

Reprinted by arrangement with the
Food and Agriculture Organization of the United Nations
by
Daya Publishing House®
A Division of
Astral International Pvt. Ltd.
New Delhi - 110 002

First Indian Edition, 1997
Reprinted, 2016
ISBN: 978-81-7035-198-6

Published by : **Daya Publishing House®**
A Division of
Astral International Pvt. Ltd.
– ISO 9001:2008 Certified Company –
4760-61/23, Ansari Road, Darya Ganj
New Delhi-110 002
Ph. 011-43549197, 23278134
E-mail: info@astralint.com
Website: www.astralint.com

Printed at : **Replika Press Pvt. Ltd.**

PREPARATION OF THIS DOCUMENT

This is the first comprehensive review of reservoir fisheries of India, based on the analysis of published and unpublished documents. Its preparation is in response to the recognition by the Indo-Pacific Fisheries Commission Working Party of Experts on Inland Fisheries of the need to assess the potential for culture-based fisheries enhancement of reservoirs in the region. The review should serve as a baseline for any future studies as it will allow fisheries specialists and administrators to evaluate the impacts of future culture-enhancements on reservoir fisheries production and yields. The study may also serve as an example for similar studies in the region, either planned or already in preparation. A study on Sri Lanka reservoir fisheries (by S.S. DeSilva) was published in 1988 (FAO Fisheries Technical Paper No. 298), and Reservoir fisheries in the countries of the Commonwealth of Independent States (by E.I. Karpova, Isaev, A.I. and Petr, T.) will appear in 1995, also in the FAO series of Fisheries Technical Papers.

Distribution

FAO Fisheries Department
FAO Regional Fisheries Officers
Directors of Fisheries, Indian States
FAO Representatives in Asia

Sugunan, V.V.
Reservoir fisheries of India.
FAO Fisheries Technical Paper. No. 345. Rome, FAO. 1995. 423 p.

ABSTRACT

The existing literature on limnology and fisheries of Indian reservoirs has been reviewed by covering more than 100 reservoirs located in various parts of the country. An assessment of environment-mediated production functions of reservoirs has been attempted. Since the ecosystem processes in reservoirs belonging to different geo-climatic regions exhibit wide variations depending on meteorological, morphometric and hydro-edaphic features of the impoundments, an effort has been made to gauge the influence of these abiotic variables on the production dynamics. Authentic information on water areas under different categories of reservoirs has been collected and interpreted in respect of all the Indian States. An attempt has also been made to resolve the anomalies pertaining to classification and nomenclature to the extent possible. Various fisheries management norms followed in the reservoirs of the country including the selection of species for stocking, stocking rate and introduction of exotic species have been reviewed. Indian reservoirs have been stocked with the Indo-Gangetic carps for many decades and the impact of this stocking has been assessed in terms of fish production and the indigenous faunistic diversity. Different modes of exploitation and the types of craft and gear employed in different reservoirs of the country are discussed. The reservoir fisheries resources and their utilization are dealt with separately in respect of each State along with a few case studies describing the water quality, biotic communities, craft and gear, and fisheries management practices followed. The status of reservoir fisheries in the States has been reviewed and the factors responsible for the slow rate of development identified.

CONTENTS

INTRODUCTION

Harnessing the rivers for irrigation and hydro-electric power generation has been the main focus of developmental activities in India ever since the country gained independence. Consequently, a number of small, medium and large river valley projects have come into existence during the last four and a half decades with the primary objectives of storing the river water for irrigation, power generation and a host of other activities. One of the direct results of these projects is the creation of a chain of man-made lakes, dotting the Indian landscape from Kashmir to Kanyakumari and Bengal to Gujarat. That the man-made lakes hold tremendous potential for inland fisheries development in India has long been recognised. However, this vital resource is not contributing to the inland fish production of the country to the extent it should. Unlike the rivers, which are under the increasing threat of environmental degradation, the reservoirs offer ample scope for fish yield optimisation through adoption of suitable management norms. Any attempt to increase productivity in inland fisheries has to rely heavily on the reservoirs. Again, the sheer magnitude of the resource makes it possible to secure a substantial increase in production by effecting even a moderate improvement in the yield rate.

The fish production from inland waters increased from 0.22 million t in 1951-52 to 1.7 million t in 1991-92. Despite this impressive growth rate and a matching contribution from the marine sector, the per capita availability of fish in the country remains at 3.2 kg, against a world average of 12.1 kg. In the freshwater sector, comprising both inland fisheries and aquaculture, the country has targeted a production of 4.5 million t by 2000 AD. The projections in aquaculture sub-sector are not very rosy. The current level of utilization of culturable water bodies is 22% (Sarma, 1990) and it has been estimated that even at an ambitious 50% level of utilization of culturable water bodies by the turn of the century, the contribution from freshwater aquaculture would not exceed 1.24 million t (Jhingran, 1990). Inevitably, the country has to depend heavily on the inland capture fisheries resources, among which the reservoirs constitute the mainstay.

In the developed world, fisheries of inland lakes largely cater to the recreational needs, whereas in a highly populous developing country like India, these resources can play a vital role in augmenting food production for human consumption and mitigating the protein deficiency. Nearly 70% of the 0.71 million fishermen in India fish in rivers, reservoirs, lakes and other inland waters. On account of underemployment, these inland fishermen live in abject poverty and constitute one of the most vulnerable sections of the rural society. Compared to the capital-intensive aquaculture enterprises, where a good part of the profit goes to the investors, development of open water fisheries is highly labour-intensive, having the potential to provide gainful employment to the weaker sections.

Notwithstanding the overriding importance of reservoirs, documentation of this vital resource is grossly inadequate. Dearth of a firm database and lack of requisite research support are often cited as the main inhibiting factors for speedy development of reservoir fisheries in India. A reliable estimate on the fish production from the man-made lakes is still eluding us due to a number of reasons. The common property nature of the resource and the varying modes of exploitation followed by different State Governments make the centralised data collection a difficult job. Moreover, the reservoirs and their fish landing centres are highly scattered, remote and often inaccessible. The problems are further confounded by the involvement of unauthorised money lenders in the fish trade, who do not part with the catch figures due to obvious reasons. Surprisingly, no serious efforts have been made to build up a database on the reservoir fisheries resources of India, taking into account the total water area under different categories of man-made lakes, their biogenic production propensites and fish yield potential.

The biological potential of the various reservoirs was not evaluated to any reliable extent until the 1970s, when fish yield from them stood at low level of 5 to 8 kg ha^{-1} y^{-1}. The investigations conducted till then were isolated attempts to unravel the ecology of individual reservoirs in one or two States. Organised research on reservoir fisheries was initiated in India only in the year 1971 with the launching of an All India Coordinated Project on Reservoir Fisheries (AICPRF). Five large reservoirs, representing distinct geoclimatic regions of the country were under its fold. The Project attempted to delve into all determinants of reservoir productivity, including climatic, morphometric and edaphic variables and the dynamics of biotic communities of the reservoir ecosystem in India. Consequently, limnology has come to be used as a mold to cast location-specific management strategy for reservoirs. The limnology-mediated management norms evolved by the AICPRF have come to be known as the ecosystem-oriented management (Natarajan, 1979). Application of these norms have resulted in a remarkable increase in fish yield in some of the Indian reservoirs. A three-pronged strategy comprising selection of appropriate mesh size of the fishing gear, increased effort and stocking support has paid rich dividends in Bhavanisagar and Gobindsagar, where the annual fish yields increased from 25.56 to 80.5 kg ha^{-1} and 25.0 to 76.5 kg ha^{-1} during the period from early 1970s to mid-1980s. The average fish yield in the country also registered an increase up to 15 kg ha^{-1} by the mid 1980s. The present yield is estimated at 20 kg ha^{-1}.

As a sequel to increased productivity, a study on levels of employment and income in respect of 7 reservoirs was also taken up by CIFRI in 1984, so as to corroborate the findings in terms of increased income of fishermen. Further, it was opined that fishing operations in reservoirs should be remunerative enough to sustain fishermen who toiled on waters. Even apprehensions were voiced that stagnant levels of productivity might lead to occupational shift in favour of other unskilled vocations. Therefore, it becomes imperative to have a contingent plan to enhance productivity at least in those reservoirs whose productive potential has been well-researched.

An objective assessment of the fish production potential of Indian reservoirs has to take into account the diverse geo-climatic regions of the country, in which the water bodies are situated. Productivity of reservoirs depends, to a large extent, on the synergetic effects of a number of geochemical, meteorological, morphometric and hydrographic variables and the biotic communities present in the ecosystem. The biomass and species number of various communities show wide variations within the country, depending on the micro- and macro environment. Although Indian rivers are known for their rich fish genetic resources, the composition of fish stock in man-made lakes is subject to a series of faunistic changes brought about during the impoundment. The fish fauna is further altered through stocking and introduction of exotic species. Thus, a database on abiotic and biotic parameters, and the fish stock of the reservoir ecosystem is an essential prerequisite for a meaningful management of the resource.

Fish yield optimisation from man-made lakes has become the focus of attention in recent years and consequently, many development projects have been initiated at the State and national level to exploit the fish production potential of this resource. While the reservoir fisheries in India is well-poised for a steady growth, this document attempts to meet the long-felt need for a comprehensive status paper on the reservoir fisheries in India. This desk review is an endeavour to compile systematically all relevant information on Indian reservoirs from a fisheries perspective. The existing literature on limnology and fisheries has been reviewed, covering more than 100 reservoirs located at different parts of the country. An assessment of environment-mediated production functions of reservoirs has also been attempted. Authentic information on water areas under different categories of reservoirs has been collected and interpreted for the first time in respect of all the Indian States. In the process, the anomalies pertaining to the classification and nomenclature of reservoirs have been resolved to the extent possible.

Stocking is *sine qua non* for reservoir fisheries management. The Indian experience of transplanting fish species from one river basin to another and introduction of exotic species has been very revealing. The performance of stocked and introduced fishes and their impact on the indigenous ichthyofauna have been examined.The Indian rivers are known for their rich faunistic diversity and many of the native species are considered to be superior to the introduced ones in their growth performance. While there are some instances of successful transplantations where the stocked species established good breeding populations in the reservoirs leading to increased yield rates, in many cases, stocking attempts were rarely governed by any ecosystem considerations. These irrational stocking efforts were not only wasteful exercises, but they also proved to be detrimental to the native fish stocks from the conservation and yield optimisation points of view.

The three main sources of data and information for this document consist of the published research papers on Indian reservoirs, many technical reports prepared by various Government agencies and the specific information provided by the Fisheries Departments of various State Governments. The material has been presented in 14 chapters. The Chapter I gives a national perspective which includes the size and distribution of the resource, the limnological attributes of reservoirs and their production trends. Highlights of reservoir fisheries management in India, including the stocking and the role of exotic fishes have also been discussed.

The Chapters 2 to 14 present a State-wise description of reservoirs and their fisheries comprising nine States *viz.*, Tamil Nadu, Kerala, Karnataka, Andhra Pradesh, Maharashtra, Madhya Pradesh, Orissa, Gujarat, Rajasthan, Himachal Pradesh, Uttar Pradesh and Bihar, and the Northeast. The States of West Bengal and Haryana, their reservoir areas being very small, have not been discussed under separate Chapters. The reservoirs in these two States have been accounted for in Chapter I, under the State-wise distribution of man-made lakes. Each Chapter in Section II contains a description of the reservoir fisheries resources, a few case studies on the reservoirs that have been subjected to scientific investigations and the status of reservoir fisheries in the respective States. The States have been arranged from south to north, beginning with Tamil Nadu which, apart from being the southern most State of India, has the distinction of pioneering the reservoir fisheries research in India.

Data received from some of the States were neither uniform nor comprehensive to enable a detailed analysis of various aspects of reservoir fisheries and this accounts for the variations in indepth coverage in respect of a few States. Similarly, more reservoirs have been investigated by various workers in Tamil Nadu and Madhya Pradesh than in offer States and thus there is a preponderance of information on reservoirs of these two States, compared to others. Data on morphometric, hydrographic and physico-chemical features of reservoirs provided at the end of each Chapter are intended to assist those who are inclined to build database on reservoirs.

A need has long been felt by those engaged in research, education, development and planning for a consolidated information on reservoirs, including the resources available, level of technologies and the extent of their utilisation. This document purports to fill the void by bringing together the up-to-date information available on the subject.

1. RESERVOIR FISHERIES RESOURCES OF INDIA

Despite the overwhelming importance of reservoirs in the inland fisheries of India, a reliable estimate of the area under this resource is still elusive, causing serious constraints to the R & D activities. The available estimates made by various agencies are conflicting and wide off the mark. The National Commission on Agriculture (NCA) has estimated the total area under reservoirs at 3 million ha during the mid-sixties and projected its growth to 6 million ha by 2000 AD (Anon., 1976). Bhukaswan (1980) put the figure at 2 million ha. Srivastava *et al.* (1985) compiled a list of 975 large and medium reservoirs in the country with an estimated area of 1.7 million ha. One of its major shortcomings is the exclusion of small reservoirs, especially those in Tamil Nadu, Karnataka and Maharashtra.

1.1 TANKS AND SMALL RESERVOIRS

Enumeration of the medium and large reservoirs is relatively easy, as they are less in number and the details are readily available with the irrigation, power and public works authorities. However, compilation of data on small reservoirs is a tedious task as they are ubiquitous and too numerous to count. The problem is further confounded by ambiguities in the nomenclature adapted by some of the States. The word *tank* is often loosely defined and used in common parlance to describe some of the small irrigation reservoirs. Thus, a large number of small man-made lakes are designated as tanks, thereby precluding them from the estimates of reservoirs. There is no uniform definition for a tank. In the eastern States of Orissa and West Bengal, pond and tank are interchangable expressions, while in Andhra Pradesh, Karnataka and Tamil Nadu, tanks refer to a section of irrigation reservoirs, including small and medium sized water bodies. In fact, some of the tanks in Tamil Nadu and Karnataka are much bigger than Aliyar and Tirumoorthy reservoirs.

David *et al.* (1974) defined the peninsular tanks as *water bodies created by dams built of rubble, earth, stone and masonry work across seasonal streams, as against reservoirs, formed by dams built with precise engineering skill across perennial or long seasonal rivers or streams, using concrete masonry or stone, for power supply, large-scale irrigation or flood control purposes*, which is obviously tedious and inadequate. Irrespective of the purpose for which the lake is created and the level of engineering skill involved in dam construction, both the categories fall under the broad purview of reservoirs, *i. e.*, man-made lakes created by artificial impoundment of surface flow. From limnological and fisheries points of view, the distinction between small reservoirs and tanks seems to be irrelevant. Moreover, numerous small reservoirs fitting exactly into the description of the south Indian tanks are already enlisted as reservoirs in the rest of the country. Therefore, the large tanks have been treated at par with reservoirs for the purpose of this study.

In Andhra Pradesh, the tanks and small reservoirs are segregated either arbitrarily or based on yardsticks that have no limnological relevance. For instance, all the small reservoirs in the State, created before independence and those without a masonry structure and spillway shutters are called tanks. Tanks in Andhra Pradesh are classified as *perennial* and *long seasonal.* Of the 4 604 perennial tanks, 1 804 in Srikakulam, East Godavari and Krishna districts, having average size less than 10 ha, are not considered as reservoirs in this study. The remaining 2 800 tanks covering a total area of 177 749 ha have been reckoned as reservoirs.

In Tamil Nadu, the tanks are classified as *short seasonal* and *long seasonal.* The latter, also known as *major irrigation tanks*, have an average size of 34 ha and retain water for 9 to 12 months a year. Major irrigation tanks of Chengalpattu MGR and Salem districts are larger with average area 222 and 156 ha respectively. A total of 8 837 major irrigation tanks of Tamil Nadu with water surface area of 300 278 ha have been included under small reservoirs. Similarly, 4 605 perennial large water bodies in Karnataka, listed as *major irrigation tanks* are brought under the ambit of reservoirs.

1.2 CLASSIFICATION OF RESERVOIRS

Fish Seed Committee of the Government of India (1966) termed all water bodies of more than 200 ha in area as reservoirs. David *et al.* (1974) while classifying the water bodies of Karnataka State, considered impoundments above 500 ha as reservoirs and named the smaller ones as irrigation tanks. In the former USSR, reservoirs up to 10 000 ha area are assigned the status of small reservoirs, whereas in USA they may range from 0.1 ha to several ha (Bennet, 1970). In China, where reservoirs are classified on the basis of storage capacity (Lu, 1986), those holding more than 100 million m^3 of water are classified as large reservoirs, 10 to 100 million m^3 as medium and 0.1 to 10 million m^3 as small reservoirs. Assuming an average depth of 10 m, small reservoirs of China are in the size range 10 to 1 000 ha.

Reservoirs are classified generally as small (<1 000 ha), medium (1 000 to 5 000 ha) and large (> 5 000 ha), especially in the records of the Government of India (Sarma, 1990, Srivastava *et al.*, 1985), which has been followed in this study. All man-made impoundments created by obstructing the surface flow, by erecting a dam of any description, on a river, stream or any water course, have been reckoned as reservoirs. However, water bodies less than 10 ha in area, being too small to be considered as lakes, are excluded.

After removing the anomalies in nomenclature, especially with regard to the small reservoirs, by bringing the large (above 10 ha) irrigation tanks under the fold of reservoirs, India has 19 134 small reservoirs with a total water surface area of 1 485 557 ha (Table 1.1). Similarly, 180 medium and 56 large reservoirs of the country have an area of 527 541 and 1 140 268 ha respectively. Thus, the country has 19 370 reservoirs covering 3 153 366 ha (Table 1.2).

The State of Tamil Nadu accounts for maximum number (8 895) and area (315 941 ha) of small reservoirs, followed by Karnataka (4 651 units and 228 657 ha) and Andhra Pradesh (2 898 units and 201 927 ha). Medium reservoirs constitute less than 1% of the total number of units and 17% of the total area. Madhya Pradesh with 169 502 ha tops the table in respect of medium reservoirs. Andhra Pradesh, Rajasthan, and Gujarat have more medium reservoirs than Madhya Pradesh, though the water area in these States is much less. Karnataka has a preponderance in number (12) of large reservoirs. Nevetheless, the 7 reservoirs in Andhra Pradesh in this category are much larger and have a water area of 190 151 ha (Table 1.2).

The predominance of reservoirs in the peninsular States, *viz.*, Tamil Nadu, Karnataka, Andhra Pradesh, Kerala, Orissa and Maharashtra is elucidated by Figs 1.1 to 1.4. These six States account for more than 56% of the total reservoir area in the country. Of the 19 134 small reservoirs, 17 989 (94%) are located there, contributing 63% of the total water area. Similarly, 34% of the medium reservoirs is distributed in these States.

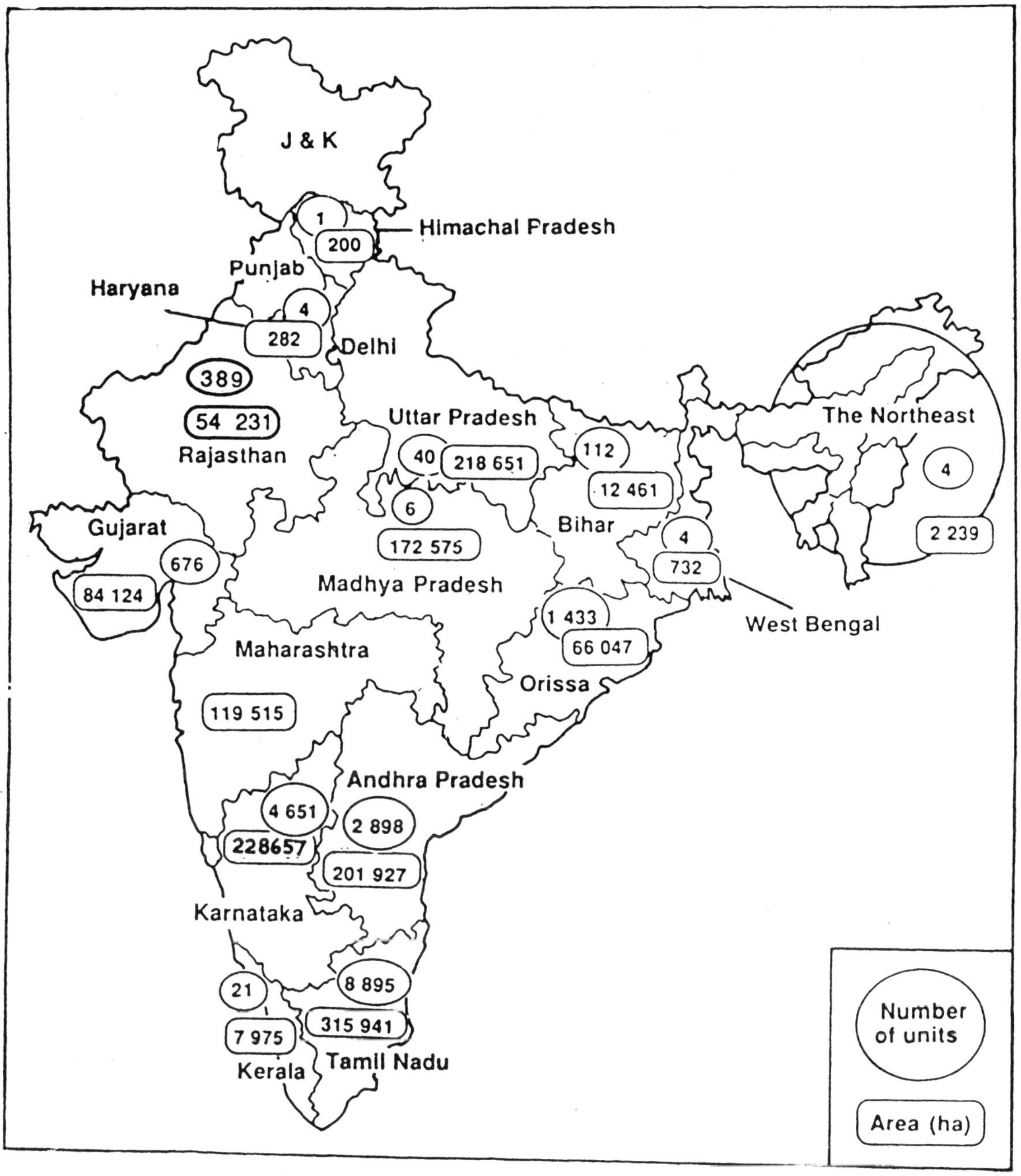

Figure 1.1. Distribution of small reservoirs in India

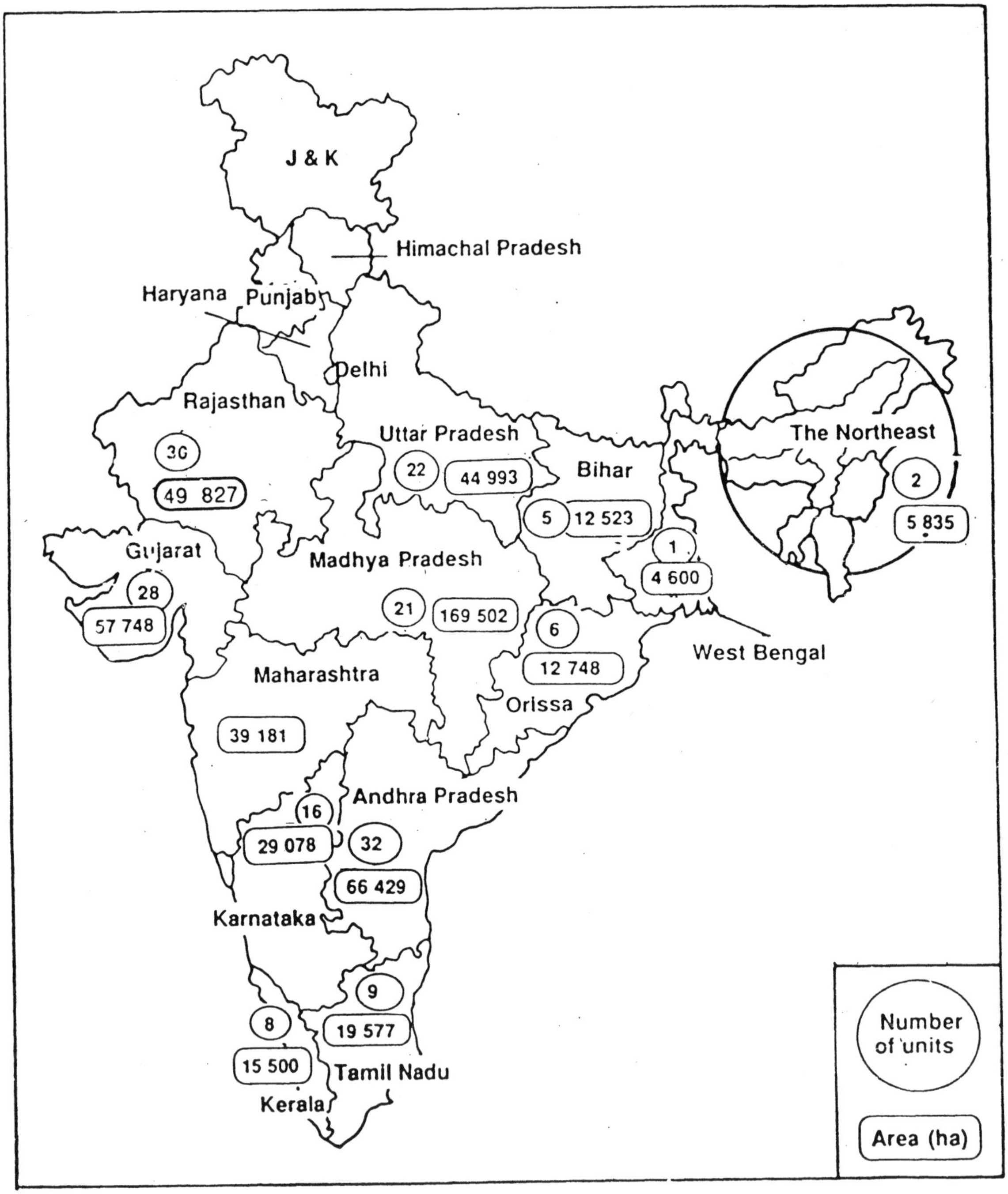

Figure 1.2. Distribution of medium reservoirs in India

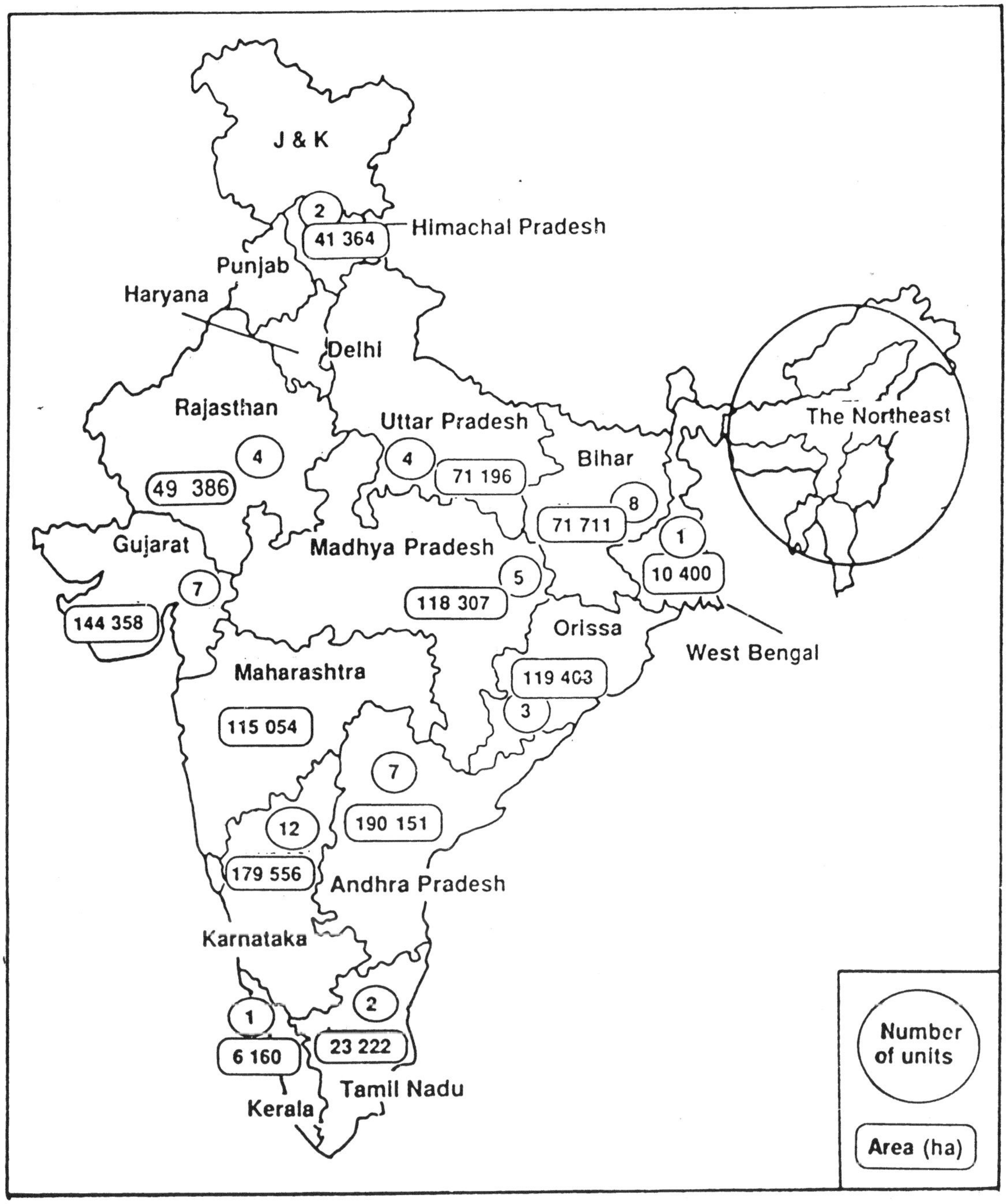

Figure 1.3. Large reservoirs in India

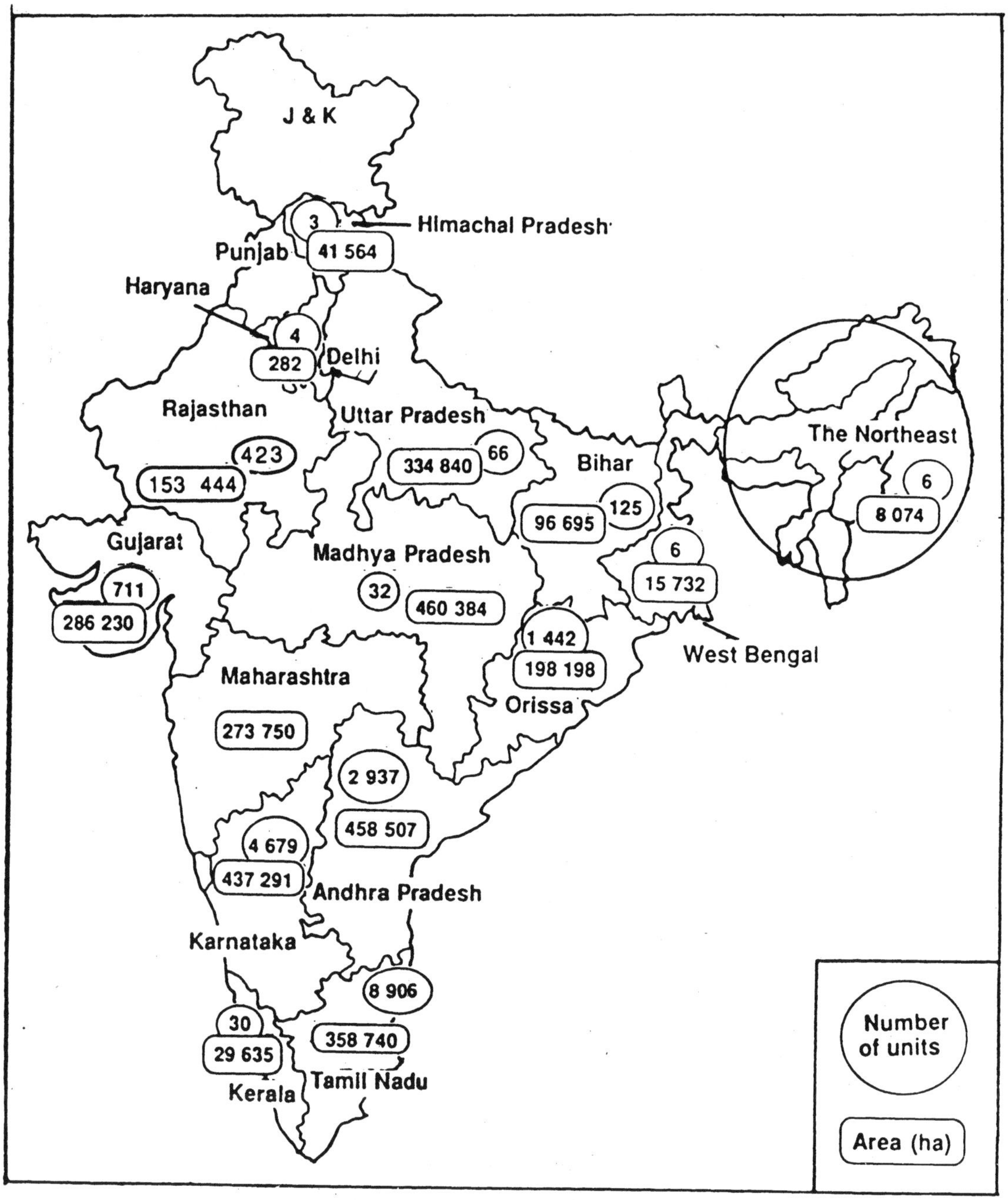

Figure 1.4. Distribution of reservoirs (all categories) in India

Table 1.1. Distribution of small reservoirs and irrigation tanks in India

States	Small reservoirs		Irrigation tanks		Total	
	Number	*Area (ha)*	*Number*	*Area (ha)*	*Number*	*Area (ha)*
Tamil Nadu	58	15 663	8 837	300 278	8 895	315 941
Karnataka	46	15 253	4 605	213 404	4 651	228 657
Andhra Pradesh	98	24 178	2 800	177 749	2 898	201 927
Gujarat	115	40 099	561	44 025	676	84 124
Uttar Pradesh	40	20 845	-	197 806	40	218 651
Madhya Pradesh	6	172 575	-	-	6	172 575
Maharashtra	-	-	-	-	-	119 515
Bihar	112	12461	-	-	112	12 461
Orissa	1 433	66 047	-	-	1 433	66 047
Kerala	21	7 975	-	-	21	7 975
Rajasthan	389	54 231	-	-	389	54 231
Himachal Pradesh	1	200	-	-	1	200
West Bengal	4	732	-	-	4	732
Haryana	4	282	-	-	4	282
North East	4	1 639	-	600	4	2 239
Total	2 331	551 695	16 803	933 862	19 134	1 485 557

Table 1.2. Distribution of small, medium and large reservoirs in India

States	Small		Medium		Large		Total	
	Number	Area (ha)	Number	Area (ha)	Number	Area (ha)	Number	Area (ha)
Tamil Nadu	8895*	315 941*	9	19 577	2	23 222	8906	358 740
Karnataka	4 651*	228 657*	16	29 078	12	179 556	4 679	437 291
Madhya Pradesh	6	172 575	21	169 502	5	118 307	32	460 384
Andhra Pradesh	2 898*	201 927*	32	66 429	7	190 151	2 937	458 507
Maharashtra	-	119 515	-	39 181	-	115 054	-	273 750
Gujarat	676*	84 124*	28	57 748	7	144 358	711	286 230
Bihar	112	12 461	5	12 523	8	71 711	125	96 695
Orissa	1 433	66 047	6	12 748	3	119 403	1 442	198 198
Kerala	21	7 975	8	15 500	1	6 160	30	29 635
Uttar Pradesh	40**	218 651	22	44 993	4	71 196	66	334 840
Rajasthan	389	54 231	30	49 827	4	49 386	423	153 444
Himachal Pradesh	1	200	-	-	2	41 364	3	41 564
Northeast	4**	2 239	2	5 835	-	-	6	8 074
Haryana	4	282	-	-	-	-	4	282
West Bengal	4	732	1	4 600	1	10 400	6	15 732
Total	19 134	1 485 557	180	527 541	56	1 140 268	19 370	3 153 366

* Including large irrigation tanks ** Not exhaustive

1.3 DETERMINANTS OF RESERVOIR PRODUCTIVITY

The production propensity of a reservoir is determined by a set of key environmental parameters, especially the water and soil quality which, in turn, are functions of the geo-climatic conditions under which it exists. Thus, the geography, climate, topography and a number of physiographic parameters play a vital role in bestowing the reservoirs their intrinsic productive potential. India, being a country of continental proportions, its reservoirs are spread over various types of terrains, and soil types exposed to diverse climatic conditions and they receive drainage from a variety of catchment areas.

1.3.1 The geoclimatic features

The land area of India covers 3 287 728 km^2, half of which lying above the Tropic of Cancer and the rest in the tropics. The southern limit is as close to equator as 8^0 4' N. The climate varies from the warm tropical in the south to the temperate in the north. The landscapes include some great mountains, extensive alluvial plains, riverine wetlands, plateau lands, deserts, coastal plains and deltas. The main soil types are alluvial, deep and medium black, red and yellow, laterite, saline and desert, and forest and hill (Fig. 1.5). Almost all conceivable forms of vegetation, including tropical evergreen, littoral and swamp, tropical moist deciduous, tropical thorn, montane sub-tropical, Himalayan and alpine are present in various parts of the country.

The major physiographic divisions of the country are the Himalayas, the Indo-Gangetic plains, the Vindhyas, the Satpuras, the Western Ghats, the Eastern Ghats, coastal plains, the deltas and the riverine wetlands (Fig. 1.6). The alignments of hills and their elevation have profound influence on the prevailing winds and thereby the distribution of rainfall in the country. India receives, on an average, 105 cm of rainfall every year, which is one of the highest in the world for a country of comparable size. Total amount of rainfall received annually is estimated at 400 million hectare meters (mhm), out of which 230 mhm goes back to the atmosphere as evapotranspiration, leaving 170 mhm to impregnate the rivers through surface flow (110 mhm) and regeneration (60 mhm). The temporal and spatial distribution of rainfall exhibits wide variations within the country.

More than one million km^2 of the country's geographical area receives inadequate rainfall. This includes the deserts, the semi-arid regions of north India and the rain shadows of the Western Ghats (Fig. 1.7). Large rivers like the Godavari, the Krishna, the Pennar and the Cauvery pass through extensive tracts of low rainfall areas and hence carry much less water than the rivers passing through high rainfall areas like the northeast and the west coast. In the northeast, the Khasi and Jaintia hills receive a bountiful 1 000 cm of rainfall annually and the Brahmaputra valley gets precipitation to the tune of 200 cm. Rainfall up to 1 142 cm recorded in Cherrapunji and Mawsyngram in the region is one of the highest in the world. In the west coast of India, heavy rainfall occurs along the slopes of the Western Ghats, where during the southwest monsoon, rainfall of very high order is recorded on the windward side which rapidly decreases towards the leeward side. The Indo-Gangetic plains and the Himalayas also receive rainfall above the national average.

Seasonal distribution of rainfall in India is worth noticing. The Western Ghats, Assam, parts of sub-Himalayan West Bengal and some higher elevations of Himalayas up to Punjab have more than 100 rainy days a year, while in extreme west Rajasthan the number of rainy days are less than 10 (Rao, 1979). The south-west monsoon season extending from June to September is the principal rainy season in the country as a whole when 75% of the annual rainfall is received. In more than one third of the country, 90% of the rainfall and thereby the surface flow is limited

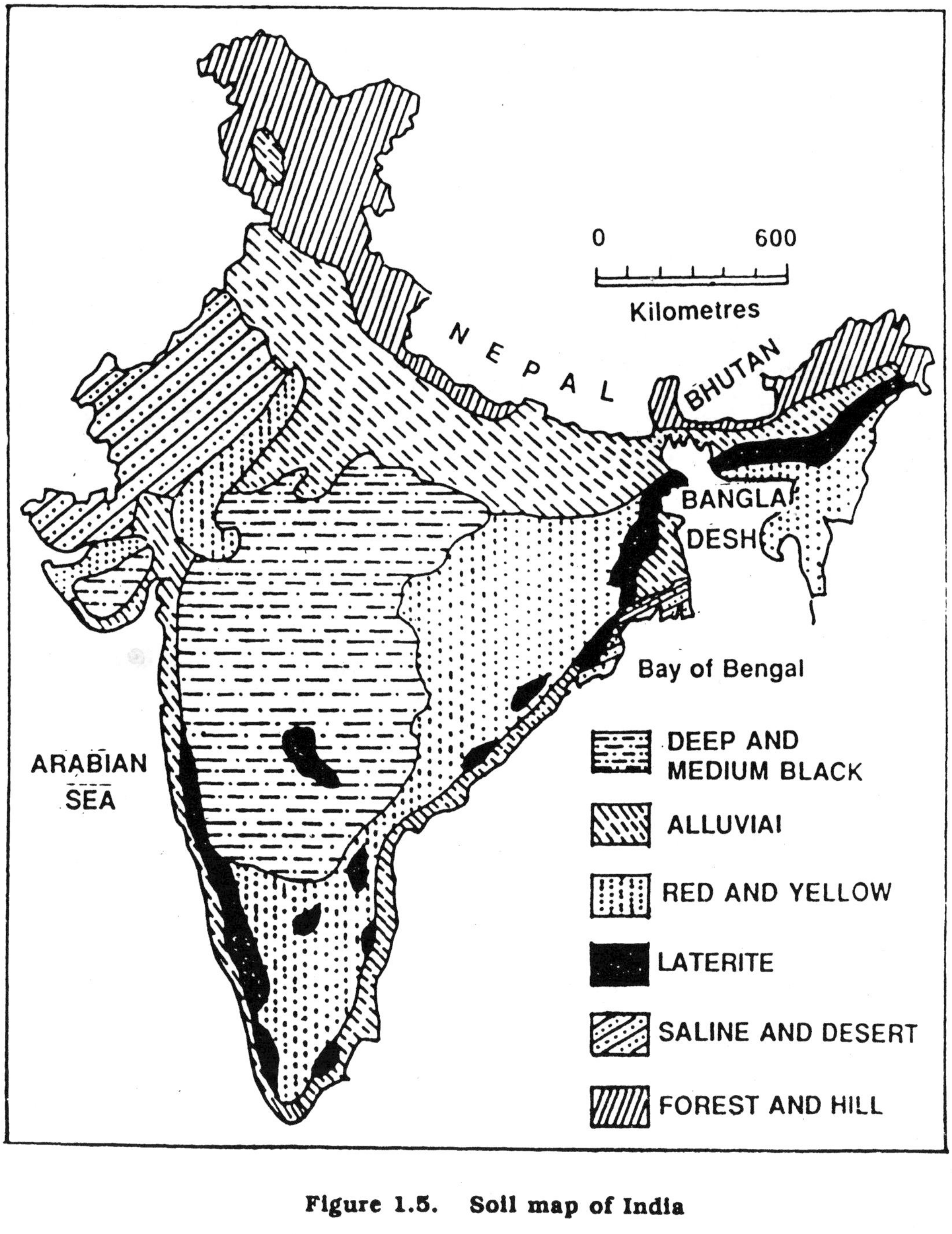

Figure 1.5. Soil map of India

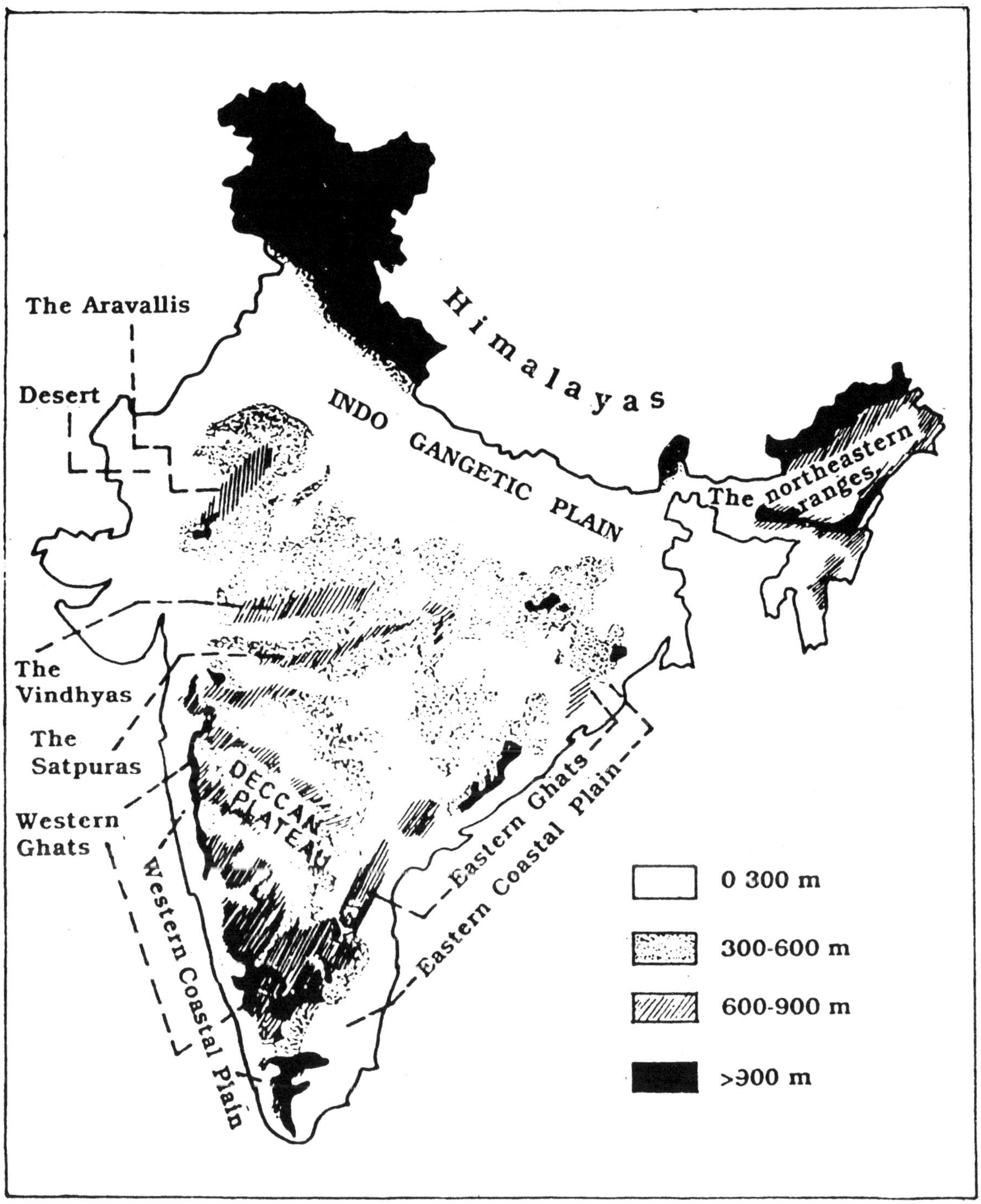

Figure 1.6. Physiographic features of India

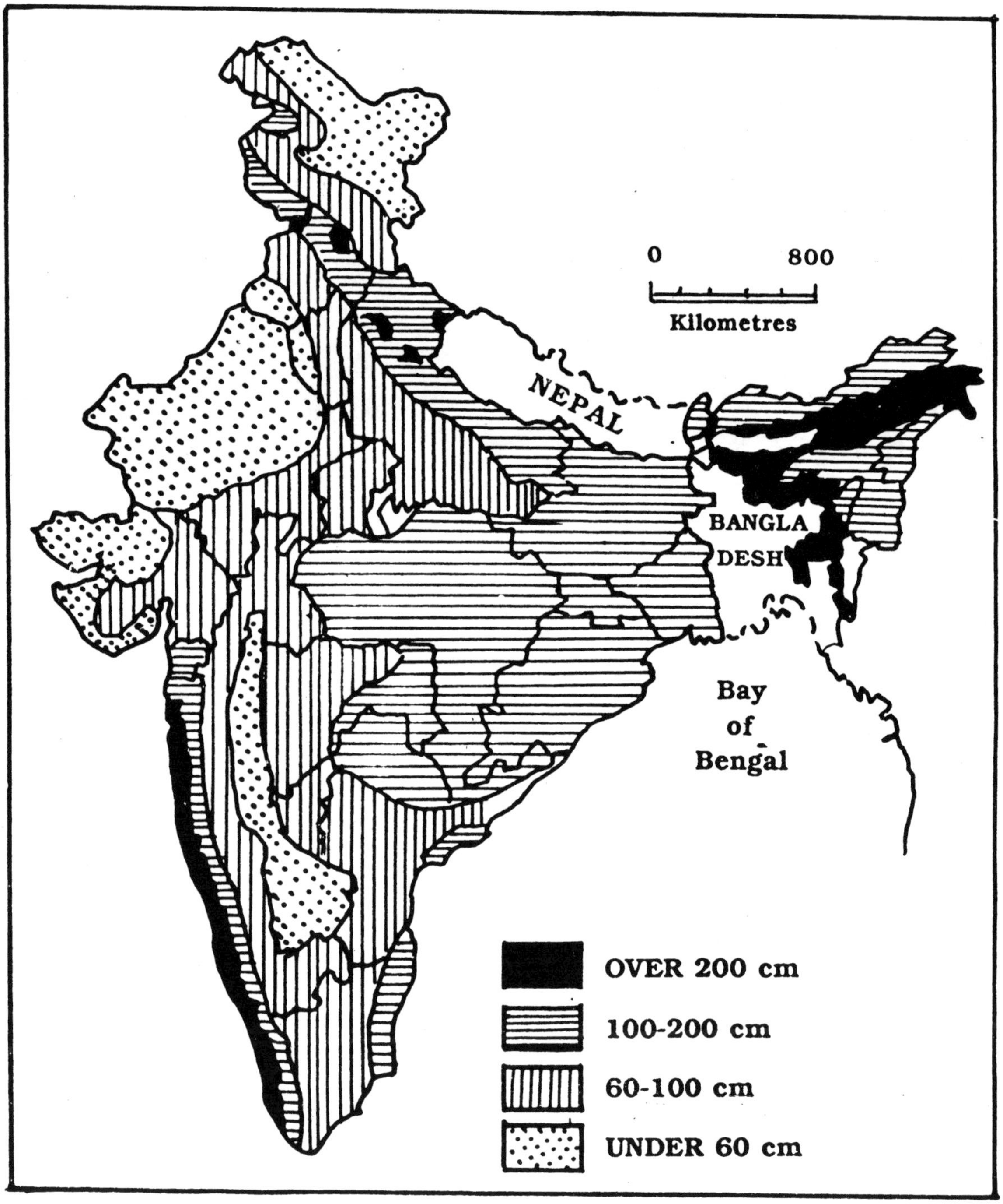

Figure 1.7. Rainfall map of India

to a very brief period of 2 to 3 months. This extreme seasonality in rainfall distribution makes the irrigation reservoirs a *sine qua non* for agriculture in India, especially in the rainshadows of the peninsular India. People inhabiting this area learned to store water by erecting barricades across minor stream and rivulets from time immemorial. In recent years, with the advent of modern hydraulic structures, larger and more complex dams came into being. The steep gradient and heavy discharge of water in the mountain slopes of Western Ghats, the northeast and the Himalayas offer ideal opportunities for hydro-electric power generation. A large number of such projects have come up in these regions in recent years. Thus, the reservoirs have become a common feature in the Indian landscape, dotting all river basins, minor drainages and seasonal streams.

1.4 RESERVOIR ECOSYSTEM

Reservoir is a man-made ecosystem without a parallel in nature. Though essentially a combination of fluviatile and lacustrine systems, a close examination of the biotope reveals that it has certain characteristic features of its own. The riverine and lacustrine characters coexist in reservoirs, depending on the temporal and spatial variations of certain habitat variables. For example, the lotic sector of the reservoir sustains a fluviatile biocoenos, whereas the lentic zone and the bays harbour lentic communities. During the months of heavy inflow and outflow, the whole reservoir mimics a lotic environment whereas in summer, when the inflow into and outflow from the reservoir dwindle, a more or less lentic condition prevails in most parts of it. Another unique feature of reservoirs that makes them distinctly different from their natural counterparts is the water renewal pattern marked by swift changes in levels, inflow and outflow.

In India, most of the precipitation takes place during the monsoon months which contribute substantially to the surface flow. During this period, due to a heavy inflow of water into the system, all outlets of the dam are usually opened, resulting in total flushing. This process dislodges a considerable part of the standing crop of biotic communities at the lower trophic level and disturbs the natural primary community succession. The sudden level fluctuations also affect the benthos by exposing or submerging the substrata. Factors determining the water and soil quality in reservoirs are different from those of natural lakes. In the latter, the basin soil plays a predominant role in determining the chemical water quality through soil water interphase. In the reservoirs, on the other hand, the nutrient input from the allochthonous source often determines the water quality, nutrient regime and the basic production potential. This is because of the fact that the catchment of parent rivers is very often situated far away from the reservoir, under totally different geoclimatic conditions. Deep drawdown, wind-mediated turbulence, locking up of nutrients in the deep basins, etc., are but some of the factors that impart the uniqueness to the reservoir ecosystem. Besides, the varying purpose and design of the dams make the reservoirs different in their hydrographic and morphoedaphic characteristics, with implications on the production potential. Some salient features of the reservoir ecosystem are depicted in Table 1.3.

The ecology of reservoirs is radically different from that of the parent rivers. Dams alter river hydrology both up- and downstream of the river. The obstruction of river flow and the consequent inundation trigger off sudden transformation of lotic environment into a lentic one. The riverine community is subjected to changes akin to the secondary community succession. A number of organisms perish, some migrate to more hospitable environs, and the more hardy ones adapt themselves to the changed habitat. There is usually an initial spurt of plankton and benthic communities due to the increased availability of nutrients released from the decay of submerged vegetation. This trophic burst is also on account of the saproxenic lacustrine species filling the vacant niches created by the disappearance of saprophobic riverine taxa. As the effects of trophic burst wean away, the reservoir passes into a phase of trophic depression and the final fertility is regained after a few years. Habitat variables responsible for a reservoir's productivity can be summed up into climatic, morphometric, and hydro-edaphic factors.

Table 1.3. Abiotic and biotic factors affecting productivity of reservoirs at various trophic levels

Positive/augmentative factors	Major effects unknown	Negative/reductive factors
High shoreline development (coves, bays, bays etc.)	Sedimentation of inorganic material	Low transparency in floods due to inorganic turbidity.
Low mean depth (less than 18 m)	High rate of evaporation	High mean depth
Existence and extent of marginal vegetation	Contributions of autochthonous nutrients	Erosion in the reservoir water shed area
Optimum nutrient levels	High surface temperature during summers (In northern India)	Reduction of quantity of water flowing into reservoir
Nutrient enrichment during floods	Low water temperature during winter (In northern India)	Large water level fluctuations creating large aridal (barren littoral)
Moderate to long growing season	Aquatic community interrelationship	Low level of dissolved oxygen in parts of hypolimnion
High frequency of phytoplankton blooms		Pollution in the reservoir water-shed
Moderately developed macrophyte community		Phytoplankton biomass mainly blue greens
Periphyton abundant		Relative low fish species diversity indicating low stability and a potentially low resilience to stresses
Well established plankton and benthos		Unbalanced fish populations favouring predatory and trash species
Tree and bush cleared		Low abundance and diversity of terrestrial vegetation hence early successional stage
Conditions permitting passage of migratory fish		Relatively low environmental heterogeneity
Introduction of fishes adapted to lentic conditions		Low diversity of plankton and benthos
Employment of modern fishing gear and optimization of fishing effort		Low diversity of aquatic macro-phytes
Enforcement of fishery regulations		Exposure of fish nests during drawdowns

(After Jhingran, 1988)

1.4.1 Climatic factors

The Indian reservoirs are exposed to a wide range of climates from the temperate Himalayas in the north to the extreme tropical in the southern peninsula. From Gobindsagar in Himachal Pradesh to Chittar in Tamil Nadu, they spread over the southern slopes of Himalayas, the Indo-Gangetic plain, the Vindhyas, the Satpuras, the Western and Eastern Ghats and the Deccan plateau.

Apart from being the main factor influencing the prevailing climate of the region, the latitudinal location is important in determining the quantum of solar radiation available at the water surface for primary productivity. Natarajan and Pathak (1983) estimated the amount of solar radiation available at four reservoirs within 31° 25' N and 11° 28' N and the rate at which the solar energy was converted into chemical energy. Incident solar energy available at the surface varied from 213 x 10^4 cal m^{-2} yr^{-1} in Bhavanisagar (10° 28' N) to 172 x 10^4 cal m^{-2} y^{-1} in Gobindsagar (31° 25' N). Jhingran (1990) observed that 0.2 to 0.68% of the incident solar energy was fixed as chemical energy by the primary producers in five reservoirs, *viz.*, Gobindsagar (Himachal Pradesh), Ramgarh (Rajasthan), Rihand (Uttar Pradesh) and Bhavanisagar (Tamil Nadu). It is often the qualitative and quantitative abundance of the producer communities that determines the photosynthetic efficiency rather than the actual amount of solar energy available. For instance, Nagarjunasagar despite receiving solar energy at the rate of 205 x 10^4 cal m^{-2} yr^{-1}, fixes chemical energy to the extent of 0.29%, whereas in Gobindsagar 0.68% of the 172 x 10^4 cal m^{-2} yr^{-1} is being fixed by the producers (Table 1.4).

Table 1.4. Latitude-induced variations in availability of energy in Indian reservoirs

Reservoir	Area (ha) at FRL	Latitude	Available light energy (cal m^{-2} yr^{1})	Available chemical energy (cal m^{-2} yr^{1})
Bhavanisagar	7 285	11° 25'	213 x 10^4	8 781 (0.41%)
Nagarjunasagar	2 8474	16° 4'	205 x 10^4	5 959 (0.29 %)
Rihand	4 6538	24°	188 x 10^4	3.970 (0.20 %)
Ramgarh	1 265	27° 12'	183 x 10^4	8 236 (0.49 %)
Gobindsagar	16 867	31° 25'	172 x 10^4	11 696 (0.68 %)

(After Jhingran, 1990)

Prevailing climatic factors including air temperature, wind velocity, rainfall, *etc.* play an important role in the biological productivity of a water body. The wide seasonal variations in air temperature is the predisposing factor in the thermal features of the north Indian and peninsular reservoirs. In contrast to the reservoirs of the north, their southern counterparts are characterised by the narrow range of fluctuations in water and air temperature during different seasons, a phenomenon which prevents the formation of thermal stratification. Normally, the thermal gradient occurs when high air temperature during summer warms up the upper layer. In peninsular India, there is no winter worth its name and the air temperature remains comparatively high during the whole year. During summer, when surface water gets heated up, the prevailing high temperature at the bottom does not offer any scope for thermal resistance by the warm upper layers. Thermal stratification is limnologically important because in thermally stratified lakes, the water above and below thermocline does not mix up and thereby rich nutrients at the bottom layer get locked up. A warm bottom layer also facilitates rapid decomposition of organic matter, thereby accelerating the process of nutrient release.

The deep basin of Nagarjunasagar, despite 40% of its capacity being dead storage, does not favour the formation of thermocline. Apart from the high water temperature throughout the year, the continous drawdown from deeper layers and the wind and wave mediated turbulence facilitate mixing up of water column. This is true to most of the reservoirs in the States of Andhra Pradesh, Karnataka, Kerala, Tamil Nadu, Orissa and Maharashtra. However, seven reservoirs in the upper peninsula comprising south Bihar, Gujarat and Madhya Pradesh undergo transient phases of thermal stratification during the summer stagnation, depending upon the other parameters such as the depth of the basin, water abstraction pattern and the wind. Konar reservoir situated beyond the Tropic of Cancer has distinct epi-and hypolimmon during the summer months. Similarly, a well-defined thermocline is reported from Gobindsagar. Apart from the solar warming of the top layer, which remains as a separate thermal regime, the inflowing Beas water that joins the reservoir at the lotic sector does not get mixed up, retains its cool character and remains as a separate layer at the bottom.

The amount of rainfall determines the rate of inflow into the reservoir, and hence plays a crucial role in bringing in the water replenishment and nutrient enrichment. More often, rainfall in the catchment of the river situated hundreds of km away from the reservoir affects the inflow rate. Another important climatic factor with implications on thermal and chemical regimes of the reservoir is the wind. It helps distribution of heat and equalisation of temperature in the water column. Wind velocity is very high in monsoon and premonsoon months in most reservoirs in India (Natarajan, 1979). Wind-induced turbulence is important in churning of the reservoirs and thereby facilitating the availability of nutrients at the trophogenic zone.

1.4.2 Morphometric factors

Reservoir morphometry is a function of the height of the dam and the topography of the impounded areas. Apart from the nature of the basin and the characteristics of the terrain, it is the design of the dam and the water use pattern that decide the influence of morphometric and hydrographic features on the aquatic productivity. Most of the hydel reservoirs on the mountain slopes of Western Ghats, Himalayas and the other highlands are deep, with steeper basin walls than the irrigation impoundments.

One of the most important morphometric considerations is the mean depth, that is believed to determine the productivity of reservoirs (Hayes, 1957, Rawson, 1952). This is based on the well known dictum that the shallower lakes have greater part of their water in the euphotic zone, facilitating greater mixing and circulation of heat and nutrients and hence higher productivity. A large portion of water in deep lakes serves as a *nutrient sink* at the bottom, where organic matter accumulates and thus the nutrients become unavailable at the photosynthetic zone. Mean depth calculated from the capacity and area varies from 5.2 in Panchet to 58 m in Gobindsagar among the large reservoirs. Medium reservoirs in the country have mean depth ranging from 2.3 m (Poondi) to 24.0 m (Bhatghar). Hope Lake in Tamil Nadu has an exceptionally deep basin of 37.7 m, while the small reservoirs in India have a mean depth range of 2.1 m (Vidur) to 14.57 (Badua). Mean depth in case of hydel reservoirs of the mountain slopes is invariably high, compared to the irrigation reservoirs of the plains and plateaus. The two largest impoundments in the country *viz.*, Hirakud and the Gandhisagar have very low mean depths of 11.3 and 11.7 m respectively, while Idukki, the hydel reservoir in the Western Ghats, which is one tenth of Gandhisagar in area has a mean depth of 32 m.

The mean depth, however, does not show any direct correlation with productivity, either at primary or fish level. Vidur, despite being one of the shallowest (mean depth 2.1 m) of all reservoirs in the country, does not support rich plankton community. Likewise, Kulgarhi and Govindgarh reservoirs in Madhya Pradesh exhibit propensities towards oligotrophy in spite of their shallowness. On the other hand, Gobindsagar, the deepest reservoir, has the highest productivity among large reservoirs. Medium deep reservoirs like Amaravathy (13.7 m), Aliyar (16.8 m), and Tirumoorthy (11 m) develop regular blooms of plankton.

Shoreline and volume development indices

A highly crinkled shoreline, as indicated by the high values of shoreline development index, is believed to be indicative of productive nature of the water body. An irregular shoreline encompasses more littoral formations and areas of land and water interface. High shoreline indices of Hirakud (13.5), Gobindsagar (12.26), Tilaiya (9.12), Konar (8.78), Nagarjunasagar (7.89) and Rihand (7.04) are accompanied by a moderate to rich plankton community.

Ratio between the maximum depth and mean depth, often described as volume development index denotes the depth of basin in relation to the nature of basin wall. An index value less than 1 suggests basin wall convex towards water. No perceptible relation exists between this parameter and the productivity of Indian reservoirs.

Hydrographic changes have a direct bearing on productivity, as sudden changes in water level, inflow and outflow directly affect the biotic communities. It has been observed that plankton, benthos, and periphyton pulses coincide with the months of least level fluctuations and all these communities are at their ebb during the months of maximum level fluctuations and water discharge. Percentage of shallow areas (littoral formation) which varies at different levels, depending on the contour, is also an indicator of productive nature of the lakes.

Storage and release of water from dams are dictated by the requirements of irrigation, power generation and other primary purposes of the dam, rather than any considerations related to fisheries. The spillway discharge, apart from dislodging the standing crop of plankton, removes the oxygenated clear water at the top layer, leaving the oxygen-deficient, turbid bottom water. Similarly, the deep drawdown removes the decomposing material including nutrients.

1.4.3 Hydro-edaphic factors

The oligotrophic tendencies shown by some of the reservoirs are mainly due to the poor nutrient status and other chemical deficiencies. In most of the cases, poor water quality is a direct reflection of the catchment soil. All reservoirs in Kerala portray a low status in terms of specific conductivity (<50 μmhos) and total alkalinity (<50 mg l^{-1}) with the attendant low primary productivity and plankton. The rivers of Kerala drain the hills of Western Ghats with lateritic and humus soils deficient in N, P and Ca. The eastern slopes of these hills drain the rivers feeding Hope Lake, Manimuthar, Pechiparai and Peruchani, all deficient in ions. Sathanur, Krishnagiri and Vidur reservoirs receiving drainage passing predominantly through cultivated area have higher levels of alkalinity and hardness. Similarly, in Madhya Pradesh the water is soft to medium soft with less mineral salts, due to geo-chemical reasons. Even small lakes with shallow bottoms, more often than not, do not show signs of productivity due to the poor chemical make up of the catchment. Soils in Madhya Pradesh are normally deep black, medium black, shallow black and mixed red and skeletal, low in nitrogen and phosphorus.

Catchment of Ravishankarsagar comprises rocky, denuded forests and upstream rivers are intercepted by impoundments, which further deprive the water of the suspended matter. Allochthonous enrichment with minerals and nutrients of the reservoir is very low, resulting in low standing crop of plankton. Limestone and other calcareous rocks underlying the water course in the Deccan plateau are responsible for the predominantly hard water character of many of the reservoirs on the Krishna and Cauvery in Andhra Pradesh and Tamil Nadu. The acidic nature of the water of the north eastern reservoirs Kyrdemkulai, Nongmahir and Barapani is attributable to the acidic soil of the reservoir bed and in the catchment.

Nutrient budget

Most of the Indian reservoirs are characterised by low levels of phosphate and nitrate. Phosphate very seldom exceeds 0.1 mg l^{-1} in reservoirs free from pollution. However, the reservoirs of Rajasthan have particularly high levels of phosphate, ranging from traces to 0.929 mg l^{-1}. They receive phosphate from the rain washings derived from soils types like brown hills, grey brown hills, red and yellow and desert soils. In the highly eutrophic reservoir of Mansarovar in Madhya Pradesh, phosphate levels of 4 to 13 mg l^{-1} were recorded.

Nitrate nitrogen in water in Indian reservoirs is mostly in traces and seldom exceeds 0.5 mg l^{-1}. Lack of nutrients in water, especially the nitrate nitrogen and phosphate, does not seem to be indicative of low productivity. In many cases, despite their virtual absence, the production processes are not hampered. In Amaravathy, Bhavanisagar, Gandhisagar, Ravishankarsagar and many other reservoirs, moderate to very high primary productivity is reported, although the phosphate in water is either absent or present in a very low concentration. In the tropical reservoirs, phosphate level in water has limited scope as an indicator of productive traits. This phenomenon is attributed to rapid turnover of nutrients (Ehrligh, 1960; Abbot, 1967) and their quick recycling, as seen from the high metabolic rates. Hayes and Phillips (1958) showed that 95% of the phosphorus could be taken up by the phytoplankton within 20 minutes, while some algae could convert inorganic phosphate into organic state in less than one minute.

Unlike the nutrients like phosphate and nitrate, the measure of total dissolved solids in the form of total alkalinity and the specific conductivity reflects the production propensities of reservoir satisfactorily, with the exception of Amaravathy which, despite very low levels of specific conductivity (38 to 63 μmhos), total alkalinity (7 to 84 mg l^{-1}) and total hardness (18 to 50 mg l^{-1}), supports a very rich plankton community and a good stock of fish.

The ranges of notable abiotic factors indicating their productivity status are given in Table 1.5. A close examination of physico-chemical data pertaining to more than 100 reservoirs in the country leads to the conclusion that none of the morphometric, edaphic, and water quality parameters can be used as a dependable yardstick to predict the organic productivity to any degree of accuracy, production propensities of each reservoir being determined by a variety of factors. The vertical gradient of some of the chemical parameters, especially the dissolved oxygen, however, conveys the status with a higher level of accuracy.

Table 1.5. Physico-chemical features of Indian reservoirs (range of values)

Parameters	Overall range	Productivity		
		Low	*Medium*	*High*
A. WATER				
pH	6.5-9.2	< 6.0	6.0-8.5	> 8.5
Alkalinity (mg l^{-1})	40-240	< 40.0	40-90	> 90.0
Nitrates (mg l^{-1})	Tr.-0.93	Negligible	Upto 0.2	0.2-0.5
Phosphates (mg l^{-1})	Tr.-0.36	Negligible	Upto 0.1	0.1-0.2
Specific conductivity (μmhos)	76-474		Upto 200	> 200
Temperature (^{o}C)	12.0-31.0	18	18-22	> 22
		(with minimal stratification : *i.e.*, > 5°C)		
B. SOIL				
pH	6.0-8.8	< 6.5	6.5-7.5	> 7.5
Available P (mg/100 g)	0.47-6.2	< 3.0	3.0-6.0	> 6.0
Available N (mg /100 g)	13.0-65.0	< 25.0	25-60	> 60.0
Organic carbon (%)	0.6-3.2	< 0.5	0.5-1.5	1.5-2.5

(After Jhingran, 1990)

The bacterial decomposition of organic matter at the bottom is well reflected by the high rate of oxygen consumption. A corresponding increase in oxygen at the trophogenic upper zone gives clues of the high rate of photosynthesis. Almost all productive reservoirs in the country, irrespective of their geographic location, have a klinograde oxygen curve.

In most of the cases, the oxycline is accompanied by a vertical stratification of other chemical parameters such as pH, carbon dioxide, total alkalinity and specific conductivity. The tropholytic zone has a steady supply of free carbon dioxide, which reacts with carbonate to produce bicarbonates. This results in an increase of bicarbonates towards the bottom. Similarly, due to the increase in the hydrogen ions, the pH drops rapidly. Thus, the increase in total alkalinity, specific conductivity and CO_2 and the decrease in pH values towards the bottom layers act as useful indicators of productivity.

Rate of primary productivity in reservoirs is very high due to the warm tropical conditions available in most parts of the country. Many workers consider 1% of the total carbon produced at the phytoplankton phase as the potential fish production from a water body, although almost all reservoirs produce much less fish than their potential.

1.4.4 Biotic communities

The highly seasonal rainfall and heavy discharge of water during the monsoons result in high flushing rate in most of the reservoirs which does not favour colonisation by macrophytic communities. Similarly, inadequate availability of suitable substrata retards the growth of periphyton. Plankton, by virtue of drifting habit and short turnover period, constitutes the major link in the trophic structure and events in the reservoir ecosystem. A rich plankton community with well-marked seral succession is the hallmark of Indian reservoirs.

Plankton

Blue-green algae form the mainstay of plankton community in vast majority of the man-made lakes studied. The overwhelming presence of *Microcystis aeruginosa* in Indian reservoirs is remarkable. The productive waters of Gangetic plains, Deccan plateau, south Tamil Nadu and Orissa invariably have good standing crop of *Microcystis*. A common feature of all these reservoirs is the bright sunshine, isothermal water column, klinograde oxygen curve and an extensive catchment area, draining a calcium rich, forested or cultivated land. The species is almost ominipresent in the southern peninsula, except in the reservoirs of Karnataka and Kerala, which tend to be oligotrophic and have poor plankton count with desmids and other green algae as the main constituents. Reservoirs of Rajasthan receiving scanty rainfall and poor flushing rate favour macrophytes and despite being productive do not harbour blooms of *Microcystis*. The oligotrophic lakes of the northeast have a desmid-dominated plankton community (Fig. 1.8).

Altitude plays a decisive role in the distribution of *Microcystis*. Gobindsagar, the highly productive high altitude temperate reservoir, supports a rich community of *Ceratium* sp. instead of *Microcystis*. Similarly, the tropical Markonahalli situated at an altitude of 731 m above MSL, has *Ceratium* sp. as the major constituent of plankton.

Most of the reservoirs have three plankton pulses coinciding with the post-monsoon (September to November), winter (December to February) and summer (March to May) seasons. The monsoon (June-August) flushing disturbs and often dislodges the standing crop of plankton. However, no sooner the destabilising effects wean away (as the dam outlets are closed), the allochthonous nutrient input favours an accelerated plankton growth. As the post-monsoon merges into winter, the turbulence decreases and water becomes clean, the plankton community progresses through a series of seral successions to culminate in a peak. The summer maxima coincide with the drastic drawdown, bringing the deep, nutrient-rich areas into the fold of tropholytic zone. The temperature, bright sunlight and rapid tropholytic activities also accelerate the multiplication of plankton during summer. In some cases, only two pulses (*i. e.*, the post-monsoon and summer) are seen. However, the shallow, nutrient-rich reservoirs in the southern tip of the peninsula, by virtue of the fast turnover of nutrients and availability of sunshine and warmth, sustain a permanent bloom of plankton.

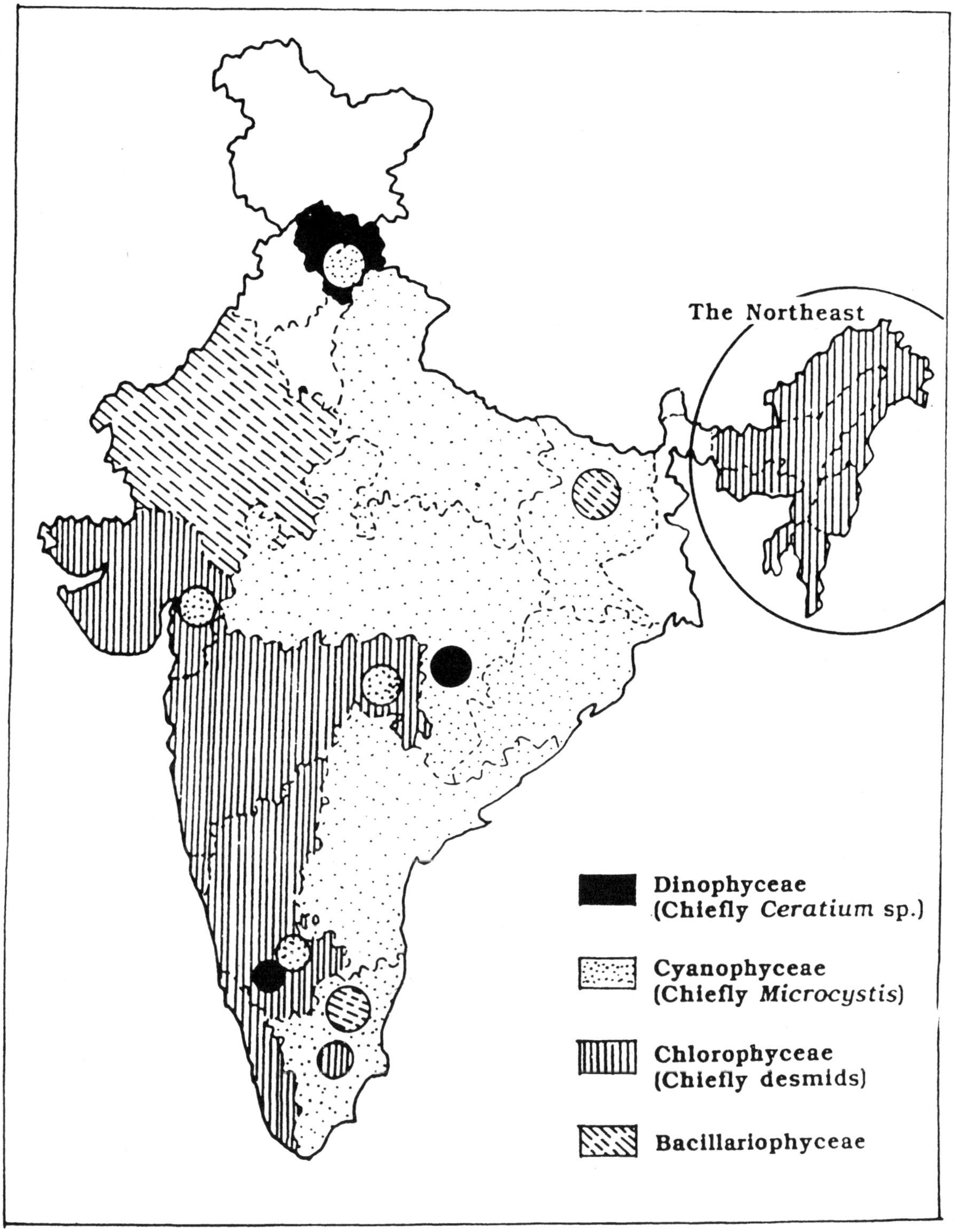

Figure 1.8. Dominant phytoplankton in reservoirs (State-wise)

The ubiquitous blooms of *Microcystis* in reservoirs in peninsular India are an example of a lacustrine biocoenose giving way to fluviatile ones in an impoundment. Studies have indicated that Chlorophyceae and Bacillariophyceae constituted the main components of riverine plankton. On reservoir formation and the consequent transformation of lotic environment into the lentic system, saprophobes disappear from the scene giving room for the rapid multiplication of saproxenes. *Microcystis*, finding a favourable note with the new environment, bursts into blooms, outnumbering all other forms into insignificance. In many reservoirs, orientation of lacustrine and fluviatile plankton can be clearly discerned from the composition of plankton in lotic, lentic and the cove sectors. The fluviatile lotic sector, although recording a lower plankton density, often shows better diversity and evenness indices, compared to the lentic and bay sectors, the still waters of which are characterised by higher concentration of dominance (C) and low evenness (J') (Sugunan, 1991).

Macrovegetation

Aquatic macrophytes do not figure prominently in the community structure and trophic events of the reservoirs in India, barring some exceptional circumstances, such as low water renewal, ageing of reservoir and pollution stress. In most of the reservoirs they are totally absent or their population is too insignificant to be taken into account. Mostly, they are restricted to isolated patches of *Vallisneria, Hydrilla* and mats of *Spirogyra*, found in the protected bays and coves. Yerrakalava in Andhra Pradesh, Ramgarh in Rajasthan and Sayajisarovar in Gujarat are the examples of macrophytic growth due to low flushing rate. Small irrigation reservoirs in Uttar Pradesh, *viz.*, Bachhra and Baghla are also known for the luxuriant growth of macrophytes.

Hussainsagar in Andhra Pradesh and Mansarover in Madhya Pradesh harbour thick vegetation which thrives due to the hyper-eutrophication. These reservoirs are well-advanced on their way to change to swamps. Age of the reservoir seems to have an influence on the macrophyte community. Vanivilas Sagar, and Markonahalli reservoir in Karnataka formed in 1901 and 1939 have a luxuriant growth of macrophytes. Similar age-related macrophyte growth can be observed in Yerrakalava, Ramgarh, Jaisamand, Hussainsagar and Fatehsagar reservoirs. Reservoirs of Rajasthan exhibit a seasonal rhythm in aquatic weeds, their population peaking in summer and declining during the monsoon season.

In the irrigation tanks of Karnataka, the qualitative and quantitative distribution of macrovegetation depends, to a large extent, on the physiographic divisions and the soil types. Thick vegetation comprising the littoral, submerged and emergent types is the characteristic feature of the tanks of coastal plains and the Malnad region. In the transitional zone between plateau and hills, marked by the presence of laterite and red soil, the tanks are fertile and plankton-rich. Weeds are sparse and whenever present, do not choke the water. In the tanks in the black soil zones, the weeds are mostly submerged and emergent types such as *Hydrilla, Chara, Nitella* etc. In most of the tanks in the red soil area, which are seasonal, the vegetation is limited to littoral areas.

Macrophytes offer substrata for an array of insects, molluscs and other invertebrate fauna, and thereby contribute to the species diversity of a water body. Nevertheless, the presence of weeds is considered to be undesirable from fisheries point of view. They accumulate large quantities of inorganic nutrients early in the season, depriving the phytoplankton of their share of nutrients. The floating vegetation utilizes the incident solar radiation for its photosynthesis and makes it unavailable to the phytoplankton communities. Submerged weeds provide shelter for minnows and weed fishes which compete with major carps for food. Excessive growth of macrophytes cause high rate of decomposition of dead plants at the bottom, creating anaerobic conditions. Problems are further confounded, if the water surface is matted by the floating vegetation which prevents light penertration. Instances of fish mortality in summer under such circumstances are reported from Hussainsagar and the reservoirs of Rajasthan. A major deleterious effect of weeds is its physical obstruction they cause to a variety of fishing gear.

The submerged plants commonly encountered in Mansarovar are *Ceratophyllum demersum*, *Potamogeton crispus*, *P. perfoliatus*, *P. pectinatus*, *P. natans*, *P. nodosus*, *Najas graminaea*, *N. minor*, *Hydrilla verticillata*, *Vallisneria spiralis*, *Chara intermedia*, *Nitella gracilus*. Floating weeds comprising *Eichhornia crassipes*, *Lemna* sp., and *Spirodela* and the emergent *Ipomoea aquatica*, *I. reptans*, *Polygonum glabrum*, *Typha angustata*, and *T. elephantia* are reported. Hussainsagar is choked mainly with the water hyacinth, while the reservoirs of Rajasthan harbour a rich flora, dominated mainly by *Hydrilla*, *Ceratophyllum*, *Potamogeton*, *Najas*, *Azolla* and *Ipomoea*. The marginal areas of Yerrakalava reservoir in Andhra Pradesh have *Hydrilla* and *Chara*, while the deeper zones harbour *Vallisneria* and *Spirogyra*

Benthos

Benthic invertebrates fauna show an erratic distribution in Indian reservoirs. The main factors that retard this community are the predominantly rocky bottom, frequent water level fluctuation and the rapid deposition of silt and other suspended particles. In spite of this, a number of reservoirs harbour rich communities of benthic invertebrates. The sequence of dominance of benthic communities closely follows the soil fertility pattern, the pre-impoundment debris often providing suitable habitats. The benthic community succession especially that of chironomids is sometimes used to characterize habitat changes. High shoreline development, variable slopes and vegetation act as favourable factors for the development of a rich assemblage of benthic organisms.

Small, shallow reservoirs of the Gangetic basin, such as Bachhra and Baghla are particularly rich in benthic fauna mainly due to the favourable substratum, rich in organic matter and the absence of swift changes in water level. In Bachhra reservoir, the standing crop of benthos registered a steady growth from 490 to 1 894 individuals m^{-2} during the last 10 years. Baghla, another small reservoir in Uttar Pradesh, has a population of benthic invertebrates represented by *Chironomus*, annelids and molluscs. The deep Rihand reservoir in the Ganga basin has a poor benthic community.

Reservoirs of Karnataka, such as Tungabhadra, Markonahalli, Hemavathy, Vanivilas Sagar and Krishnarajasagar have impressive populations of benthic organisms. So are the reservoirs of Himachal Pradesh and Rajasthan. Trends in Tamil Nadu and Madhya Pradesh are erratic. The local conditions, rather than a general geo-climatic features of the area, determine the density of benthic populations in their reservoirs.

Chironomid larvae, being a saprophobic, quickly fill the niches vacated by the saproxenes during the tranformation of habitats. They form the most important constituent of benthos, reported from all soil types and geographic locations and depths. Gastropods and annelids form the next important groups. *Viviparus bengelensis* enjoys country-wide distribution.

Among the biotic communities of the reservoir ecosystem, periphyton is the least reported upon. It constitutes an important component of food for the browsing fishes which contribute substantially to the total fish biomass of the tropical reservoirs. Apart from the limited littoral region in reservoirs, it is the frequent level fluctuations that prevent the growth of periphyton on natural substrata. Significantly, rich periphyton, whenever reported, coincides with relatively stable reservoir levels. There are reports of rich periphyton deposits on anchored boats, rafts, *etc.* that move down along with the receding water level. The fixed substrata either get totally exposed when water level decreases or they are submerged too deep for the communities to survive when level goes up. Propensities for rich settling rates of periphyton have been established through experiments with the artificial substrata, such as glass slides (David *et al.*, 1975; Jha, 1979; Sugunan and Pathak, 1986).

Ichthyofauna

Despite the cataclysmic faunistic changes associated with the impoundment, Indian reservoirs preserve a rich variety of fish species. The ichthyofauna of a reservoir basically represents the faunal diversity of the parent river system. On the basis of studies conducted so far, large reservoirs, on an average, harbour 60 species of fishes, of which at least 40 contribute to the commercial fisheries. The fast-growing Indo-Gangetic carps, popularly known as Indian major carps, occupy a prominent place among the commercially important fishes. More recently, number of exotic species have contributed substantially to commercial fisheries. Broad categorisation of the species is as follows:

The Indian major carps : *Labeo rohita, L. calbasu, L. fimbriatus, Cirrhinus mrigala, Catla catla.*

The mahseers : *Tor tor, T. putitora , T. khudree, Acrossocheilus hexagonolepis*

The minor carps including snow trout and peninsular carps : *Cirrhinus cirrhosa, C. reba, Labeo kontius, L. bata, Puntius sarana, P. dubius, P. carnaticus, P. kolus, P. dobsoni, P. chagunio, Schizothorax plagiostomus, Thynnichthyes sandkhol, Osteobrama vigorsii,*

Large catfishes : *Aorichthys aor, A. seenghala, Wallago attu, Pangasius pangasius, Silonia silondia, S. childrenii,*

Featherbacks : *Notopterus notopterus, N. chitala,*

Airbreathing catfishes : *Heteropneustes fossilis, Clarias batrachus,*

Murrels : *Channa marulius, C. striatus , C. punctatus, C. gachua,*

Weed fishes : *Ambassis nama, Esomus danrica, Aspidoparia morar, Amblypharyngodon mola, Puntius sophore, P. ticto, Oxygaster bacaila, Laubuca laubuca, Barilius barila, B. bola, Osteobrama cotio, Gadusia chapra.*

Exotic fishes: *Oreochromis mossambicus, Hypophthalmichthys molitrix, Cyprinus carpio specularis, C. carpio communis, Gambusia affinis, Ctenopharyngodon idella.*

Most of the catfishes, featherbacks, air breathing fishes, murrels and the weed fishes enjoy a country-wide distribution, while that of the major carps, minor carps and mahseers (*Tor putitora, T. tor, Acrossocheilus hexagonolepis*) varies according to river basins. The Indian major carps, catla (*C. catla*), rohu (*L. rohita*) and mrigal (*C. mrigala*) constitute the important native ichthyofauna of the rivers of the Gangetic system. These rivers also harbour *Labeo bata*, *P. sarana*, *P. chagunio*, and *C. reba*. *Tor putitora*, *Labeo dero* and the snow trouts (*Schizothorax* spp.) form the dominant riverine fish fauna of Indus system. Mahseers, especially the chocolate mahseer, *Acrosscheilus hexagonolepis* are also found in the streams associated with all the major river systems of the country. Indigenous fishes of the peninsular rivers include *Cirrhinus cirrhosa*, *C. reba*, *Labeo kontius*, *L. fimbriatus*, *P. dubius*, *P. sarana*, *P. carnaticus*, *P. kolus*, *P. dobsoni*, *T. tor*, *T. sandkhol* and *O. vigorsii*.

Fish faunistic diversity of a reservoir at a given time is the result of the impact of a series of man-made and natural changes on the native fauna of the parent river. Riverine fish fauna is subjected to a series of habitat changes such as water current, turbidity levels, fishing pressure, loss of breeding grounds and the changes in fish food organisms due to lake formation. The original fauna changes and hardy fish species take advantage of the vacant niches. In many reservoirs, transplantation of fishes from other basins and introduction of exotic species have led to further radical changes in the species set up.

The three Indian major carps have been stocked extensively in reservoirs all over the country for many decades and, in many instances, they have established themselves in reservoirs far away from their original habitat. Sathanur reservoir in Tamil Nadu has a naturalised population of catla that contribute 80 to 90% of the total catch. This Indo-Gangetic carp has eclipsed all indigenous fish fauna including *Labeo fimbriatus*, which dominated the scene by contributing 36% of the catch during the mid 1960s. Similarly, introductions of the silver carp in Gobindsagar, common carp in Krishnarajasagar and tilapia (*Oreochromis mossambicus*) in Amaravathy are examples of man induced changes in fish communities.

1.4.5 Impact of reservoir formation on the native ichthyofauna

Formation of reservoirs have affected especially the following indigenous fish stocks:

1. The mahseers, snow trouts and *Labeo dero* and *L. dyocheilus* of the Himalayan streams,
2. The anadromous hilsa, the catadromous eels, and freshwater prawns of all major river systems,
3. *P. sarana*, *T. tor*, *Tor mahanadicus*, *T. mosal*, *L. fimbriatus*, *L. calbasu*, and *Rhinomugil corsula* of the Mahanadi river,
4. *P. dobsoni*, *P. dubius*, *P. carnaticus*, *C. cirrhosa* and *Labeo kontius* of the Cauvery basin,
5. *P. kolus*, *P. dubius*, *P. sarana*, *P. porcellus* *L. fimbriatus*, *L. calbasu*, *L. pangusia* and *Tor kudree* of the Krishna river system, and
6. The mahseers, eels and *Osteobrama belangiri* of the northeast (Fig. 1.9).

The pristine streams of the river Sutlej harboured at least 51 species of fishes including the (exotic) trout, *Salmo truta fario*, the snow trouts, *Schizothorax* spp. and several species of hillstream fishes (Anon., 1989b). Most of them were unique due to the sub-temperate climate and the zoogeographic affiliations to the Himalayan region. The upper reaches of the Sutlej and its tributaries were particularly rich in *T. putitora*, *L. dero*, *L. dyocheilus* and *Schizothorax* spp. A decline in the number of species and their populations has been reported on account of changed ecological

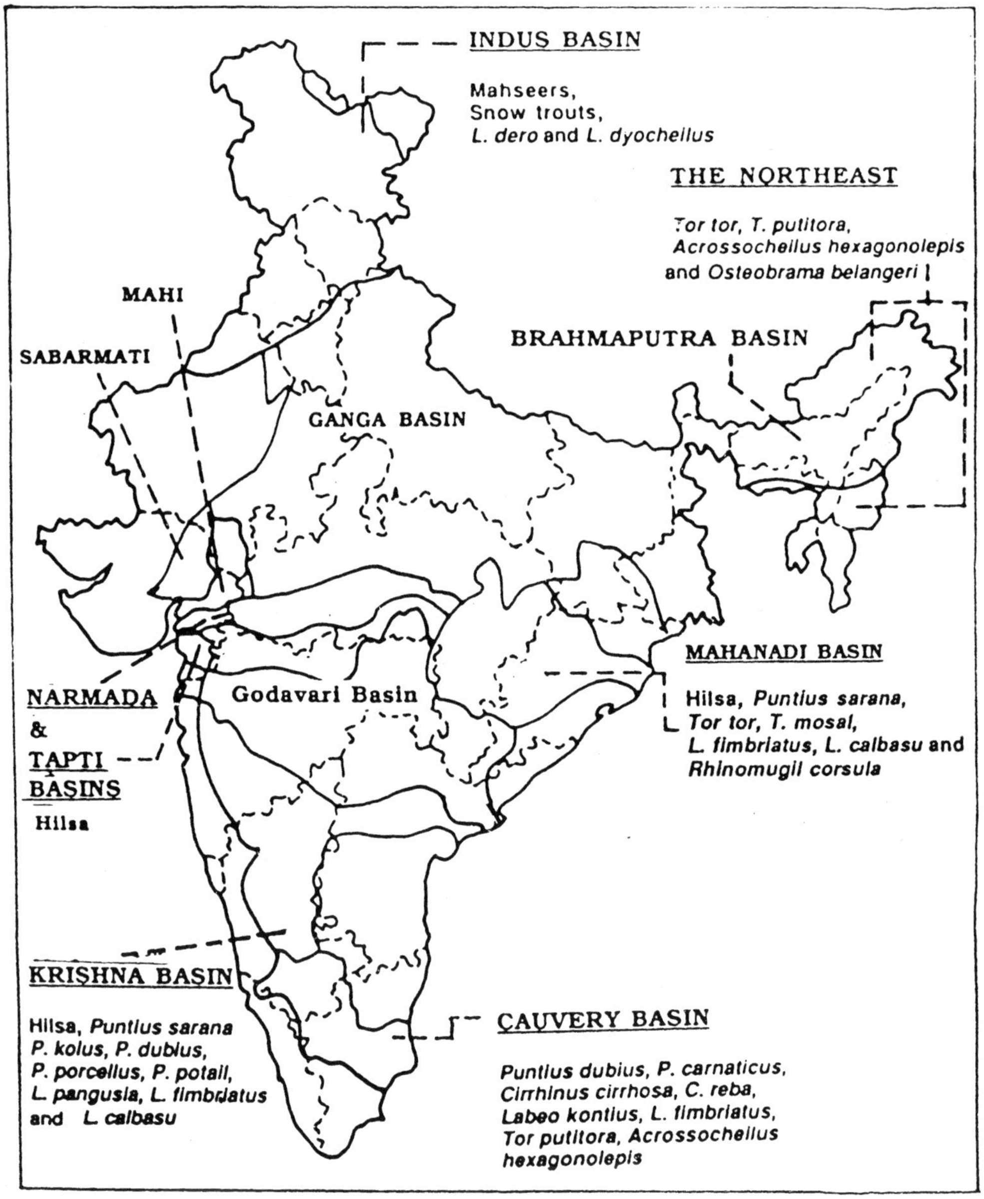

Figure 1.9. Indigenous fish species affected by reservoir formation in India

conditions especially the silt deposition at the bottom and the stratification of water body. Apart from minnows, many native species, such as, *Schizothorax plagiostomus*, *T. putitora*, *L. dero*, *L. dyocheilus* are on the decline. Proliferation of the exotic carps, *H. molitrix* and *C. carpio* in recent year has further posed threat to the local species.

Before the creation of Hirakud reservoir, the parent river Mahanadi had a rich fish fauna of 103 species, comprising both plain and sub-montane forms with sizeable representation of carps and catfishes. The common species of the river were *P. sarana*, *T. tor*, *T. mahanadicus* *T. mosal*, *L. fimbriatus* and the Indo-Gangetic major carps. The endangered *T. mosal* and *T. mahanadicus* were protected in the temple tanks that were submerged during reservoir formation. Presently, the number of species has declined to 40, of which many may disappear. The worst affected are *Tor mosal*, *Rhinomugil corsula* and the freshwater prawn *Macrobrachium malcolmsonii*.

Fishes affected in the two reservoirs in the Krishna river system *viz.*, Tungabhadra and Nagarjunasagar are *P. sarana*, and *Labeo* spp. Soon after the impoundment, the Tungabhadra reservoir harboured a good population of indigenous *P. kolus* that contributed up to a third of the total fish landings. *P. dubius*, *P. sarana*, *P. porcellus*, *P. potail* and *Labeo pangusia* were also present in large numbers. Most of these native species have eiher disappeared or declined drastically due to the absence of fluviatile environment and the changed trophic structure. The vacant niche has been filled by minnow-predator combination, due to the management lapse of not inducting the fast-growing species into the system.

Nagarjunasagar reservoir on the mainstream Krishna harboured rich populations of *Labeo fimbriatus*, *Labeo calbasu* and *T. khudree* in the earlier years of impoundment. On account of recurring breeding failure and habitat loss, these species have declined over the years, giving way to minnows which have shared the common niche with them. In the absence of any commercial fish to utilize the rich planktonic resources of the reservoir, they are mainly channelled through the detritus-molluscs chain to favour *Pangasius pangasius* and through grazing-predator chain to help the *Silonia childreni* populations. Thus, finally, these two catfishes have established a firm hold on the fish fauna.

Cauvery river system is the original abode of a number of fish species including *P. dubius*, *P. carnaticus*, *C. cirrosa*, *C. reba*, *L. kontius*, *L. fimbriatus*, *Tor putitora* and *Acrossocheilus hexagonolepis* which have been affected in various degrees by the impoundments, chiefly, Krishnarajasagar, Mettur, Bhavanisagar and Amaravathy reservoirs. During the last 25 years, the indigenous economic carps in Krishnarajasagar *viz.*, *Labeo* spp., *P. dubius* and *P. carnaticus* have suffered setback due to the changed ecological conditions, especially the components of fish food biotic communities. These species, contributing more than 60% of the total catch during the 1950s, have given way to the transplanted exotic common carp, *C. carpio*, which found a favourable environment in the reservoir *vis-a-vis* feeding and breeding. At present, most of the energy transfer is channelled through the detritus/benthos chain, giving considerable edge to the common carp which is a prolific breeder and competitor to *Cirrhinus* spp. for food.

One of the earliest casualties of hydraulic structures is the Indian shad *Tenualosa ilisha* (the hilsa) which was affected as early as mid 19th century when the Upper and Lower Anicuts were constructed on Cauvery. These barrages had severly restricted the migration of hilsa by obstructing their pathways and construction of Mettur dam (Stanley reservoir) in 1935 completely stopped the hilsa run in Cauvery. Several fishes were affected by the Mettur dam. *Puntius* spp. which used to form 28% of the landings in 1943-44 faded out in the mid 1970s. Although the indigenous *Cirrhinus cirrhosa* took some initial advantage, it also could not survive. Low water levels during July for three consecutive years have most probably caused its decline. Similarly, *Labeo kontius* which was next only to *C. cirrhosa* in Cauvery also disappeared from the reservoir. The Gangetic carps transplanted into the reservoir also did not find roots there. *C. mrigala* was stocked in 1950-51 and appeared in catch during 1957-58, contributing up to 13.9% in 1966-67, only to fade into insignificance later. The some fate met *Labeo rohita*. Recruitment failure, water level changes, and predator pressure are the main reasons for the failure of Indian major carps in Stanley reservoir. In 1993, the total catch of 115 t in the reservoir comprised *L. rohita* (19%), *Wallago attu* (15%), other catfishes (14%), *Puntius* spp (14%) and *C. catla* (10%).

Bhavanisagar is the only reservoir in the Cauvery basin, where the indigenous species like *Puntius* spp., *Tor putitora*, *T. tor*, *A. hexagonolepis*, *P. dubius*, *P. carnaticus*, *L. kontius* and *C. cirrhosa* still hold together well. Their survival is mainly due to the uninterrupted breeding activities at Moolathurai and Nellithurai, especially when water is released from the upstream Pilloor reservoir. *P. dubius* ascends the river Moyar during the northeast monsoon and lays eggs in batches of 1 000 to 2 000 on the gravel beds. Similar breeding success has been confirmed in case of *Cirrhinus reba*, *Labeo fimbriatus*, *Labeo calbasu*, *L. kontius* and *Puntius carnaticus*.

Amaravathy and Sathanur with their prime fishes of tilapia and catla respectively are the examples of introduced fishes finding a favourable environment and propagating themselves into a dominant position.

Positive impact of reservoirs on fish fauna

Many species of fish not only manage to adapt to the reservoir ecosystem but also find it congenial and flourish there, which is the main reason for the rise in the biomass of reservoir in the early stage of impoundment. However, most of the fishes that manage to multiply in the reservoir system are not very high in priority from the commercial and ecological point of view. Stocks of the small clupeid, *Salmostoma phulo phulo* and *O. vigorsii*, which support a flourishing dry fish trade in Nagarjunasagar and Tungabhadra reservoirs, multiply in a much higher scale than they do in the riverine ecosystem. The catfish, *P. pangasius* which was believed to be a catadromous migrant, has not only adapted itself to become a resident population in Nagarjunasagar, it has also become a very important component of the population. Ramakrishniah (1994) described many instances where reservoirs acted as sanctuaries by citing examples of *Barilius bola* in Tilaiya (Damodar), *Mystus krishnensis*, *Osteobrama vigorsii*, and *Pseudeutropius taackree* in Nagarjunasagar (Krishna), *T. sandkhol* in Nizamsagar (Godavari), *Tor khudree* and *T. mussullah* in Shivajisagar (Krishna), *A. seenghala* and *T. putitora* in Pong (Beas) and Vallabhsagar (Tapti).

1.5 RESERVOIR FISHERIES

A realistic evaluation of fish production from reservoirs in India is elusive. Compared to the impressive volume of data generated by the individual research workers and various institutions on limno-chemical variables and biotic communities, the estimates on fish catch remain grossly inadequate. Reliable fish catch statistics and yield estimates remain as the weakest link in the database on reservoirs. More discomfitting is the fact that the production figures available on most of the reservoirs are inaccurate and unreliable. This lacuna is imputable to a large number of factors, chiefly:

1. the multiplicity of agencies owning the fishing rights that pose difficulties in some States to gather data,

2. highly scattered and unorganised market channels, mostly under the clutches of illegal money lenders,

3. ineffective cooperative set up,

4. diverse licensing/royalty/crop sharing systems practised by different State Governments, some of which include a *free for all* system, providing little scope for recording catch statistics, and

5. inadequate and poorly trained manpower at the disposal of State Governments/Cooperatives to collect catch data, follow statistically sound sampling procedures, unable to cover the whole reservoir.

The All India Coordinated Project on Reservoir Fisheries took the lead in evolving a collection methodology on the basis of stratified random sampling during 1971 to 1985. However, many State Governments were not able to follow the procedure and to continue with recording of catch data after the project wound up in 1985. Nevertheless, some States like Tamil Nadu and Madhya Pradesh, despite the enormity of the resource size, have a streamlined machinery to record catch statistics. Similarly, Himachal Pradesh has good documentation on catch. On the contrary, Karnataka, has very little information on the catch structure of its reservoirs. Figures of Andhra Pradesh seem to be off the mark, due to the free fishing system and inadequate methods of data collection. For instance, system followed in Nagarjunasagar virtually allows anybody to catch fish in the reservoir and freely sell it anywhere. Considering the size of the lake and the large number of remote landing centres, an effective monitoring of catch is almost impossible. This is equally true with regard to many reservoirs of Kerala, Maharashtra, Orissa, Karnataka and Uttar Pradesh.

Based mainly on the data obtained from various State Governments, the fish production particulars from 422 reservoir have been presented in Table 1.6. Fish yield figures of small reservoirs of Andhra Pradesh, as given by the State Fisheries Department are very impressive (188 kg ha^{-1}), followed by those of Kerala, Madhya Pradesh, Tamil Nadu and Rajasthan in the range of 46.43 to 53.5 kg ha^{-1}. Medium reservoirs of Rajasthan, on an average, produce fish at the rate of 24.47 kg ha^{-1}, while Tamil Nadu, Maharashtra, Madhya Pradesh and Orissa record about half this yield. The two large reservoirs in Himachal Pradesh produce 35.55 kg ha^{-1} which is remarkably high, compared to other States.

The estimated fish yield of 291 small, 110 medium and 21 large reservoirs of the country are 49.90, 12.30 and 11.43 kg ha^{-1} respectively. Based on the catch of 422 reservoirs belonging to 10 states during 1992-93, the national fish production rate of Indian reservoirs is estimated as 20.13 kg ha^{-1}. Applying this national average yield rates into the 1 485 557 ha of small, 527 541 ha of medium and 1 140 268 ha of large reservoirs in the country, their current production rate can be estimated as 74 129, 6 488 and 13 033 t respectively. A modest increase in yield rate up to 100, 75 and 50 kg ha^{-1} in respect of small, medium and large reservoirs, would ensure production of 148 556, 39 565 and 57 013 t. This would increase the production by 2.5 times *i. e.*, from the present 93 650 t to 245 134 t (Table 1.7).

1.5.1 Reservoir fisheries management in India

The present low level of fish production in Indian reservoirs can be attributed to inadequate management inasmuch as many of them have high propensities of production from a limno-chemical point of view. In many of the reservoirs, the high rate of the primary and secondary productivity is not channelled to fish production. Insufficient understanding of the reservoir ecosystem often comes in the way for adopting effective management measures.

Table 1.6. Fish production in different categories of reservoirs in India

State	Small reservoirs			Medium reservoirs			Large reservoirs			Pooled		
	Number	Production (t)	Yield (kg ha^{-1})	Number	Production (t)	Yield (kg ha^{-1})	Number	Production (t)	Yield (kg ha^{-1})	Number	Production (t)	Yield (kg ha^{-1})
Tamil Nadu	52	760	48.50	8	269.0	13.74	2	294.0	12.66	62	1323.0	22.63
Uttar Pradesh	31	168	14.60	13	156.0	7.17	1	50.0	1.07	45	374.0	4.68
Andhra Pradesh	37	2224	188.00	29	306.0	22.00	3	800.0	16.80	69	4330.0	36.48
Maharashtra	6	72	21.09	12	313.5	11.83	4	794.0	9.28	22	1179.6	10.21
Rajasthan	78	970	46.43	17	599.7	24.47	2	120.0	5.30	97	1690.0	24.89
Kerala	7	118	53.50	2	17.3	4.80	-	-	-	9	135.0	23.37
Bihar	25	22	3.91	3	7.2	1.90	1	0.8	0.11	28	30.0	0.054
Madhya Pradesh	2	24	47.26	20	624.9	12.02	3	1184.0	14.53	25	1833.1	13.68
Himachal Pradesh	-	-	-	-	-	-	2	1453.0	35.55	2	1453.0	35.55
Orissa	53	349	25.85	6	163.0	12.76	3	925.0	7.62	62	1437.0	9.72
Total	291			110			21			422		
Average			49.90			12.30			11.43			20.13

Table 1.7. Present and potential production from reservoirs of India

Category	Yield ($kg\ ha^{-1}$)	Area (ha)	Present production	Potential production
Small	49.90	1 485 557	74 129	148 556
Medium	12.30	527 541	6 488	39 565
Large	11.43	1 140 268	13 033	57 013
Total		**3 153 366**	**93 650**	**245 134**

Since fish production from reservoirs is essentially extractive in nature, the essence of management strategy lies in exploitation of natural stocks. Nevertheless, the ecosystem management provides different degrees of freedom for stock maniputation, depending on the size and class of the water body (Table 1.8). One of the possible criteria that can be used to differentiate between capture and culture fisheries is the extent of human intervention in the ecosystem management. While aquaculture systems provide maximum avenues for the man to monitor and change the habitat variables and the biotic communities at will, this freedom attenuates as we proceed from aquaculture to the culture-based and capture fisheries. In a large water body, managed on capture fishery norms, there is little room for altering the habitat variables and the scope for effecting change in biotic communities is limited to stocking and ranching, which have uncertain chances of success.

Relative contribution of culture and capture norms in management vary, depending on the category of the reservoir. Medium and large reservoirs are predominantly capture fisheries systems and the management norms are based on the principle of stock manipulation, adjustment in fishing effort, observance of conservation measures and gear selectivity. Selective stocking is resorted to for correcting imbalances in species spectrum and to fill the vacant ecological niches. The small reservoirs, on the other hand, are generally managed as culture-based capture fisheries, akin to extensive aquaculture, where the main accent is on stocking, fattening and harvesting. An imaginative stocking and harvesting schedule and right species mix hold the key for effective management of small reservoirs.

1.5.2 Stocking

Inducting fast-growing extraneous species into the ecosystem to colonise the diverse niches is a necessary prerequisite of reservoir management. Since one of the primary aims of stocking is to ensure utilization of the enhanced food reserves, the ideal time to stock new species is the period of trophic burst. Any lapse in this important management measure causes the proliferation of trash fishes by taking advantage of the increased availability of fish food organisms, which in turn, may provide forage base for catfishes. Nagarjunasagar, Tungabhadra, Hirakud and a number of other large reservoirs in India are examples where the minnows, catfishes, murrels and other uneconomic fishes gained grounds in the early years, leading to establishment of long food chains. These reservoirs harbour good standing crops of plankton and benthos, which are not reflected in the fish output. Even intensive stocking at a later stage has failed to reverse the situation.

Table 1.8. The broad distinguishing features of small and large reservoirs

Small reservoirs	Large reservoirs
Single-purpose reservoirs mostly for minor irrigation.	Multi-purpose reservoirs for flood-control, hydro-electric generation, large-scale irrigation, etc.
Dams neither elaborate nor very expensive. Built of earth, stone and masonry work on small seasonal streams.	Dams elaborate, built with precise engineering skill on perennial or long seasonal rivers. Built of cement, concrete or stone.
Shallow, biologically more productive per unit area. Aquatic plants common in perennial reservoirs but scanty in seasonal ones.	Deep, biologically less productive per unit area. Usually free of aquatic plants. Subjected to heavy drawdowns.
May dry up completely in summer. Notable changes in the water regime.	Do not dry up completely. Changes in water regime slow. Maintain a conservation-pool level (= dead storage).
Sheltered areas absent.	Sheltered areas by way of embayments, coves, etc. present.
Shoreline not very irregular. Littoral areas with a gentle slope.	Shoreline more irregular. Littoral areas mostly steep.
Oxygen mostly derived from photosynthesis in these shallow, non-stratified reservoirs, lacking significant wave action.	Although photosynthesis is a source of dissolved oxygen, the process is confined to a certain region delimited by vertical range of transmission of light (euphotic zone). Oxygen also derived from significant wave action.
Provided with concrete or stone spillway, the type and size of the structure depending on the size of the runoff.	Provided with more complex engineering devices.
Major carps breed in the reservoirs.	Breeding mostly observed in the headwaters or in other suitable areas of the reservoir.

Cont.

Table 1.8 (cont.). The broad distinguishing features of small and large reservoirs

Small reservoirs	Large reservoirs
Can be subjected to experimental manipulations for testing various ecosystem responses to environmental modifications.	Cannot be subjected to experimental manipulations.
Trophic depression phase can be avoided through chemical treatment and draining. Cycle of fish production can be repeated as often as the reservoir is drained.	Trophic depression phase sets in.
The annual flooding during rainy season may be compared to overflowing of floodplains. Inundation of dry land results in a release of nutrients into the reservoir when it fills up, resulting in high production of fish food through decomposition of organic matter, predominantly of plant origin, leading to higher fish growth and survival.	Loss of nutrients occurs as they are locked up in bottom sediments. Rapid sedimentation will reduce benthos production.
Through complete fishing or overfishing of seasonal reservoirs, no brood stock is left. Fish stock has to be rebuilt through through stocking. There is thus established a direct relationship between stocking rate and catch per unit of effort.	Prominent annual fluctuations in recruitment occur and balancing of stock number against natural mortality requires high density stocking of fingerlings. Their capture requires efficient capture methods.

(Jhingran, 1988)

Since large and medium reservoirs are to be developed on the principles of capture fisheries, it is desirable to stock the species that may breed and ultimately get naturalised in the system through autostocking. This is imperative to meet the long-term objective of obtaining a sustained yield rate. Management involving persistent stocking in large water bodies not only pushes up the input cost, such systems also create many practical difficulties in raising the stocking material in adequate quantities. However, naturalisation of introduced species is quite often beset with many problems. The alien species should find the habitat conducive to its biological and physiological requirements. It should have an edge in the competition for food and finally, the environmental conditions should favour its requirements of spawning and larval development. Recruitment failure due to the erratic hydrographic conditions that break the breeding rhythm have been found to be the single major factor responsible for the failure of stocked fishes to hold out in reservoir.

In Konar, strong currents wash down the carp eggs into the deep zones of the reservoir leading to their destruction and resultant recruitment failure (Parameswaran *et al.*, 1969). But more often, the fishes do not find suitable spawning grounds as in the case of many south Indian reservoirs. Since the breeding of major carps is governed to a great extent by the magnitude of monsoon floods, annual variations in this parameter affect their breeding and recruitment.

Selection of species for stocking

Jhingran (1988) has summarized the principles to be followed in the selection of species for stocking as:

1. The species should find the environment suitable for growth and reproduction.
2. It should be quick growing, ensuring high efficiency in food utilization.
3. A fishery comprising herbivores with a short food chain is preferable, as they have a better conversion of primary production to fish flesh.
4. The stocking density should be such that the food resources of the ecosystem are fully utilized and optimum population maintained, consistent with normal growth.
5. The size of the fingerlings to be stocked should be so chosen to get the desired results.
6. Seed should be readily available with minimal transportation cost .
7. Cost of stocking and managing the species must be such that the operation is economically viable.

One of the important considerations is to know the amount of food available in the new environment. This factor has a considerable bearing in determining stocking rates and hence production. Fish production from unit area is a product of individual growth rate and population density. From a study of growth rate of various species in a particular body of water, it is possible to assess their optimal stocking density. Efficient utilization of fish food communities increases the carrying capacity and therefore calls for a higher population density which is achieved by addition of species to the original fish populations. In addition, information on differences, if any, in the growth rates of the endemic and introduced species, and the time taken by the introduced species in attaining harvestable size would provide insight into the production dynamics of the system.

Competition

Both intra- and inter-specific competitions are to be considered in the stocking programme. Situations where two or more species use a similar resource, such as food or space, lead to overcrowding and poor growth rate. Under a higher than optimum stocking rate, though production may be high, the individual growth rate will be so small that to attain a marketable size a long growing period will be needed. On the other hand, if bigger fishes are needed, the rate of stocking should be lowered and a low production will have to be accepted. Similarly, when a marketable size is to be attained in a shorter period, stocking rate will have to be lowered to allow faster growth. Thus, a desired balance among stocking rate, population density and growth is to be maintained with enough flexibility so as to swing it to suit the changes in enviromental factors. Such a plan must determine tentative stocking rates and population thinning, depending on the need(Jhingran, 1988).

The sizes of fingerlings that are stocked in Indian reservoirs come under the pre-recurit phase and so, up to the size of entering the exploited phase they are prone only to natural mortality. Therefore, a knowledge of the natural mortality rate is essential as due compensation can be provided for it while computing optimum stocking rates. Jhingran and Natarajan (1969), while evaluating the stocking rates of Damodar Valley Corporatin (DVC) reservoirs, arbitrarily recommended a compensation factor for natural mortality @ 25% for reservoirs with no large predaceous fishes and 50% for those harbouring large predators. While estimating optimum stocking rates for such populations, about which no reliable estimates of natural mortality are available, it is felt that assumption of higher natural mortality rate would be desirable as a little overstocking would be less harmful than understocking. Fish Seed Committee of the Government of India (1966) recommended the stocking rate for reservoirs at about 500 fingerlings ha^{-1} of the size range 40 to 150 mm.

Stocking measures adopted in Indian reservoirs

The policies hitherto adopted in Indian reservoirs mainly consisted of stocking fingerlings of a species or a combination of species without any definite density levels or ratios based on the biogenic capacity of the reservoir. Rate of stocking and the species-mix are often determined by their availability, as can be seen from the case studies in the chapters ahead.

Basic productivity of the reservoir is dependent on the amount of solar energy available and the efficiency of the system to transform it into chemical energy. Besides, the energy conversion efficiency at trophic levels of consumers differs considerably from one reservoir to another, depending on the qualitative and quantitative variations in the biotic communities. Any conversion rate above 1% can be considered as good. In an ideal situation, the commercial species share the ecological niches in such a way that trophic resources are utilised to the optimum. At the same time, the fishes should belong to short food chain in order to allow maximum efficiency in converting the primary food resources into harvestable materials. But in reservoirs, such conditions seldom prevail.

Indian reservoirs, by and large, have a wide ranging representation of biotic communities. Phytoplankton comprising Cyanophyceae, Chlorophyceae, Dinophyceae and Bacillariophyceae dominate over the zooplankton such as copepods, cladocerans, rotifers and protozoans. Benthos is represented by insect larvae and nymphs, oligochaetes, nematodes and molluscs. There is a rich growth of periphyton on the submerged objects. The large magnitude of water level fluctuations does not favour the establishment of aquatic macrophytic communities. Significantly, many of the above niches with the exception of insects, Cyanophyceae and molluscs are shared between Indo-Gangetic major carps and trash fishes, focussing the need for controlling carp minnows and weed fishes. The ecosystem-oriented management policy places due emphasis on trophic strata in terms of shared, unshared and vacant niches. Two main pathways through which primary energy finds its way to fish flesh are the grazing chain and the detritus chain. Contribution by both the pathways to the total availability of the energy needs to be assessed for determining the species combination, most suited to the ecosystem. A large number of Indian reservoirs exhibit the detritus chain of energy transfer.

Prior to the development of carp seed production technology in India, natural spawn collected from rivers were stocked in reservoirs. Thus, the seed of *Puntius* spp., *Cirrhinus* spp. and *Labeo* spp. collected from the Cauvery were extensively stocked in the reservoirs of Tamil Nadu along with euryhaline species such as *Chanos chanos*, *Etroplus suratensis* and *Megalops cyprinoides*. *Labeo fimbriatus*, *Cirrhinus cirrhosa*, tilapia and *Etroplus suratensis* were the common stocking material in Kerala during the early years. With the advent of induced breeding, most of the States in India were able to raise the carp seed in large numbers by the 1970s and this resulted in a shift in species-mix in favour of catla, rohu and mrigal. Today, reservoir fisheries in India largely centre on development of carp fisheries. Their unmistakable role has been demonstrasted in the Gangetic as well as peninsular reservoirs. Major carps, by virtue of their feeding habits and fast growth rate are indispensable in reservoir management. However, the Indian major carps are ill-suited to utilize phytoplankton, the most dominant fraction of plankton. The remarkable ability of silver carp in efficient conversion of phytoplankton into fish flesh has been demonstrated in Kulagarhi and Getalsud reservoirs, despite the persisting doubts about the digestibility of *Microcystis*. However, introduction of exotic fishes in open waters is still a subject of controversy due to its possible deleterious effects on indigenous populations.

Development of endemic candidates as stocking material has not made much headway in the country although some of them have a proven track record in ensuring an efficient energy transformation rate. *P. pangasius*, subsisting on a molluscan diet is a species to be considered in the detritus-based, mollusc-rich reservoirs of the country. *Puntius pulchellus*, the peninsular species is a well-known macrophyte feeder and *Thynnichthys sandkhol* consumes *Microcystis*, the common alga in Indian waters. Diversification of stocking material is essential for establishment of a multi-species fish stock that utilize all food niches of the ecosystem. In reservoirs, where annaul drawdown is not pronounced and water level fluctuations are not steep, phytobenthos and macrovegetation develop in various degrees. The grass carp, *Ctenopharyngodon idella* can be considered for such water bodies. The common carp is being stocked in many reservoirs. This sluggish fish does not survive normally in the warm, deep-basin reservoirs of the south, especially when infested with predators. However, this prolific feeder could carve out a place for itself in the reservoirs of the northeast, in Gobindsagar and in some of the peninsular reservoir like Krishnarajasagar. The fish being a mud-stirrer, is considered to be unsuitable for already turbid waters.

Tilapia, due to its records of rapid proliferation and consequent stunted growth in pond ecosystem, does not find favour with many fishery managers, although in Amaravathy and Malampuzha it has performed well. Stocking of prawn, *Macrobrachium malcolmsonii* has been tried in Tungabhadra and Konar (Natarajan, 1979b), where they could not survive and contribute to the commercial fisheries. Ahmed (1993) emphasised the need for extensive stocking of this prawn, by quoting instances of its self-sustaining populations in many reservoirs. However, the prawn's ability to complete its full life cycle in the freshwater phase is yet to be proved. *T. putitora*, *L. dero*, and the exotic species such as mirror carp, silver carp, grass carp, *Tinca tinca* and *Carassius carassius* are advocated for the high altitude reservoirs (Natarajan, 1979b).

Impact of stocking

The most important objective of stocking, *i.e.*, to augment the yield, can be achieved only if the stocked fishes survive, grow and get caught in the fishing gear. This is achieved, to a large extent, in small reservoirs where the management centres round the stocking and recapture system. However, in larger water bodies, the recapture is uncertain on account of many reasons as mentioned earlier.

Impact of stocking in medium and large reservoirs

Experience in a number of medium and large reservoirs prompts us to conclude that the stocking programme can be termed as successful, only when the stocked fishes breed in the reservoir and contribute towards autostocking. In many cases, despite persistent stocking, the transplanted species did not show up in the catch, thereby rendering the expenditure incurred in stocking as waste. Only in a few instances the resources mobilised for stocking operation were compensated by generation of income through recapture of the stocked fishes.

Sreenivāsan (1984) reviewed the impact of stocking in 10 reservoirs of Tamil Nadu. The stocked catla built up a naturalised population in Mettur reservoir. Just 10 000 fingerlings were stocked during 1922 to 1935, which formed the nucleus of a self-propagating stock and dominated the catch during the 1960s. Catla fisheries, however, suffered periodic setback due to breeding failures. The current contribution is as low as 10%. Recapture of two other stocked fishes *viz.*, *L. rohita* and *L. calbasu* is reported to be adequate (Sreenivasan, 1984). However, stocking of *L. fimbriatus* (over 2 million), common carp (1 million), *L. kontius* (0.4 million), *P. carnaticus* (0.4 million), *C. reba* (several hundred thousands) and *P. dubius* (several hundred thousands) is believed to be wasteful, since they were never recaptured in any appreciable quantity.

In Bhavanisagar, although the transplanted catla showed up in the catch, they could not make any impact on the catch structure. As opposed to this, *L. calbasu*, another transplant, found the environment congenial for breeding and established itself in the reservoir, supporting a lucrative fishery for the last few decades. More than 2 million common carp fingerlings were stocked in the reservoir, of which only a few hundreds were recaptured. Intensive stocking of *L. fibriatus* did not make any dent on the fisheries, despite the fact that the fish was native to the system.

In Sathanur, a breeding population of catla has been successfully established through stocking and the fish contributes more than 80% of the total catch. The species has also eclipsed the indigenous *L. fimbriatus* by reducing its contribution from 36% to 1%. It is pertinent to note that the stocking done over 12 years involving 2 million fingerlings of indigenous species such as *C. cirrhosa*, *L. kontius*, *C. reba* and *L. fimbriatus* could not make any impact on restoration of their fisheries, primarily due to the inability of the fishes to breed and propagate themselves.

In Krishnagiri, although the increase in percentage of the stocked major carps has resulted in some initial increment in their contribution in the catch, the recapture was not commensurate with the stocking rate. Stocking of Gangetic carps, common carp and *L. fimbriatus* done in Vaigai, so far, has been described as wasteful (Sreenivasan, 1984), where presently, catla contribute 20% of the catch.

In Malampuzha and Peechi, the two medium reservoirs in Kerala not register any substantial increase in yield rate despite sustained stocking with Gangetic major carps, mainly on account of their failure to breed. In Malampuzha, despite a sharp increase in stocking density from 52 ha^{-1} in the 1970s, and a corresponding increase in the percentage of major carps in the stocked fishes, the yield rate remains at a low level of 5.0 kg ha^{-1}. Although catla grows to an impressive size, the contribution of major carps never exceeded 20% of the total catch. Considering the low yield rate, the quantity of major carps harvested does not commensurate the stocking effort. Similarly, in Peechi, 90% of the fingerlings stocked belong to the Indian major carps, especially *L. rohita*. But they are not reflected adequately in the catch and the yield rate remains at a low level of 4.5 kg ha^{-1}.

In Nagarjunasagar, Andhra Pradesh, regular annual stocking at the rate of 50 000 to 833 000 fingerlings comprising catla, rohu and mrigal during the 1970s had little impact on the catch structure, as none of the stocked fishes could breed and contribute to recruitment. Similarly, stocking as a management option has failed in Tungabhadra reservoir situated on the same basin and Krishnarajasagar in the Cauvery basin. In Tungabhadra, the stocking rate during 17 years ranged from 1 to 11 ha^{-1}, while it varied from 0.15 to 67 ha^{-1} in Krishnarajasagar. In both the reservoirs, the selection of species has always been arbitrary, the stocking density was inadequate and the stocked fishes failed to breed in the reservoir.

Gandhisagar in Madhya Pradesh is an example of the stocked fishes contributing to fish catch in a sustained manner through breeding and recruitment. During the 1950s through 1970s, on account of steady stocking of catla (2 million), rohu (1.3 million), and mrigal (1.1 million), the fish yield rose to 20.33 kg ha^{-1} from the initial 0.51 kg ha^{-1}. Among the stocked fishes, catla contribute 60 to 70%, while mrigal and rohu form only 1 to 20%. Mrigal, despite being a part of the indigenous ichthyofauna, is on the decline. Ravishankarsagar in the same State was stocked, on an average, with 1.77 million fingerlings every year and yet the yield rate could not be raised above 8.3 kg ha^{-1}, primarily due to the non-establishment of the carps. The amount spent on stocking has been much higher than the cost of recaptured fishes. Rihand reservoir in U. P. could build up a breeding population out of the initial stocking. Although the yield rate of 0.58 kg ha^{-1} is not impressive, 73 to 99% of the catch comprises *C. catla*.

The success of stocked Indian major carps in Ukai reservoir in Gujarat can also be attributed to their breeding in the reservoir. Apart from augmenting the reservoir fisheries through recruitment, the young ones of Indian major carps are also reported to escape through the outlet of the dam and contribute to stocking of downstream impoundments. In all the DVC reservoirs, *viz.*, Konar, Tilaiya, Maithon and Panchet, stocking did not have any impact (Sreenivasan, 1984). Some breeding activities in respect of catla and mrigal have been reported, though low survival rate of the spawn and frequent breeding failure due to erratic monsoon prevented proper recruitment of the planted species.

Indian experience of stocking medium and large reservoirs suggests that by and large, the stocking becomes effective only when the stocked fishes propagate themselves. Moreover, this breeding population can be built up only if the stocking is resorted to during the early phase of the reservoir formation.

Impact of stocking in small reservoirs

In sharp contrast to the large and medium reservoirs, stocking has been more effective in improving the yield from small reservoirs as success in the management of small rerservoirs depends more on recapturing the stocked fish rather than on their building up a breeding population. The smaller water bodies have the advantage of easy stock monitoring and manipulation. Thus, the smaller the reservoirs, the better are the chances of success in the stock and recapture process. In fact, an imaginative stocking and harvesting schedule is the main theme of fisheries management in small, shallow reservoirs. The basic tenets of such a system involve:

1. Selection of the right species, depending on the fish food resources available in the system.

2. Determination of a stocking density on the basis of production potential, growth and mortality rates.

3. Proper stocking and harvesting schedule including staggerred stocking and harvesting, allowing maximum growout period, taking into account the critical water levels.

4. In case of small irrigation reservoirs with open sluices the season of overflow and the possibilities of water level falling too low or completely drying up, are also to be taken into consideration.

Aliyar is a standing testimony to the efficacy of the management based on staggered stocking. The salient features of the management options adopted in Aliyar are:

1. stocking is limited to Indian major carps (earlier, all indigenous, slow-growing carps were stocked)

2. Increasing the size at stocking to 100 mm and above,

3. reducing the stocking density to 235-300 ha^{-1} (earlier rates were erratic ranging between 500-2 500 ha^{-1}),

4. staggering the stocking, and

5. regulating mesh size strictly and banning the catch of Indian major carps < 1 kg in size.

A direct result of the above management practice was an increase in fish production from 1.67 kg ha^{-1} in 1964-65 to 194 kg ha^{-1} in 1985.

Effective recapture of the stocked fishes renders the stocking more remunerative in small reservoirs. Successful stocking has been reported from a number of small reservoirs in India. In Markonahalli, Karnataka, on account of stocking, the percentage of major carps has increased to 61% and the yield increased to 63 kg ha^{-1}. Yields in Meenkera and Chulliar reservoirs in Kerala have increased from 9.96 to 107.7 kg ha^{-1} and 32.3 to 275.4 kg ha^{-1} respectively through sustained stocking. In Uttar Pradesh, Bachhra, Baghla, and Gulariya reservoirs registered steep increase in yield through improved management with the main accent on stocking. An important consideration in Gulariya has been to allow maximum growout period between the date of stocking and the final harvesting, *i. e.*, before the levels go below the critical mark, The possible loss due to the low size at harvest was made good by the number. Bundh Beratha in Rajasthan, stocked with 100 000 fingerlings a year (164 ha^{-1}) resulted in a fish yield of 94 kg ha^{-1}, 80% of which constituting catla, rohu and mrigal (Table 1.9).

Table 1.9. High yields obtained in small reservoirs due to management based on stocking.

Reservoir	State	Stocking rate (number ha^{-1})	Yield (kg ha^{-1})
Aliyar	Tamil Nadu	353	194
Meenkara	Kerala	1 226	107
Chulliar	-do-	937	316
Markonahalli	Karnataka	922	63
Gulariya	Uttar Pradesh	517	150
Bachhra	-do	763	140
Baghla	-do-	?	102
Bundh Beratha	Rajasthan	164	94

Instances where intensive stocking of Indian major carps became ineffective in small reservoirs are very rare. In Govindgarh, despite stocking of Indian major carps at the rate of 19 to 390 fingerlings ha^{-1}, the yield remained at 15.92 kg ha^{-1}, during the '60s and the '70s. Large-scale escapement of fishes through the open weir is believed to be the main reason for this low fish yield. An estimated 1.1 million fingerlings of catla, rohu and mrigal were stocked into Badua reservoir, Bihar during the period 1975 to 1979. However, the fish yield from the reservoir during the period remained within 4 to 7 kg ha^{-1}. Other management measures taken in the reservoir are not known. Sreenivasan (1984) reported disappointing recapture of major carps after their heavy stocking in Manjalar reservoir (Tamil Nadu). Proliferation of the tilapia, *O. mossambicus* is the main factor that prevented the major carps from getting a foothold in the fishery, probably due to their competition for food.

1.5.3 Removal of predators and weed fishes

Presence of predatory and weed fishes poses impediments in survival and growth of economic species in many Indian reservoirs. Keeping these unwanted population under check is a very difficult management problem, especially in large reservoirs. A small population of predators helps to crop the trash fishes which compete for food with the economic species. A small predator population of Gobindsagar which keeps the minnows under check is a good example. However, no scientifically sound methods are available to keep a limited population of predatory species. Repeated use of gill nets of appropriate mesh size, use of long lines, traps, etc. are suggested for control of the uneconomic and undesirable populations. Manipulation of reservoir level with a view to checking the breeding and destruction of the young ones of predators and the minnows has been tried in several countries. However, this is not practicable in many Indian reservoirs since water release pattern is dictated by priority sectors like irrigation and power generation. Poisoning of selected sheltered areas arms and coves as practised abroad has also limited use in India due to the multiple use of water and objections from the other water users. David and Rajagopal (1969) reported that non-selectivity of shore seines helped in reducing catfish population in Tungabhadra reservoir by 76 to 81%. *Alivi*, the giant shore seine of Tungabhadra also removes the trash fish in large numbers. Judicious use of this gear, with a condition that the juveniles of economic species are released back, can go a long way in containing the trash fish population.

Recent findings of Kartha and Rao (1990) with regard to the efficacy of trawling in checking predators and trash fishes are of interest. Bottom trawling in Gandhisagar was found to catch 64 to 91% of the unwanted fishes and this has been recommended as a method to crop the predators and carp minnows. Nevertheless, applicability of this method is limited to places where the bottom is free from obstructions. Natarajan (1979b) suggested biological control of trash fishes by stocking two euryhaline species *viz*., *Megalops cyprinoides* and *Lates calcarifier*. Since these predatory fishes do not breed in freshwater, they cannot go out of control.

1.5.4 Exotic fishes and their role in the reservoir fisheries of India

In spite of an already rich and diverse fish genetic resource of India, more than 300 exotic species have been introduced into the country so far (Jhingran, 1989a). While a vast majority of them are ornamental fishes which remain, more or less, confined to the aquaria, some others have been introduced in aquaculture and open water systems with varying degrees of success. Three larvicidal fishes *viz*., *Lebistes reticulatus*, *Nothobranchus* sp. and *Gambusia affinis* were introduced for containing the insect larvae in confined waters. Silver carp and the three varieties of common carp were brought into the country with the objectives of broadening the species spectrum in aquaculture and increasing the yields through better utilization of trophic niches. In recent years, the bighead carp *Hypophthalmichthys nobilis* and *O. niloticus* have been reported form the culture systems of eastern India. After unauthorised introduction, these two fishes are becoming popular among the aquaculturists of the region. While a few of the introduced species proved to be a boon in aquaculture and acted as an instrument for yield optimisation from ponds, the accidental and deliberate introduction of some of the exotic fishes into the open-waters has generated a lot of debate in recent years. There is a growing concern in India about the possible deleterious impact of the exotic fishes on the fish species diversity of the Indian rivers.

Oreochromis mossambicus, Hypophthalmichthys molitrix, Ctenopharyngodon idella, Cyprinus carpio communis, C. carpio specularis and *C. carpio nudus* have gained entry into the reservoir ecosystem through accidental or deliberate introduction. Among them, tilapia, silver carp and common carp could make a negative impact on the fisheries in various reservoirs in the country. Instances of *Gambusia affinis* getting naturalised in reservoirs are rare. In Markonahalli reservoir, the fish has established itself as a breeding population and reported to be affecting the larval stages of commercially important fishes.

Tilapia

The tilapia, *O. mossambicus* was first introduced into the pond ecosystem of the country in 1952 and soon it was stocked in the reservoirs of south India. By the end of 1960s, most of the reservoirs in Tamil Nadu and those in the Palakkad and Trissur districts of Kerala were regularly stocked with tilapia. Performance of tilapia in ponds of south India has been discouraging mainly due to its early maturity, continuous breeding, over-population and dwarfing. It is reported to mature at 6 cm length at an age of 75 days and to breed at an interval of one month under the tropical conditions.

Performance of tilapia in reservoirs

The warm waters of the tropical reservoirs in India have provided a conducive habitat for the tilapia and it has established a secure position in a number of south Indian reservoirs. The fears of its stunted growth have been allayed as the average size of tilapia did not decline as much as it did in ponds. Sreenivasan (1967) stated that the fluctuating water levels affected the breeding pits of the fishes and the predators took a heavy toll of their young ones. These two factors are believed to keep check on the excessive proliferation of tilapia in reservoirs.

Size of tilapia in the commercial catches of reservoirs has been very good, as opposed to the unmarketable size reported from the ponds. The average size of tilapia from Tamil Nadu reservoirs has been 1.5 kg during the 1960s, with the minimum size of 500 g. Similarly, tilapia weighing 2.5 kg was very common in Malampuzha reservoir, Kerala during the 1960s, with an average size of 1.5 to 1.75 kg. The present size of 0.5 to 0.7 kg in Malampuzha and 0.68 kg in Tamil Nadu reservoirs are well within the limits of market preference, their continuous slide in size over the years is a cause of concern as it is feared that if the fall in size continues, it may become unmarketable.

Tilapia has dominated and virtually eliminated all other fishes including the stocked Gangetic carps in a number of reservoirs in Tamil Nadu. Vaigai, Krishnagiri, Amaravathy, Uppar and Pambar reservoirs in Tamil Nadu are harbouring sizeable populations of tilapia since 1960s, contributing substantially to commercial catches. While its contribution has declined since 1979-80 in Vaigai, it continues to form a major fishery in all the other reservoirs. In Krishnagiri, the fish has a changing fortune on account of competition with the mullet, *Rhinomugil corsula*. From the predominant position in the 1960s the percentage of tilapia came down to 4.3% in 1983-86, only to increase in the year 1989-90 to 69% (Jhingran, 1991). At present, tilapia forms 24% of the catch.

Chulliar, Meenkara, Peechi and Malampuzha reservoirs in Kerala have been stocked with tilapia in the early sixties and the fish contributes substantially to the catch. In the relatively large (2 313 ha) Malampuzha reservoir, tilapia is reported to have registered a fast growth rate and large size, contributing 10 to 70% of the catch in different years. Tilapia also found its way into Kolleru lake in Andhra Pradesh and Sondur reservoir in Raipur district, Madhya Pradesh. Jhingran (1991) examined the yield of tilapia in different classes of reservoirs in the size range 50 to 10 000 ha and came to the conclusion that, by and large, small reservoirs in the size range of 50 to 200 ha showed a better yield (44 to 101 kg ha^{-1}), than the larger ones. Relatively large reservoirs such as Malampuzha and Amaravathy have a better size of tilapia, although documentation is not sufficient to attempt any correlation between the size attained by the fish and the area of reservoirs.

Distribution of tilapia is more or less restricted to the tropical belt as the fish is constrained with slow growth and winter mortalities in the higher latitude. Attempts to introduce the fish in Baghla, a small irrigation impoundment in the Gangetic plain have not succeeded (Jhingran, 1991). The ongoing debate on the introduction of tilapia centres round its:

1. suitability to enhance yield through niche utilization,
2. propensity for affecting or even replacing the native ichthyofauna, and
3. consumer preference.

One of the main considerations in determining introduction of exotic fishes is their feeding habits. Since none of the Indian culturable carps feeds on Cyanophyceae blooms like *Microcystis aeruginosa*, tilapia is often cited as a welcome addition to the blue-greens-dominated water bodies. Although Sreenivasan (1967) and many workers abroad expressed their reservations about the ability of tilapia to digest and assimilate *Microcystis*, Jhingran (1991) believed that the sodium-calcium ratio in alkaline reservoirs broke down the cell walls of blue-greens, facilitating their digestion by tilapia. *Oreochromis mossambicus* has wider omnivorous food spectrum, compared to many other species of tilapia.

Tilapia versus indigenous carps

The apprehensions about tilapia affecting the native ichthyofauna seem to be valid, going by its track record in India and abroad. In a number of reservoirs in India, introduction of tilapia has resulted in poor growth rate or even elimination of the indigenous species and the transplanted Gangetic carps. Sreenivasan (1967) found that the growth rates of *C. catla*, *L. fimbriatus* and *C. mrigala* were adversely affected by tilapia in Ayyamkulam pond. He also observed that growth of *Chanos chanos* was restricted to less than 100 g yr^{-1}, against the usual 500 g yr^{-1} in many water bodies of Tamil Nadu due to its co-existence with tilapia. In Kabini reservoir tilapia has adversely affected the indigenous *Cirrhinus reba*. During the period from 1980-81 to 1984-85, tilapia has caused decrease of *C. reba*'s share in the catch from 70% to 20% (Murthy *et al.*, 1986). Tilapia and the Indian economic species sharing a common food niche, the success of one in competition with the other is determined by the ability to breed and propagate. Given the propensities of tilapia for autostocking, the indigenous species, which are prone to breeding failure, have a definite disadvantage in its struggle to coexist with the former.

Introduction of *O. mossambicus* and *Tilapia zillii* for weed and insect control in Californian reservoirs has affected the native ichthyofauna (Moyle, 1976). A more devastating effect on indigenous fish fauna has been reported from Kyle reservoir in Zambia, where a valuable local species *Paretopus petite* has been eliminated from the ecosystem (Lamarque *et al.*, 1975).

Consumer preference

Tilapia of right size has a good consumer preference. Tilapia is also known as a species affordable by the poor. In Palakkad and Trissur districts of Kerala, where the fish is an important component of reservoir catch, tilapia enjoys a consumer preference over the Indian carps even when sold at an equal price (Jhingran, 1991).

Relevance of tilapia in reservoir fisheries

Tilapia (*O. mossambicus*) has entered the Indian scene, when the inland fisheries contributed negligibly to the total fish production in the country and the ecosystem management was in its infancy. Today, a number of indigenous species are available for stocking to broaden the species spectrum, bridge the gaps in niche utilization and increase the yield. Barring a very few reservoirs, tilapia-dominated fishery invariably leads to low yields. In many reservoirs like Krishnagiri and Vaigai, the production has been found to be erratic due to the unpredictive behaviour of tilapia population due to competition from other fishes. Fishery managers of India are striving hard to change the fish from its dominant position, wherever, they occupy one. Thus, in the present context, tilapia does not figure among the species preferred for stocking in Indian reservoirs.

Oreochromis niloticus has not yet entered the reservoir ecosystem in India. Confined to the estuarine and freshwater wetlands of the eastern India, the fish has registered an impressive growth of 250 g in 6 months. Since this fish is not reported to have problems of stunted growth and prolific breeding, it may probably have a more positive role to play in Indian reservoirs, compared to *O. mossambicus*.

Silver carp

Silver carp, *Hypophthalmichthys molitrix* was introduced in India in 1959 and unlike tilapia, it has not strayed into many reservoirs. However, silver carp has attracted more attention from the ecologists and fishery managers, generating a more animated debate. Importance of silver carp in reservoirs emanates mainly from:

1. its reported ability to utilise *Microcystis*
2. the impressive growth rate, and
3. its propensities for affecting the indigenous species, especially *Catla catla*.

The silver carp has a specialised structure of gill rakers adapted to microplankton feeding (Inaba and Nomura, 1956). Gut analysis of the fish carried out at the Central Inland Fisheries Research Institute revealed a wide feeding range including Chlorophyceae, Cyanophyceae, Chrysophyceae, Bacillariophyceae, Dinophyceae, Protozoa, Rotifera, Cladocera, Ostracoda, and Copepoda (Jhingran and Natarajan, 1978).

Performance of silver carp in reservoirs

An experimental consignment of 239 fingerlings of silver carp was stocked in Kulgarhi reservoir (Madhya Pradesh) in 1969. Based on the recapture of 8 specimens, growth rates ranging from 597 mm in 783 days (0.76 mm day^{-1}) to 404 mm in 293 days (1.4 mm day^{-1}) have been recorded (Rao and Dwivedi, 1972). The fish was also introduced in Getalsud reservoir, Bihar in 1974, where it has recorded growth rates ranging from 2.20 to 5.79 g day^{-1} (Table 13.4). In both reservoirs, the fish did not breed. There are reports about the stocking of silver carp in a number of reservoirs across the country ranging from Gumti in the northeast to Aliyar in Tamil Nadu. However, it did not get established as a breeding population anywhere.

The most spectacular performance of silver carp has been reported from Gobindsagar reservoir, where after an accidental introduction, the fish formed a breeding population and brought about a phenomenal increase in fish yields. (see the chapter on Himachal Pradesh). Silver carp is instrumental in enhancing the fish production from the reservoir from 160 t in 1970-71 to 964 t in 1992-93 (Fig. 1.10). Many workers have suggested the over-intensity of feeding in respect of silver carp, as reflected by the *gorged* and *full* conditions of gut all through the year. Jhingran and Natarajan (1978) pointed out that the silver carp, being a cold-water fish, when introduced into a warm regime, consumed food much in excess and grew faster as expected of a true poikilotherm. A similar latitude- induced change worth noticing is the age at maturity. The fish mature when they are 5 to 6 years old in North China, 4 to 5 years in Central China and 2 to 3 years in south China. In India, it breeds just at the age of one year under optimum conditions.

Silver carp versus catla

Pond culture experiments conducted in the country have unmistakably established the superior performance of silver carp when cultured along with catla (Sukumaran, *et al.*, 1968). Jhingran and Natarajan (1978) expressed the view that this need not hold good for large water bodies like reservoirs. They argued that the silver carp short-circuited the food-web, resulting in the poor performance of catla. For instance, silver carp showed preference for smaller zooplankters especially rotifers and nauplii. It was quite likely that too much of grazing on copepode nauplii by silver carp could disturb the life cycle of copepodes in a small water body like pond, causing poor performance of catla. While advocating a cautious approach, they advocated stocking of silver carp in closed reservoirs like Gobindsagar and Nagarjunasagar which are blocked by dams, both down- and upstream and no connected with the Ganga river system, the original abode of *C. catla*. The fish was not stocked in either of the reservoirs. However, in Gobindsagar, stocking of silver carp has since become irrelevant as the fish has already carved out a niche for itself in the reservoir. In the process, the lurking fear that the exotic fish is deleterious to the populations of indigenous catla has been proved beyond any doubt. Ever since silver carp gained a stronghold in the reservoir, catla which constituted an annual fishery of the magnitude of 200 to 300 t has declined considerably (Fig. 1.10)

Karamchandani and Mishra (1980), while evaluating the co-existence of silver carp and catla, established that the two fishes shared a common niche and compete with each other for food in a reservoir ecosystem. Percentage composition of phytoplankton in the guts of both the fishes caught during the same time from Kulgarhi reservoir was more or less the same. Zooplankton, the favourite menu of catla, formed 21% of the gut contents of silver carp. The authors concluded that silver carp hampered the growth of catla in the reservoir and advocated caution before its stocking in Indian reservoirs.

It is significant to note that despite its entry into a number of Indian reservoirs, by accident or otherwise, silver carp failed to get naturalised anywhere except Gobindsagar. Considering that the reservoir, with its temperate climate, is closer to the original habitat of the fish and has a distinctly cold water hypolimnion due to the discharge from Beas, the silver carp seems to have found a congenial habitat for growth and propagation. Although introduction of silver carp was never cleared by the Committee of Experts constituted by Govt. of India, the fish is being

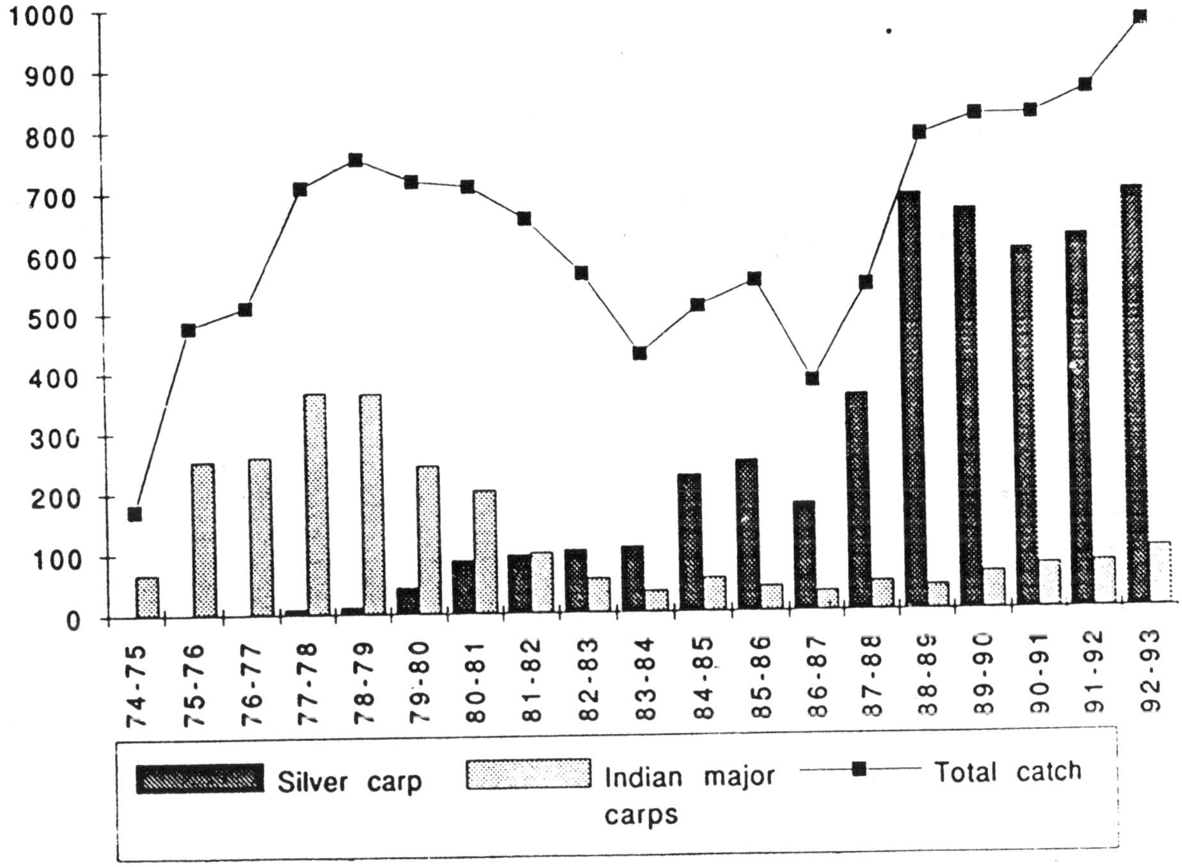

Figure 1.10. Catch of silver carp in relation to the total fish catch in Gobindsagar (in t)

stocked in a number of reservoirs in the country. Nowhere did the fish make an impact as it did in Gobindsagar. Therefore, fears regarding the threat of extinction of catla from the Gangetic and peninsular India posed by silver carp are perhaps misplaced.

Common carp

The three varieties of the Prussian strain of common carp, *viz.*, the scale carp (*Cyprinus carpio communis*), the mirror carp (*C. carpio specularis*) and the leather carp (*C. carpio nudus*) were introduced in India during 1939. They were stocked in several high altitude ponds and lakes during the1950s. Later, in the 1957, the Chinese (Bangkok) strain of the common carp was brought into the country, primarily for aquacultural purposes, considering its warm water adaptibility, easy breeding, omnivorous feeding habits, good growth and hardy nature.

Like tilapia, common carp soon found its way to all types of reservoirs in the country. Relative ease at which the fish could breed in controlled conditions prompted the departmental fish farms throughout the country to produce the seed of common carp in large numbers and to stock them in the reservoirs. However, such stocking attempts were devoid of any ecological reasoning. The Bangkok strain of common carp has been stocked in a large number of reservoirs in the plains and European strain was introduced in the reservoirs of temperate zones and high altitudes. But their performance in reservoirs is erratic, despite heavy stocking.

Common carp is not a suitable fish for stocking in Indian reservoirs, especially the larger ones, for diverse reasons. Being a sluggish fish, its chances of survival in a predator-dominated reservoir are very poor. They are not frequently caught in a passive fishing gear like gill net, due to its slow movement and bottom dwelling habit. It is no wonder, despite a regular stocking for 13 years (involving 537 000 fingerlings), not a single common carp was ever caught from Nagarjunasagar. Obviously, the stocked fishes failed to survive among the marauding predators. This has been the fate of common carp stocked in all the deep reservoirs, with a few exceptions such as Krishnarajasagar. A more important disqualification is its propensities to compete with some important indigenous carps like *Cirrhinus mrigala* and *C. cirrhosa* and *C. reba* with which common carp shares food niche. Instances where the presence of common carp has resulted in the decline of *Cirrhinus* sp. are available in Girna and Krishnarajasagar.

The mirror carp has a dubious distinction of jeopardizing the survival of a number of native fish species, after its introduction in some of the upland lakes of Kumaon Himalayas, the Dal lake in Kashmir, Gobindsagar, and the reservoirs of the northeast. In the Dal lake, common carp found a favourable environment by virtue of the shallow lake basin, extensive submerged vegetation, and rich food resources. By virtue of the specific ecological advantage, the fish propagated itself profusely to the peril of indigenous snow trouts like *Schizothoraichthys niger, S. esocinus*, and *S. curvifrons*. The snow trouts had the twin disadvanges of low fecundity and the stream breeding behaviour. Mirror carp has caused similar damage to the snow trouts in Gobindsagar reservoir and *Osteobrama belangeri* in Loktak lake of the northeast. Analogy of events related to the common carp and snow trouts sends out enough signals regarding the potential harm the former can do to the ichthyofauna in the plains.

Other exotic species

Three exotic carps, being considered for introduction in the country are the bighead carp, *H. nobilis* (already gained entry unofficially) the mud carp, *Cirrhinus molitorella* and the snail carp, *Mylopharyngodon piceus*, feeding on zooplankton, detritus and molluscs respectively. Natarajan (1988), after a thorough probe into the ecological implications of their introduction, considered their introduction as an irrational step, as all of them infringe on the food niche of economic carp species of India and the alien species possessed all propensities for causing extinction of their native counterparts.

1.5.5 Artificial eutrophication

Fertilization of reservoirs as a means to increase water productivity through abetting plankton growth has not received much attention in India. Multiple use of the water body and the resultant conflict of interests among the various water users are the main factors that prevented the use of this management option. Surprisingly, fertilization has not been resorted to even in reservoirs which are not used for drinking water and other purposes. Documentation on fertilization of reservoirs in India is scarce. Sreenivasan and Pillai (1979) attempted to improve the plankton productivity of Vidur reservoir by the application of super phosphate with highly encouraging results. As soon as the canal sluice was closed, 500 kg super phosphate with P_2O_5 content of 16 to 20% was applied in the reservoir when the waterspread was 50 ha with a mean depth of 1.67 m. As an immediate result of fertilization, phosphate content of water increased from nil to 1.8 mg l^{-1} and that of soil from 0.242 to 0.328%. Similar improvements in organic carbon and Kjeldal nitrogen has been reported from soil and water phases on account of fertilization. Experiments were also conducted with urea in the same reservoir.

Application of lime was tried in some upland natural lakes for amelioration of excessive CO_2 and acidity at the bottom (Sreenivasan, 1971). This measures, together with the application of superphosphate in Yercaud lake, raised the pH of water from 6.2 to 7.3 and decreased the CO_2 in bottom water from 38 to 6.5 mg l^{-1}. There was a corresponding increase in species number and biomass of plankton.

The basic objective of fertilization is to increase the plankton density and thereby accelerating primary productivity. Fertilization in Vidur reservoir resulted in a marked increase in benthic and plankton communities and doubling of the primary production rate. After two successive applications of fertilizers, significant limnological changes took place including the presence of free carbon dioxide and decrease in pH and dissolved oxygen at the bottom layer of water. The methylorange alkalinity increased from 44 to 108 mg l^{-1} from the surface to bottom, indicating a high organic productivity. Phosphate fertilization triggered the tropholytic activities mineralising the organic matter and producing carbon dioxide. As a direct benefit from the fertilization, a 50% increase in fish production, along with three-fold increase in the size (average weight) of catla, rohu, mrigal, *L. fimbriatus* and *L. calbasu* were achieved.

Experiments on fertilization is in progress in the 90 ha Naktra reservoir in Madhya Pradesh, under a research project of the Central Inland Capture Fisheries Research Institute. Organic and inorganic fertilisers are being applied to improve the water and soil quality status. Artificial eutrophication as a decisive management option was tried in India for the first time in Kyrdemkulai (80 ha) and Nongmahir (70 ha) reservoirs of the northeast (Sugunan and Yadava, 1991a, b) by applying poultry manure (10 t ha^{-1}), urea (40 kg ha^{-1}) and single superphosphate (20 kg ha^{-1}).

Fertilization can play a key role in many small reservoirs of India, which require correction of oligotrophic tendencies. A number of reservoirs in Madhya Pradesh, the northeast and the Western Ghats, receiving drainage from poor catchments show low productivity, necessitating artificial fertilization. Chinese experience in fertilizing the small reservoirs for increasing productivity has been reassuring (Yang *et al.*, 1990). In Shishantou reservoir, a management strategy comprising fertilization by organic and inorganic manures and feeding resulted in phenomenal production hike from 1 500 kg ha^{-1} to 6 000 to 7 000 kg ha^{-1} during

1985 to 1989. Before fertilization, the plankton biomass in Shishantou was 1.5 mg l^{-1}, which was raised to 6.5 mg l^{-1} through application of organic fertilizers at the rate of 6.375 t ha^{-1}. The plankton biomass, after dropping during the peak precipitation period, picked up to 20.51 mg l^{-1} during the post-rainy season months, with corresponding increase in fish production.

1.5.6 Pollution

Ecodegradation of reservoirs has been on the increase due to the rapid pace of industrialisation, poor environment management in the catchment and a variety of other factors. Apart from the direct entry of industrial, municipal and thermal wastes, the pollution load carried by the upstream rivers is also accumulated in the reservoirs. The environmental degradation in reservoirs is caused mainly by the waste discharge from industrial, municipal and agricultural sources and the thermal power plants (Table 1.10). High rate of siltation due to poor catchment management also affects the biological productivity.

Table 1.10. Pollution in reservoirs

Reservoir	Name of river	Sources of pollution
Getalsud	Subarnarekha	Heavy engineering, chemicals and sewage.
Gandhisagar	Chambal	Textile, chemicals, trade effluents from Indore, Ujjain and Kota.
Tungabhadra	Tungabhadra	Paper, iron and steel, rayon, chemicals and sewage.
G. B. Pantsagar	Rend	Thermal power plant, coal washery, chemicals.
Bhavanisagar	Bhavani	Viscose factory effluent.
Hussainsagar	Musa	Trade effluents and sewage from Hyderabad city.
Hirakud	Mahanadi	Paper mill
Byramangala	Vrishabhavati	Industrial effluents and city sewage
Sandynulla	-	Animal products

(Modified from Joshi, 1990)

Thermal pollution

A number of reservoirs have been selected, of late, as sites for thermal power plants due to their dual utility as perennial source of water supply and disposal point for heated effluents. Thermal power generation capacity of the country has been registering a steady growth of 8% per annum and by the turn of the century, the installed capacity is expected to reach 84 000 MW. Various thermal plants in the country are estimated to generate 10 billion m^3 of hot water (40^0 to 5 ^{0}C) and 17 million t of fly ash every year. Fly ash is known to contain heavy metals such as Zn (6%), Ba (12.2%), Cu (1.3%), As (0.02%), V (0.08%), Ti (0.02%) and Mn (0.23%), which may find their way to the nearest river stretch or a reservoir.

Rihand is a large man-made lake of 46 000 ha, into which converge cooling waters from four super thermal power plants under the public sector *viz.*, Singrauli (2 000 MW), Vindhyachal (2 260 MW), Anpara (3 130 MW) and Rihand (3 000 MW), besides the private sectors Renusagar thermal power plant with a capacity of 210 MW. All these power generating plants are located within a small area of 30 km^2. Chandra *et al.* (1985) reported adverse effects of heated discharge on resident aquatic organisms. They recorded mortality of fish and decrease of aquatic life within 50 km of the discharge point, owing to high temperature (46 to 52 ^{0}C) of the effluent. Deposition of fly ash has been reported up to 500 m downstream of the outfall point. Cooling waters of Renusagar power plant discharged into Rihand reservoir are acidic and high in chlorides. Although an increase in water temperature is known to cause deoxygenation, a rise within reasonable limit enhances photosynthetic activities resulting in supersaturation of water with oxygen.

The main ecological consequences of a heated water discharged into the aquatic ecosystem are increase in water temperature, change in chemical composition and change in metabolism and life history of aquatic communities. The heated discharge may elevate the water temperature by 8 to 10 ^{0}C which may cause mortality of fish and fish food organisms. Temperature also exerts direct influence on toxicity. Apart from the rise in temperature, discharged waters are often altered chemically during the cooling processes. Davies (1966) showed that cooling tower discharge has lower ammonia level, higher concentration of nitrate and TDS, and lower levels of organic nitrogen, when water is abstracted from a polluted water source.

Temperature above 40 ^{0}C has been reported to negatively affect the plankton and benthic communities. Generally, fishes avoid heated effluents discharge points by swimming away to safer places. They can also withstand wide fluctuation of temperature (8 to 10 ^{0}C). However, the reproduction of fish is affected due to deposition of fly ash in the marginal areas of the river/reservoir which act as their breeding grounds. All the power plants around Rihand reservoir are located near the intermediate and lotic sectors, where the fishes are known to congregate (Desai, 1993). The most deleterious among the impacts of thermal pollution is the blanketing effect on the reservoir bed. Fly ash covers extensive areas of the bottom, blanketing off the substratum, resulting in retardation or total elimination of benthic communities. Thick mat of fly ash deposit at the bottom bed over the years may seal the nutrients away from the water phase and thereby affect productivity.

Domestic wastes

A number of reservoirs contiguous to towns and cities face threat from sewage pollution. Although from the fisheries point of view, organic loading within certain limits does not hamper the productivity, sewage load in excess can cause aseptic conditions, adversely affect the biotic communities, retard productivity and render the fish unfit for human consumption. Moreover, the problem needs to be addressed from public health and aesthetic points of view. The acute cases of hyper-eutrophication due to city sewage discharge in Hussainsagar, Mansarovar, Byramangala and Sandynulla reservoirs cause serious impediments in ecosystem management. Cases of heavy fish mortality is reported in Byramangala (Raghavan *et al.*, 1977) and Hussainsagar (Hingorani *et al.*, 1977).

The major adverse impact of sewage pollution can be assessed from deoxygenation, high BOD load, rapid eutrophication and accumulation of heavy metals in the environment. Sharp fall in dissolved oxygen in water puts the biotic communities under severe stress. While some species can tolerate a wide range of dissolved oxygen, many communities are highly sensitive to this parameter. As a chronic effect of oxygen depletion, some of the component populations are eliminated from the riverine community, causing far reaching changes in the trophic cycle. For instance, complete absence of zooplankton during January to August and its reappearance in September represented by *Keratella* sp., associated with abundance of phytoplankton like *Microcystis* sp. *Oscillatoria* sp., *Hormidium* sp., and *Nitzschia* sp., have been observed downstream of the sewage effluent outfall on the Ganga and Yamuna. The outfall area is dominated by *Chironomus*, followed by oligochaetes (*Tubifex* and *Nais*) in both the rivers, while areas below the outfall are characterised by the dominance of *Chironomus* followed by gastropods and bivalves.

Apart from affecting the organisms at lower trophic levels, intensive rate of pollution from municipal sources often causes direct fish kill, especially in small reservoirs where the problem gets aggravated due to reduced water flow rate. There are potential problems relating to the use of chlorine for disinfecting the sewage effluents for public health purposes. Owing to the increased use of synthetic detergents for domestic purposes, their incidence in the sewage effluents are on the increase. Synthetic detergents being absorbed into the body system of fish impair their growth and reproductive capacity. Detergents mixed with oil may be 60 times more toxic than the oil alone. Synergistic action of detergents with insecticides has also been recorded. Its sub-lethal concentration causes thinning and elongation of respiratory epithelial cells. Sodium lauryl sulphate is more toxic to freshwater teleosts, compared to alkyl benzene sulphonate (13-60 mg l^{-1}). Impact of heavy metals is discussed later on in this document.

Industrial effluents

Wastes emanating from an array of industries such as chemical plants, textile mills, heavy engineering plants, paper mills, iron and steel factories, rayons, etc., cause pollutional hazards in Indian reservoirs. Several instances of ecosystem damage and fish kill due to industrial effluents have been documented. Effluents from the Kanoria chemicals discharged into Rihand reservoir are alkaline (pH 9.2) and high in total alkalinity (4 770 mg l^{-1}), specific conductivity (12 816 μmhos), chlorides (5 173 mg l^{-1}) and free chlorine (1 924 mg l^{-1}) (Chandra *et al.*, 1983). These wastes have shown severe toxic effects on phyto - and zooplankton. Direct fish kills have also been reported in this reservoir due to high chlorine bearing wastes (Arora, *et al.*, 1970). Effluents from a paper mill at Brajrajnagar are discharged into Hirakud reservoir and Sugar mill wastes finding their way to a small reservoir in Gorakhpur were the cause of complete replacement of carps by the uneconomic fishes (Natarajan, 1979b). Discharge of industrial wastes consisting of dissolved and insoluble solids, free chlorine and lime of Mettur chemical factory into the surplus water channel of Stanley reservoir is reported to have resulted in large-scale mortality of carps and catfishes in summer.

An industrial unit near Sandynulla reservoir manufacturing gelatin from animal bones and the effluents from this factory bearing high BOD and phosphate are discharged into the reservoir, which is already eutrophic on account of sewage wastes from the city of Ooty. The south India Viscose, manufacturing viscose rayon and staple fibre, discharges 16 000 m^3 of waste material into the river Bhavani at

Sirumughai, causing pollutional hazard in Bhavanisagar reservoir, causing increase in bicarbonate and carbon dioxide and fall in dissolved oxygen and pH along with occasional fish kills. Another synthetic fibre manufacturing unit, Harihar Polyfibre discharges wastes into the tributaries of the river Tungabhadra, resulting in ecosystem degradation and fish kills (Joshi and Sukumaran, 1987). Several major and minor industries located on the river Bhadra in the industrial town of Bhadravati in Shimoga district are the source of heavy metals discharge in the inflowing water of Tungabhadra reservoir (Singit *et al.*, 1987).

Impact of industrial pollution

Industrial effluents, though comparatively less in volume, may cause considerable harm to the aquatic environment and the biotic communities including fish and ultimately affect man through food chain. Non-biodegradable and persistent types of pollutants like heavy metals, chlorinated hydrocarbon pesticides, oil components having high boiling points and radionuclides get more concentrated at higher trophic levels through biomagnification and pose threat to human health. Industrial effluents include a wide variety of chemical toxicants and heavy metals, apart from those contributing substantially to the BOD load such as pesticides which are used in processing the raw materials in many industries. In addition to the sub-lethal chronic effects on the environment, certain direct impacts are also discernible.

Natarajan (1979b) stressed the importance of protecting the upstream zones which are biologically sensitive areas. So are the head zones (*i. e.*, that part of the reservoir into which river flows) of the reservoirs where fish concentration is much higher.Discharge of effluents into the upstream can throw up a chemical barrier for breeding migration of economic carps, apart from causing considerable mortality to spawn and hatchlings.

Chronic effects of effluent discharge

There are two general classes of effects of pollutants on water uses. Some of the dramatic effects of toxicity, including fish kills are often well-publicised. But the other class of effluxion which involves continuous chronic sublethal degradation needs a more demanding consideration. This degradation goes unnoticed except by ecologists, taxonomists and sometimes by fishermen. It is neither dramatic nor well-publicised. No gory pictures of heaps of dead fishes or large water areas covered with oil spill or debris, but the killer masquerades in the form of reduction in the rate of reproduction by aquatic species or subtle changes in the food chain pattern on which the fish populations depend. The contaminants get accumulated in the water, soil and detritus phases of the environment and get biologically magnified, as they enter into fish tissues.

The harmful industrial and municipal effluents are as diverse as they are obnoxious. There is a diversity of harmful chemical toxicants that emanate from different industrial units, the nature of substances varying, depending on the products, production processes and the raw materials used. Similarly, the agricultural runoff carry heavy load of non-biodegradable pesticides. Domestic wastes also contain a variety of chemicals, detergents and organic load. Unfortunately, the impact of these toxicants on the biotic communities is very complex and our knowledge in this regard is grossly inadequate. It is not even possible to prescribe a precise *safe* limit in respect of any of the chemical pollutants. The pre- 1960 literature in this regard was based on short-term studies and used mortality as an end point. This is no longer valid, as the emphasis has now shifted to a balanced ecosystem rather than prevention of fish kills. To prevent the ecosystem from gradual degradation, we must provide criteria that will protect the entire life cycle of the desirable species as well as the food chain on which these species depend. A significant reduction in available food or reproductive success will result in a condition similar to that after a fish kill. Criteria must, therefore, be based on chronic or life cycle studies that may also permit extrapolation to untested species or toxicants.

Where multiple discharges exist, many chemical reactions may occur in the receiving water that intensify or reduce the toxic effect of the original materials. Thus, apart from the information concerning the effluents, specific knowledge on the potential chemical and physical changes involved is imperative to estimate the effects of multiple effluents on the environment. Moreover, the potential of combined stress on aquatic life cannot be explained on the basis of a single contaminant. The problems involving pH and metal toxicity are common where toxicity increases due to decrease in pH values. Similarly, the environments barely acceptable with regard to dissolved oxygen become totally unacceptable if the temperature is permanently increased, resulting in an increase in oxygen demand by the aquatic life.

The validity of applying the existing *safe concentration limits* is rather limited, as they are generally determined under controlled laboratory conditions. In the laboratory, fish are fed *ad libitum*. They are treated prophylactically, if needed; there are no predators, no competition for spawning areas, and no exposure to extremes of natural water quality. The effect of 2, 4-D on fishes adequately illustrates the gradual unspectacular decline in the quality of aquatic life. A study conducted at the Bureau of commercial fisheries pesticide laboratory at Gulf Breeze, Florida (Brungs, 1972) indicated that fish exposed to 2, 4-D for 1 to 5 months grew and survived as well as the control animals. However, the exposure, apparently lowered the general body resistance to a microsporidian parasite and a massive invasion of the central nervous system of the fish resulted.

Pesticides and heavy metals

Hazardous and toxic substances such as pesticides and heavy metals are carried to the reservoirs through the effluents and the rain washings from the catchments. These substances are highly persistent and thereby contaminate the entire biogeochemical cycle of static systems like reservoirs. The problems are aggravated due to the capacity of toxic substances to get biomagnified in fish tissues, which otherwise exist in water in extremely low concentrations. Such a situation, apart from resulting in low fish, transports toxic metals and pesticides into the human body through the contaminated fish. Heavy metal accumulation in water, sediments and plant tissue has been reported from Byramangala, (Table 1.11), Stanley (Table 2.8), Hussainsagar (Table 5.12) and Rihand (Agarwal and Kumar, 1978) reservoirs and the Tungabhadra river (Joshi and Sukumaran, 1987).

Pesticide residues have been detected in the riverine ecosystems in all the river basins of the country (Joshi, 1990). High levels of BHC, methyl parathion, endosulfan and DDT and their biomagnification in biotic communities like plankton, benthos and fish have been reported from Cauvery, Ganga and Yamuna rivers. However, such observations from reservoirs are rare. Joshi (1990) recorded significant presence of residues of different isomers of BHC and DDT and its metabolites (DDE, DDD) in fish and plankton of Rihand reservoir. Considering that the reservoir is situated in a relatively remote place far away from agricultural and industrial activities, the observation assumes importance.

Studies conducted in Panchet reservoir, Bihar (Gopalakrishnan *et al.*, 1966) showed adverse effects of effluents from coal washings on the recruitment of Indian major carps. Sinha (1986) reported rapid eutrophication in man-made lakes due to drainage from coal fields in south Bihar.

Table 1.11. Status of heavy metals in Byramangala reservoir

Metal	Concentration		
	Water *($\mu g\ l^{-1}$)*	**Sediment** *($\mu g\ g^{-1}$)*	**Plant** *($\mu g\ g^{-1}$ dry wt.)*
Zn	87-130	50-197	76.5-207.8
Cu	28-52	38-64	33-143
Cd	nd-15	32-106	1.4-2.1
Cr	nd	0.88-1.32	0.42-0.7
Pb	16.22	53.4-101.2	5.3-9.0
Hg	0.08-0.12	0.14-0.4	0.29-0.63

(After Joshi, 1990)

Siltation

Excessive siltation leading to drastic decrease in the water holding capacity and even damage to concrete hydraulic structures is a common problem in reservoirs. Siltation also hampers the productivity of water body by affecting the life processes of biotic communities. Erosion of top soil in the catchment area is the main man-made factor that leads to increased sediment load in rivers. Vegetation cover on the slopes acts as an adherent of top soil during the surface runoff. Removal of forest cover through logging, grazing, road construction or for urban needs makes the soil susceptible to erosion. The entire suspended and bed load materials carried by the rivers, however, are not exclusively the contribution by man. The catchment areas, especially those of the Ganga river are characterised by a prolonged dry season followed by a turbulent monsoon, with river discharges up to 85 000 m^3 per second. Therefore, heavy erosion and high sediment load are characteristic of Indian rivers. However, the tampering of environment in catchment areas adds considerably to the sediment load and the problem needs to be addressed through appropriate conservation measures.

Suspended particles tend to settle down in the lentic waters of the reservoir causing many problems. It is estimated that in India 5 334 million t of soil is eroded every year from the cultivable land and forests. The Indian rivers carry about 2 050 million t of silt, of which nearly 480 million t is doposited in the reservoirs and 1 572 million t is washed away into the seas. Loss of storage capacity of the reservoirs due to siltation is one of the most serious consequences of soil erosion. Many of the reservoirs recorded siltation rates, much in excess of what was envisaged during the planning stage of the project, due to increased rate of sediment load in the incoming waters (Table 1.12).

Table 1.12. Rate of siltation in some selected Indian reservoirs

RESERVOIR	Rate of silting (in ha m 100 km^{-2} yr^{-1}) *Assumed*	*Actual*
Gobindsagar	4.29	6.00
Nizamsagar	0.29	6.57
Tungabhadra	4.29	6.11
Hirakud	2.52	3.98
Shivajisagar	3.24	15.24
Gandhisagar	3.61	10.05

(After, Joshi, 1990)

Apart from diminishing the water holding capacity of the reservoir and cutting its life, siltation also affects the biota by blanketing the benthic and periphytic community. It also hampers the recruitment by destroying the breeding grounds and retards the overall productivity of the ecosystem.

1.5.7 Cage and pen culture

The unconventional production systems, such as cage and pen cultures have not become very popular in India, although they have a definite role to play in augmenting fish production from open water, especially the reservoirs. It is now widely accepted that the pen enclosures erected in the reservoir margins can be used as nurseries to raise stocking material to obviate the necessity for constructing concrete nursery farms which are cost-intensive. Similarly, the rearing of fish in cages and pens up to marketable size enables easier stock manipulation and total harvesting. However, non-standardization of farm practices and the materials to be used in the operation still acts as a major retardant for large-scale adoption of these culture systems in Indian reservoirs.

Species selection

Main criteria for the choice of candidate species for cage and pen culture are:

1. *fast growth rate,*
2. *adaptability to the stresses in enclosures due to crowded conditions,*
3. *ready acceptance of artificial feeds consisting mainly of cheap agricultural byproducts,*
4. *high feed conversion rates,*
5. *resistance to diseases, and*
6. *good market demand.*

The candidate species should preferably not breed in the cages and upset the population balance. Under the Indian conditions, the Gangetic major carps (*C. catla, L. rohita, C. mrigala*), the Chinese carps (*Hypophthalmichthys molitrix, Ctenopharyngodon idella*), common carp (*Cyprinus carpio*), the magur (*Clarias batrachus*) and tilapias satisfy these requirements to a great extent. Murrels (*Channa* spp.) also can be cultured in maritime States, where marine trash fish is available at a discount. Selection of species, however, is mainly dictated by the local demands and availability of quality seed and other inputs in adequate quantities.

Site selection

Appropriate site selection is important for successful enclosure aquaculture. Sheltered, weed-free, shallow bays are the ideal locations for installing pens and cages. The sites should have adequate circulation of water, with wind and wave action within moderate limits. Excessive turbulence may lead to wastage of fish energy for stabilizing themselves and loss of feed. The other major considerations are that the water should be pollution-free, availability of seed in the vicinity, easy accessibility to the site and a ready market for fish. Flowing waters with a slow current of 1.0 to 9.0 m per minute are considered ideal for cage siting. It is desirable to install cages a little away from the shore to prevent poaching and crab menace.

Water level fluctuation is the most important consideration in site selection for the pen culture operations in reservoirs. A scrutiny of the contour map and the monthly fluctuation patterns of reservoir levels will enable the location of suitable sites, which retain sufficient water for the required period of time. Sites which dry out during summer will be ideal, as it is easier to erect pens on dry land, to be inundated later as the water level increases. Similarly, some bays of the reservoir retaining water for sufficient period can be identified and cordoned off by erecting barricades.

Cage culture

Experiments on cage culture conducted in India have been exploratory in nature and the yields obtained, so far, are not impressive. The supplemental feeds given are oilcakes, ricebran, soy bean flour and silkworm pupae, which have great demand in cattle, poultry, pig rearing and other animal husbandry practices and hence command a good price in the market. The food quotient obtained in the cage culture of various species has not been high, except in the case of tilapia, making conventional supplemental feeding unremunerative. The low production and feed conversion rates are mainly due to the relatively low stocking density and many deficiencies in the feed. The feed is often not in a water-stable form and nutritionally balanced to promote growth. There is need for evolving suitable complete feeds for individual species of fish from the locally available raw materials, by experimentation.

One of the major constraints of the cage culture system is the lack of suitable cage designs to withstand severe wave action, common in Indian reservoirs. Mukherjee (1990) suggested a number of flexible, floating barriers, sheet barriers and rigid floats to protect the cage structures from wave action. The floats dampen the wave thrust and absorb the wave energy before the wave can propagate and strike the cage and cause damage. Kumaraiah and Parameswaran (1985) proposed a circular cage that could be used in reservoir with moderate wave action for culture of carps, tilapia and air breathing fishes. The cages can float at the surface, remain just submerged or rest at the bottom. Floating cages are considered to be most appropriate for Indian conditions and all the experiments conducted so far in the country for seed rearing, growout, nutrition and biomonitoring have been in such enclosures.

Cage materials

Floating fish cages can be constructed out of a variety of materials including metal, wood, bamboo and netting. Fairly fine-meshed nylon netting is used for nursery purposes. Cages made of monofilament woven material of 1.0 to 3.0 mm mesh size are light and easy to handle but last only for six months to one year, depending on their thickness. Knotless nylon webbing of 3 to 6 mm mesh size and knotted nylon webbing of 7 to 15 mm mesh have been found to be very durable as cage material. A battery of cages can be buoyed up within a bamboo catwalk which will serve as a working platform, floated by sealed empty barrels. Circular and box-like cages of varying diamensions on conduit pipe structures which can be easily assembled, and suitable flotation systems have been designed in India. Similarly, self-floating cage with HDPP pipe structure has also been experimented with succesfully.

In Jari tank near Allahabad, nylon cages (20 mesh cm^{-1}; size 2.2 x 1.6 x 1.45 m) were stocked at a density of 8 500 hatchlings m^{-2} (size 6.5 to 7.8 mm). These grew in 21 to 28 days to 30.2 to 45.6 mm with a survival of about 25% (Anon., 1979). In fry rearing, the stocking rate in the cages (mesh size 3 mm) was 700 to 2 500 m^{-2} and within 90 days they attained a size of 103.6 to 121.8 mm. The feed given was powdered soybean, groundnut cake and rice bran in equal proportions. Rearing of carp fry was done in Getalsud reservoir, where they (10 to 31 mm in size) were stocked in 2.4 x 1.5 x 1.5 m cages at the rate of 300 to 700 m^{-2}. The growth rate per month was 17, 25 and 20 mm in mrigal, catla and rohu respectively. The stock was fed with mustard and groundnut cake and rice bran in the ratio 3:1:1 at 30% of the body weight (bw) of the stock for 4 days and thereafter at 20% for the rest of the period. Summary of cage culture experiments conducted is presented in Table 1.13.

A series of cage culture trials have been reported from a 12 ha impoundment in Bangalore (Parameswaran, 1993). In an experiment conducted with monofilament cloth cages of size 10.5 m^2, common carp and silver carp fry were reared at a ratio of 40:1 at a stocking density of 225 m^{-2}. Put on a diet of powdered rice bran, defatted silkworm pupae, groundnut cake and soya flour in 12:5:2:1 ratio at 10 to 20% bw day^{-1}, the survival obtained at the end of 4 months rearing was 97.5% in common carp and 88% in silver carp. The stock attained average final weights of 20 and 8.6 g respectively. However, experiments conducted on catla gave erratic results with survival rate varying from 9 to 71.4%. In another trial, 30 000 spawn obtained from cage grown common carp parents were reared in 4.5 m^3 auto-floating (PVC frame) monofilament cloth (mesh: 15 cm^{-1}) cages. In 35 days, the fry attained a size of 25.4 mm with 38% survival. Restocked in 3.5 m^3, 8 mm mesh knotless nylon netting cages, at a density of 475 m^{-3}, they grew to 54.8 mm/4.9 g in 75 days with a survival rate of 88.5%.

In a cage culture experiment reported from Tamil Nadu (Parameswaran, 1993),10 days old fry (size 10 mm) stocked at a density of 500 m^{-2} were raised to a size of 50 to 60 mm in 40 days, with survival rates ranging from 45 to 85%. Department of Fisheries in Tamil Nadu has been undertaking rearing of spawn and fry of major carps in floating cages during July to September every year. However, data on the stocking density, nutrition, growth and survival are not available.

Rearing of the fry of Indian major carps was tried in Tungabhadra reservoirs in the year 1984-85 (Singit et al., 1985). Four floating cages, made of 16-P velon screen fitted on rectangular bamboo frame of 10 x 4 x 1 m, were stocked with rohu and mrigal. Although survival rates ranging from 37.5 to 87.5% were obtained at the end of the 3 months rearing period, the experiment was vitiated due to the destruction of cages due to heavy winds. At stocking densities ranging from 2 to 5 million ha-1, growth of about 100 mm (33 g) was obtained.

Dependent on the type of management input, fish production rates obtained for growout in cages vary greatly. Unlike the hi-tech system of saturated stocking and feeding on enriched formulated diets, the production recorded in cage culture of common carp is 35, 37.5 and 25 kg m^{-3} $month^{-1}$ respectively in Japan, Germany and the Netherlands. In Asia, in general, only semi-intensive and low cost technologies are adopted, mainly due to economic considerations. In India, the growing season is almost year round, except for December-January in northern parts, where the temperature is low during these winter months.

Table 1.13. Summary of growout experiments conducted in cages in India

Species cultured	Cage volume (m^3)	Stocking density (m^{-2})	Stocking size (g)	Mean harvest size (g)	Culture period (months)	Production (kg m^{-2} $month^{-1}$)	Feed	Feeding rate (% bw)	FCR	Reference
Cyprinus carpio	15.75	30-38	40-50	325	6	1.55-2.22S	WP, GNC, RB (8:9:3)	10-20	8.3-10.4	Govind (MS.) 1983
Catla catla	15.0-15.75	13-49	8-50	544-772	6-8	0.83-1.30	GNC, RB (1:1)	5-10	5.6-6.6	Govind *et al.*, 1988
Hypophthalmichthys molitrix	10	15	61	472	10	0.7	SWP, RB, GNC (1:2:3)	3-5	3.1	Kumaraiah *et al.*, 1991
Labeo calbasu	10	5	16.5	208	8	0.1	GNC, RB (1:1)	2	2.9	Kumaraiah *et al.*, (unpublished)
Ctenopharyngodon idella	3	33-67	7-10	350-400	6	2.0-3.3	Lemna, Hydrilla	80	-	Bandopadhyay *et al.*, 1991
Oreochromis mossambicus	5-10	100-200	6.0-7.6	32-62	2-5	09-1.6	RB, GNC, CFP (1:1:1)	3-5	1.8-2.3	Kumaraiah et al., 1986
Channa marulius	5	40	25.8	177	5.3	0.8	Trash fish	10-12	2.5	Kumaraiah and Parameswaran; (unpublished)
Clarias batrachus	2	100	7.4	36.9	3	1	-	-	-	Murugesan and Kumaraiah; 1972

SWP=Silkworm pupae; GNC= Groundnut cake; RB= rice bran; CFP=cattle feed pellets; FCR= food conservation ratio

Pen culture

Pen culture has a special relevance in reservoir management, since it has been widely recognised as a means to rear, *in situ*, the fingerlings for stocking. The number of fingerlings required for stocking the reservoirs in the country is so enormous that it is impossible to raise all of them in land-based nursery farms which makes pen nurseries *sine qua non* for reservoir management. Nevertheless, pen culture on a regular basis has not been practised anywhere in India except at Tungabhadra reservoir. The factors that hamper the standardisation of pen culture technique are:

1) the steep level fluctuations,
2) wind and wave action,
3) lack of suitable pen materials,
4) weed infestation and the related harvesting problems, and
5) nonsynchronisation of suitable water levels and the spawn availability.

The water retention time is important, since the rearing has to be completed before the water level in the pen goes down the critical limit. In reservoirs with high drawdown, the water retention time is very limited. Sometimes the filling takes place so late that no spawn of desirable carps will be available when the water level attains the desirable limit. The pen walls limiting the water circulation to some extent, the accumulated feed and fertilizers case eutrophication leading to weed infestation fouling of water and fish kills.

Pen culture in Tungabhadra reservoir

Despite all the limitations, pen nurseries are used with remarkable success in Tungabhadra reservoir for the last 12 years. During 1992-93, 21 pens were erected in Ladakanabhavi, 25 km away from the dam site, covering a total enclosure of 3.3 ha. The pen site is situated at an elevation of 496 m above MSL and the installation was completed in the month of July, when the site was still exposed. Later, when the water level increased, the pen got inundated.

The pen area was pre-treated with organic manure that resulted in a rich growth of plankton after the filling. A total of 15 million spawn were stocked in the pens, comprising 6.75 million *Labeo rohita*, and 8.25 million *Cirrhinus mrigala*. After a rearing period of 90 days, 2.41 million fingerlings were collected from the pen and released into the reservoir. This included 1.085 million *L. rohita* and 1.325 million *C. mrigala*, worth Rs. 495 875. Pen culture operations on similar lines are being in Kyrdemkulai and Nongmahir reservoirs of the northeast (see Chapter on the Northeast).

Seed rearing experiments were conducted in a split bamboo pen enclosure of 247.5 m^2 reinforced with a nylon netting in Punjar swamp, adjoining the Bhavanisagar reservoir (Abraham, 1980a). The pen was stocked with the spawn of *C. mrigala* (size 7 mm) and *L. fimbriatus* (size 5 mm) at the rate of 4.6 million ha^{-1} and usual farm practices were followed. In 30 days, mrigal attained a size of 38 mm and L. fimbriatus, 28 mm. At the time of conclusion of the study after 3 months, the former had attained a size of 88 mm and the latter, 75 mm. The overall survival obtained was 27.8%.

A pen culture experiment for raising catla and rohu in Manika maun, a floodplain lake in Gandak basin yielded a (computed) production of 4 t fish ha^{-1} in six months. The experiment was conducted in a bamboo screen pen (1000 m^2) and the stock was fed with mixture of rice bran and mustard cake, apart from a feed formulated from the aquatic weeds collected from the lake.

1.5.8 Crafts and gear

Gear

The presence of underwater obstacles restricts the use of active gear in reservoirs and the choice is often limited to passive gear such as simple gill nets. The most common among them is the *Rangoon* net, an entangling type of gill net without a foot rope. Another entangling type of net used in reservoirs is *uduvalai*, which has a reduced fishing height and is usually operated in shallow marginal areas to catch small fish. Shore seines of various dimensions and mesh sizes are employed in many reservoirs. Although a number of other fishing gear such as long lines, hand lines, pole and line, cast nets, dip nets, *etc.* are in use, their contribution to the total catch is very insignificant.

Unlike the marine fisheries, very little attention has been paid over the years towards improvement of gear in the inland sector, barring an attempt to upgrade the reservoir fishing gear by two experts under the aegis of FAO/UNDP programme during the 1960s (Gulbadamov, 1962). A number of improvements have been suggested by the experts to the fishing techniques followed by the reservoir fishermen. Apart from the introduction of frame net, they have suggested improvements on the design of gill nets, beach seines and long lines, Unconventional methods such as electrical fishing, use of light as fish lure, and the use of echo-sounder for fish detection and survey of bottom topography were also suggested.

The main emphasis of gear improvement was the modification of gill nets. Gulbadamov (1962) designed two nets *viz.*, Sebgul I and Sebgul II, which were gill nets with modified rigging pattern. While ensuring a proper fixing of webbing on head and foot ropes, a uniform hanging coefficient was ensured. Similarly, the sideways movement of the webbing was checked to maintain effective area of the net. Ranganathan and Venkataswamy (1967) conducted experiments in Bhavanisagar reservoir to check the efficacy of the new design and found no appreciable superiority for Sebgul nets over the *Rangoon* nets.

The Central Institute of Fisheries Technology has experimented with gill nets of various colours and found yellow and orange coloured nets yielding better catch than the white coloured ones(Table 1.14). It has been observed that 77% of the carps were caught by entangling and the rest by gilling. The usual method of increasing the entangling capacity of gill net is by decreasing the slackness of webbing, which can be achieved by suitable modifications in the hanging of the net (Nayar, 1979).

Alivi, the giant drag net of Tungabhadra reservoir is described in the chapter on Karnataka. This giant shore seine catch fishes of all hues in large numbers, including the juveniles of commercially important carp species. Similar nets are used in Rihand and Keetham reservoirs, where they are known as *mahajal*.

Table 1.14. Fish catch by gill nets of different colours

Colour of net	Catch kg 1000 m^{-2}
White	38.9
Blue	55.1
Green	66.9
Orange	83.4
Yellow	85.5

(After Nayar, 1979)

Trawling

Trawling has been attempted only in two reservoirs *viz.*, Gandhisagar and Hirakud. The findings of Kartha and Rao (1990) with regard to species selectivity of different type of trawling are interesting. The findings suggest efficacy of various types of trawling in increasing the productivity of commercial carps and checking the predator and weed fish populations. After experimental trawling with *single-boat bottom trawling, two-boat bottom trawling* and *two-boat mid-water trawling* under different speeds, it has been found that more than 92% of the total catch consisted of economic varieties such as catla, rohu, murrels, mullets and featherbacks in two-boat mid-water trawling. This was in sharp contrast to the bottom trawling, of both single boat and two boat variety, which yielded mostly (64 to 91%) the non-commercial species of fish. The two-boat mid-water trawling at a speed of 3 to 4 knots have been recommended for exploitation of commercial species and single and two-boat bottom trawling at 2 to 3 knots for eradication of uneconomic species of fishes.

Fishing crafts

Coracle, a saucer shaped country craft, is the major fishing craft used in the reservoirs of peninsular India. It is made of a split bamboo frame, covered with buffalo hide. Apart from being simple and inexpensive, coracle is durable and has very good manoeuvrability in choppy waters. It is also a versatile craft used for laying and lifting of nets, besides navigation and transport of fish and other material. Coracles of Krishnarajasagar are prepared by wrapping HDPP over the bamboo frame with the help of coal tar as an external covering in place of hide. This modified version of coracle is cheaper and more durable (Anon., 1984b; Parameswaran and Murugesan, 1984).

Unlike Gobindsagar, where all the fishermen possess their boats, reservoir fishermen, in general, are too poor to own boats. In many reservoirs like Vallabhsagar and Hirakud, the fishermen could get their boats with the help of subsidy and other financial assistance from the Government or fundings agencies. In Vallabhsagar, boats are distributed by the State among the fishermen at a subsidy of 50%, while in Hirakud, they get it from various schemes under NABARD and NCDC.Wooden boats are used for fishing in a number of reservoirs, especially in the North India. Flat bottomed, locally fabricated boats ranging in length from 3 to 7m are used in Kyrdemkulai, Hirakud, Malampuzha, Gobindsagar, Gandhisagar and Rihand. A plank-built, flat bottomed canoe, 2 to 3 m in length is the most popular fishing craft of Gandhisagar. In the same reservoir the repatriates from the erstwhile East Pakistan used the Bengal type dinghy, which is 5 to 7 m in length and have the additional facility of setting sails for wind propulsion.

Mechanised boats are not used in reservoir fishing in any appreciable extent. A 9.1 m long wooden, mechanised boat has been introduced by the CIFT in Hirakud reservoir, but they are too expensive for the fishermen. It is significant to note that large water bodies like Nagarjunasagar, Tungabhadra and Krishnarajasagar have no motorised craft neither for fishing nor for fish transport.

Dugout canoes, carved out of palm trees are used in Yerrakalava reservoir. In most of the reservoirs in the country the fishermen rely on improvised materials. Reservoir fishermen show considerable ingenuinity in fabricating makeshift rafts out of discarded old tyres, logs, used cans *etc.* In a vast majority of Indian reservoirs, where the catch is not very remunerative, no boats are used and the fishermen depend entirely on these improvised devices.

1.5.9 Other management measures

Pre-impoundment surveys

Until recently, pre-impoundment surveys conducted in India in connection with dam construction invariably lacked a fisheries perspective. Faunistic surveys of the river stretches carried out before dam construction in Tungabhadra (Chacko and Kuriyan, 1948), Bhavanisagar (Chacko and Dinamani, 1949), Pipri and Rihand (Hora, 1949), Damodar Valley Corporation reservoirs (Job *et al.*, 1952), Hirakud (Job *et al.*, 1955) and Gandhisagar (Debey and Mehra, 1959) were not comprehensive and did not help in any way the conservation and developmental efforts. A complete survey comprising the fish and fisheries, together with inventories of fishing villages and fishermen populations, and fishing craft gear for planned development is lacking in most cases.

The pre-impoundment surveys provide the framework for future development policies and should encompass:

i) the native ichthyofauna in the river stretch above and below the dam, and their likely chances of survival,

ii) breeding habits of fishes and the possible impact of impoundment on their recruitment,

iii) survey of breeding grounds in relation to submergence, both above and below the dam,

iv) hydrobiological characteristics of water and soil with special emphasis on the nutrient and thermal regime,

v) needs for creating infrastructure such as, hatcheries, nurseries, ice plants etc.,

vi) site selection for pen nurseries, cages etc., and

vii) possibilities for cleaning the area of submergence of trees and other obstructions.

Holistic pre-impoundment survey for fisheries development is a new concept in India. A beginning in this direction has been made by the Narmada Control Authority. A recently concluded socio-economic survey of the Narmada basin tried to address the problem of fisheries development with a holistic approach.

Timber clearance

Opinion is divided on the wisdom of removing timber from the reservoir bed. While it is mostly appreciated that a reservoir bed free from obstructions facilitates the use of active fishing gear and leaves room for many other management options, many workers feel the necessity to leave at least the non-commercial timber intact for a variety of purposes such as, reducing wave action, flocculating the colloidal clay turbidity, providing habitat for fishes and substrata for periphyton deposition (Bhukaswan, 1980). Timber clearance has been tried in a number of reservoirs in India, both before and after the impoundment. In Chillar and Benisagar reservoirs of Madhya Pradesh, trees were cut from the lake bed and auctioned before the reservoirs were filled. Harsi, Jamoia and Ghatera reservoirs are examples of complete clearance of date palm trees from the marginal areas during the summer months. Forest area of about 61.4 km^2 were cleared in Hirakud, when the bed was exposed during drawdown.

Exploitation systems

Fisheries being a state subject, management of reservoir fisheries vests with the State Governments. There is a great deal of divergence in the management practices followed by individual States which vary from outright auctioning to almost free-fishing. Cooperative societies (primary and apex) and the State level Fisheries Development Corporations are also involved in the fishing and marketing operations. Involvement of the above agencies and their role in fishery operations and market interventions often vary from one reservoir to another within the same State. Some sort of uniformity in fishery regulations among various categories of reservoirs as well as the need to monitor the socio-economic aspects of reservoirs fisheries.

Commercial exploitation systems followed in different States can be broadly classified under four headings *viz.*, 1) departmental fishing, 2) lease by auctioning, 3) issue of licences to cooperative societies or individuals and 4) fishing on a royalty basis (crop sharing). Direct departmental fishing being not an economic proposition, is followed only in a very few reservoirs. In some reservoirs like Hirakud, Nagarjunasagar, and DVC, this practice is partially resorted to for experimental and exploratory purposes. In most of the cases, the Department exerts its control over the exploitation by acting as a marketing link and controlling the fishing effort. In Rajasthan, Madhya Pradesh and Uttar Pradesh, the small reservoirs are mostly auctioned on an yearly basis. In a number of large reservoirs, free licences are issued to fishermen without any limit.This virtual *free for all* system has been found to be detrimental to the interests of the ecosystem and fishermen in Nagarjunasagar, Yerrakalava and a large number of other reservoirs in Andhra Pradesh. Crop sharing is a very popular mode of exploitation in Tamil Nadu, where, the fishermen are provided with all fishing implements, in return of which they pay a royalty (sometimes up to 50% of the catch) to the Government.

Input-output relationship

Evaluation of the socio-economics of reservoir fisheries is a very tedious task due to the multiplicity of agencies involved in reservoir management. Reservoir fisheries is developed basically on capture fisheries lines, following the common property norm. Like the rivers, lakes and the seas, the biological wealth is considered as a nature's endowment and the State's intervention in developmental activities benefits the poor fishermen who toil in water. The investment made in developing reservoir fisheries shall be viewed in the light of the social benefits it accrues in the form of:

1) *rehabilitating the displaced population,*
2) *improving the living conditions of fishermen, and*
3) *providing employment opportunities.*

Capture fisheries activity of the reservoirs is akin to the extractive industries like coal, oil, iron ore *etc.*, where the yield depends on the state of technology involved and the quantum of labour and capital deployed. But the renewable nature of the resource and the intricate biological principles involved in the ecosystem management imparts a heavy element of challenge into the reservoir fisheries management. Human intervention being less intense in reservoirs, compared to the aquaculture operations, yields often display violent fluctuations, even if the effort in terms of labour and capital is kept constant. Therefore, production-function relationship is bound to be intricate and less precise. A certain measure of stability needs to be imparted in production by affecting sustained improvement in yield along with remunerative returns to fishermen by narrowing the price spread between the producer and the consumer.

In aquaculture, it is estimated that 77.23% of the price paid by the consumer is received by the producer (Paul, 1990). As opposed to this, a major chunk of the price is siphoned away by the wholesalers and other market intermediaries in reservoir fisheries. A study of seven reservoirs (Paul and Sugunan, 1990) for a period of six years has brought to light the major factors that determine the remunerativeness of fishing in reservoirs. Reservoir fisheries is a sector, where the chief input is labour, besides a marginal depreciation of crafts and gear. Even if the costs of stocking and other developmental measures are taken into account, this area does not call for heavy investment as in the case of pond culture.

A factor that can bring serious distortions in the income level of fishermen is the over-concentration of fishermen in the wake of a low fish productivity. For instance, in Ukai reservoir, with an area of 36 525 ha, 306 boats with 3 400 gill nets (50 m each) were operated during 1985-86 to 1982-83. In a fishing year comprising 260 days, 1 836 fishermen netted out 174 t of fishes. In sharp contrast to this, 520 fishermen of Nagarjunasagar shared a catch of 170 t in a year. In lower Aliyar, 17 t of fishes were harvested by 14 fishermen, each of them, after meeting the royalty obligations, could take home only Rs. 1 000 to 1 400 a year. A better picture, however, emerged in Bhavanisagar, where 80 fishermen after sharing 150 to 300 t of fishes, earned an annual individual income of Rs. 8 175/ (Paul and Sugunan, 1983).

There is a need to dovetail the twin objectives of conservation and yield optimisation in reservoir fisheries management. While the fishermen and the fish merchants strive to increase the production for economic considerations, it is the responsibility of the State to ensure that economic expediency of development does not mar the ecological reasoning. Virtual free fishing, as followed in Andhra Pradesh is counter-productive to the norms of conservation and yield optimisation. Although there are fair possibilities of linking reservoir fisheries development with poverty alleviation programmes, the progress made so far in this direction is not very encouraging. Chances of creating additional employment are not much in majority of Indian reservoirs. On the contrary, many reservoirs have surplus manpower which can be diverted to others which can absorb more men without eroding the income level of the existing fishermen (Paul and Sugunan, 1990).

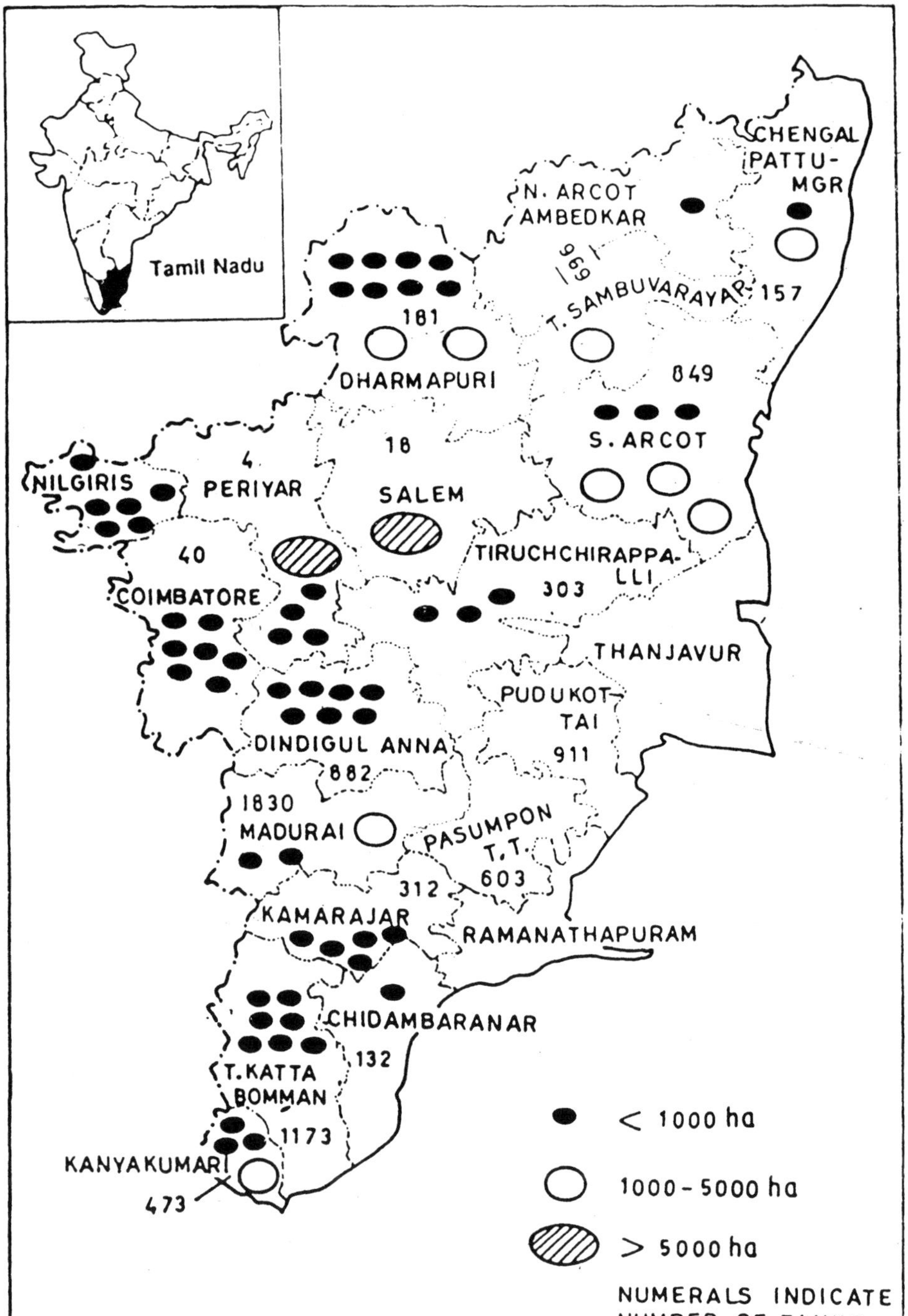

Figure 2.1. Distribution of reservoirs and tanks in Tamil Nadu

2. TAMIL NADU

The south-eastern State of Tamil Nadu is an important geo-political, economic and cultural constituent of India. From the reservoir fisheries point of view, the State occupies a pre-eminent position, inasmuch as it has an ancient tradition of creating impountments for irrigation. Tamil Nadu is also in the forefront of research and development of reservoir fisheries in the country. Some of the pioneering research studies on limnology and fisheries of reservoirs were conducted in the State.

Situated in the rain shadow behind the Western Ghats, most part of the State is semi-arid and dry, drained mostly by the east-flowing seasonal, rain-fed streams. Northeast monsoon is more important to the State than the southwest, thus making Tamil Nadu unique among the Indian States. The districts of Chengalpattu MGR, South Arcot, Thanjavur, Madurai, Ramanathapuram and Tirunelveli Kattabomman receive rainfall mainly from the north-east monsoon, while North Arcot, Salem, Coimbatore and Tiruchirapally in the central region depend on both monsoons. North Arcot and Thanjavur districts receive rich rainfall of 102 to 114 cm per annum, while the range of annual rainfall in the rest of the State is 640 to 890 mm. Despite the limitations in water resources, the State has made great strides in agriculture. Tamil Nadu's rice yield of 2.5 t ha^{-1} is among the highest in India and its sugar cane yield of 100 t ha^{-1} is a world record.

People of Tamil Nadu learned to conserve water and store it for irrigation since time immemorial. The landscape of the State, especially in Madurai, North and South Arcot, Dindigul Anna, Tirunelveli Kattabomman and Pudukottai districts is dotted with thousands of small impoundments (locally called *tanks*) created to store the surface flow for irrigation. Ganapati (1956a), quoting Madeley (1914) reported as many as 141 *minor irrigation tanks* or reservoirs in the 363 km^2 catchment of Red Hill reservoir. Madurai and Ramnad districts of Tamil Nadu have the maximum number of tanks and these areas are called *the land of irrigation tanks* (Pandian, 1987). The Cauvery river was tamed in the first century A. D., when the Grand Anicut was built of stone and mud, covered with an outer facing of dressed granite, set in lime mortar. Upper and Lower Anicuts, the two barrages on the Cauvery downstream, were constructed by Sir Arthur Cotton in 1836 and the Mettur dam came into being in the 1930s. Even in the recent past, a large number of concrete masonry and earthen dams were built in the State, both before and after independence, resulting in the creation of 58 small and 11 medium and large reservoirs.

Fisheries Department of Tamil Nadu, with one of the best resource data bases in the country, has surveyed almost all the man- made lakes in the State, most of them being small irrigation reservoirs. Fifty-eight, out of sixty-nine reservoirs in the State fall under the category of small reservoirs (<1 000 ha). Even among them, the reservoirs less than 500 ha in size dominate. However, the small reservoirs form only 27% of the total water surface area. One reservoir in the category of 5 000-10 000 ha (Bhavanisagar) and one in the >10 000 category (Mettur) form 40% of the total water surface area. There are 9 reservoirs in the medium category (1 000 to 5 000 ha), covering an area of 19 577 ha (Table 2.1 ; Fig. 2.1).

Most of the earlier accounts on reservoir fisheries of the State seem to have overlooked the water bodies designated as *tanks*. In the southern States of Karnataka, Tamil Nadu and Andhra Pradesh, many small irrigation reservoirs in the country side, created by erecting small earthern dams to obstruct stream flow, are generally referred to as tanks. Many of them have a very small catchment, restricted to the immediate surroundings and the others receive water from small streams and

Table 2.1. Number and area of reservoirs in Tamil Nadu (by category)

Category	No. of units	Area (ha)	
		At FRL	*Average*
Small reservoirs			
<500 ha	*46*	*6 858*	*3 594*
500 - 1 000 ha	*12*	*8 805*	*5 034*
Total	**58**	**15 663**	**8 628**
Medium reservoirs			
1 000 - 5 000 ha			
Total	**9**	**19 577**	**6 374**
Large reservoirs			
5 000 - 10 000 ha	*1*	*7 876*	*3 208*
>10 000 ha	*1*	*15 346*	*9 324*
Total	**2**	**23 222**	**12 532**
Grand total	69	58 462	27 534

(Classification on the basis of level at FRL)

Small irrigation reservoirs (tanks)

rivulets. Tanks also include some community ponds of the villages and those attached to the temples. Nevertheless, vast majority of them are man-made impoundments created by obstructing surface flow, falling under the broad definition of reservoirs. Although some lakes of tectonic origin, meteorite craters, *etc.*, must be masquerading among them, it is difficult, for the fisheries officials of the lower and middle rank, to tell the actual reservoirs from other lentic formations.

The Fisheries Department of Tamil Nadu has a village-wise inventory of all tanks in the State, and that is categorised into *short seasonal* and *long seasonal* (*major irrigation*) tanks. The short seasonal tanks having an average size less than 10 ha, retain water only for a period less than 9 months in an year, while the long seasonal ones have an average size of 34 ha and retain water for 9 to 12 months. In some of the districts, the major irrigation tanks have an average size of 156 to 222 ha. Total waterspread under tanks in 22 districts of the State is estimated at 740 182 ha, of which 439 904 ha are short seasonal, with the size of individual units less than 10 ha in most cases. These small tanks have been tagged with the ponds. There are 8 837 long seasonal (major irrigation) tanks, covering 300 278 ha, situated in 17 districts, which are *de facto* reservoirs and so treated in this work. Madurai, Tirunelveli Kattabomman, Pudukottai and South Arcot districts have maximum number of perennial tanks (Table 2.2), while the average size of tanks in Chengalpattu MGR, Salem and Pasumpon districts is higher.

2.1 DISTRIBUTION OF RESERVOIRS BY DISTRICTS

Reservoirs of Tamil Nadu are scattered in 16 districts. The two large reservoirs (> 5 000 ha) *viz.*, Stanley (Mettur Dam) and Bhavanisagar are situated in Salem and Periar districts respectively. Dharmapuri has the maximum number of reservoirs comprising 8 small and 2 medium reservoirs, with a total area of 3 600 ha; followed by Coimbatore, Dindigul Anna and Tirunelveli Kattabomman districts with 7 reservoirs each. South Arcot district with just 3 each of small and medium reservoirs has a waterspread of 9 933 ha (Table 2.3). While taking irrigation tanks also into consideration, Pasampon Thevar Thirumagan District has the maximum water area (48 734 ha), closely followed by Pudukottai and Madurai districts with 46 278 and 41 508 ha respectively. The State has 8 906 water bodies in total, spread over 358 740 ha at FRL (Table 2.4). Eleven medium and large reservoirs in 8 districts and 58 small reservoirs in 14 districts along with their annual fish production and major fisheries are listed in Tables 2.5 and 2.6 respectively.

Table 2.2. Distribution of major irrigation tanks in Tamil Nadu

District	No.	Area (ha)	Average of a unit
1. Chengalpattu MGR	457	34 931	222
2. N. Arcot Ambedkar 3. T. Sambuvarayar	969	20 328	21
4. Dharmapuri	181	13 960	77
5. Soth Arcot	849	17 695	21
6. Salem	18	2 805	156
7. Thiruchirapalli	303	10 063	33
8. Periyar	4	275	69
9. Coimbatore	40	1 792	45
10. Dindigul Anna	882	12 915	15
11. Madurai	1 830	41 508	23
12. Kamarajar	312	17 328	55
13. T. Kattabomman	1 173	26 683	23
14. V. O. Chidambarar	132	1 990	15
15. Kanyakumari	473	2 993	6
16. Pasumpon	603	48 734	81
17. Pudukottai	911	46 278	51
Total	**8 837**	**300 278**	

Table 2.3 Distribution of reservoirs in Tamil Nadu districts

District	No. of reservoirs		Total area (ha)	
	Small	Medium and Large (L)	At FRL	At. Av. Level
Chengalpattu MGR	1	1	4 163	1 931
North Arcot Ambedkar	1	0	678	339
T. Sambuvarayar	0	1	2 010	56
Dharmpapuri	8	2	3 600	1 444
South Arcot	3	3	9 933	2 977
Salem	0	1 (L)	15 346	9 324
Thiruchirapally	3	0	227	102
Periyar	4	1(L)	8 786	3 766
Coimbatore	7	0	3 172	1 768
Nilgiris	6	0	486	253
Dindigul Anna	7	0	829	546
Madurai	2	1	2 736	1 768
Kamarajar	5	0	933	400
Tirunelveli Kattabomman	7	0	1 722	883
V. O. Chidambarar	1	0	657	328
Kanyakumari	3	1	3 184	1 649
Total	**58**	**11**	**58 462**	**27 534**

Table 2.4. Number and area of reservoirs and perennial irrigation tanks in Tamil Nadu districts

Districts	Reservoirs		Irrigation Tanks		Total	
	No.	Area (ha) (FRL)	No.	Area (ha)	No.	Area (ha)
Chengalpattu MGR	2	4 163	157	34 931	159	39 094
North Arcot Ambedkar	1	678	969	20 328	970	21 006
Dharmapuri	10	3 600	181	13 960	191	17 560
South Arcot	6	9 933	849	17 695	855	27 628
Salem	1	15 346	18	2 805	19	18 151
Thiruchirappally	3	227	303	10 063	306	10 290
Periyar	5	8 786	4	275	9	9061
Coimbatore	7	3 172	40	1 792	47	4 964
Nilgiris	6	486	-	-	6	486
Dindigul Anna	7	829	882	12 915	889	13 744
Madurai	3	2 736	1 830	41 508	1 833	44 244
Pudukottai	-	-	911	46 278	911	46 278
Pasumpon	-	-	603	48 734	603	48 734
Kamarajar	5	933	312	17 328	317	18 261
Tirunelveli K. B.	7	1 722	1 173	26 683	1 180	28 405
V. O. Chidambarar	1	657	132	1 990	133	2 647
Kanyakumari	4	3 184	473	2 993	477	6 177
T. Sambuvarayar	1	2 010	-	-	1	2 010
Total	**69**	**58 462**	**8 837**	**300 278**	**8 906**	**358 740**

Table 2.5. Reservoirs in Tamil Nadu (> 1000 ha)

Name of Reservoir	Area (ha) (*at FRL*)	(*Average*)	Fish landing (*t* yr^{-1})	Major fisheries
Salem District				
1. Mettur	15 346	9 324	115	Rohu (19%), *Wallago attu* (15%), catfishes (14%), *Puntius* spp. (14%), catla (10%)
Periyar District				
2. Bhavanisagar	7 876	3 208	179	*L. calbasu* (22%), *L. bata* (19%), *M. aor* (16%), *Puntius dubius* (10%)
South Arcot District				
3. Veeranam	3 885	38	36	Catla (31%), tilapia (44%), rohu (17%), silver carp (8%)
4. Perumaleri	2 590	1 295	-	--
5. Wellington	1 554	500	9	Catla (31%), common carp (8%), *rohu* (17%), tilapia (44%)
Chengalpattu MGR District				
6. Poondi	3 263	1 402	15	Tilapia (41%), *common* carp (9%), mrigal (9%), *Puntius* spp. (7%)
Madurai District				
7. Vaigai	2 419	1 554	24	Catla (56%), mrigal (22%), silver carp (8%), tilapia (6%)
T. Sambuvarayar District				
8. Sathanur	2 010	56	126	Catla (51%), mrigal (27%), common carp (12%), tilapia (6%),
Kanyakumari District				
9. Pechiparai	1 515	700	9	*Puntius* spp. (45%), catla (19%), mrigal
Dharmapuri District				
10. Krishnagiri	1 248	768	47	Misc. (73%), tilapia (24%), catla (20%)
11. Vaniyar	1 093	61	3	Misc. (48%), rohu (27%), catla (13%)

Compiled from: *1. Director of Fisheries, Tamil Nadu*
2. Srivastava, et al., 1985
3. Anon., 1987

Table 2.6. Reservoirs in Tamil Nadu (<1000 ha)

Name of reservoir	Area (ha) at *FRL*	*Av.*	Fish landings ($t\ yr^{-1}$)	Major fisheries
Chengalpattu MGR District				
Kolavai eri	900	529	30	
N. Arcot Ambedkar District				
Goddar	678	339	3	Tilapia (57%), common carp (25%), catla (10%), rohu (6%)
Dharmapuri District				
Thumbalahalli	193	89	11	Tilapia (79%), catla (9%), silver carp (4%)
Kasarikulihalla	105	63	4	Tilapia (73%), common carp (13%), mrigal (6%)
Sicclagiri Chinnar	54	27	8	Misc. (58%), tilapia (28%), catla (5%), common carp (5%)
Pambar	243	60	26	Misc. (41%), tilapia (35%), c. carp (8%), rohu (7%)
Barur	256	128	24	Tilapia (90%), misc. (7%), catla (2%)
Nagavathy	118	71	5	Tilapia (76%), *Puntius* spp. (9%), catla (6%)
Thoppaiyar	120	60	15	Tilapia (77%), catla (30%), rohu (5%), mrigal (4%)
Chinnar	170	117	6	Tilapia (26%), *Puntius* spp. (26%), common carp (13%)

Table 2.6 (cont.). Reservoirs in Tamil Nadu (<1000 ha)

Name of reservoir	Area (ha) at FRL	Av.	Fish landings ($t\ yr^{-1}$)	Major fisheries
Coimbatore District				
Pillur	400	233	3	-
Tirumoorthy	388	240	13	-
Amaravathy	850	544	112	-
Aliyar	650	384	27	-
Sholaiyar	526	312	5	-
Nirar (upper & lower)	68	35	-	-
Peruvaripallam	290	20	-	-
Nilgiris District				
Sandynulla	300	156	5	-
Pykara	50	25	3	
Glenmargam	50	25	1	-
Ooty lake	35	16	4	-
Kundah	40	23	1	-
TR Bazar	11	8	-	-

Table 2.6 (cont.). Reservoirs in Tamil Nadu (<1000 ha)

Name of reservoir	Area (ha) at FRL	Av.	Fish landings ($t\ yr^{-1}$)	Major fisheries
Dindigul Anna District				
Kuthirayar	-	-	5	Tilapia (50%), catla (16%), mrigal (9%)
Palarporanthalar	518	376	95	-
Maruthanathi	72	42	8	Tilapia (37%), common carp (24%), catla (23%), mrigal (9%)
Manoor	10	10	2	Common carp (48%), catla (33%), silver carp (19%)
Parappalar	114	60	6	Tilapia (37%), common carp (24%), catla (23%)
Kombai (Peria & Chinna)	35	19	7	Tilapia (71%), catla (14%), common carp (4%), silver carp (4%)
Varathamanathi	80	39	7	Tilapia (71%), catla (14%), common carp (4%), silver carp (4%)
South Arcot District				
Vidur	798	479	6	Tilapia (55%), common carp (9%), mrigal (9%), *Puntius* sp. (7%)
Gomukhi	360	218	6	*Puntius* sp. (35%), catla (41%), mrigal (11%)
Manimuktha	746	447	12	Catla (47%),common carp (17%), tilapia (6%),

Table 2.6 (cont.). Reservoirs in Tamil Nadu (<1000 ha)

Name of reservoir	Area (ha) at FRL	Area (ha) Av.	Fish landings ($t\ yr^{-1}$)	Major fisheries
Kamarajar District				
Vembakottai	467	160	18	Tilapia (22%), mrigal (17%), catla (13%), misc. (35%)
Periyar	76	26	17	Tilapia (53%), mrigal (23%), catla (14%), silver carp (6%)
Kovilar	74	24	12	Tilapia (48%), mrigal (24%), silver carp (11%), catla (10%)
Kullur Santhai	316	190	48	Tilapia (70%), mrigal (15%), catla (7%)
Anaikuttam	-	-	1	Tilapia (40%), misc. (34%), mrigal (10%), common carp (9%)
Tirunelveli Kattabomman Dt.				
Karuppanathi	50	25	4	Tilapia (91%), common carp (4%)
Ramanathi	39	20	4	Tilapia (57%), common carp (22%), *Puntius* spp. (15%)
Gundaru	21	10	1	-
Gadana	80	48	2	*Puntius* spp. (65%), common. carp (14%), tilapia (12%)
Manimuthar	940	470	7	*Puntius* spp. (56%), mrigal (16%), tilapia (15%), *Labeo fimbriatus* (6%)
Hope lake	580	300	3	-
Sreemoolaperi	12	10	-	-

Table 2.6 (cont.). Reservoirs in Tamil Nadu (<1000 ha)

Name of reservoir	Area (ha) at FRL	Av.	Fish landings ($t\ yr^{-1}$)	Major fisheries
Madurai District				
Manjalar	197	154	28	-
Sathiar	120	60	6	Tilapia (48%), catla (30%), common carp (7%), rohu (6%)
Tiruchirapalli District				
Kannathudai	50	26	3	Catla (56%), mrigal (22%), tilapia (6%), common carp (12%)
Ponnaniyar	60	16	4	Catla (51%), mrigal (27%), tilapia (6%) common carp (12%)
Uppar	117	60	-	-
Periyar District				
Uppar	453	405	45	-
Varattupallam	89	53	6	Tilapia (56%), silver carp (17%), catla (17%)
Vattamalai (Karaiodai)	307	66	4	-
Gunderipallem	61	34	15	Silver carp (25%), tilapia (12%), catla (6%), misc. (54%)
V. O. Chidambaranar District				
Kadama	657	328	51	Tilapia (85%), catfish (1%), misc. (13%)
Kanyakumari District				
Chittar-1	293	175	12	*Puntius* spp. (41%), catla (25%),
Chittar-2	414	248		tilapia (13%), silver carp (8%)
Perumchani	962	526	9	*Puntius* sp. (71%), tilapia (28%)

2.2 SCIENTIFIC INVESTIGATIONS OF TAMIL NADU RESERVOIRS

At least twenty reservoirs in Tamil Nadu have been studied and the literature on the subject is rich in descriptive limnology and production dynamics. No other State in the country has generated scientific information of this magnitude on the reservoir ecosystem. Ganapati (1940) through his epoch-making paper on Red Hills lake, a water supply reservoir near Madras, laid the foundation for reservoir limnology in India. The paper dealt with the plankton cycles in relation to the water quality and meteorological parameters. He followed up the studies with investigations on Stanley reservoir (Ganapati, 1955; Ganapati and Alikunhi, 1949), Errakuppam (Ganapati, 1956b), Hope lake (Ganapati, 1956a), an irrigation tank near Madras, and Mukerti reservoir (Ganapati, 1957) in addition to his numerous papers on the pond ecosystem. Since 1960s, Sreenivasan, undertook a number of studies on a large number of reservoirs in Tamil Nadu. His studies on three lakes *viz.*, Stanley (Sreenivasan, 1966, 1969) Bhavanisagar (Sreenivasan,1964a, Sreenivasan *et al.*, 1964) and Amaravathy (Sreenivasan,1965a) set the pace for other limnological studies for the rest of India. His contributions include some comparative studies on the reservoirs of South India (Sreenivasan,1970a & b, 1976, and 1979). Sounder Raj et al., (1971) studied the limnology and productivity of Poodi reservoir.

The following account provides information on fish stocks and fisheries of 5 major reservoirs of Tamil Nadu.

2.3 STANLEY RESERVOIR (Fig 2.2)

Mettur dam, situated at 11° 49' N, is often described as an engineer's delight. A straight gravity structure, 1 615 m long, rising 54 m above the Cauvery river bed, the dam is constructed across two hills of the Eastern Ghats. At the time of its construction, Mettur was the highest masonry dam in Asia and the largest in the world. The dam, constructed primarily to stabilise the irrigation in Thanjavur delta, caters to about one third of the irrigated area of Tamil Nadu, besides generating hydroelectric power of 240 MW. Commenced in 1925, the dam was completed in 1934. Stanley is the largest reservoir in Tamil Nadu.

Stanley reservoir was commissioned in 1939 and it continues to be the only reservoir in the State representing the >10 000 ha category. Situated at an elevation of 243 m above MSL in the Salem district, Stanley reservoir has a waterspread of 15 346 ha and capacity of 2 646 million m^3 at FRL, the average area being 9 324 ha. It receives water from the Cauvery river basin of 42 217 km^2. Morphometric indicators such as high shore development index (6.7) and a volume development index more than unity (1.4) point towards productive nature of the reservoir.

Information on Stanley reservoir comes mainly from Sreenivasan (1966, 1969, & 1976). The highest and lowest water temperatures recorded during 1962-63 being 32 °C (in May) and 24.2 °C in (January) respectively. Dissolved oxygen in the surface water layer in most of the months was above 4 mg l^{-1} with the maximum supersaturated conditions at the surface, accompanied by a depletion of oxygen in the hypolimnion. The water was alkaline, the pH values ranging between 7.5 to 8.8. High values of methyl orange alkalinity (140 mg l^{-1} to 278 mg l^{-1}) and specific conductivity (170-350 µmhos) indicated a fairly high ionic concentration, suggesting that the water body is reasonably productive. Total hardness as $CaCO_3$ varied within a range of 86-128 mg l^{-1} ranking the lake as a hard water one. Calcium had a negligible contribution (11-20 mg l^{-1}) to hardness .

Despite being a deep basin, the water column was not stratified thermally, in conformity with the situations reported from elsewhere in peninsular India. Even during the so called winter, air temperature in this part of the country does not drop below 15 °C keeping the water relatively warm. With the onset of summer, when the

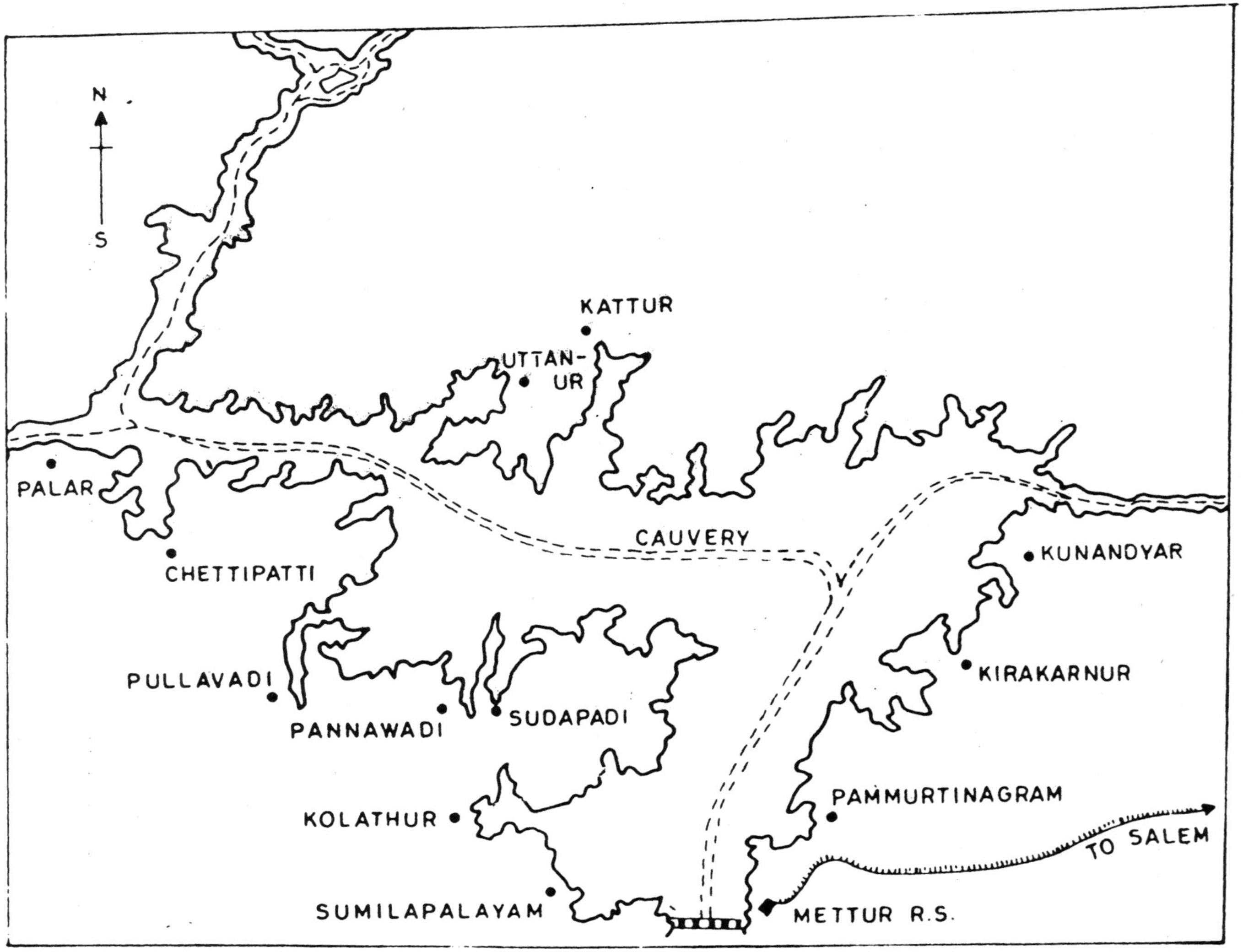

Figure 2.2. Stanley Reservoir (Mettur Dam), Tamil Nadu

top layer warms up, there is no cool water below to offer any thermal resistance. Moreover, the release of cooler water from the bottom layer through the outlets of the dam remove any disparity in temperature between the top and bottom layers.

There was a distinct klinograde distribution of dissolved oxygen in the reservoir. Oxygen recorded from the tropholytic layer varied from nil to 6.8 mg l^{-1} and the deficit of oxygen at the hypolimnion ranged from 0.64 mg l^{-1} to 5.84 mg l^{-1} during various seasons. Oxycline unaccompanied by a thermal stratification is a positive indication of the productive nature of the reservoir. On the one hand, the warm waters at the bottom hastened the bacterial decomposition of organic matter and on the other, the high dissolved oxygen levels in the trophogenic layer indicated high rate of photosynthesis. The low oxygen at the bottom indicated high oxygen demand for decomposing organic matter.

Oxygen depletion at the bottom was accompanied by a concomitant stratification of pH, CO_2, total alkalinity and other related parameters. Inorganic soluble phosphate and nitrate were rarely recorded. Silicate ranged from nil to 19.9 mg l^{-1}. Unavailability of nutrients is a common feature of the South Indian reservoirs.

With Secchi disc visibility up to 1 m and a photosynthetic zone extending down to 5.0 m, the primary production rate was generally high within a range of 1.05 to 10.97 g O_2 m^{-2} day^{-1}. Sreenivasan (1976) estimated the photosynthetic efficiency of Stanley waters at 1.04%, based on the amount of solar radiation and the rate of carbon synthesis by primary producers. Only 0.0927% of the carbon fixed by primary producers was being harvested as fish.

Phytoplankton community of Stanley consisted predominantly of diatoms represented by the ubiquitous *Nitzschia* and *Navicula*, with occasional presence of blue greens like *Merismopedia*, and *Oscillatoria*. *Microcystis*, the common form in the warm water ponds and lakes of South India was conspicuous by its near absence in Stanley.

Net plankton was less important in the trophic chain of Stanley reservoir as they were outnumbered 10 to 1 by nannoplankton. This explains the rather unusual species spectrum of phytoplankton, despite the positive indicators of productive nature of the water body.

Fish and Fisheries

Indigenous ichthyofauna of the river comprised *Acrossocheilus hexagonolepis*, *Puntius carnaticus*, *P. dubius*, *Tor putitora*, *Labeo kontius*, *L. fimbriatus*, *Cirrhinus cirrhosa*, *C. reba*, *Aorichthys aor*, *A. seenghala*, *Pangasius pangasius*, *Silonia silundia* and *Wallago attu* (Chacko *et al.*, 1955). The three downstream barrages on Cauvery *viz.*, the Lower Anicut, the Great Anicut and the Upper Anicut restricted the upriver migration of the anadromous fish, *Tenualosa ilisha* and after the construction of Mettur dam, this fish was reported to disappear completely from upstream stretches of Cauvery. Sreenivasan (1976) reported the disappearance of *Puntius* spp., which used to form 28% of the landings in 1943-44. The indigenous carp *C. cirrhosa* showed an initial increase in catches, but later declined. Low water levels during July in the preceding two or three years were blamed for the spawning failure of *C. cirrhosa*. *Labeo kontius*, which was common and next only to *C. cirrhosa* in the Cauvery, also disappeared from the reservoir.

Table 2.7. Fish production (t) trends in Stanley reservoir during 1960-61 to 1964-65 (Sreenivasan, 1966)

Species	1960-61	1961-62	1962-63	1963-64	1964-65
Catla catla	151 (37%)	138 (40%)	169 (39%)	126 (31%)	39 (17%)
Cirrhinus cirrhosa	156 (39%)	123 (36%)	92 (21%)	82 (20%)	55 (25%)
C. mrigala	21 (5%)	21 (6%)	35 (8%)	42 (10%)	17 (8%)
Catfishes	73 (18%)	60 (17%)	131 (30%)	141 (35%)	112 (50)
Other species	4 ((1%)	2 (1%)	7 (2%)	16 (4%)	0.4 (-)
Total	**405**	**344**	**434**	**407**	**223**

During 1960-61 to 1965, *C. catla* and *C. cirrhosa* dominated the fish landings. However, there was progressive decrease in the percentage of *C. cirrhosa* with a gradual increase in catfish populations (Table 2.7). The annual total fish landings from 1967-68 to 1976-77 ranged from 230 t to 555 t (Sreenivasan, 1976), the average production during this period being 344 t, equivalent to the yield of 22.76 kg ha^{-1} (based on reservoir area at FRL). From 1969-70 to 1972-73, catla was reduced to 10%.

Mrigal, introduced into the Stanley reservoir during 1950-51, appeared in catches during 1957-58 contributing up to 13.9% in 1966-67 and later on, it declined and reduced to insignificance. Rohu also showed some initial increase but later declined. Recruitment failure from the vicissitudes of water level fluctuations and predator pressure have been attributed by Sreenivasan (1966) for the erratic behaviour of the introduced carp species and the emergence of predatory catfishes as the most stable component of fishery in the reservoir. The latest report received from the State Fisheries Department in 1993 put the total catch at 115 t comprising *Labeo rohita* (19%), *Wallago attu* (15%), other catfishes (14%), *Puntius* spp. (14%) and *Catla catla* (10%).

Fishing gear

Mettur being one among the earliest dams to be commissioned in the State of Tamil Nadu, Stanley reservoir has been considered as a testing ground for evolving reservoir fishing practices in India. The use of gill nets, now used widely in reservoirs all over the country, was developed by the fishermen of Stanley by trial and error methods. They named the floating gill nets as *Rangoon* nets, after the migrant fishermen from Myanmar. This gear is the most suitable for reservoir conditions, inasmuch as most of them have rugged bottom with steep slopes. The fishermen have even evolved the best mesh size to suit each fishery. For instance, during 1964 to 1965, there were separate nets targetted against catla of 25 kg, mrigal of 16 kg.

Pollution

Factory effluents from the Mettur Chemical and Industrial Corporation were reported to be discharged into the river course below the dam. During dry seasons, when the outflow surplus channels were closed, the effluents caused pollution in the deep pools of the river (Ganapati and Alikunhi, 1949). Bioaccumulation of mercury had been reported in many fishes (Jayachandran and Raj, 1975; Table 2.8).

Table 2.8. Bioaccumulation of mercury in Stanley reservoir

Species	length (mm)	Wt (g)	Hg (mg kg^{-1})
Aorichthys aor	370	350	0.250
Notopterus notopterus	340	270	0.086
Cirrhinus cirrhosa	335	450	0.047
Labeo calbasu	410	720	0.090
Rhinomugil corsula	330	275	0.085
Puntius sarana	255	250	0.072
Labeo rohita	430	1570	0.027

(From Jayachandran and Raj, 1975)

2.4 BHAVANISAGAR (Fig. 2.3)

Chacko and Dinamani (1949) prepared a preliminary report on the fishery resources of the Bhavani river, three years before Bhavanisagar reservoir was formed. After the impoundment was completed in the year 1953, Dorairaj and Pankajam (1956), and Menon and Chari (1959), provided the first insights into the limnology and fisheries of the reservoir. Pankajam (1956), Sreenivasan(1962 and 1964b) and Sreenivasan *et al.* (1964) studied plankton communities in the light of environmental changes due to impoundment. Franklin (1969) documented the phytoplankton species. Bhavanisagar is also mentioned in the comparative studies of Tamil Nadu reservoirs made by Sreenivasan (1970a and 1976). Biology of commercially important species of fish *Puntius dubius* (Ranganathan *et al.*, 1962), *Aorichthys aor* (Ranganathan and Radha, 1966), *Labeo calbasu* (Natarajan, 1971a) and *Channa marulius* (Devaraj, 1973) was also studied. Bhavanisagar was covered under the All India Coordinated Research Project on Reservoir Fisheries, launched in 1971. The final report contains the results of investigations carried out in the reservoirs during 1971 to 1981 and a set of recommendations for the management of its fisheries (Natarajan *et al.*, 1981).

Bhavanisagar dam is constructed on the river Bhavani at 11° 30' N lat., below its confluence with the river Moyar in the Cauvery basin in the Periyar district (Fig. 2.3). The reservoir has a maximum waterspread of 7 876 ha at FRL which is 280.2 m above the MSL, and its net capacity is 908 million m^3. The average reservoir area is 3 695 ha. During the driest years, the reservoir shrinks to 745 ha in summer (Natarajan, *et al.*, 1981). Receiving water from catchment of 4 200 km^2, the reservoir serves irrigation and flood control. The calculated mean depth of Bhavanisagar is 11.5 m, the maximum depth being 33 m. The shore development and volume development indices are 4.0 and 1.0 respectively.

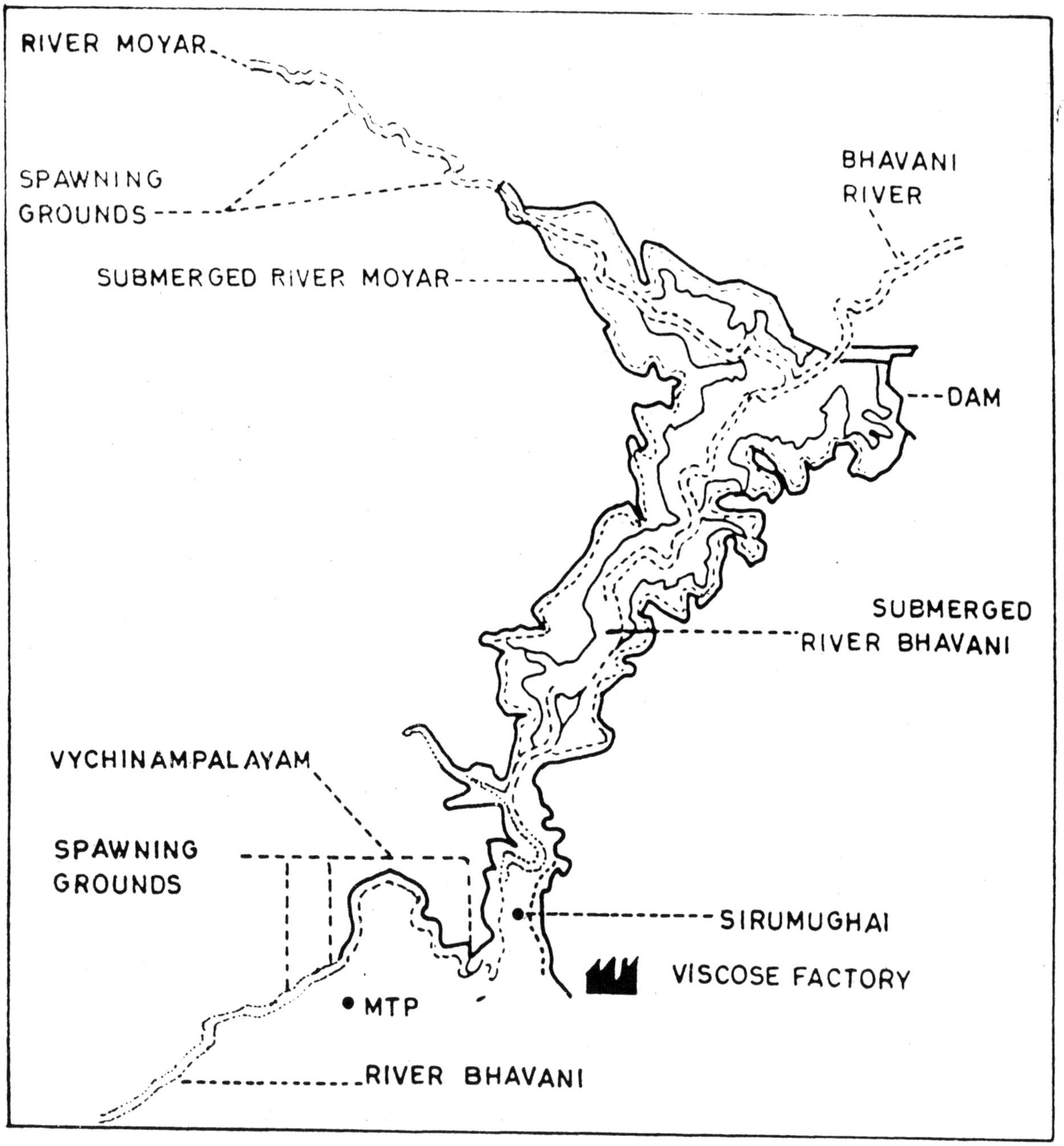

Figure 2.3. Bhavanisagar reservoir, Tamil Nadu

Soil of Bhavanisagar reservoir bed is silty clay loam and mostly acidic in reaction. Judged by the presence of available nitrogen, available phosphorus and organic carbon, the soil was categorised by Gupta (1979) as conducive to aquatic productivity. During pre-monsoon months, the pH was found to increase towards neutrality. Water level fluctuations in Bhavanisagar are erratic. In the year 1961 and 1962, annual level fluctuations were of 15.2 and 7.9 m respectively. During 1971 to 1981, water level fluctuations varied from 5.46 m to 22.6 m yr^{-1}, with the accompanied changes in area and capacity of the reservoir. Air temperature regime seems to have undergone a change during the last few years. The maximum air temperature recorded during 1957-62 was 38 °C, while the summer temperature during 1971 to 1981 was on an average 40.5 °C. The highest temperature exceeded 45.0 °C during 1974-75 and 1975-76. However, the range of water temperature during 1956-61 and 1971-82 (23.7 °C to 30.2 °C) did not differ much.

Water temperature in Bhavanisagar is not dependent on the prevailing air temperature at the reservoir site. The inflowing water from Moyar and Bhavani rivers determines the water temperature of the reservoir (Sreenivasan, 1962). Absence of thermal stratification, oxygen depletion in the bottom water layer, and increase in total alkalinity, CO_2 and specific conductivity towards the depth indicate the productive nature of the water body.

The reservoir has a rich plankton community dominated by Cyanophyceae (*Microcystis, Oscillatoria* and *Anabaena*), diatoms (*Nitzschia, Melosira*) and Chlorophyceae (*Pediastrum, Mougeotia*). During the early years (1956-61), *Nitzschia* and *Melosira* formed an important constituent of the phytoplankton community, along with Cyanophyceae (Sreenivasan, 1964b; Sreenivasan *et al.*, 1964). During 1970's and 80s, however, the diatoms were not very important. Initially, Bhavanisagar had a very high phytoplankton species diversity (Sreenivasan *et al.*, 1964; Franklin, 1969). By the 1970s blooms of *Microcystis* (occasionally *Mougeotia*) have become a common feature accompanied by substantial reduction in diversity (Abraham, 1980b). This fall in diversity and increase in dominance of lentic species, apart from indicating organic enrichment, suggested a transient eutrophication stage through which the reservoir was passing. The rich benthic community of Bhavanisagar was represented by oligochaetes, *Chironomus, Chaoborus* and mayfly nymphs, with a population density of 483 organisms and 3.8 g m^{-2} respectively (Abraham, 1979).

Pollution

Bhavanisagar reservoir receives effluents from the South India Viscose Factory at Sirumughai, ↄ km downstream of Moolathurai, which is considered as the breeding ground for indigenous and introduced fish species. There is a sharp fall in pH and dissolved oxygen at the discharge point with sharp increase in carbon dioxide and alkalinity. Free CO_2 recorded at the site is 27.6 mg l^{-1}, compared to 1.5 mg l^{-1} upstream. However, further downstream in the lotic zone and the main reservoir sector, the water quality became normal (Natarajan *et al.*, 1981). Fish mortality due to decrease in pH and oxygen depletion was reported occasionally (Sreenivasan, 1979b).

Ichthyofauna of the reservoir comprises 51 species belonging to 11 families (Natarajan *et al.*, 1981). Among them, 18 species contribute to the commercial fisheries. Impoundment did not alter the fish faunistic spectrum to any appreciable extent as most of the indigenous species (Chacko *et al.*, 1955) except *Tor tor* continue to be caught in the reservoir.

Puntius spp., *T. putitora, T. tor, Accrossocheilus hexagonolepis, Puntius dubius* and *Puntius carnaticus*, along with *Labeo kontius* form the traditional fisheries of river Cauvery and the stretch above the reservoir are believed to be their breeding grounds (Wilson, 1920). One reassuring feature of the reservoir is the continued presence of these indigenous forms especially *P. dubius, P. carnaticus, Labeo kontius, Cirrhinus cirrhosa* and *A. hexagonolepis*. Their survival is attributed to the uninterrupted breeding activities at Moolathurai and Nellithurai, especially in the former when water is released from Pilloor. *P. dubius* which has a fecundity of 66 000, ascends the rivers Cauvery and Moyar during the north east monsoon for spawning and lays eggs in batches of 1 000 to 2 000 on the gravel bed (Ranganathan *et al.*, 1962). It is due to the continued availability of breeding grounds that the indigenous fishes survive in the lake and therefore there is a need to conserve this biologically sensitive environment. *Cirrhinus reba* breeds during June-July at Moolathurai in Bhavani and Hogainakkal in the main river Cauvery (Rao and Gopinathan, 1972). Evidence for recruitment of *Labeo fimbriatus, L. calbasu, L. kontius* and *P. carnaticus* has also been obtained along with successful natural breeding of introduced carps such as catla, rohu and mrigal (Natarajan *et al.*, 1981). Sharp variations in catch per unit of effort in respect of *P. dubius* indicated inconsistencies in annual recruitment (Natarajan *et al.*, 1981). Water level fluctuations seem to be the major factor behind the fish catch fluctuations, rather than fishing pressure.

L. calbasu, which feeds mainly on bottom detritus, is reported to mature at a size of 400 mm. The fish spawns in Moolathurai breeding grounds during south west monsoon (Natarajan, 1971a and b). The mean length in catch varies between 465 and 530 mm among the > 3 year group. Other southwest monsoon breeders are the Indo-Gangetic major carps *Catla catla, Labeo rohita and Cirrhinus mrigala*. The indigenous *Puntius dorsalis* breeds intermittently with at least three annual peaks in its spawning intensity (Natarajan *et al.*, 1981). Commercial fish landings from the reservoir during 1964 to 1980 ranged from 85 t to 293 t at an average of 156.5 t. This is rather low, in view of the good primary production rate. Sreenivasan (1976) estimated that 0.0775% of the carbon synthesised and 0.000513% of the solar energy were converted into fish flesh in Bhavanisagar. Community metabolism is mainly through the detritus chain. In the absence of an adequate population of planktiphagous fish, most of the energy at plankton phase is added to detritus. *Labeo calbasu* (95%), *Cirrhinus mrigala* (87.5%), *P. dubius* (94.9%) and *Catla catla* (55%) had a predominantly detritus-based diet. *L.calbasu*, and *C. cirrhosa* are increasing in catches while *P. dubius* and *C. catla* show a decline (Table 2.9).

Conservation

Conservation measures to be adopted in Bhavanisagar should be relevant to the feeding, breeding and other habits of the target fishes. *P. dubius* has a natural distribution in the upper reaches of Bhavani and Cauvery in the sub-mountainous Tamil Nadu. This fish feeds mainly on benthic chironomid larvae, oligochaetes, insects and nymphs at the depths of 2 to10 m. It breeds once a year during October to December, and undertakes a distinct, short upstream spawning migration. The eggs are laid on gravel by the side of rapids and pools.

In Bhavanisagar, the breeding grounds of *P. dubius* are reported to be intact and the fish still thrives in the reservoir. In Stanley, the Hogainakkal fall is a natural barrier obstructing the upstream migration of *P. dubius*. Similarly in Amaravathy, the river stretch above the reservoir has Duvanam Falls where the ready to spawn breeders congregate in large numbers from December. Ranganathan *et al.* (1962) noted that the failure of migration caused atrophy of gonads and breeding failure. Stocking of fry above waterfalls is suggested by the above authors as a conservation measure. However, Sreenivasan (1976) felt that the natural cascades associated with Davanam Falls allowed upstream migration of this fish. Ranganathan *et al.*, (1962) recommended stocking of *P. dubius* in reservoirs situated in the sub-montane areas, provided there was a suitable river stretch above and no barrier for upstream migration. Conservation measures should also include:

1) total closure of fishing activities during the breeding season, and

2) ban on gill nets < 6 cm in mesh size (stretched) so that fish of three years and less will not be caught.

Table 2.9. Percentage composition of fish landings in Bhavanisagar Reservoir

Year	*W. attu*	*A. aor*	*P. dubius*	*L. calbasu*	*L. rohita*	*C. catla*	*L. bata*	*C. mrigala*	*L. fimbriatus*	Total catch (t)
1964	10.19	21.30	27.16	12.67	1.74	8.20	5.91	2.56	3.50	-
1965	14.03	39.84	14.40	11.35	0.70	11.81	2.80	0.68	1.98	-
1965-66	14.45	41.48	11.70	12.00	0.76	12.47	1.59	0.58	1.76	-
1966-67	19.04	20.60	5.54	18.61	1.95	23.98	4.86	1.03	0.57	-
1967-68	10.95	18.02	3.88	42.24	0,59	10.86	-	0.81	-	109.6
1968-69	9.26	16.01	5.81	47.74	0.37	14.91	-	0.36	-	112.9
1969-70	12.78	13.73	11.75	42.36	0.68	14.14	-	3.54	-	134.4
1970-71	12.49	16.06	4.87	27.05	1.14	27.24	-	10.09	-	150.4
1971-72	11.04	14.74	6.35	48.96	0.59	9.67	-	8.08	-	96.7
1972-73	11.09	17.98	11.19	34.89	1.17	10.02	-	10.63	-	120.8
1973-74	7.20	16.96	4.17	40.41	1.22	5.44	-	12.52	1.16	146.5
1974-75	7.00	30.60	3.60	40.30	0.90	2.60	-	11.30	1.96	205.0
1975-76	12.27	19.99	4.04	39.76	2.77	3.39	1.22	13.28	1.02	195.1
1976-77	8.16	13.46	5.29	42.07	7.71	6.86	1.77	12.09	0.94	294.3
1978-79	5.61	15.30	5.89	21.45	1.22	3.17	21.55	3.26	0.64	-
1979-80	5.04	16.28	9.62	21.71	2.61	5.10	19.42	6.83	1.47	-

From Sreenivasan 1976 and Natarajan 1981

2.5 AMARAVATHY (Fig. 2.4)

The river Amaravathy, a tributary of the Cauvery was dammed in 1958 at 10^0 15' N, 5 km below the point, where the river debouches into the plains to create the Amaravathy reservoir. The reservoir water surface area is 850 ha at the full level of 358 m above MSL. Though small, Amaravathy is one of the most productive reservoirs in the State. The lake has a net capacity of 112.37 million m^3 at FRL and the mean depth is 13.7 m . Volume development and the shore development indices are 1.2 and 2.3 respectively. The catchment area of the reservoir is 832 km^2. The reservoir was described by Sreenivasan (1965a, 1965b, 1970a, 1976 and 1979a, b, and c.).

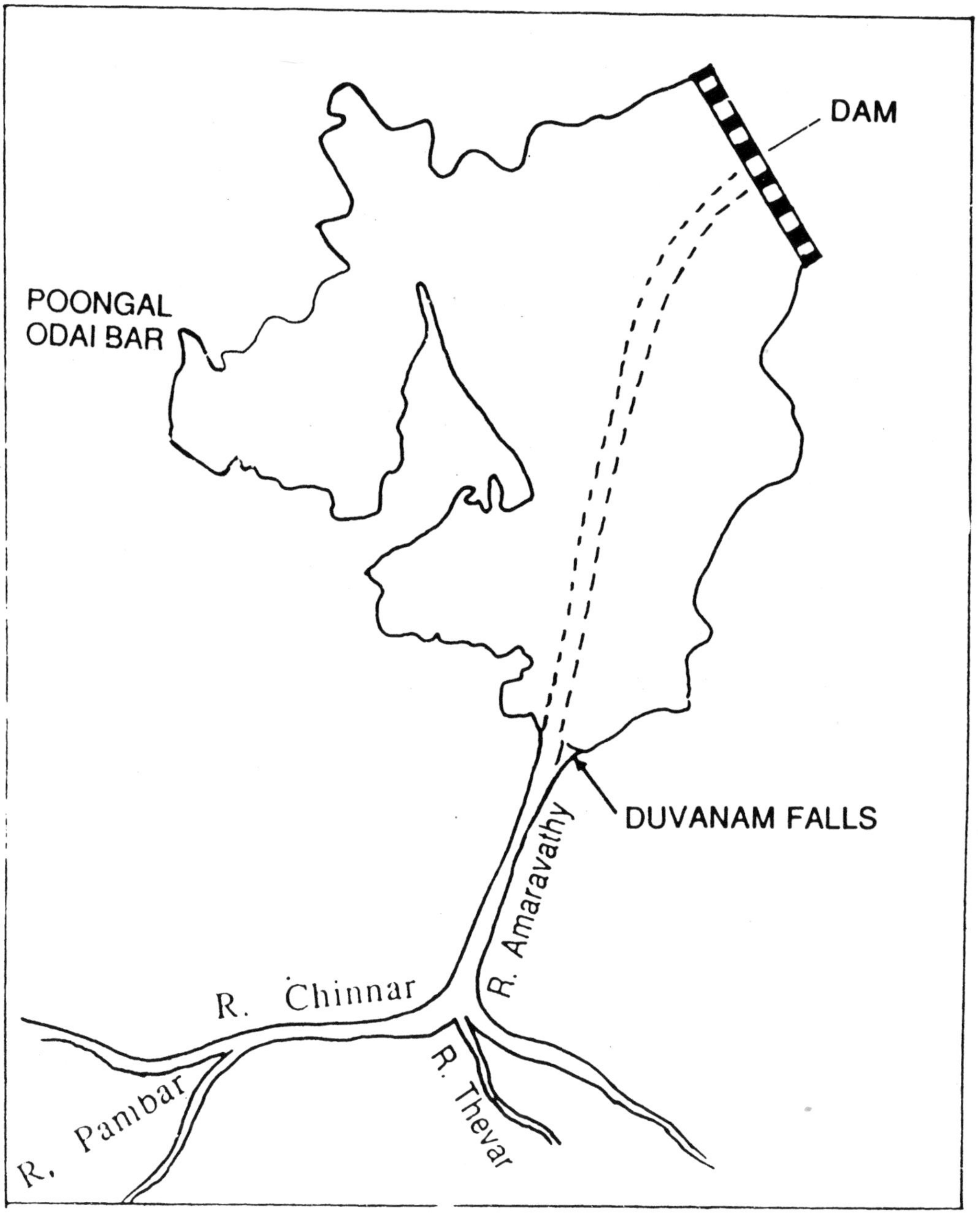

Figure 2.4. Amaravathy reservoir, Tamil Nadu

Bottom soil of the reservoir is sandy loam with organic nitrogen @ 0.156%, organic carbon 0.397% and phosphate 0.362 g per 100 g. The catchment of Amaravathy is the thickly wooded, highly rain-fed Munnar hills of the Western Ghats. Geologically, the catchment comprises rocks of biotite granite gneiss and schistose gneiss. Hard, coarse, crystalline metamorphic rocks are highy resistant to weathering. The river originates at an altitude of 1 515 m and passes through forests. The inflowing water does not stratify in the reservoir. However, there is a distinct horizontal gradient in temperature between the water at the lentic zone (28 °C) and the fluviatile sector (23.8 °C), just 8 km distant from one another.

The reservoir has all the features of an oligotrophic lake such as low total hardness, virtual absence of nutrients like nitrate and phosphate and low electrical conductivity and total alkalinity. Morphometric indicators like shore development index and the deepness of water column also point towards oligotrophy. Yet, Amaravathy has the highest rate of primary productivity among the reservoir of Tamil Nadu. This shows the efficacy of tropical impoundments to turn eutrophic, taking advantage of the abundant sunlight and the warm unstratified water column (Table 2.10). The inflowing water is poor in nutrients and dissolved solids. The ionic build-up and nutrient loading are a local phenomenon happening through metabolic activities.

Table 2.10. Physico-chemical characteristics of water in Amaravathy reservoir

Parameters	Range	Average
Specific conductivity (µmhos)	38-63	48
Total alkalinity (mg l^{-1})	7-84	25.86
Total hardness (mg l^{-1})	18-50	38
Chlorides (mg l^{-1})	0.4-1	0.7
Organic C (mg l^{-1})	12.5-21.6	18.2
Silicate (mg l^{-1})	1.4-38.5	-
Phosphates (mg l^{-1})	0.0-0.01	-
pH	6.7-9.1	-
Gross primary productivity (O_2 m^{-2} d^{-1})	0.30-18.3	13.7

(Sreenivasan, 1965a; 1979c)

Methyl orange alkalinity of 7.0 mg l^{-1} at the surface increases to 40 mg l^{-1} below 8 m depth. The water turns acidic as the depth increases. At the top layer, apart from using up all the free CO_2 for photosynthetic activities, CO_2 is also derived from the bicarbonate, thus reducing its concentration, whereas in the non-photosynthetic zone, with the abundance of free CO_2, bicarbonates are high. Amaravathy is a testimony to the fact that the conventional indicators of productivity in terms of morphometric and catchment parameters have limited validity in the warm tropical reservoirs, where the water body can develop intrinsic eutrophy. The high rate of metabolic activities due to conducive temperature hastens both photosynthetic production as well as the organic decomposition.

Amaravathy has one of the highest primary productivity rates, next only to Tirumoorthy. Carbon fixation rate, measured in terms of oxygen released, was estimated @ 11.320 cal m^{-2} yr^{-1} x 10^{-6} at a photosynthetic efficiency of 1.442% from sunlight to carbon. An unusually high rate of oxygen release at the rate of 59.1 g m^{-2} day^{-1} has been reported from the reservoir by Sreenivasan (1965a). Plankton community of Amaravathy is dominated by *Microcystis* which blooms round the year.

Alterations in the plankton community due to the reservoir formation is believed to have contributed to radical changes in fish stocks. The indigenous fishes could not thrive on a *Microcystis*-dominated plankton. Almost all indigenous species of commercial value have disappeared. *Puntius dubius*, which ranked first till 1965-66, forming 55% of the fish landings, dwindled to less than 2% from 1968-69 and disappeared later (Sreenivasan, 1976). *P. carnaticus* was another casualty. Today, the commercial fishery is entirely based on the transplanted fishes including some exotic species. Tilapia, *Oreochromis mossambicus*, which was stocked in the year 1957-58, gradually established a foothold and by mid-seventies, they formed more than 90% of the catch. The common carp, *Cyprinus carpio* has increased since 1975-76 (Table 2.11).

Table 2.11 Changes in percentage composition of fishes in the commercial catches of Amaravathy reservoir

Year	Total landing (t)	Percentage composition						
		Oreochromis mossambicus	*Cyprinus carpio*	*Labeo fimbriatus*	*Catla* catla	Rohu & mrigal	*Puntius dubius*	*Puntius carnaticus*
1978-79	139	88.19	4.41	5.21	0.42	0.30	-	-
1977-78	85	78.93	13.33	6.13	0.39	1.25	-	-
1976-77	64	88.96	6.76	1.63	0.73	1.23	-	-
1975-76	90	89.04	4.21	3.11	-	-	-	-
1974-75	121	91.33	0.91	3.80	-	1.04	-	-
1973-74	132	86.70	2.47	3.03	-	-	-	-
1972-73	86	88.02	-	3.51	-	-	0.15	0.62
1971-72	72	79.47	3.88	3.50	0.70	1.46	0.27	0.91
1970-71	76	79.98	6.91	6.58	0.33	3.80	0.85	1.12
1969-70	68	84.12	0.71	5.60	0.57	0.80	1.86	1.28
1968-69	101	85.41	-	2.34	0.66	1.78	33.75	4.61
1967-68	87	61.75	0.11	1.39	0.86	1.46	n. a	n. ax
1966-67	48	48.17	0.01	1.06	0.83	0.52	33.75	4.61
1965-66	33	13.54	0.01	3.30	3.03	1.11	n. a.	n. a
1964-65	18	16.26	0.02	3.61	2.18	3.72	54.94	5.97

(Modified from Sreenivasan, 1976)

Performance of tilapia in Amaravathy, over the years was phenomenal. Apart from adding 48 to 132 t to the production (average fish production 90 t yr^{-1}), the individual fish grew to *c.*1.5 kg in 1968 (later reduced to 0.68 kg). This allayed the fears of stunted growth, a common complaint against this particular tilapia (*Oreochromis mossambicus*) in pond environment. Tilapia also had the highest catch per unit of effort (CPUE) of 11 950 to 22 569 kg yr^{-1} (CPUE in Tamil Nadu reservoirs is calculated on the basis of units comprising 20 gill nets of 50 m length each, operated from a coracle as described by Ranganathan and Venkataswamy, 1967). Moreover, the fishery based on tilapia could efficiently convert energy into fish flesh. The energy conversion rates from primary producer to fish and light energy to fish were 0.2081% and 0.003003% respectively, both being very high and next only to Sathanur reservoir.

2.6 SATHANUR (Fig. 2.5)

Sathanur reservoir was created in 1957 on the River Ponniar at Sathanur Village in Tiruvannamalai Sambuvarayar district (12^0 12' N). It covers 2 010 ha (Fig. 2.5) at the FRL of 222.2 m. The reservoir has a capacity of 228.91 million m^3 at full level and a mean depth of 11.4 m. Shore and volume development indices are favourable for productivity. Sathanur is a hard water reservoir (total hardness range from 112 - 254 mg l^{-1}), with a matching high specific conductivity (320-800 μmhos), and total alkalinity (145-616 mg l^{-1}). Values of all the three parameters mentioned above are some of the highest recorded in Tamil Nadu reservoirs. Nitrate nitrogen in water which is rarely found in Tamil Nadu waters is quite high (4.4-5.6 mg l^{-1}). There is no regular plankton bloom as in the organically rich reservoir, Amaravathy. Dissolved oxygen drops from 8.5 mg l^{-1} to nil towards the bottom. However, the fall in pH is not steep, due to the buffering action and poor tropholytic activities (Sreenivasan, 1976). Alkalinity increases by 14.0 to 78.0 mg l^{-1} from surface to bottom.

Both morphometric as well as metabolic indicators point to the productive nature of the reservoir There is a high rate of carbon fixation. The primary chemical energy fixed by producers is estimated at 8.1 x 10^6 cal m^{-2} yr^{-1} at a photosynthetic efficiency of 1.047%. The fish, *Wallago attu, Cirrhinus reba, Labeo kontius, L. fimbriatus, Notopterus notopterus and Mastacembelus armatus* are indigenous species, while *Catla catla, Cirrhinus mrigala, C. cirrhosa, L. calbasu, L. rohita* and *Cyprinus carpio* (Bangkok strain) are introduced.

Drifting gill nets of mesh bar 5, 6 and 7.5 cm are employed for fishing. A fishing unit comprises 10 gill nets, each 50 m in length, a coracle (a saucer-shaped country craft made of split bamboo and hide) and two men. Annual fish landings recorded during 1964 to 1977 ranged from 48.1 t (in 1967-68) to 214.4 t (in 1976-77) with an annual average of 128.2 t. During this period, catla has established firmly in the reservoir. The indigenous *L. fimbriatus*, which was dominant (36%) during 1964-65, represented only 3% of the total catch by 1976-77. *W. attu*, the main predator, is less in percentage, though steady in its occurrence (Table 2.12). Presently, catla along with other Indian major carps form more than 78% of the catch (1992-93).

It is encouraging to note that *C. catla*, the prime fish species of Sathanur, is represented in the catch by the 3 to 4 year class. The presence of *W. attu* has not affected the carp fishery. Prabhavati and Sreenivasan (1979) believe that small numbers of this predator are welcome as a means to check the weed fish populations. The decline in populations of *L. fimbriatus* and *L. calbasu* is attributed to their breeding failure. Until 1977, fish catches in Sathanur were steadily increasing every year, despite a steady increase in fishing effort (Sreenivasan, 1979a). The catch per unit effort (as described by Rangathan and Venkataswamy, 1967) still remained high at 7 318 to 14 295 kg yr^{-1}, thus suggesting scope for further increase in effort.

2.7 ALIYAR RESERVOIR (Fig. 2.6)

Aliyar dam was constructed in 1962 across the Aliyar river. Aliyar reservoir, situated between 10° 15' and 10°30' N and 76° 50' and 77^0 10' E, covers 646.0 ha at the FRL of 320.04 m above MSL. Apart from a catchment area of 468.8 km^2, the lake receives water from Upper Aliyar reservoir, through a hydro-electric power station at Navamali and the Parambikulam reservoir (Kerala) through a contour canal. The reservoir with its maximum depth of 36.5 m and mean depth of 18.2 m, exhibits annual water level fluctuations ranging from 13.15 m to 31.03 m at an average of 21 m per year (Table 2.13). Aliyar reservoir was studied by the Central Inland Capture Fisheries Research Institute for eleven years from 1982 to 1992.

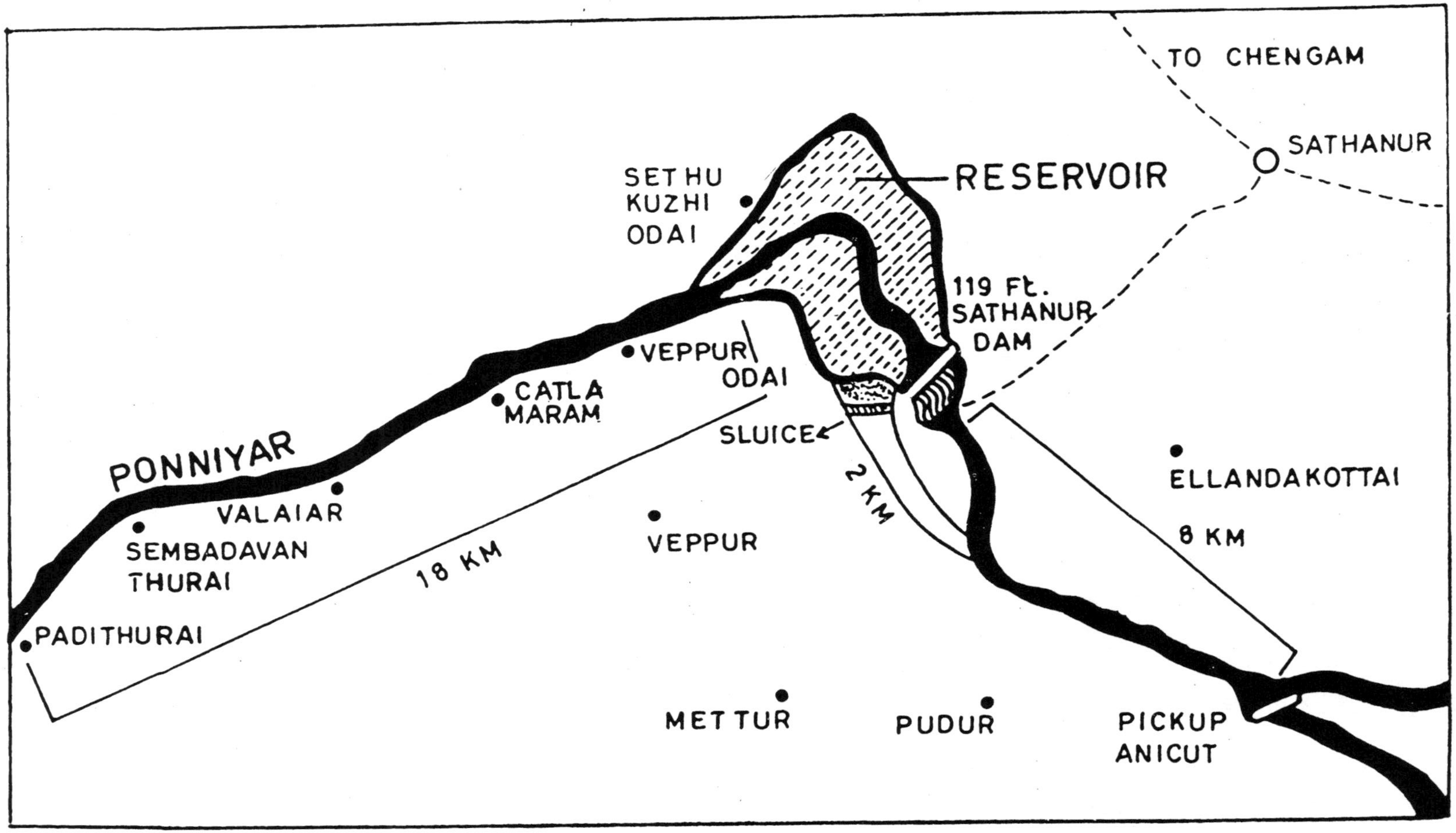

Figure 2.5. Sathanur reservoir, Tamil Nadu

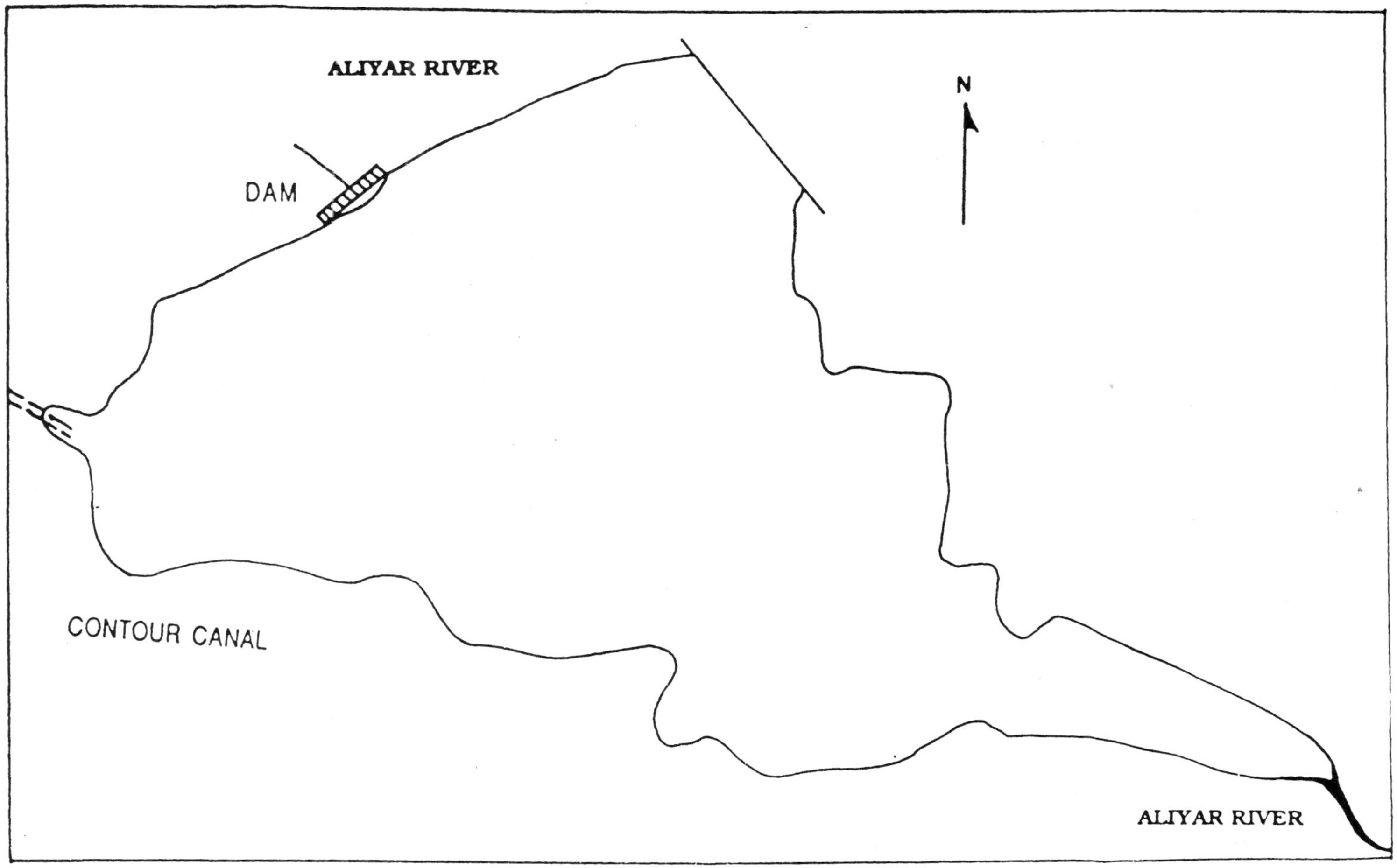

Figure 2.6. Aliyar reservoir, Tamil Nadu

Table 2.12. Fish catch composition of Sathanur reservoir

Year	Total fish catch (t)	Percentage composition							
		Catla	Rohu	Mrigal	*C. cirrhosa*	*L. fimbriatus*	*L. calbasu*	*W. attu*	Tilapia
64-65	na	20.40	7.53	6.86	-	35.96	-	1.20	12.40
65-66	na	27.60	9.65	12.35	7.95	19.35	-	10.40	0.90
66-67	na	51.70	5.40	17.50	4.40	6.30	3.50	10.80	-
67-68	48.1	46.20	5.80	16.80	3.80	12.70	8.30	6.20	-
68-69	54.1	51.10	5.70	15.90	2.10	7.30	7.00	8.30	-
69-70	103.2	59.44	2.67	11.27	0.81	5.58	12.38	7.45	-
70-71	109.3	78.25	2.76	6.91	0.78	0.94	2.20	6.97	-
71-72	113.2	80.72	2.58	1.96	0.48	1.62	1.07	8.70	-
72-73	116.2	82.62	2.31	1.89	1.78	2.20	3.03	7.68	-
73-74	139.0	86.38	2.87	1.07	1.23	1.08	0.38	6.57	-
74-75	175.4	85.65	1.36	1.38	0.43	1.93	1.06	5.62	-
75-76	209.1	91.47	1.16	0.90	0.05	0.89	0.34	4.76	-
76-77	214.4	87.70	0.86	1.93	0.67	3.02	0.89	4.69	-
Mean	*128.2*								

Compiled from Sreenivasan (1976) and Prabhavati & Sreenivasan (1979)

Table 2.13. Morphometric features of Aliyar reservoir

Catchment area	:	468.8 km^2
Length of shoreline at FRL	:	16.0 km
River bed level (MSL)	:	283.6 m
Full reservoir level (MSL)	:	320.04 m
Maximum depth	:	36.58 m
Dead storage level	:	293.0 m
Mean depth	:	18.2 m
Average water level fluctuation	:	21.0 m
Area of dead storage level	:	2.5 ha
Maximum area at FRL	:	646.0 ha
Average area (DSL + FRL)/2	:	324.25 ha
Capacity at FRL	:	10 942.8 ha m
Volume development	:	1.2

(CICFRI, Barrackpore)

Physico-chemical quality of water and soil

The basin soil is acidic (pH 6.42), comprising sand (24.1%), silt (29.0%) and clay (46.9%) with levels of available phosphate, available nitrogen and C/N ratio indicating medium level of productivity.

The catchment area of Aliyar mostly comprises the hill ranges of the Western Ghats with limited forest cover and negligible agriculture, resulting in low nutrient leaching into the reservoir. Water of Aliyar is warm (temperature at the surface within the range of 21.2 and 32.5 °C) and clear (transparency 108 - 182 cm) with a klinograde oxygen distribution and a discernible increase in the values of specific conductivity, bicarbonate, total alkalinity and free carbon dioxide from surface to bottom. The vertical stratification is more pronounced during the first and last quarters of the year due to lesser turbulence from wind and inflow. Higher values of specific conductivity, total alkalinity and dissolved organic matter are mostly noted in the lentic sector. Ammonia is present only at the bottom of the profundal region of the lentic sector. Carbonate alkalinity is absent during most of the times. The water is generally poor in nitrate (nil - 0.04 mg l^{-1}), phospate, calcium and magnesium, but moderate to high values of silicates are recorded (Table 2.14).

While temperature, pH, transparency and dissolved oxygen levels depict a congenial environment for productivity, the alkaliphobic condition of the water and the low values of calcium and magnesium are retardants to productivity. The low to moderate levels of phosphate and nitrate in Tamil Nadu reservoirs probably indicate their constant utilisation. There is no sign of eutrophication in Aliyar reservoir. With the maturation of the reservoir, the dissolved oxygen concentration has been increasing. There is also a decrease in oxygen demand as the flooded vegetation is already decomposed (Sreenivasan, 1970 c). The alkalinity, specific conductivity and the pH do not show significant change in values over the years. Based on the water characteristics, the reservoir continues to be oligotrophic even after three decades of its formation.

Table 2.14. Water quality of Aliyar reservoir (1983-87)

Parameteres	Range
Water temperature (°C)	21.2 - 32.50
Transparency (cm)	108.0 - 182.00
pH	6.6 - 8.80
Dissolved oxygen (mg l^{-1})	4.2 - 11.60
Specific conductivity (µmhos)	38.7 - 109.60
Carbonate alkalinity (mg l^{-1})	nil - 14.00
Bicarbonate alkalinity (mg l^{-1})	16.0 - 58.00
Total alkalinity (mg l^{-1})	16.0 - 58.00
Free Carbon dioxide (mg l^{-1})	nil - 10.00
Ammonical Nitrogen (mg l^{-1})	nil - 0.20
Nitrate-Nitrogen (mg l^{-1})	Tr. - 0.04
Phosphate (mg l^{-1})	Tr. - 0.02
Silicate (mg l^{-1})	2.8 - 24.00
Calcium (mg l^{-1})	4.0 - 18.00
Magnesium (mg l^{-1})	1.2 - 6.20
Dissolved organic matter (mg l^{-1})	0.9 - 3.60

(CICFRI, Barrackpore)

Plankton

The mean standing crop of plankton is estimated at 10.8 ml m^{-3} within a range of 6.4 to 26.5 ml m^{-3} (observation period 1983 to 1986). The characteristic feature of the reservoir is the *Microcystis sp.*, blooms during low water levels. The blue-greens constitute 84.32% of the total plankton of the reservoir (Fig. 2.7).

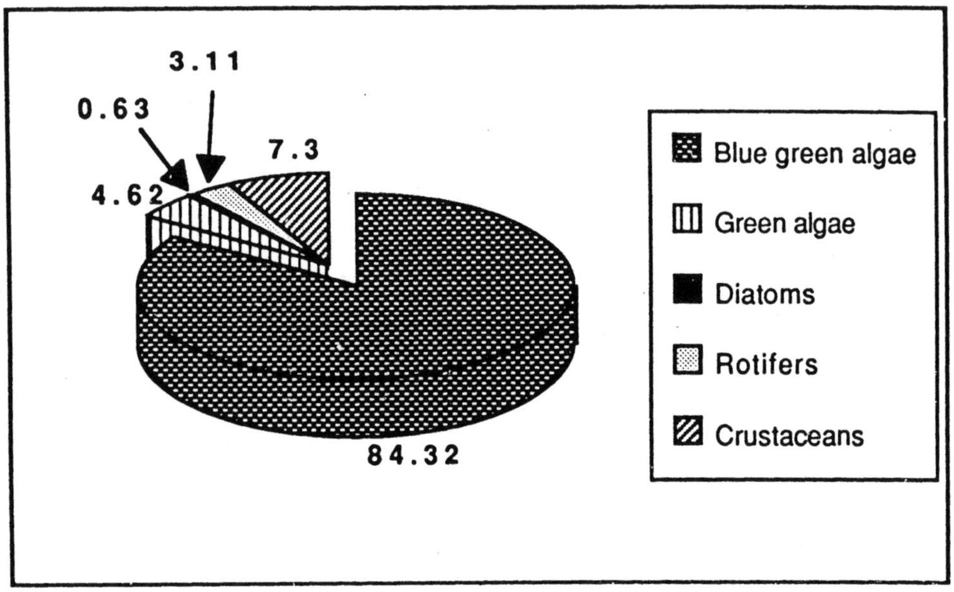

Figure 2.7. Percentage composition of plankton in Aliyar reservoir

The abundance of plankton in Aliyar reservoir is related to the water level; the low mean depth coinciding with a higher density of plankton, mainly contributed by *Microcystis*. The low mean depth of 2.6 m during March to June 1983 resulted in a *Microcystis* bloom of 95.95 ml m^{-3} . During 1984, when the mean depth remained high (6.1 to 14.9 m), there was no bloom. The blooms recorded in 1985 and 1986 also coincided with the lowest mean depth. Correspondingly, seasons having higher mean depth matched with low density of plankton population in the reservoir. The relatively high water temperature of the reservoir favours the dominance and abundance of the *Microcystis*. Since blooms can prove detrimental to the health and survival of the fish, the maintenance of minimum water level has to be reckoned as a management tool to ensure the health of the ecosystem and fish stock. For the common planktonic organisms, see Table 2.15.

Table 2.15. Common planktonic organisms in Aliyar reservoir

Chlorophyceae	:	*Mougeotia* sp., *Pediastrum* sp. *Staurastrum* sp.
Myxophyceae	:	*Microcystis* sp., *Anabaena* sp.
Bacillariophyceae	:	*Asterionella* sp., *Synedra* sp.,*Navicula* sp.
Dinophyceae	:	*Ceratium* sp.
Protozoa	:	*Actinospherium* sp., *Cyclotella* sp.
Rotifera	:	*Keratella* sp., *Brachionus* sp., *Filinia* sp.
Cladocera	:	*Daphnia* sp.
Copepoda	:	*Cyclops* sp., *Diaptomus* sp., nauplii

Numerically, the plankton population is dominated throughout the year by *Microcystis* sp. The only other biota worth mentioning are *Mougeotia*, the filamentous alga, *Cyclops* (2.9%) and *Keratella* (2.5%). The diversity of species in the reservoir is poor with negligible sectoral variation. The year-round distinct dominance of *Microcystis* suppresses development of a species-rich plankton population in the reservoir.

Benthos

Bottom macrofauna : The average standing crop of bottom macrofauna in Aliyar, recorded for four years from the depths of 2 to 10 m, is estimated at 258 ind. m^{-2} (3.18 g m^{-2}). Oligochaetes dominate both in terms of numbers (47.8%) and weight (86.8%). Shallows have a low density of macrobenthos and their concentration generally increases with depth. The maximum concentration of benthic organisms (2 331 organisms m^{-2}/48.45 g m^{-2}) was recorded at 30 m depth.

Primary productivity

During 1984 to 1986, the gross primary production rate in the reservoir ranged from 0.278 to 0.699 g C m^{-2} day^{-1} (net production: nil to 0.385 g C m^{-2} day^{-1}). The average rate of energy transformation is estimated at 13 580 cal m^{-2} day^{-1}. The incident solar energy of visible light at Aliyar reservoir works out at 2 150 000 cal m^{-2} day^{-1}. Thus, 0.64% of available light energy is fixed by the producers in the reservoir. Considering that only about 40 to 62% of the energy fixed by the producers is stored by them and the rest is utilised for respiration and metabolic activities, the quantum of energy finally converted into fish flesh depends on the qualitative and quantitative abundance of biotic communities within the impoundment. The peak yield from the reservoir @ 193.58 kg ha^{-1} in 1989-90 reveals a conversion efficiency of 0.47% of the available primary energy in the reservoir. The rate of energy conversion at primary producer level as well as at the fish production level at Aliyar is considered higher than in any other Indian reservoir.

FISH FAUNA

The indigenous ichthyofauna of the reservoir consists of 40 species belonging to 13 families, besides the seven stocked species (Table 2.16).

Table 2.16. Ichthyofauna of Aliyar reservoir

Family : CYPRINIDAE
sub-family : Cyprininae

1. *Catla catla* (Hamilton-Buchanan),
2. *Cirrhinus mrigala* (Hamilton-Buchanan)
3. *Ctenopharyngodon idella* (Valenciennes)
4. *Cyprinus carpio communis* Linnaeus
5. *Labeo rohita* (Hamilton-Buchanan)

sub-family : Leuciscinae

6. *Hypophthalmichthys molitrix* (Valenciennes)

Family : CICHLIDAE

7. *Oreochromis mossambicus* (Peters)

Family : NOTOPTERIDAE

8. *Notopterus notopterus* (Pallas)

Family : ANGUILLIDAE

9. *Anguilla bengalensis bengalnesis* (Gray)

sub-family : Cyprininae

10. *Cirrhinus cirrhosa* (Bloch)
11. *Cyprinus carpio* var. specularis Lacepede
12. Puntius *dubius* (Day)
13. *Labeo boga* (Hamilton)
14. *Labeo calbasu* (Hamilton)
15. *Labeo fimbriatus* (Bloch)
16. *Labeo kontius* (Jerdon)
17. *Puntius carnaticus* (Jerson)
18. *Puntius chola* (Hamilton)
19. *Puntius denisonii* (Day)
20. *Puntius dorsalis* (Jerdon)
21. *Puntius filamentosus* (Valenciennes)
22. *Puntius mahecola* (Valenciennes)
23. *Puntius sarana* (Hamilton)
24. *Puntius ticto* (Hamilton)
25. *Puntius punctatus* (Day)
26. *Puntius melanampyx* (Day)
27. *Tor khudree malabaricus* (Jerdon)
28. *Salmostoma chela untrachi* (Day)

Sub-family : Rasborinae

29. *Amblypharyngodon melettinus* (Val.)
30. *Barilius gatensis* (Valenciennes)
31. *Danio aequipinnatus* (McClelland)
32. *Perluciosoma daniconius* (Hamilton)

Sub-family : Garrinae

33. *Garra mecclellandi* (Jerdon)

Family BAGRIDAE

34. *Mystus malabaricus* (Jerson)

Family : SILURIDAE

35. *Ompok bimaculatus* (Bloch)
36. *Ompok malabaricus* (Valeniciennes)

Family : SISORIDAE

37. *Glyptothorax housei* Herre

Family : CLARIDAE

38. *Clarias batrachus* (Linnaeus)

Family HETEROPNEUSTIDAE

39. *Heteropneustes fossilis* (Bloch)

Family : BELONIDAE

40. *Xenentodon cancila* (Hamilton)

Family : NANDIDAE

sub-family : Pristolepidinae

41. *Pristolepsis (malabaricus) marginata* (Jerdon)

Family : CICHLIDAE

42. *Etroplus canarensis* (Day)
43. *Etroplus maculatus* (Bloch)

Family : GOBIIDAE

44. *Glossogobius giuris* (Hamilton)

Family : CHANNIDAE

45. *Channa marulius* (Hamilton)

Family : MASTACEMBELIDAE

46. *Macrognathus guentheri* (Day)
47. *Mastacembelus armatus* (Lacepede)

No. 1 to 7 - Species stocked

Feeding habits of fishes

The stomach contents analysis of *Cirrhinus mrigala, Labeo rohita, L. fimbriatus, L. calbasu and Cyprinus carpio* showed that they feed on bottom sediments (Table 2.17). *C. catla* is known to be zooplankton feeder. However, in Aliyar, the stomach contents of this species comprise blue green algae to the extent of 44 - 80%, followed by organic detritus (12.7%) and zooplankton (12 - 24%). The feeding orientation of fishes in Aliyar reservoir indicates the flexibility in food preference and adaptability of various fishes to the reservoir system.

Blue-green algae constitute almost 40% of the food of catla, mrigal, common carp, *L. calbasu* and mahseer (*Tor*). This is not surprising since 85 to 90% of the plankton population in the reservoir belongs to blue-green algae. The higher share of blue-green algae among the food components of these fishes may not mean a preference for blue-green algae, especially as the digestibility of the *Microcystis* found in the guts of the fishes is probably low. One could conclude that the fishes of Aliyar reservoir rely more on the detritus for their nourishment. Organic detritus at the bottom, constantly enriched by the *plankton rain* from the water column, is a rich food resource of the reservoir. Most of the plankton forms, especially *Microcystis* sp., recorded from the stomachs of detritivorous fishes are, in all probability, ingested from the bottom sediment phase.

A substantial proportion of the energy transformation from plankton phase to fish flesh seems to be channelled through the detritus chain. Thus, *C. catla*, among the plankton feeders and mrigal and common carp among the bottom feeders should form a substantial proportion of the regularly stocked fingerlings.

Table 2.17. Stomach contents of fish in Aliyar reservoir

FISH		STOMACH CONTENTS
Cirrhinus mrigala	:	**Blue-green algae**, dominated by *Microcystis sp.*; **detritus; decaying organic matter**
Labeo rohita	:	**Detritus** (56.1 - 58.8%); **Blue-green algae** mainly *Microcystis* sp (14.6%); **decaying organic matter** (10.2%)
Catla catla	:	**Blue-green algae** (43.8 - 80 %); **detritus** (21.3 - 24.7%); **zooplankton** (12.7 - 24.5%) consisting of copepods and rotifers.
Cyprinus carpio	:	**Algae** (36.6 - 62.6%); **detritus** (9.2-31.6%); **zooplankton** and **diatoms** as minor components.
Labeo fimbriatus	:	**Detritus**(42.9-59.0%); **decayed organic matter**, sand and silt.
Labeo calbasu	:	**Blue-green algae** (36.2 -55.0%); **detritus** (30.0 - 42.0%), **decaying organic matter** (15.0 - 15.1%)

Maturation and fecundity of commercial carps

Commercially important fishes do not breed in the reservoir, thereby necessitating continuous stocking. The investigations made by the Central Inland Capture Fisheries Research Institute (CICFRI) reveal that the gonads of the major carps reach only up to IV stage of maturity in the reservoir, after which they undergo resorption, leading to breeding failure, with the exception of a few specimens of *C. mrigala*, in the Vth stage of maturity. The failure of major carps to reach the ripe stage in the reservoir is further corroborated by the low gonado-somatic index (GSI) and fecundity. The maximum GSI observed is only 6.2 for catla, 9.8 for mrigal and 10.5 for rohu. Common carp, however, has a higher GSI (16.0). The relative fecundity (the number of ripe ova per kg body weight) is as low as 0.39 to 2.10 x 10^5 for mrigal and 0.62 - 2.20 10^5 for common carp.

Fisheries management

Current fish yield rate of Aliyar reservoir is one of the highest in the country. This reservoir acted as testing ground for the field trials of the scientific management package developed by the CICFRI and the yield optimisation achieved thereof is a standing testimony to the validity of the package. The State Fisheries Department launched a stocking programme in the reservoir immediately after its formation in 1962 (Selvaraj and Murugesan, 1990). The stocking density was erratic in the earlier years with a high preponderance of medium and minor carps including *Puntius carnaticus, P. dubius, Labeo kontius, L. fimbriatus, L. calbasu, Cirrhinus reba, C. cirrhosa, Acrossocheilus hexagonolepis* and *Tor khudree malabaricus*. Apparently, availability of fish seed in the vicinity and other logistics of seed supply governed the stocking density and the species-mix, rather than any ecological considerations. The size of fry stocked was usually so small that they fell an easy prey to predators.

The exploitation system followed during the early phase also left much to be desired. The fishing was done, both by the Department of Fisheries and private fishing units, on a royalty basis, without any restriction on mesh size. This resulted in indiscriminate exploitation of stocked fishes much below the desirable size. A cumulative effect of a number of irrational reservoir management practices resulted in very low yields, ranging from 2.67 to 54.7 kg ha^{-1} yr^{-1}, at an average of 26.21 kg ha^{-1} yr^{-1} (Table 2.18).The reservoir was subjected to scientific investigations since 1983 and a number of measures were initiated to develop the fisheries. The results of scientific management started reflecting in the yield rate from 1985. The management options used (Table 2.19) for enhancing fish yield were:

1) Stocking was limited to Indian major carps
2) Size at stocking was increased to advanced fingerlings (>100mm)
3) Stocking density was reduced to 236-312 ha^{-1} (Av. 637 ha^{-1})
4) The stocking schedule was spread over different months of the year
5) A strict mesh regulation, banning nets < 50 mm in mesh bar (knot to knot),
6) Ban on catching Indian major carps less than 1kg in size.

A direct outcome of the adoption of scientific management practices in Aliyar was a substantial hike in fish yield to the tune of 193.58 kg ha^{-1}. The yield ranged during 1985-86 to 1989-90, from 77.75 to 193.58 kg ha^{-1}, with major carps constituting 84.04 to 96.25% (Table 2.20).

Table 2.18. Fish yields from Aliyar reservoir during 1962 to 1985

Year	Total landings (Kg)	Yield (kg ha^{-1})	Percentage contribution Major carps	Miscellaneous
1964-65	864.25	2.67	2.43	97.57
1965-66	1 323.75	4.09	11.28	82.72
1966-67	2 238.50	6.91	9.31	90.69
1967-68	3 341.75	10.31	36.22	63.78
1968-69	5 000.00	15.43	38.99	61.01
1969-70	8 913.75	27.51	28.45	61.55
1970-71	11 268.75	34.78	47.78	52.22
1971-72	9 312.50	28.74	61.92	38.08
1972-73	9 720.75	30.00	27.27	72.73
1973-74	15 776.75	48.69	25.59	74.41
1974-75	15 790.75	48.74	27.27	72.72
1975-76	12 413.25	38.31	17.40	82.60
1976-77	11 019.00	34.01	18.35	81.65
1977-78	7 994.45	24.67	25.17	74.83
1978-79	3 061.50	9.45	44.01	55.99
1979-80	3 386.75	10.45	48.15	51.85
1980-81	8 066.60	24.90	71.64	28.36
1981-82	10 896.50	33.63	73.10	26.90
1982-83	***	-	-	-
1983-84	17 636.85	54.43	89.77	10.23
1984-85	11 837.50	36.54	90.05	9.95
Average	**8 493.20**	**26.21**	**46.27**	**53.73**

*** Data not available

(*After Selvaraj and Murugesan, 1990*)

Table 2.19. Stocking in Aliyar reservoir during 1985 to 1989 (total number of fingerlings)

Fish	1985-86	1986-87	1987-88	1988-89	Average
Catla	25,729	23,777	71,030	30.041	37,644
Rohu	18764	12,248	37,679	26,561	23,813
Mrigal	16,608	14,556	47,475	23,998	25,659
Common carp	15,243	18,409	45,300	20,547	20,252
Silver carp	131	4,509	5,031	-	2,418
Total	**76,475**	**73,499**	**206,515**	**101,147**	**114,409**
no ha^{-1}	***236***	***227***	***637***	***312***	***353***

(*After Selvaraj and Murugesan, 1990*)

Table 2.20. Fish yields from Aliyar reservoir during 1985-86 to 1989-90

	1985-86	1986-87	1987-88	1988-89	1989-90
Major carps					
Qty.	21 926	36 637	21 966	51 150	59 313
%	87.04	99.01	84.04	96.25	94.6
Misc. Fishes					
Qty.	3 625	370	4 175	2 032	3 406
%	12.96	0.99	15.96	3.75	5.40
Total *(kg)*	**25 192**	**37 007**	**26 141**	**54 182**	**62 720**
Yield (kg ha^{-1})	***77.75***	***114.22***	***80.68***	***167.23***	***193.58***
CPUE *(kg)*	5.64	8.02	9.05	16.91	-

(After Selvaraj and Murugesan, 1990; CICFRI Barrackpore)

2.8 STATUS OF RESERVOIR FISHERIES IN TAMIL NADU

Tamil Nadu is harnessing its water resources to the fullest extent by creating impoundments all over the State and this has resulted in the location of reservoirs in a variety of places ranging from the uplands of the west to the coastal deltas in the east. They are situated in diverse morpho-edaphic and geo-climatic environs, receiving drainage from varying types of catchment areas. All these factors add an element of diversity to the environmental conditions under which the reservoirs exist. Scientific data on various environmental variables and biotic communities are available on at least 21 reservoirs in the State. After a close scrutiny of 17 of them, Sreenivasan (1976) concluded that the production processes and the community succession in these reservoirs are so complex and interrelated that classifying them, based on the conventional methods of morphometric and edaphic production indices, was almost impossible. Most of the limnological information with predictive value has been found to be of limited application in these reservoirs.

Most of the man-made lakes in Tamil Nadu, except Pykara, Sandynulla and Mukerti, are situated in medium to low altitude, the elevation ranging from 38 to 483 m above MSL. Thus, almost all the reservoirs are exposed to the warm tropical climate with air temperature never dropping below 15 °C. A high temperature regime throughout the year, coupled with plentiful sunshine has compensated for much of the adverse attributes in the shallow reservoirs of Tamil Nadu, which explains the non-conformity with morphometric and chemical indicators of productivity. From the two mountain reservoirs *viz.*, Sandynulla and Pykara, situated at + 2 120 and +2 036 m altitude respectively, the former is organically rich and highly productive, harbouring a permanent bloom of *Microcystis*, whereas the latter is truly oligotrophic, with low alkalinity, weak oxygen stratification and an acidic water with the presence of dissolved carbon dioxide even at the top layer (Sreenivasan, 1968). Manimuthar and Pechiparai, two small reservoirs in the plains remained oligotrophic mainly due to nutrient-starved catchments.

Shallow lakes are often considered to be more productive, as larger proportion of the water volume remains in the euphotic zone. This maxim does not hold good in respect of Tamil Nadu reservoirs. Neither mean depth nor the maximum depth has any correlation with productivity. Vidur reservoir, despite being the shallowest of all reservoirs (mean depth of 2.1 m), did not develop a rich plankton community, even after seven years of its impoundment. Deeper reservoirs like Amaravathy (13.7 m), Aliyar (16.8 m) and Tirumoorthy (11m), on the other hand, developed algal blooms as soon as the dams were sealed. Other morphometric indices like shoreline development and volume development also did not convey the status of productivity with any degree of accuracy. Sreenivasan (1976) examined the area, mean depth and volume, of a number of reservoirs in Tamil Nadu and found no correlation among the above variables and productivity.

Total dissolved solids, measured in terms of specific conductivity or the bicarbonate alkalinity, is an index of biogenic production potential. Eventhough this indirect measure of edaphic quality reflected the status, to an appreciable extent, there are many exceptions. For instance, Amaravathy and Tirumoorthy, despite their low values of total alkalinity, specific conductivity and hardness, have high organic productivity.

Thus, the productive nature of each reservoir is determined by different factors. For example, the low productivity of Manimuthar and Pechiparai is due to morphometric features (deep basin and low catchment area respectively), while the high rate of productivity in Sandynulla is due to the rapid eutrophication from urban wastes. However, a close examination of the reservoirs reveals certain definite indicators of productivity. The nutrients load of the inflowing waters being dependent on the nature of catchment, quality of the catchment area plays an important role in determining the nutrient status of a reservoir. The percentage of cultivated area is often taken as a criterion for assessing the catchment quality. The high bicarbonate alkalinity and hardness of Sathanur, Krishnagiri and Vidur reservoirs can be directly attributed to the geochemical status of its catchment (Sreenivasan, 1976). Similarly, the low values of alkalinity, conductivity and hardness of Hope lake, Manimuthar, Pechiparai and Perumchani are a reflection of the poor nutrient status of their catchment area.

Thermal regime and oxygen distribution, determining the vertical profile of chemical parameters give a more reliable measure of productivity status in the reservoirs under study. The chemical stratification is the measure of metabolic and organic processes and gives a more direct indication about the trophic status. Absence of proper thermocline is the characteristic feature of all reservoirs in the State. In many cases, the inflowing river water is much cooler than the standing water, as the former flows down from heavily wooded sheltered forests. This cooler incoming water, unlike in the temperate region, readily mixes with the lentic water. Other reasons for the ready mixing are the withdrawal of water from the bottom and the wind-induced turbulence.

Oxygen depletion at the bottom is a very reliable indicator of productivity in tropical waters, especially because it is unaccompanied by thermal stratification. The bacterial decomposition of organic matter at the bottom is hastened by high temperature. Similarly, the photosynthetic processes at the euphotic zone keep up a steady release of oxygen. All the productive reservoirs in the State have the klinograde oxygen distribution, while the oligotrophic Pykara, Uppar and Pambar reservoirs have the orthograde oxygen distribution. Where oxygen depleted water is discharged downstream, it gets reoxygenated within 100 meters downstream (Sreenivasan, 1970a) and its impact on riverine aquatic communities is negligible.

It is often believed that total alkalinity above 50 (Northcote and Larkin,1956) and 90 mg l^{-1} (Spencer,1964) is indicative of high productivity. In Amaravathy, the inflowing water is poor in total alkalinity and hardness. However, the standing crop of plankton provides a continuous plankton rain to enrich the bottom, which in turn decomposes to release the nutrients and keep the carbon production going. Thus, it is the vertical gradient of these chemical parameters, rather than the ambient concentration in inflowing water, that indicates the status of productivity.

Changes in biotic communities

Construction of dams and the subsequent impoundment brings in drastic changes in the biotic communities at all levels. These changes at the micro-level have far reaching impact on the trophic structure and events of the ecosystem, affecting the survival of higher organisms, especially fishes. Sandynulla reservoir, despite its deep basin and high altitude, sustains a rich plankton community, represented mainly by *Microcystis* in blooms. Drastic reduction in diversity of planktonic organisms with a high index of *concentration of dominance* is characteristic of reservoirs which receive rich supply of nutrients. Most of the reservoirs in the State such as Stanley, Bhavanisagar, Amaravathy, Aliyar, Tirumoorthy, Vaigai, Vidur and Poondi have blooms of blue-greens round the year.

Apart from light penetration and nutrient loading, water renewal plays an important role in the abundance of planktonic organisms. Abraham (1980b) pointed out that plankton density in Bhavanisagar reservoir was rich during the months of low level fluctuations. In reservoirs with violent level fluctuations, the annual flood water sweeps away not only the nutrients but even physically dislodges all the passive communities like plankton from the system. It is after the closure of outlets that the plankton community starts developing again. The same is true with benthic and macrophytic communities.

Fish fauna

Fish fauna of the reservoirs in Tamil Nadu is comprised indigenous species, those transplanted from other river systems and the exotics. In most cases, the last two categories support commercial fisheries while the indigenous ones, are on the decline. The gradual phasing out of indigenous fishes is a matter of concern. Reasons for their decline can be summarised as:

1) loss of habitat,
2) changes in fish food organisms
3) obstruction of migratory pathways,
4) loss of breeding grounds, and
5) loosing out in competition with the transplanted species.

The anadromous Indian shad (*Tenualosa ilisha*), the fishery of which has collapsed in the Cauvery above the Mettur Dam, is the most serious loss of indigenous fishes. The hilsa from Cauvery- Coleroon system used to ascend the River Cauvery for breeding. The three lower barrages *viz.*, the Upper Anicut, the Grand Anicut and the Lower Anicut restricted hilsa runs even before the Mettur dam came up. After commissioning the dam, hilsa disappeared from the upstream stretches.

Another species that should attract the attention of conservationists is *Puntius dubius*, indigenous to Mettur, Bhavanisagar, Amaravathy and Krishnagiri reservoirs. In Amaravathy, it contributed 55% of the catch in the mid-960s and declined to 0.15% in the 1970s and thereafter disappeared completely. In Bhavanisagar, it still constitutes 5 to10% of the total catch. Decline of *P. dubius* in the reservoirs is mainly due to recruitment failure. The genus *Puntius* had the maximum species diversity in the region which, in addition to 5 commercially important species, comprised a number of uneconomic fishes.

Mahseers, comprising *Tor tor*, *T. putitora* and *Acrossocheilus hexagonolepis* are also threatened. *Cirrhinus cirrhosa*, although still caught in Bhavanisagar, Stanley and other reservoirs is on the decline. This fish faces severe competition from *C. mrigala* that has been introduced into the reservoirs. Tilapia *Oreochromis mossambicus* has been the most successful among the introduced fishes. At least 24 reservoirs in Tamil Nadu have tilapia either as the major component of fish catch or represent a dominant fishery as in Amaravathy, where tilapia outnumbers other fishes by 90 to 10. Catla is the major fishery only in five reservoirs *viz.*, Sathanur, Ponnaniyar, Kannathudai, Manimukta and Gomukhi. The indigenous fishes still dominate in Stanley, Bhavanisagar (*P. dubius*), Manimuthar, Gadana, Chittar and Perunchani (*Puntius* spp.).

Fish Production

Sixty-nine reservoirs of Tamil Nadu produce 1 323 t of fish annually. Figures from the small reservoirs designated as *irrigation tanks* are not available. Small (<1 000 ha in size) reservoirs' share to this fish production is 760 t. Thus 57.4% of the production comes from small reservoirs which form just 26.8% of the total area. Production from unit area is also high in respect of the small reservoirs. The small reservoirs yield 48.5 kg ha^{-1}, calculated for FRL, 88.01 kg ha^{-1}, if calculated for mean area. Medium reservoirs have fish yield of 13.74 kg ha^{-1} (FRL) and 42.20 kg ha^{-1} (mean surface area). Large reservoirs have the lowest yields of 12.66 and 23.46 kg ha^{-1}, calculated at FRL and mean surface area respectively (Table 2.21). If all categories of reservoirs are pooled together, the Tamil Nadu reservoirs produce 22.6 kg ha^{-1} (at FRL) and 48.04 kg ha^{-1} (at mean surface area level).

Table 2.21. Fish production and yields from different categories of reservoir in Tamil Nadu

Category	Size (ha)	Total fish catch (t)	Yield (kg ha^{-1}) At FRL area	At mean surface area
I	<1 000	760	48.50	88.10
II	1 000-5 000	269	13.74	42.20
III	>5 000	294	12.66	23.46
Total/Mean	-	*1323*	*22.63*	*48.04*

Data pertain to 1993 *Source: Govt. of Tamil Nadu*

This is one of the highest yields in the country, but yields vary widely among the districts. Dindigul Anna district produces 156.81 kg ha^{-1} from its 130 reservoirs. Thiruvannamalai Sambuvarayar, Kamarajar and Coimbatore districts have average yields rates above 50 kg ha^{-1} (Table 2.22).

Fish yield potential

Fish yield potential, based on the primary production rates, has been estimated in respect of seven reservoirs (Sreenivasan, 1976). Tirumoorthy is the most efficient in converting solar energy (1.803%), followed by Amaravathy (1.442%) and Bhavanisagar (1.050%). The energy harvested as fish is estimated at 0.02931% to 0.2152% of the energy fixed by the primary producers. The best conversion of Sathanur is 0.2152 % and considering the fact that up to 1% of the carbon can be harvested, there is sufficient room for increasing the fish yield. In Amaravathy, there is scope for a five-fold increase in the fish production. The mono-species fisheries of catla and tilapia in Sathanur and Amaravathy facilitate better conversion of primary production to fish flesh and consequently, these two species are better converters of energy (Table 2.23).

Table 2.22. Fish production and yields from reservoirs in Tamil Nadu, by districts

Districts	Total fish production (t)	Yield ($kg\ ha^{-1}$) At FRL	At mean surface area
Chengalpattu MGR	45	10.81	23.30
North Arcot Ambedkar	3	4.42	8.85
Dharmapuri	149	41.39	103.18
South Arcot	69	6.95	23.18
Salem	115	7.49	12.33
Thiruchirapalli	7	30.84	68.63
Periyar	249	28.34	66.12
Coimbatore	160	50.44	90.50
Nilgiris	14	28.81	55.33
Dindigul Anna	130	156.81	238.09
Madurai	58	21.20	32.80
Kamarajar	96	102.89	240.00
Tirunelveli	21	12.19	23.78
Kattabomman			
V. O. Chidambaranar	51	77.6	155.40
Kanyakumari	30	9.42	18.19
T. Sambuvarayar	126	62.68	2 250.00
Total/Mean	*1457*	*53.39*	*25.13*

Table 2.23. Production in terms of energy in some reservoirs of Tamil Nadu

	Stanley	Bhavani-sagar	Sathanur	Krishna-giri	Amara-vathy	Sandy-nulla	Tiru-moorthy
Total radiation $g\ cal\ m^{-2}\ yr^{-1} \times 10^5$	7 775	7 775	7 738	7 720	7 847	8 578	7 845
Photosynthetic oxygen production $cal\ m^{-2}\ yr^{-1} \times 10^6$	8.147	8.162	8.100	5.610	11.320	8.121	14.15
Photosynthetic efficiency %	1.047	1.050	1.047	0.727	1.442	0.947	1.803
Fish production $g\ m^{-2}\ yr^{-1}$	6.29	3.32	14.52	2.05	19.60	4.54	3.456
$cal\ m^{-2}\ yr^{-1}$	7550	3972	17430	2460	23560	5454	4147
Conversion efficieny: Photosynthesis to fish %	0.0927	0.0775		0.0449		0.0671	0.02931
Light energy to fish %	0.000971	0.000513	0.000366	0.000319	.003003	0.000636	0.000529

(50% of the area at full reservoir level taken as the average water surface area

(After *Sreenivasan, 1976*)

Increase in productivity

Nearly 60% of the reservoirs in the State of Tamil Nadu and all the tanks are less than 500 ha in size, making them ideal for extensive aquaculture. Management strategy in very small reservoirs and the irrigation tanks should centre around an imaginative stocking and harvesting schedule to allow the stocked fish maximum time to grow. In many cases, the reservoirs nearly or completely dry up during summer, eliminating chance for fish reproduction. This calls for *put and take* system of management, where the selected species of fishes are stocked and harvested, all within less than one year. Such systems can also be fertilized to enhance plankton production. The small reservoirs also offer opportunities to integrate aquaculture with animal husbandry and livestock farming. However, the possible conflicts with other water users may have to be reconciled.

Pollution

The reservoirs in Tamil Nadu do not face problems from pollution on a very large scale. In the absence of any organised sewage disposal system in small towns, the community wastes often find their way to the reservoir. From fisheries point of view, this organic loading within certain limits is acceptable. Nevertheless, the problem needs to be addressed from a public health and aesthetic point of view. Organic wastes pose serious eutrophication problems in Sandynulla, which receives city sewage from Ooty. Non-point sources, mainly agricultural wastes, are potentially harmful. Accumulation of pesticide residues in water or sediment phase, their bioaccumulation in organisms and biomagnification in different trophic levels have been not yet investigated in Tamil Nadu and need urgent attention.

Some such instances of pollution from industrial sources have been reported from many reservoirs with varying degrees of impact on the environment and biotic communities. Fish mortalities in Mettur dam area due to effluent discharge from the Mettur Chemical and Industrial Corporation were reported as early as 1949. The effluents contained high concentrations of dissolved and insoluble solids, chlorides and free chlorine and were highly alkaline. However, the effluents were dangerous only during low discharge from the dam when the offensive fluids remained entrapped in rock pools where they affected fish. Apart from this, a series of chemical plants at Mettur discharge their effluents into the Ellis surplus side, but none of them is known to affect the fish life downstream.

Waste discharge into the Bhavani river from the Seshasayee Pulp and Board Limited, Pallipalayam, manufacturing writing paper and boards using bamboos, wood and bagasse adopting the sulfate processes, has been studied and found to be innocuous by Sreenivasan (1979a and b). The South India Viscose, manufacturing viscose rayons and staple fibre, discharges 16 000 m^3 of wastes into the river Bhavani at Sirumughai every year. The sulfite process of manufacturing is highly polluting. There have been reports of fish mortality in Bhavanisagar reservoir due to this effluent discharge. Low dissolved oxygen level and high carbon dioxide concentrations were recorded in affected areas. Combining effluents with the septic wastes from the residential colony and treatment of the combined wastes in stabilization ponds before discharging them into river have been suggested by Sreenivasan (1979b).

Exotic fishes

Among the fishes transplanted into reservoirs of the State, the tilapia, *Oreochromis mossambicus* has been the most successful. It has become so ubiquitious that no water body in Tamil Nadu is free from it. After the initial enthusiasm with which the people have accepted it, the fish has become very unpopular due to the stunted size. In ponds, myriad of tilapia flocking the entire water body is a very common sight. The small-sized tilapia does not fetch a remunerative price and only way to utilize them is conversion into fish meal. However, in the reservoirs, the problem of stunted growth has not yet arisen. In Mettur, where the fish forms the bulk of the catch, the individual size is more than 0.68 kg, which is readily accepted by the market. However, it is significant to note that the average size of tilapia decreased from 1.5 kg in 1968 to 0.68 kg over a period of 25 years. Excessive proliferation of tilapia in reservoirs could be kept under check by predators like *Notopterus cinotala*, *Wallago attu*, *Channa* spp. and eels, and frequent level fluctuation that exposes the breeding pits.

After heavy stocking of tilapia in Vaigai, Sathanur, Amaravathy, Krishnagiri and Manimuthar reservoirs in the early 1960s, fairly large sized fish were caught in gill nets. In Vaigai they formed 25-50% of the catch by weight soon after stocking. Presently *O. mossambicus* is the single largest component of the fishery in the reservoir. Opinion is divided on the utility of this fish for yield optimization in Indian waters.

Impact of stocking on the species spectrum and productivity

Catla catla, *Cirrhinus mrigala*, *C. cirrhosa*, *Labeo rohita*, *L. calbasu*, *L. fimbriatus* and *Cyprinus carpio* are the species being stocked in reservoirs. In Mettur, catla started appearing in the fisheries in 1943-44 and gained the third rank in ten years to form16% of the catch. During the next decade, *i. e.*, from mid 1950s to 1960s catla was the most dominant component of the catch. It had a fluctuating fortune, since then. Today, catla form nearly 10% of the catch in the reservoir. This is the fate of catla in most of the reservoirs, where it does not reproduce. The exception is Sathanur, where self-reproducing catla contributes up to 91% of the total catch, while the fish forms only 10 - 20% of the catch in many reservoirs with regular stocking. *C. mrigala* and *L. rohita* are the other two species which could not establish themselves in any appreciable extent. *L. calbasu* was successfully introduced in Stanley and Bhavanisagar.

In most of the reservoirs, stocking is practised on an ad-hoc basis. Introduction of new species should be done with due consideration of long-term and short-term objectives. In larger reservoirs, the first option should be to stock suitable fishes which have a reasonable chance to establish self reproducing stocks. The main criterion for selection should be to fill a vacant or underutilized niche. On the other hand, a small reservoir with no breeding facilities needs to be managed on a continuous stocking and harvesting basis. This is a culture-based fishery, where what is stocked is harvested. The stocking material is selected after an assessment of trophic structure *i.e.*, fish feed availability. Such approach is adopted successfully in Aliyar.

The organically rich Aliyar reservoir has a permanent bloom of *Microcystis* which is not directly consumed by any species of fish, but it is added to the detritus. The detritivorous *Cirrhinus mrigala* (37%) and *Cyprinus carpio* (24%) together form 61% of the catch, followed by catla (21%) and *Labeo rohita* (16%). The Aliyar experience stressess the importance of utilising the detritus as a fish food resource, where direct plankton feeders are not available or cannot utilize blue-green algae. Job and Kannan (1980) corroborated this by studying the caloric value and chlorophyll content in detritus phase in Sathiar reservoir, where the introduced carps account for 98% of the total fish harvested from the lake. In both reservoirs, the fisheries is totally dependent on the stocked fish. The stocking and harvesting schedule is carefully planned and the fish is removed after attaining a certain size. Smaller fish caught accidentally are released back into the reservoir. This management approach has increased the yield in Aliyar from the initial 35 kg ha^{-1} up to 192 kg ha^{-1}.

Craft and gear

The most common gear used by the fishermen in the reservoirs of Tamil Nadu is the gill net of entangling type. The *Rangoon* net is the most popular among them, which has been introduced into the state from Andhra Pradesh (Nayar, 1979). Another variety of gill net used extensively in reservoirs is *udu vala*, a narrow bottom set gill net, operated usually in the shallow areas of the reservoir. Gill nets of varying mesh sizes are used to catch different species of fish. The small-meshed nets are used for *Oxygaster phulo, Osteobrama vigorsii* and other small fish. Large-meshed nets are called *catla nets* in Tamil Nadu.

There is no uniform standard length or height for gill nets. For the sake of computing catch per unit of effort in Bhavanisagar, Ranganathan and Venkataswamy (1966) reckoned 20 pieces of *Rangoon* nets comprising three different mesh sizes and a coracle as a standard unit. Mesh sizes were 6.3 cm (50 m length), 5 cm (40 m length) and 10 cm (100 m length). The All India Coordinated Project on Reservoir Fisheries has considered 50 m hung length as a unit and the effort was identified in terms of cluster of 50 m units.

Apart from gill nets, simple long lines, pole and line and cast nets (both stringed and unstringed) are occasionally employed. Common coracle, a saucer shaped country craft made of split bamboo mat covered with buffalow hide is the most popular craft in reservoirs.

Post harvest arrangements

Most of the ice plants and cold storages in the State cater to the marine sector. Mettur, Bhavanisagar, Sathanur, Krishnagiri and Amaravathy reservoirs have ice plants and cold storages for storing fish before its despatch to the Howrah railway station in Calcutta. Various market channels are in vogue in respect of reservoir fishes. The agencies involved are primary cooperative societies, Tamil Nadu Fisheries Development Corporation (TNFDC), Department of Fisheries and the private traders. Networks involving one or more of the above agencies are also not uncommon. For instance, the TNFDC collects fish from Aliyar, Sathanur, Bhavanisagar and Tirumoorthy and sells it directly to the public or to merchants. There are also arrangements between the Department of Fisheries and TNFDC to dispose the catch in the event of a glut. A substantial part of the catch from Sathanur, Mettur, Bhavanisagar and Aliyar is despatched to the Howrah market.

Cooperatives

Tamil Nadu has 598 primary fishermen's cooperatives with a total membership of 157 400, of which 250 are inland fishermen societies. These societies are not as efficient as their counterparts in agriculture, marketing, consumer and housing sectors, primarily because the fishermen cooperatives confine themselves mainly to the provision of cheap credit (long, medium and short-term) and working capital to fishermen (Parasuraman, 1979). They do not attract institutional finance due to legal restrictions. Fishermen cooperatives in the State are financed solely by the State Government in the form of share capital and loans in the long, medium and short-term categories. They normally do the following functions:

1) Distribution of essential commodities to their members with the working capital accommodations granted by State Government,
2) Taking small reservoirs on lease and let them on sub-lease to members on yearly basis,
3) Fish marketing, and
4) Distribution of yarn under subsidy schemes.

Table 2.24. Salient features of some reservoirs in Tamil Nadu

Parameters	Stanley	Bhavanisagar	Poondi	Vaigai	Sathanur	Krishnagiri
Year of sealing	1939	1953			1957	
Area at FRL (ha)	15 346	7 876	3 263	2 419	2 010	1 248
Area at DSL (ha)	-	120	-	-	-	-
Average area (ha)	-	3998	-	-	-	-
Volume (million m^3)	2 646	908	77.87	193.1	229	68.25
Maximum depth (m)	37.5	33	10.1	29.3	30.2	70.0
Mean depth (m)	17	11.5	2.3	8.1	11.4	5.2
Catchment area (km^2)	16 300	4 200	-	2 253	10 826	5 428.43
Elevation (m)	243	280	38	279	222	483
Length of shoreline (km)	-	125	-	-	-	-
Shore dev. index	6.7	4.0	-	-	6.2	1.5
Volume dev. index	1.4	1.0	0.69	0.8	1.0	0.94
Annual level fluctuations (m)	-	22.6	-	-	-	-
Inflowing rivers	Cauvery	Bhavani Moyar	Kortaliyar Nagari	Vaigai	Ponnair	Ponnair
Outflowing rivers	Cauvery	Bhavani	-	Vaigai	Ponnair	-
Latitude (N)	11^0 49'	11^0 30'	-	10^0 01'	12^0 12'	10^0 30'
Soil						
pH	-	5.2-6.4	-	-	-	-
Organic carbon (%)	-	1.83-2.68	-	-	-	-
Total nitrogen (%)	-	0.14-0.26	-	-	-	-
Available nitrogen (mg 100 g^{-1})	-	31.0-50.8	-	-	-	-
Available phosphorus (mg 100 g^{-1})	-	1.0-3.58	-	-	-	-
Water						
Water temp. (^oC)	24.2-32.0	25.57	27.0-31.6	-	25.0 -30.4	31.4-23.5
pH	7.5-8.8	7.8	8.8-9.6	7.5-8.5	8.3	8.0
Transparency (cm)	-	18.5-135.0	66-55	-	-	11.5-70
DO (mg l^{-1})	4.5-12.6	7.48	4.8	-	8.5	6.8
CO_2 (mg l^{-1})	-	3.90	0-2.0	trace	-	-
Total alkalinity	140- 278	40.92	229-554	79-247	145-616	146-373
Spec. cond. (µmhos)	170-350	269.96	500-400	125-300	320-800	425-575
Total hardness (mg l^{-1})	86-128	26.11	98-66	65-280	112-254	76-146
Calcium (mg l^{-1})	tr.	-	33.5-11	16.5-17.5	-	-
Nitrates (mg l^{-1})	tr.	tr. - 0.34	nil	-	4.4-5.6	-
Phosphates (mg l^{-1})	0-0.1	0.01 - 0.06	tr.	-	-	-
Silicates (mg l^{-1})	0-19.9	0-14.3	22.8-5	0-9.4	0.1-32	0.34-4.4
Chlorides (mg l^{-1})	19-37	9-30	23-88	10-23	18-66	35-120
Organic carbon (mg l^{-1})	1.8-12	6.6-9.0	-	-	7.8-12.6	1.26-8.58
Level fluctuations (m)	-	5.5-22.6	-	-	-	-

Table 2.24 (cont.). Salient features of some reservoirs in Tamil Nadu

Parameters	Willingdon	Pechipparai	Manimuthar	Amaravathi	Vidur	Aliyar
Year of sealing	-	-	1958	1958	-	1962
Area (ha) at FRL	1 554	1 515	940	850	798	646
Volume (m m^3)	573.4	126.4	146	112	16.93	107
Maximum depth (m)	-	37.7	36.0	33.8	9.8	36.5
Mean depth (m)	3.86	8.4	16.8	13.7	2.1	18
Catchment area (km^2)	129.5	207.19	161.61	832	1298	468.80
Elevation (m)	72.0	92	110	358	38	320
Length of shoreline (km)	-	-	-	-	-	16
Shore dev. index	-	-	-	2.3	-	-
Volume dev. index	-	0.75	1.4	1.2	0.68	1.2
Annual level fluctuations (m)	-	-	-	-	-	31
Inflowing rivers	Periya Odai	Kodayar	Manimuthar	Amaravathy	Varahanadhi	Aliyar
Outflowing rivers	-	Kodayar	Manimuthar	Amaravathy	Varahanadhi	Aliyar
Latitude (N)	-	-	-	10° 15'	12° 04'	10°15'
Longitude (E)	-	-	-	77° 15'	79° 34'	76°50'
Soil						
pH	-	-	-	-	-	6.42
Sand (%)	-	-	-	-	-	24.1
Silt (%)	-	-	-	-	-	29
Clay (%)	-	-	-	-	-	46.9
Water						
Water temp. (^{o}C)	20.3-24.5	-	-	24-28.8	-	21.2-32.5
pH	8.2-8.4	6.6-7.2	6.9-7.9	6.7-9.1	8.7-8.6	6.6 - 8.8
Transparency (cm)	20-51	-	--	--	--	108-182
DO (mg l^{-1})	6.4-8.6	-	-	8.1	-	4.2-11.6
CO_2 (mg l^{-1})	0	-	-	--	-	nil-10
Total alkalinity	120-150	28-24	12.5-28	7-84	840-570	16.0-58.0
Spec.cond. (µmhos)	240-320	32-19	29	38-63	850-150	38.7-109.6
Total hardness (mg l^{-1})	-	10	20-14	18-50	56-152	14-40
Calcium (mg l^{-1})	-	-	-	-	-	4.0-18.0
Magnesium (mg l^{-1})	-	5	9 16.7	-	14.5-26.5	1.2 - 6.2
Nitrates (mg l^{-1})	-	-	-	1.7-2.8	-	tr. 0.04
Phosphates (mg l^{-1})	tr	-	-	0.0-0.01	-	tr. 0.02
Silicates (mg l^{-1})	6-10	1.8-3.0	1.6-5.7	1.4 38.5	1-26.5	2.8-24.0
Chlorides (mg l^{-1})	18-22	10-11	14-6	0.4-1.0	104-16	4-24
Organic carbon (mg l^{-1})	-	-	-	12.5-21.6	-	-
GPP (g O_2 m^{-2} d^{-1})	-	-	-	0.3-18.3	-	0.278-0.699

Table 2.24 (cont.). Salient features of some reservoirs in Tamil Nadu

Parameters	Hope Lake	Pykara	Uppar	Tiru-moorthy	Gomukhi	Sholiar
Area (ha) at FRL	573	448	453	388	360	553
Volume (m m^3)	216.2	59.69	15	51.0	15.86	147.2
Maximum depth (m)	55	36.6	-	32	14	-
Mean depth (m)	37.7	14.5	-	11	4.6	27.4
Catchment area (km^2)	57.5	97.5	903.88	80.29	292.67	120
Elevation (m)	244	2036	276	407	183	370
Shore development index	-	-	-	--	2.6	-
Volume development index	2.0	1.2	-	1.0	1.0	1.01
Inflowing rivers	-	-	Uppar	Palar	Gomu-khinadi	Sholiar
Latitude (N)	-	-	10^0 47'	-	-	10^0 10'
Longitude (E)	-	-	77^0 25'	-	-	76^0 55'
Water Temp. (^{0}C)	-	-	22.5-29	-	25.1	22.6-24.2
pH	6.6-8.6	6.9	8.0-8.8	7.7	7.8-8.6	6.7-7.1
Transparency (cm)	-	0.77-3.78	-	-	25.5	-
DO (mg l^{-1})	-	-	4-10	7.6	5.7	8.0-8.2
CO_2 (ml l^{-1})	-	-	Nil	-	-	2.2-4.0
Total alkalinity (ml l^{-1})	10-21.3	6.0-18.0	56-182	16-124	88.6-309	22.0-32.0
Spec. cond. (μmhos)	30	14-45	-	40-200	230-550	40-70
Total hardness (mg l^{-1})	8.0	10.4-18.0	-	16-44	68-124	20.0-28
Calcium (mg l^{-1})	4	Trace	-	-	24	-
Phosphates (mg l^{-1})	-	1	-	0-2.0	Trace	nil-0.06
Silicates (mg l^{-1})	1.4-3.0	-	-	5-16	19.3-6.1	6
Chlorides (mg l^{-1})	1.0-18.0	10.0-18.2	-	2-14	24.0-54	2-12
Organic carbon (mg l^{-1})	-	-	-	4-14.5	-	8.5-10.2

Table 2.24 (cont.). Salient features of some reservoirs in Tamil Nadu

Parameters	Sandynulla	Pambar	Manimukta	Goddar	Pilloor
Area (ha) at FRL	258	243	746	678	389
Volume (m m^3)	20.28	7	20.6	153.60	44.4
Maximum depth (m)	24.7	-	-	-	-
Mean depth (m)	10.3	14	10.97	8	11.4
Catchment area (km^2)	43.5	1736	484.31	-	117
Elevation (m)	2120	321	128	-	920
Shore dev. index	4.6	-	-	-	-
Volume dev. index	1.2	-	-	-	0.46
Inflowing rivers	-	Pambar	Manimukta	-	-
Latitude (N)	-	-	-	-	11^0 16'
Longitude (E)	-	-	-	-	76^0 49'
Water Temp. (^0C)	22.6-16.4	23.2-29	23	19-28	26-29.6
pH	7.95	8	8.4	8.3-8.7	7.9-8.6
Transparency (cm)	78.5	-	32.5	40-70	-
DO (mg l^{-1})	8.3	7.2	7.4	4.4-7	7.6-8.8
Co_2 (mg l^{-1})	2.4	NIL	5	0	nil-1.4
Total alkalinity	26-62	396-398	153	274-350	36.5-85.0
Spec. cond. (μmhos)	99-280	300-950	315	550-680	60-125
Total hardness (mg l^{-1})	28-74	204-208	-	-	28
Calcium (mg l^{-1})	-	-	-	-	7-11
Nitrates (mg l^{-1})	1.5-2.1	-	-	-	-
Phosphates (mg l^{-1})	0-Trace	tr-0.2	0.02	-	nil
Silicates (mg l^{-1})	0-4.7	10	34	4-6	6.2-20.8
Chlorides (mg l^{-1})	12-48	-	15	30-40	4-12
Organic carbon (mg l^{-1})	10.05	-	-	-	4.7
GPP (g O_2 $m^{-2} d^{-1}$)	6.1	-	-	-	-

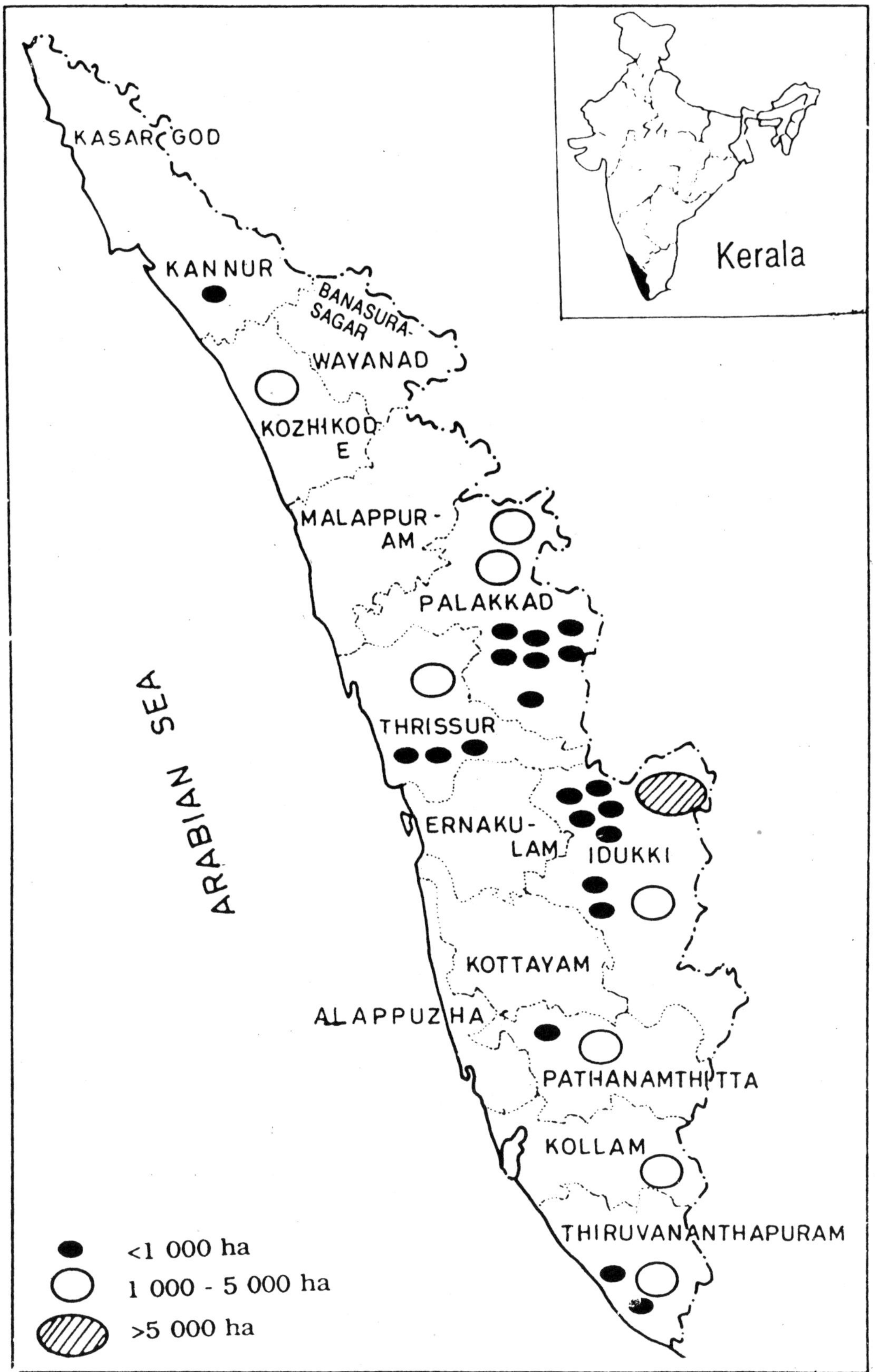

Figure 3.1. Distribution of reservoirs in Kerala

3. KERALA

Kerala is a small state, tucked away between the Western Ghats and the Arabian Sea, in the southwestern corner of India. The land is known for its picturesque landscapes and highly educated people. With a very high quality of life index and literacy, the State is one of the most densely populated in India. Geographically, Kerala is unique and its climate is often described as pseudo-tropic. The State is endowed with liberal rainfall and all round lush vegetation. Some of the last remnants of the tropical rain forests of India are situated here. It is the privilege of Kerala to usher in the southwest monsoon to the country. The tall, well-wooded hills of Western Ghats precipitate a bountiful rainfall that flows down to the Arabian sea through 41 small, west-flowing drainages. These rivers, the longest being Bharatapuzha, 267 km, carry an annual discharge of 72 673 million m^3 per year, which is much higher than the quantum of water being carried by large rivers like Cauvery, Krishna and Tapti.

3.1 DISTRIBUTION OF RESERVOIRS

Unlike its big neighbour, Tamil Nadu, Kerala does not have a history of ancient irrigation reservoirs to boast; the traditional practice of agriculture depending mostly on the local rainfall. Storage of river water for irrigation and power generation is a relatively modern concept in Kerala, where most of the dams are constructed after independence. There are 30 reservoirs in the State, spread over 29 635 ha, in the districts of Palakkad, Trissur, Thiruvanathapuram, Pathanamthitta, Kollam, Kannur and Idukki. Most of them are logistically suitable for fish yield optimisation, primarily due to their manageable size. Twenty-one of them are below 1 000 ha in area and among the rest, only the Idukki reservoir is over 5 000 ha (Fig. 3.1). Bharathapuzha is extensively dammed for irrigation purposes in Palakkad district, resulting in nine impoundments (11 351 ha). Similarly, Periyar river is tamed for hydro-electric power generation in the undulated terrains of Idukki district, creating nine reservoirs (6 633 ha; Table 3.1). The total waterspread areas under 21 small, 8 medium and 1 large reservoirs are 7 975, 15 500 and 6 160 ha respectively (Table 3.2).

Kerala is the foremost fish producing state in India. However, 97% of the State's production is from the marine sector and out of the inland fish production of 3%, only 0.3% comes from the reservoirs (Das *et al.*, 1993). Despite the existence of many small reservoirs in the State, no serious attempts have been made, till recently, to develop them on scientific lines for fish yield optimisation. This is primarily due to the inadequate attention received by the inland fisheries sector during the early years of the State's planned development, as the Keralites have a marked bias in favour of marine fish. During the last few decades, due to improved facilities for processing, storage and transport, marine fish has become available to consumers all over the State, resulting in a shortage. Thus, the people have turned towards the inland fish and they even developed a taste for it. Catla, rohu, mrigal and even common carp are no longer at a discount, in comparison to sardines and mackerels. Now, the State has recognised the importance of inland fisheries and freshwater aquaculture. However, a perusal of literature indicates that no major initiative has been taken to create a database on the ecosystem functions and production potential of reservoirs in the State. Very little information is available on their water quality and biotic communities.

Table 3.1. Reservoirs of Kerala

No.	Name of Reservoir	Area (ha)	District	Location Lat.(N)	Long. (E)	River
1	2	3	4	5	6	7
1.	Malampuzha	2313	Palakkad	10° 40'	76° 39'	Bharatapuzha
2.	Mangalam	393	do	10° 31'	76° 58'	do
3.	Meenkara	259	do	10° 38'	76° 48'	do
4.	Chulliar	159	do	10° 36'	76° 46'	do
5.	Pothundi	363	do	10° 07'	76° 10'	do
6.	Walayar	259	do	10° 50'	76° 51'	do
7.	Parambiculam	2092	do	-	-	do
8.	Thunakkadavu	283	do	-	-	do
9.	Kanjirampuzha	512	do	-	-	do
10.	Peechi	1263	Trissur	10° 30'	76° 15'	Karuvannoor
11.	Vazhani	255	do	10° 40'	76° 15'	Keecheri
12.	Sholayar	870	do	10° 17'	76° 45'	Chalakkudy
13.	Peringalkuthu	263	do	-	-	do
14.	Neyyar	1500	Thiruva-nanthapuram	08° 32'	77° 08'	Neyyar
15.	Peppara	582	do	08° 07'	76° 40'	Vamanapuram
16.	Aruvikkara	258	do	08° 28'	77° 58'	Karamana
17.	Pamba	570	Pathanam-thitta	09° 20'	76° 53'	Pamba
18.	Kakki	1800	do	-	-	do
19.	Parappar	2590	Kollam	09° 57'	77° 4'20"	Kallada
20.	Kuttiadi	1052	Kozhikode	11° 36'	75° 49'27"	Kuttiadi
21.	Pazhassi	648	Kannur	-	-	Valapatta-nam
22.	Idukki	6160	Idukki	09° 48'	76° 53'	Periyar
23.	Ponmudi	260	do	09° 55'	77° 05'	do
24.	Anayirankal	433	do	10° 0'	77° 0'	do
25.	Kundala	230	do	-	-	do
26.	Mattupatti	324	do	10° 05'	77° 05'	do
27.	Sengulam	33	do	10°00'	77° 05'	do
28.	Neriamangalam	413	do	-	-	do
29.	Bhoothathankettu	608	do	-	-	do
30.	Periyar lake	2890	do	10° 10'	76° 15'	do
	Total	**29 635**				

Table 3.2. Category-wise reservoir areas in Kerala

Category	Number	Area (ha)
Small	21	7 975
Medium	8	15 500
Large	1	6 160
Total	**30**	**29 635**

3.2 IDUKKI RESERVOIR (Fig. 3.2)

Idukki, the prestigeous hydro-electric power project of Kerala consists of three dams, *viz.*, the Idukki double arch dam across the river Periyar, the adjacent Cheruthoni concrete gravity dam and the big masonry Kulamavu dam, located 30 km upstream of Idukki on the left bank. The 6 160 ha reservoir is spread on the beds of two rivers, *viz.*, Periyar and Cheruthoni. Idukki reservoir receives water from the Periyar at Ayyappankovil, from the Erattayar tunnel at Kanjiyar and from the catchments of a number of minor streams. Total catchment area is estimated at 649.3 km^2 and gross storage capacity, 1 996.3 million m^3. At FRL, the reservoir attains an elevation of 732.62 m with average depth ranging from 50 to 80 m. The calculated mean depth (volume/area) is 32 m.

Khatri (1985) reports the physico-chemical profile of water in Idukki reservoir. The water, although warm (21.8 °C to 26.3 °C) and clean (transparency 55 to 540 cm), is poor in nutrients. Being a deep basin reservoir, a substantial part of the storage is always held up as dead storage. Among nutrients, phosphate is totally absent and nitrates and silicates are in very low concentrations (Table 3.3). Nutrients, especially phosphates, seem to be the limiting factor for primary productivity in the reservoir. Whatever nitrates and silicates are available are locked up in the deep aphotic layers.

Table 3.3. Physico-chemical profile of water in Idukki reservoir

Air temperature (^0C)	21.2-29.8
Water temperature (^0C)	21.8-26.3
Transperancy (cm)	55.0-540.0
pH	6.0-9.0
Free CO_2 (mg l^{-1})	nil-70.0
Bicarbonates (mg l^{-1})	7-42
Dissolved oxygen (mg l^{-1})	1.2-9.8
Silica (µg l^{-1})	95-199
Phosphate (µg l^{-1})	nil
Nitrate (µg l^{-1})	1.7-18.7

After Khatri, 1985

Standing crop of phyto and zooplankton in Idukki is reported to be extremely poor with the overwhelming presence of the oligotrophic species, *viz.*, *Ankistrodesmus* sp. and *Cosmarium* sp. Cyanophyceae are scanty (Khatri, 1987a) and believed to be of allochthonous origin. The reservoir has an equally poor zooplankton community (Khatri, 1988). Seventeen species of fishes, including two exotics are reported from Idukki (Gopinath and Jayakrishnan, 1984). Presently, there is no organised fishing in the reservoir. A few local people inhabiting along the reservoir margin are engaged in improvised fishing practices, using local fishing implements such as

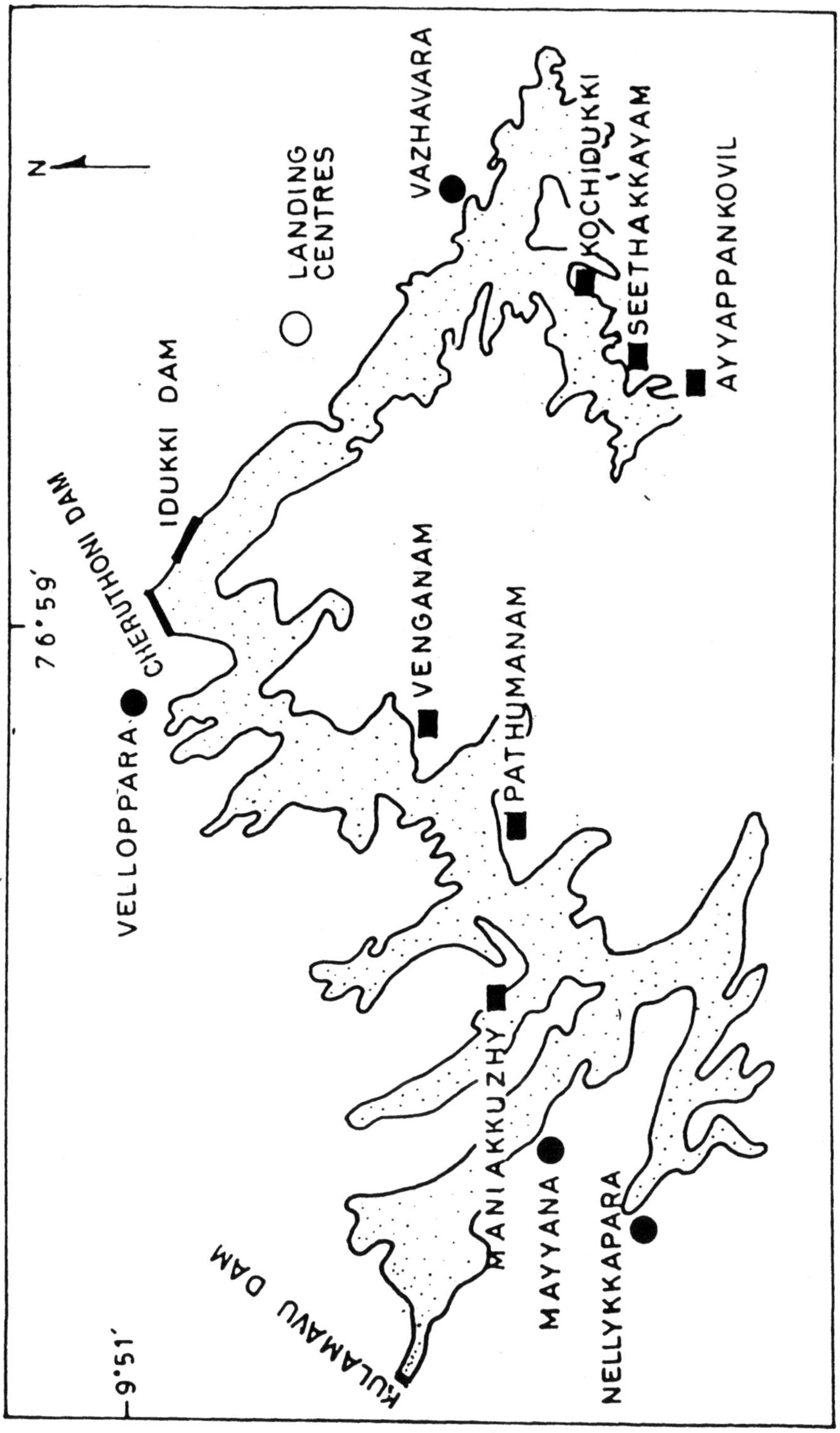

Figure 3.2. Idukki reservoir, Kerala

thandadivala or *kethevala*, which are crude gill nets with mesh size varying from 26 to180 mm. Gopinath and Jayakrishnan (1984) recorded fish catch of 705.97 kg during 11 months from September 1982 to July 1983. Six species that figured in the commercial landings were *Cyprinus carpio*, *Cirrhinus cirrhosa*, *Channa gachua*, *Mastacembelus guentheri*, *Tor khudree* and *Oreochromis mossambicus* (Fig. 3.3). *Ompok bimaculatus* and *Heteropneustes fossilis* were also present in small quantities. Seventeen species of fishes and eleven birds have been recorded from Idukki (Table 3.4 and 3.5). Nair (1988) considers Idukki to be passing through a trophic depression, as evidenced by low rate of colonisation by fish food biotic communities and the retarded growth rate of *Oreochromis mossambicus*.

Table 3.4. Fish species recorded from Idukki reservoir

ORDER	**CYPRINIFORMES**
Family	CYPRINIDAE
Sub-family	Rasborinae
	1. *R. daniconius*
	2. *B. bendelisis*
Sub-family	Cyprininae
	3. *C. carpio*
	4. *P. melanampyx*
	5. *P. bovianicus*
	6. *T. khudree*
	7. *C. cirrhosa*
Sub-family	*Garrinae*
	8. *G. lamta*
Family	COBITIDAE
Sub-family	*Neomacheilinae*
	9. *N. denisonii*
	10. *N. scaturigina*
	11. *N. rupicola*
ORDER	**SILURIFORMES**
Family	SILURIDAE
	12. *O. bimaculatus*
Family	HETEROPNEUSTIDAE
	13. *H. fossilis*
ORDER	**ATHERINIFORMES**
Family	CYPRINODONTIDAE
	14. *A. lineatus*
ORDER	**CHANNIFORMES**
Family	CHANNIDAE
	15. *C. gachua*
ORDER	**PERCIFORMES**
Family	CICHLIDAE
	16 *O. mossambicus*
ORDER	**MASTACEMBELIFORMES**
Family	CHANNIDAE
	17. *M. guentheri*

Source: *Gopinath and Jayachandran, 1984*

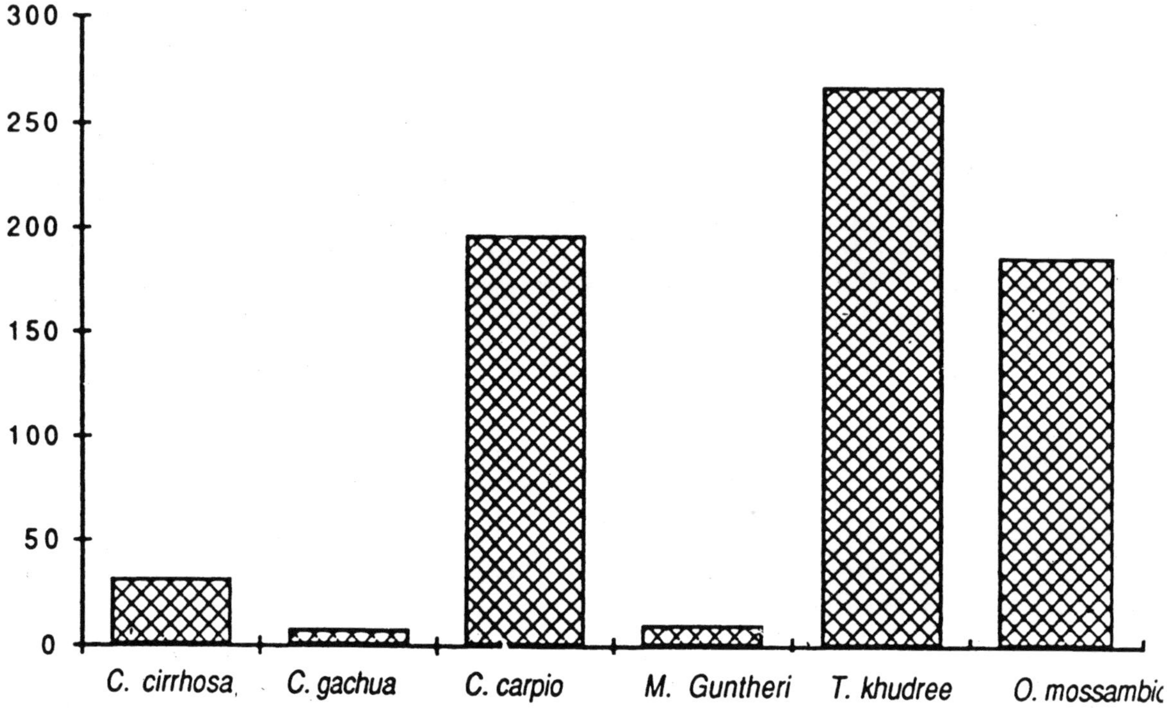

Figure 3.3. Catch by weight (kg) for the major fish species (Idukki reservoir) after Gopinathan and Jayakrishnan, 1984

Table 3.5. Common birds of Idukki reservoir

Common name	Zoological name
Dabchick	*Podiceps ruficollis*
Darter	*Anhinga rufa*
Little cormorant	*Phalacrocorax niger*
Cattle egret	*Bubulcus ibis*
Little egret	*Egretta garzetta*
Common sand piper	*Tringa hipoleucos*
Night heron	*Nycticorax nycticorax*
Pond heron	*Ardeola grayii*
White-breasted kingfisher	*Halcyon smyrtensis*
Small blue kingfisher	*Alcedo atthis*
Pied kingfisher	*Ceryle rudis*

Source: Gopinath and Jayachandran, 1984

3.3 PARAPPAR RESERVOIR

The 2 590 ha Parappar reservoir was commissioned in 1987 on the Kallada river in Kollam district at 9° 57' N, 77° 4' E. The Project comprises two hydraulic structures, *viz.*, the Thenmala dam and the Ottackal weir. The dam Thenmala is built across Kallada, just below the confluence of its three small tributaries, the Kulathupuzha, the Chendurni and the Kalthurthy, forming the Parappar reservoir at an elevation of 1 225 m above MSL. The outflowing river Kallada joins the Ashtamudi lagoon and estuary (Fig. 3.4).

Kallada river, downstream the Parappar reservoir, receives industrial effluents from the Punalur Paper Mills, causing concern among the environmentalists. The reservoir, however, is reported to be free from pollution. A pre-impoundment survey by Nair (1986) throws light on the environmental characteristics of river stretch. Sahib and Aziz (1989) examined the water quality above and below the dam, with special emphasis on the pollution load and water quality. They found an increase in water temperature and transparency in the impounded area during monsoon and post-monsoon seasons, compared to the pre-impoundment period, with a corresponding drop in hardness and dissolved oxygen levels. The reservoir water was neutral in reaction with low levels of nitrates and phosphates (Table 3.6). No appreciable level of vertical gradient was observed in respect of temperature and chemical parameters through seasons.

3.4 NEYYAR RESERVOIR

Neyyar is an irrigation reservoir created on the Neyyar river at lat. 8° 32' N and long. 77° 08' E, about 30 km south-east of Thiruvananthapuram at the foot of the holy hill, Agasthyakudam. Commissioned in 1959, the reservoir has a waterspread of 1 500 ha and receives catchment from 140 km^2. Neyyar reservoir, apart from meeting the irrigation needs of Thiruvananthapuram district, recharges the Aruvikkara reservoir to maintain uninterrupted water supply to the city of Thiruvananthapuram and its suburbs (Fig. 3.5).

Harikrishnan and Azis (1989), who made an indepth study of the reservoir, found it ideal for fisheries development. The water temperature, which varied from 28.0 to 32.0 °C during different seasons, did not show any vertical stratification. Range in values of other physico-chemical parameters is given in Table 3.7.

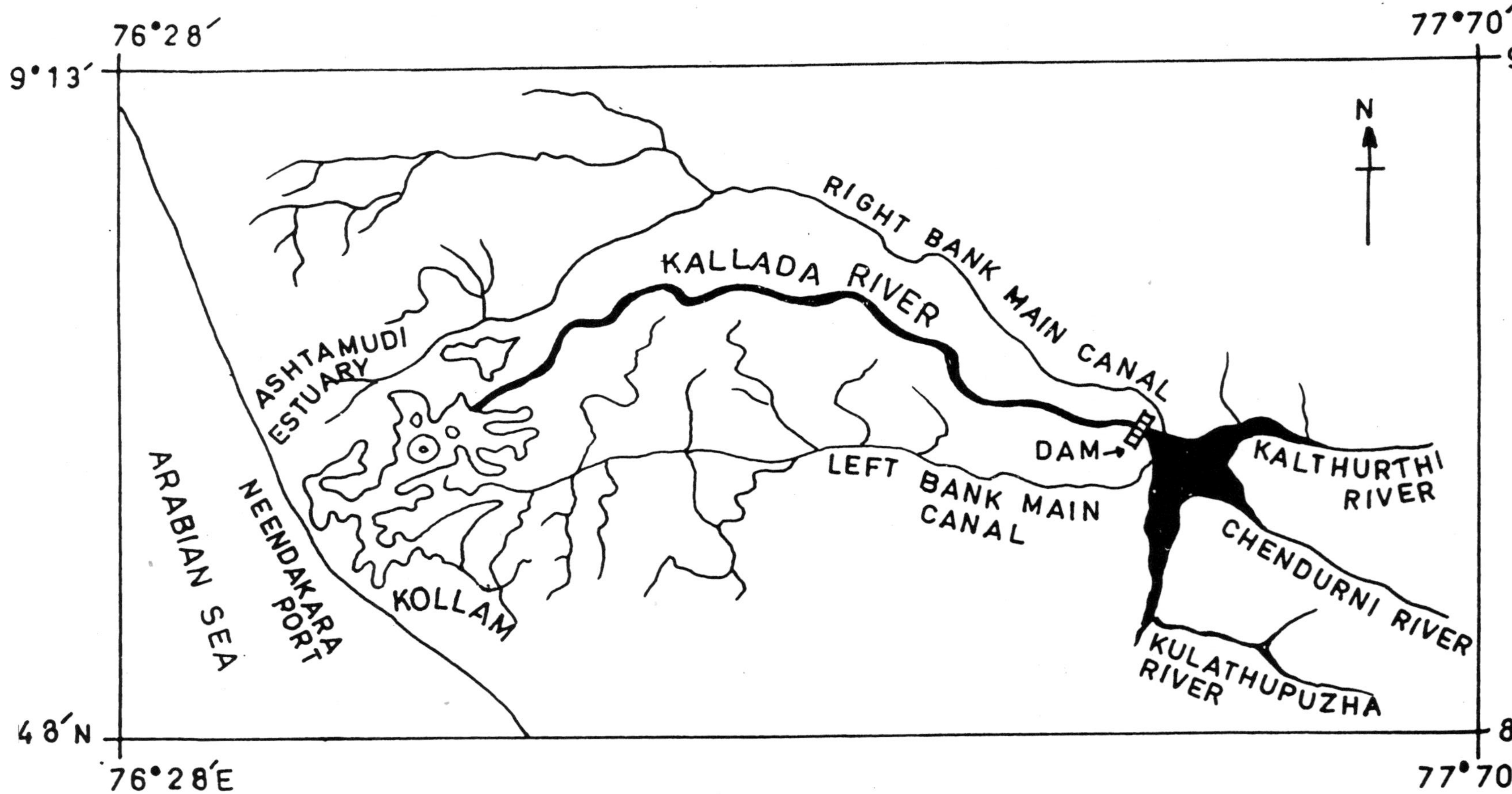

Figure 3.4. Parappar reservoir, Kerala

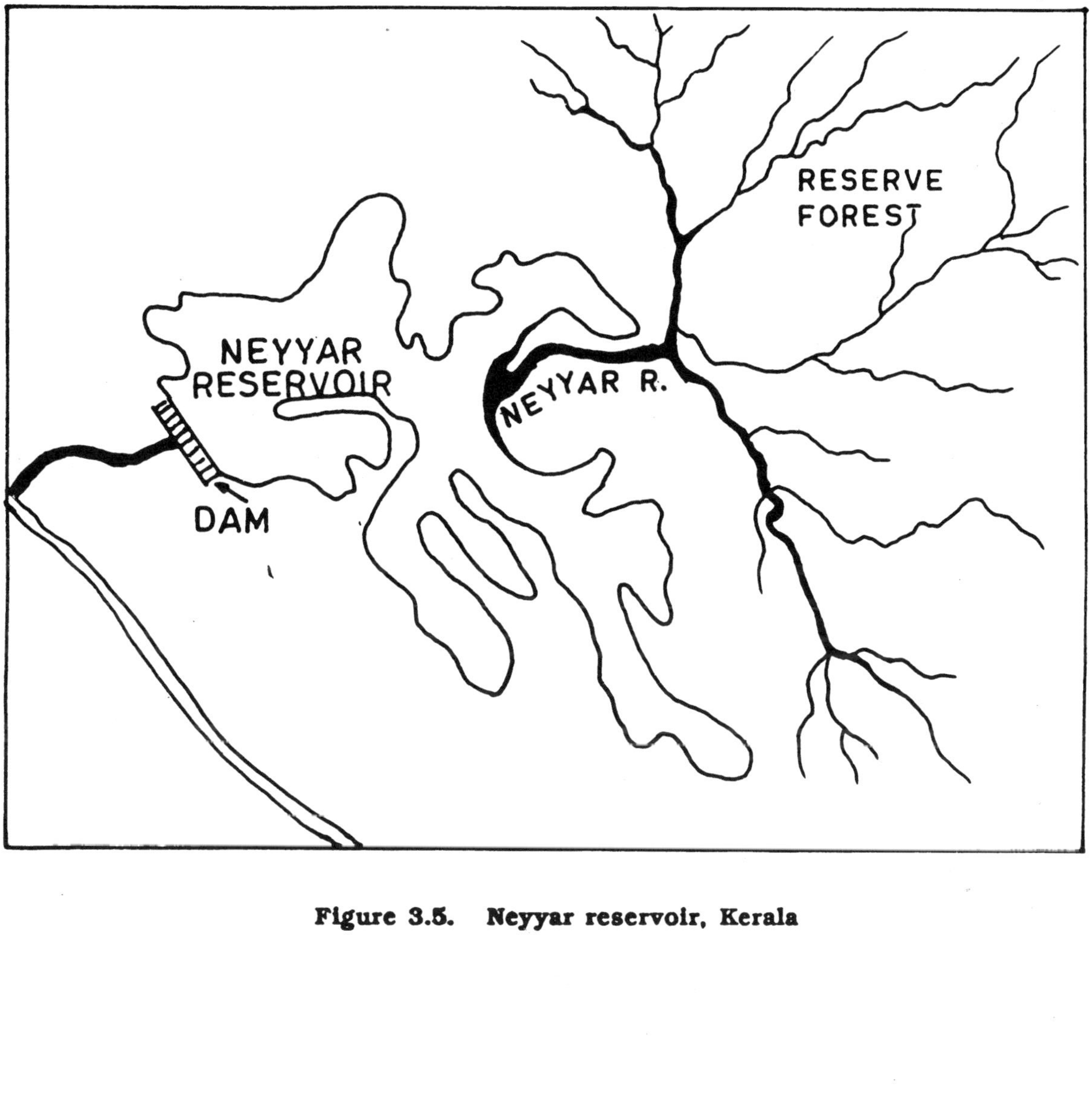

Figure 3.5. Neyyar reservoir, Kerala

Table 3.6. Some physico-chemical attributes of water in Parappar reservoir (surface values)

Parameter	Monsoon	Post-monsoon	Pre-monsoon
Water temperature (^{o}C)	27.5	26.7	28.4
pH	6.8	7.4	7.6
Transparency (cm)	155.4	166.2	184.7
Total hardness (mg l^{-1})	10.5	5.5	8.7
Dissolved oxygen (mg l^{-1})	3.4	5.5	5.2
Nitrite (μg l^{-1})	4.7	0.9	0.9
Nitrate (μgl^{-1})	-	-	0.9
Phosphates (μg l^{-1})	27.49	-	-
Silicate (μg l^{-1})	133.8	151.1	175.3

(After Sahib and Azis, 1989)

Table 3.7. Water quality parameters in Neyyar reservoir

Parameter	**Range in values**	
	Station I	*Station II*
Water temperature (^{o}C)	28.0-32.0	28.5-32.0
Transparency (m)	0.06-2.12	0.06-1.64
pH	6.7-7.9	6.8-7.8
Dissolved oxygen (mg l^{-1})	5.9-9.6	6.3-9.7
Dissolved carbon dioxide (mg l^{-1})	1.1-6.6	1.1-4.4
Total hardness (mg l^{-1})	6.0-20.0	8.0-24.0
Nitrite nitrogen (μg l^{-1})	1.23-12.25	1.23-7.35
Nitrate nitrogen (μg l^{-1})	5.28-211.2	7.92-142.0
Phosphate phosphorus (μg l^{-1})	0-16.64	0-8.32
Silicate-silica (μg l^{-1})	2.08-4.48	2.24-4.16

(After Harikrishnan and Azis, 1989)

3.5 KUTTIADI RESERVOIR

Kuttiadi river originates from Narikotta ranges in the Western Ghats, at an elevation of 1 220 m above MSL and joins the Arabian Sea at Kottackal, after traversing 74 km across the State. The straight gravity masonry dam, situated at Peruvannamuzhi, is 173 m long and 36 m in height. At the dam site, the river receives catchment from 1 095 km^2. Gross and net capacity of the reservoir are 121.2 and 113.9 million m^3. Mean annual rainfall at the watershed is estimated at 3 445 to 7 760 mm. The irrigation reservoir serves a total cultivable command area of 25 490 ha and a total irrigated area of 14 566 ha. Kunhi (1990) reports thermal stratification in the reservoir during post-monsoon months, accompanied by a cinograde oxygen distribution. Reporting very high levels of phosphorus in the water phase, the author places the lake under the eutrophic category.

3.6 STATUS OF FISHERIES DEVELOPMENT IN KERALA

Fish yield

Total fish production from the reservoirs of Kerala is not known. Although 17 reservoirs in the State are under regular stocking by the State Government, latest (1992-93) annual fish production figures are available only in respect of 9 reservoirs (Table 3.8). Normally, the data on reservoir fish production are based on sales figures from co-operative societies, which do not account for other important sources and uses of fish produced, such as illegal removal of fish by poachers, non-marketed production (subsistence consumption) and illegally marketed fish (Das *et al.*, 1993). Present total fish supply from the five reservoirs (Malampuzha, Peechi, Pothundi, Vazhani, and Chulliar) covered under the Indo-German Reservoir Fisheries Development Project is estimated at 384 t yr^{-1}, out of which 336 t (88%) is marketed. This is against the official figure of 81.6 t. The Project has estimated an annual potential fish supply of 650 t from the five reservoirs and 1 700 t from all reservoirs of Kerala.

Table 3.8. Fish production from 9 reservoirs of Kerala (1992-93)

Reservoir	Area (ha)	Total fish landings (kg)	yield (kg ha^{-1})
Trissur district			
1. Peechi	1 263	5 734	4.5
2. Vazhani	255	6 985	27.4
Palakkad district			
3. Malampuzha	2 313	11 540	5.0
4. Mangalam	393	17 494	44.5
5. Meenkara	259	17 776	68.6
6. Chulliar	159	50 334	316.6
7. Pothundi	363	7 057	19.4
8. Walayar	259	4 700	18.1
9. Kanjirampuzha	512	13 406	26.2

Source: State Fisheries Department

During the last 13 years, the yield from the managed reservoirs registered an impressive four-fold (from 34.9 t in 1978-79 to 147.7 t in 1991-92) increase, though the present yield is less than 5 kg ha^{-1}, *i.e.*, a third of the national average, when all the reservoirs are taken into consideration (Table 3.9).

Table 3.9. Fish yields of Kerala reservoirs

	1978-79	1987-88	1988-89	1989-90	1990-91	1991-92	1992-93
Total yield (t)	34.90	35.40	79.50	91.60	118.20	147.70	135.03
Average yield from managed reservoirs (kg ha^{-1} $year^{-1}$)	6.60	6.70	15.10	17.50	20.40	25.50	23.38
Average yield from total reservoir area (kg ha^{-1} $year^{-1}$)	1.20	1.20	2.70	3.10	4.00	5.00	-

Ecosystem management

Hydel reservoirs created in the hilly regions generally have deep basins and they portray oligotrophic tendencies, compared to those in the plains which are in transient phases of eutrophication. The irrigation reservoirs of the midland are shallow and spread over the rich alluvial soils. Bountiful sunlight, high temperature regime and nutrients from allochthonous sources make these reservoirs conducive to high primary productivity. Neyyar reservoir in Thiruvananthapuram District has high nutrient loading, in terms of nitrate, phosphate and silicate, indicating eutrophication. Similar nutrient status is reported for Aruvikkara reservoir in Thiruvananthapuram and Parappar reservoir in the Kollam District. Natural populations of fish recorded from Neyyar, Aruvikkara, Kallada and Peppara reservoirs have *Puntius sarana, P. filamentosus, Labeo* spp. and *T. khudree.* Among predators, catfishes *Wallago attu* and *Mystus vittatus* and the murrels are present.

Development of reservoir fisheries in Kerala on scientific lines is yet to pick up momentum. The State Government initiated steps to develop the reservoir fisheries, way back in 1951, through a three-pronged strategy comprising large-scale production of major carp seed, stocking of carps and creation of breeding and rearing facilities at the reservoir sites. Even though breeding experiments commenced at Malampuzha in 1962, encouraging results were obtained only in the year 1967. Annual seed production in the State rose from 3 million in the 1970s (Nair, 1973) to 14 million in the late 1980s (Sidhardhan and Nair, 1988). Fish farms have been constructed in eight reservoirs for raising stocking material. Available details on six of them are given in Table 3.10.

The first reservoir to be stocked with major carps in the State was Malampuzha, where the fish seed obtained from Tamil Nadu and West Bengal were introduced in 1951. During the 22 years from 1951 to 1972, 2.553 million seed was stocked at an annual rate of 52 ha^{-1}. By 1972, 12 other reservoirs in the State were also stocked at densities ranging from 35 to 805 ha^{-1} (Table 3.11). The stocking rate in the state over the years is considered to be very inadequate (Nair, 1973; Mathew and Mohan, 1990).

Table 3.10. Farm facilities available at reservoir sites in Kerala

Reservoir	Nursery area (ha)	Ponds (No.)	Rearing area (ha)	Ponds (No.)	Stocking area (ha)	Ponds (No.)
Malampuzha	1.0157	91	0.7176	22	1.1558	14
Meenkara	0.1890	12	0.3000	5	-	-
Chulliar	0.0670	5	0.3040	8	-	-
Walayar	0.0432	6	-	-	0.1028	2
Mangalam	0.0068	3	0.1116	3	-	-
Neyyar	0.0692	17	-	-	-	-
Vazhani*						
Peechi*						
Total	**1.3909**	**134**	**1.4332**	**38**	**1.2586**	**16**

*Data not available *(After Venugopal, 1993)*

Recent advances in fingerling production

Recently, an advisory mission, under the framework of Indo-German Reservoir Fisheries Development Project has examined the feasibility of organising alternate ways of producing stocking material, using low-cost and easy to manage systems (Rajts, 1993). The emphasis was on a decentralised fry/fingerlings production through target group organisations, as opposed to the present centralised and government services. The project envisaged a fingerling production system operated through fishermen under a cooperative fold, thereby contributing fundamentally to self-sustained culture-based reservoir fisheries development. The proposed production units to be set up in the five reservoirs under the Project, *viz.*, Malampuzha, Peechi, Pothundi, Vazhani, and Chulliar comprise broodfish stock, mini-hatchery and rearing components, in addition to the pens to be erected in the reservoir margin and the off-shore floating cages.

A total incubator volume of 4.0 m^3, 1.79 ha of pond area, and 7 ha of pen enclosures are involved in the operation, with an estimated production potential of 40 million hatchlings (Indian major carp, common carp and tilapia). However, an economic evaluation, based on cost and return estimates, made by the project (Schack, 1993) did not consider the pond and pen culture experiments at Pothundy and Malampuzha as viable investment packages. In both cases, the internal rates of return (IRR), calculated over a period of 10 years, was found to be below the loan interest rates charged by the banks. While pointing out the low recapture rates of Indian major carps in many reservoirs, the study strongly advocated the need for evolving alternatives with still lower-cost production techniques.

Table 3.11. Details of stocking in reservoirs of Kerala

Sl. No.	Reservoir	Period	No. of finger-ling stocked *(million)*	Stocking rate (no. ha^{-1} yr^{-1})	Reference
1.	Malampuzha	1951-72	2.553	52	Nair, 1973
		1987-88	2.870	1 297	Mathew and Mohan, 1990
2.	Mangalam	1956-72	0.913	137	do
		1987-88	0.550	1 400	
3.	Meenkara	1961-72	1.054	339	do
		1987-88	0.730	2 818	
4.	Chulliyar	1968-72	0.640	805	do
		1987-88	0.240	1 509	
5.	Pothundy	1968-72	1.241	684	do
		1987-88	0.690	1 900	
6.	Walayar	1958-72	1.883	485	Nair, 1973
7.	Vazhani	1958-72	0.310	81	Nair, 1973
		1987-88	0.200	784	Mathew and Mohan, 1990
8.	Peechi	1958-72	0.670	35	do
		1987-88	0.850	673	
9.	Neyyar	1971-72	0.315	105	Nair, 1973
10.	Kallada	1987-88	0.430	166	do
11.	Kanjirampuzha	1987-88	0.450	879	do
12.	Peppara	1987-88	0.200	344	do
13.	Kuttiadi	1987-88	0.900	855	do

Recent advances in fisheries management

Recently, 14 reservoirs in the State, having a waterspread of 17 411 ha, have been identified under a fisheries development scheme. However, stocking and management of even these reservoirs are not up to the desired level. In 1987-88, only 11 reservoirs with a total waterspread of 9 641 ha had been stocked (Table 3.11). Selection of species for stocking seems to be guided more by expediency rather than ecosystem considerations (Table 3.12). During the initial years, a substantial percentage of the seed stocked belonged to the slow growing *Labeo fimbriatus*, which was later replaced with the major carps. The common carp also found a place among them, though in limited quantities. Criteria for stocking depended entirely on availability of seed, or more precisely, the available seed was distributed among reservoirs in an arbitrary manner. The size of stocking material was invariably small (20 to 40 mm) and the stock was left at the mercy of nature and the predators. Notwithstanding all the shortcomings mentioned above, the stocking did have some positive impact on the fish production in many reservoirs.

Studies on fish population dynamics in Malampuzha reservoir

Under the Indo-German Reservoir Fisheries Development Project, considerable effort has been made by Taege *et al.* (1993) towards continuous evaluation of stock composition and annual stock assessment in Malampuzha reservoir. Taking into account the complexities arising out of the unstable hydrological, ichthyological, ecological and socio-economic conditions, they suggested continuous revision management decisions on an annual basis.

Table 3.12. Species composition of fish seed stocked in the reservoirs of Kerala (1986)

Reservoir	Area	No. of seed stocked					
		Catla	Rohu	Mrigal	Labeo fimbriatus	Common carp	Total
Malampuzha	2 213	9 525	17 700	357 630	254 335	318 492	957 682
Meenkara	259	-	-	4,425	228 000	72 000	304 425
Mangalam	393	97 700	97 150	150 400	50 000	-	350 250
Pothundy	363	99 500	71 295	79 205	50 000	-	300 000
Walayar	259	202 950	835	2 008	462	9 200	215 525
Peechi	1 263	77 464	276 735	30 475	39 630	-	424 304
Peppara	582	3 000	-	5 000	38 950	8 000	54 950
Kanjirampuzha	512	8 000	20 725	155 360	147 000	25 000	395 585
Chulliar	159	-	-	-	-	-	45 000
Kuttiadi	1 052	66 600	420 000	198 000	271 000	-	955 000
Total	7 055	564 735	1004 440	982 573	879 377	432 692	4 002 721

(After Mathew and Mohan, 1990)

Fishery resources of Malampuzha reservoir comprise indigenous species and tilapia, that multiply in the reservoir, constituting about 75% of the present yield, and Indian major carps making up the rest. At least three generations of tilapia were reflected in the catch, among which the youngest (99.08 g) was considered as underweight and uneconomical. The pearlspot, *Etroplus suratensis* was reported to be present in two generations. Pearlspot, along with the exotic member of Cichlidae- *Oreochromis mossambicus*, registered some decline in recent years, with a simultaneous increase in catches of *Puntius* spp. suggesting biological imbalance. *Puntius sarana* was represented by 3 to 4 generations, of which the youngest was considered undersized. *Gonoproktopterus curmuca*, *Parambassis thomassi*, and *Ompok bimaculatus* appeared in only single size groups. Catla and *Cyprinus carpio* caught during November to April were dominated by undersized individuals. A large population of predatory species, *Xenentodon cancila* was a cause for concern.

The management strategy evolved by the Indo-German Reservoir Fisheries Development Project comprises:

* stocking measures and stock enhancement programmes
* self-reliance on fingerling supply of Cooperative Society
* diversification of the abundant tilapia stocks,
* introduction of strict regulations on mesh sizes of fishing gear
* continuous evaluation of stock composition and annual stock assessment, and
* annually adjusted management strategies.

Impact of stocking on fish yield rate

The low productivity in most of the reservoirs can be easily attributed to undesirable species mix. A perusal of catch composition in 9 reservoirs reveals that in most cases, the fast-growing species do not get the desired level of representation (Table 3.13). Among Indian major carps, only catla made some impact in Mangalam, Meenkara, and Peechi. Tilapia forms a major component of the catch in many of the reservoirs. Taking all the reservoirs into consideration, out of 79 454 kg of fish, the shares of catla and tilapia were nearly 25 000 kg each.

Sustained stocking had the desired effect in altering the catch structure in favour of major carps in some of the lakes. In Chulliar reservoir, the percentage of major carps varied from 0.2 to 5.5% of the total production during 1969 to 1987 and it went up to 11.5 in 1988-90 (Jhingran, 1991). A similar picture emerges in Meenkara, where percentage composition of catla, rohu and mrigal touched 35.70% during 1988 to 1990 from an abysmal 1.02% till 1983. In Peechi, catla (57.3%), mrigal (9.6%) and rohu (4.8%) together formed 71.6% in 1989-90, compared to 2% in 1966-67. In Malampuzha, the Indian major carps have established a firm foothold. Specimens of *Catla catla* weighing 16 to 17 kg are a common sight. Similarly *L. rohita* grows to 4.5 kg. The largest specimen recorded, so far, is the one caught in 1982, weighing 47 kg and measuring 1 170 mm in length. However, the Indian major carps never exceeded 19.9% of the total catch. Stocking rate and fish yield in respect of four reservoirs are depicted in Tables 3.14 to 3.17.

Chulliar and Vazhani reservoirs registered an impressive increase in yields during 1988 to 1993 on account of the new stocking policy, which gave emphasis on the size and number of the seed stocked. In Chulliyar, yield has increased from 49.10 kg ha^{-1} yr^{-1} in 1988-89 to 275.50 kg ha^{-1} yr^{-1} in 1991-92. In Vazhani, the corresponding figures were 2.30 and 48.60 kg ha^{-1} yr^{-1} respectively.

Tilapia

Oreochromis mossambicus is ubiquitous in the inland waters of Kerala. After its introduction in the State in 1952, this exotic fish has found its way into the household ponds, temple tanks and reservoirs all over the State. Almost all reservoirs in Palakkad and Trissur were stocked with tilapia by the late 1950s and early 1960s (Venugopal, 1993). Employing tilapia for filling the vacant niches in reservoirs is a subject of polemics. The main controversy regarding tilapia is its habit of early maturing and stunted growth. Nevertheless, experience of Kerala reservoirs shows that the fish adapt well in the reservoir ecosystem and thrive on a variety of food items.

Table 3.13 Fish catches (kg) from Kerala reservoirs 1988-89

Fish	Malampuzha	Mangalam	Walayar	Pothundy	Meenkara	Chulliar	Peechi	Vazzhani	Kuttiadi	Total
Etroplus	1112	-	-	-	5	-	-	-	-	1417
Murrels	37	290	-	48	94	3	142	-	2	616
Mullets	-	-	-	-	-	-	-	-	-	-
Catfish	-	57		230	-	-	-	-	1	60
Tilapia	5101	546	429	1014	7927	6805	2831	29	4	24686
Labeo rohita	1109	250	9	296	3194	419	679	89	206	6251
Barbus	7387	1323	13	61	247	2	1259	77	461	10830
Miscellaneous	1570	299	-	7	36	-	18	-	1225	3155
Eels	87	20	-	3	-	-	-	-	40	150
Cyprinus carpio	1116	63	6	23	231	1	651	-	4	2095
Mrigal	1186	552	22	154	1095	571	1263	198	94	5135
Gourami	-	5	-	3	-	-	-	-	-	8
Catla	881	9039	115	144	11617	8	3293	205	51	25353
Total	19586	12444	594	1755	24446	7809	10136	598	2088	79456

Table 3.14. Stocking rates and yields in Meenkara reservoir

Year	Stocking (no. ha^{-1} yr^{-1})	Yield (kg ha^{-1} yr^{-1})
1966-72	176	9.96
1973-75	-	9.53
1975-77	-	8.86
1977-79	166	4.55
1981-83	-	4.32
1986-88	94	15.76
1988-89	142	94.40
1989-90	1 027	86.60
1990-91	1 106	53.80
1991-92	1 226	107.70

Table 3.15. Stocking rates and yields in Peechi reservoir

Year	Seed stocking (no. ha^{-1} yr^{-1})	Yield (kg ha^{-1} yr^{-1})
1985-87	184	8.30
1987-89	1 016	8.41
1989-90	58	10.50
1990-91	228	16.24
1991-92	235	13.60

Table 3.16. Stocking rates and yields in Malampuzha reservoir

Year	Seed stocking (no. ha^{-1} yr^{-1})	Yield (kg ha^{-1} yr^{-1})
1981-86	-	4.57
1986-89	136	8.50
1989-90	120	5.10
1990-91	751	3.48
1991-92	8	4.60

Table 3.17. Stocking rates and yields in Chulliar reservoir

Year	Seed stocking *(no. ha^{-1} yr^{-1})*	Yield *(kg ha^{-1} yr^{-1})*
1983-85	-	32.30
1985-87	-	11.00
1988-89	1 525	49.10
1989-90	1 289	148.10
1990-91	1 062	225.40
1991-92	937	275.40

Tilapia contributed significantly to the catch of Chulliar (27.90%) even in 1988-90. In Meenkara, the percentage of tilapia remained high (36.57%) throughout the period of 1966 to 1990. In Malampuzha, where the fish was first introduced in 1956, it is firmly established and forms 31% of the catch since 1986. Individual fish in this reservoir attain an average weight of 2.5 kg, allaying all fears of stunted growth. Tilapia is, in fact, relegating the indigenous species into unimportant positions.

Craft and Gear

A cursory look at the available information on craft and gear indicates that very little thought has been given while organising the fishing effort in reservoirs as an input for development. Quantum of fishing efforts employed in the reservoirs is much below the optimum level, probably with the sole exception of Chulliar. Details of craft and gear in five reservoirs are given in Table 3.18. Small meshed gill nets are operated in the reservoirs without any check, leading to exploitation of undersized fishes. Catching fishes at uneconomic sizes brings pressure on populations and affects the yield considerably.

Cooperatives

Fishing in the reservoirs of Kerala was done directly by the staff of the Department of Fisheries until the late 1980s. Later, the cooperative societies were established at the initiative of the Government at all reservoirs. Five cooperative societies operating in Malampuzha, Chulliar, Pothundi, Vazhani and Peechi reservoirs together have a total membership of 1428, belonging to 290 Scheduled Caste and Scheduled Tribe families. Members of these societies are organised into small groups, who go for fishing on selected days according to a roster for avoiding conflicts. However, all members of the society are not keen to fish due to economic and social compulsions.

Drewes (1993), after a critical examination of the scope and issues involved in the organisation and mangement of reservoir fisheries cooperatives in Kerala, concluded that they were not economically viable, both either as fish production or fish marketing bodies. Nevertheless, he expressed the opinion that they could be transformed into effective instruments for accelerating the socio-economic development, with necessary policy changes at Government level. The policy interventions suggested include:

- Cancellation of the royalty to the Government, which amounts to 25% of the cooperative's gross income.

- Utilization of the 25% royalty by the cooperative as operating capital (nets, boats, fingerlings).

- Discontinuation of government subsidies to meet the cooperative's operating cost (fingerlings, nets, boats, etc.). This will allow a more self-reliant management and avoid outside interference.

- Cancellation of the fixed fish sale prices which are about 75% below the prevailing wholesale market prices. This will result in higher income of fishermen/members of the cooperative and increased gross returns of the cooperative to meet operating costs.

- Introduction of fish auctioning by the cooperatives. This will ensure that a maximum possible price is realized by the cooperative.

- Re-establishing the fishermen/members' share of their cooperative's gross income at a rate which will safeguard an increased income of members and leave the cooperative with sufficient capital to meet its operating cost.

- In case of the Pothundy Cooperative, the members' share may be established at 40% of the cooperative's gross income, leaving 60% for the cooperative to purchase fingerlings, boats and nets. The sharing system for the other cooperatives has to be worked out.

Table 3.18. Craft and gear employed in some selected reservoirs of Kerala

Reservoir	Craft	Gear
Malampuzha	3 country crafts with outboard engine	3", 8" and 10" mesh gill net; 20 kg/group
Peechi	4 country boats and 2 FRP boats	Up to 18" mesh gill nets
Chulliar	8 coracles	4", 12" and 15" mesh gill nets; 8 piece/group
Pothundi	7 coracles	2", 4", 6" and 8" mesh gill net; 4 kg/group
Vazhani	4 country crafts	4", 8" and 12" mesh gill nets.

Source: Interim Report, Indo-German Collaborative Project on Reservoirs in Kerala, 1993

Marketing channels

Das *et al.* (1993) delved deep into the fish marketing and allied activities in five reservoirs. Demand for fish is always high in Kerala and there is a substantial production gap for all types of fishes. This shortfall is expected to become more pronounced in future as marine fisheries are almost fully exploited. This gap could be filled, at least partially, by increased production from reservoirs. There is a market demand for reservoir fish production of at least 700 t in Palakkad and Trissur district and 1 700 t in Kerala as a whole (Das *et al.*, 1993). Margins in reservoir fish production and marketing are reasonable, when compared to marketing of other commodities (rice, beef), except for cooperatives, some of which are in the red (*e.g.*,Malampuzha). Total marketing margin range between 30 to 38%, while wholesale and retail margins are estimated at 21%, and 15 to 38% respectively. Share of producer price in final consumer price is 52 to 59%.

Socio-economics

For most fishermen, fishing is not their only occupation. A study made by the Indo-German Reservoir Fisheries Development Project showed that at least in three reservoirs the fishermen earned as much or more from the non-fishing activities (Table 3.19). This suggests the absence of gainful employment opportunities for all members at the present level of fish yield. Nevertheless, Pothundy portrays a brighter picture, where the individual fisherman earn Rs. 14 380 annually from fishing, against Rs. 2 167 from other sources. According to the present arrangement, fishermen get 50% of the sale proceeds and the remainder is shared equally by the cooperative society and the Government.

Table 3.19. Annual income per person engaged in fishing and non-fishing activities at some selected reservoirs

	Annual income (Rs.) per person from		
Reservoirs	Fishing	Non-fishing activity	Benefit from fishing
Malampuzha	3 638	2 900	+738
Chulliar	6 841	4 272	+2 750
Pothundy	14 380	2 167	+12 213
Vazhani	3 537	811	+2 726
Peechi	4 845	5 299	-454

Source: Interim Report, Indo-German Collaborative Project on Reservoir Fisheries in Kerala.

Among the 290 households in five cooperatives surveyed by Iyer *et al.* (1993), 73% were nuclear households and 27% were joint families, with an average household size of 5 persons. The average income of a fisherman was about Rs. 12 000 a year *i. e.*, slightly above the poverty line established by Government (Rs. 11 500 a year). Capital outlay of household was described as alarming. Almost all fishermen were indebted through loans from banks and other financing agencies; the average outstanding loan amounting to 25% of a yearly household income. Living conditions of fishermen were equally precarious. Almost 90% of houses were classified as *kuccha* houses, *i.e.*, makeshift or semi-permanent, against 20% for the State as a whole.

Causes of low yield from reservoirs

Factors contributing to the low yield rates from reservoirs of Kerala can be summarised as follows:

1. *Underutilization of resource :* Only 9 out of the 30 reservoirs in the State are commercially exploited. This accounts for only 20% of the total reservoir area.

2. *Multiple ownership:* The State Electricity Board controls the reservoirs under the hydro-electric projects, which cover more than half of the total area and those situated in the reserve forests are under the jurisdiction of the Forest Department. Fisheries Department has access only to a few irrigation reservoirs. This multiplicity of ownership comes in the way of implementing development plans.

3. *Lack of scientific database:* None of the reservoirs in the State has been subjected to thorough scientific investigation and as a result, a deep insight into the eco-dynamics of the ecosystem is lacking. Kerala reservoirs are situated in unique geo-climatic conditions and there is a need to study the environment and the dynamics of biotic communities. A few reservoirs, representing various morpho-edaphic groups, should be studied in detail and the data thus generated need to be used as an input for drawing up management plans.

4. *Inadequate stocking:* Most of the man-made lakes in the State, belonging to the category of small reservoirs, are ideally suited for the system of continuous stocking and harvesting. The success of management in such cases essentially centre round an imaginative stocking and harvesting schedule. This calls for creation of adequate nursery facilities to meet the demands. A viable, cost-effective nursery rearing practice for meeting the huge demands from the reservoir sector is yet to be developed.

5. *Selection of species:* Selection of candidate species for stocking and determination of species-mix and density have to be made after due consideration of the location-specific information on the trophic status. The practice of stocking undersized fish seed (<100 mm) is undesirable.

6. *Lack of manpower development:* Upgrading of skill among the operators, creation of awareness about norms of conservation and sustainable yield are essential pre-requisites for motivating people to continue in their profession.

Table 3.20 Salient features of three reservoirs in Kerala

Parameters	Idukki	Parappar	Neyyar
Year of sealing`	1973	1987	1959
Area (ha) at FRL	6 160	2 590	1 500
Volume (m m^3)	1 996	524	106
Maximum depth (m)	80		
Mean depth (m)	32	-	-
Catchment area (km^2)	649	-	140
Elevation (m)	732.6	1 225	84.7
Inflowing rivers	Periyar	Kalthurthi	Neyyar
	Erattayar	Kulathupuzha	
		Chendurni	
Outflowing rivers	Cheruthoni	Kallada	Neyyar
Maximum discharge (m^3 sec^{-1})	8 000	2 832	7.10
Latitude (N)	$9^0 48'$	$9^0 57'$	$8^0 32'$
Longitude (E)	$76^0 53'$	$77^0 04'$	$77^0 08'$
Water temp. (^0C)	21.8-26.3	-	28-32
pH	6-9	6.8-7.6	6.7 -7.9
Transparency (cm)	55-540	155-185	6-212
DO (mg l^{-1})	1.2-9.8	3.4-5.5	5.9-9.7
Total hardness (mg l^{-1})	-	5.5-10.5	6-24

4. KARNATAKA

Karnataka, the eighth largest State in India, is situated on the western edge of the Deccan plateau. The climate and physiography of the region make the State one of the most important in the country with regard to water resources. Karnataka has been harnessing these resources to the fullest extent over the years, creating a chain of storage reservoirs all over the State. The four physiographical regions of the State are the coastal region, the Malnad, the northern plains and the southern plains. The coastal region is a 240 km wide strip of land between the Arabian Sea and the Western Ghats, covered mainly with the alluvium deposited by the short, turbulent, west-flowing rivers, receiving an annual rainfall of 250 cm. Malnad is the undulating land lying east of Western Ghats at an average elevation of 150 m, with many tall peaks (average elevation of peaks, 900 m). With a wet, hot climate and a rainfall of 250 cm yr^{-1}, the region has thick forest cover. The northern and southern plains are a part of the Deccan plateau, with the characteristic dry, semi-arid climate and scanty rainfall of 50 to 75 cm $year^{-1}$.

Karnataka has two types of drainages, *viz.*, the east-flowing large rivers Krishna and Cauvery with their tributaries (Fig. 4.1) and the short, west-flowing rivers. The Western Ghats, which remain as a wedge between the two major watersheds, feed both the east-flowing and west-flowing rivers with an impressive amount of water. The surface flow, in many rivers and their tributaries, has been interrupted for irrigation and power generation and in the process, a large number of storage reservoirs have come up all over Karnataka. Today, Karnataka occupies a prominent position in the reservoir map of the country.

The main river Krishna flowing to Andhra Pradesh carries an annual discharge of 37 000 million m^3. The river has five main tributaries in the State, of which Tungabhadra and Bhima are the largest. Most of the tributaries and their sub-tributaries have been dammed. The Cauvery, which originates in the State flows into Tamil Nadu, and being an upper riparian State, Karnataka taps a substantial quantity of the river water. The main river is dammed at Krishnarajasagar and all the tributaries upstream and below Krishnarajasagar are also obstructed to create reservoirs. While all the east-flowing drainages are harnessed mainly for irrigation, the west flowing streams are impounded to tap their hydro-electric power potential (Table 4.1).

4.1 RESERVOIR FISHERIES RESOURCES

Karnataka, like Tamil Nadu, has a traditional irrigation system of impounding surface flow by creating small earthen dams across streams, creeks and rivulets. Locally called *tanks*, these small irrigation reservoirs can be seen in all the districts of the State, except Kodagu. Based on the available data, there are 4 605 large irrigation tanks in the State covering an area of 213 404 ha, with an average area of 50 ha, as opposed to 19 673 small tanks, with an average area < 7 ha (Anon., 1989). Large tanks have been considered at par with small irrigation reservoirs for the purpose of this study, whereas the small tanks are reckoned as ponds and excluded.

The 74 reservoirs in Karnataka cover an area of 223 887 ha, nearly four times that of Tamil Nadu. Among them, 46 belong to the category of small reservoirs, *viz.*, <1 000 ha with a waterspread of 15 253 ha. After taking into account the irrigation tanks, the total surface water area of small reservoirs goes up to 228 657

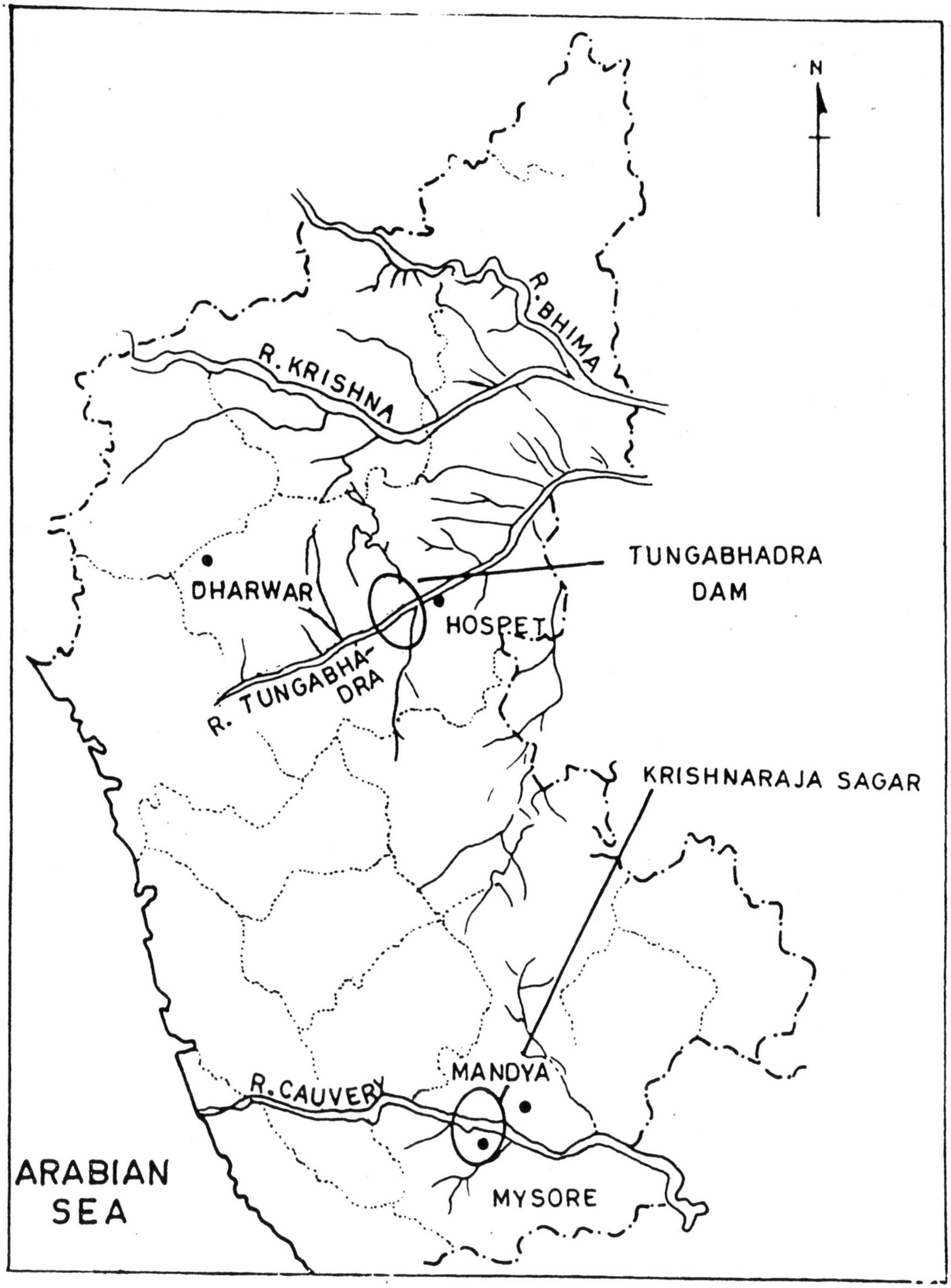

Figure 4.1. Major river systems of Karnataka

ha. The 16 medium reservoirs have an area of 29 078 ha and the large reservoirs (> 5 000 ha), over 179 556 ha. Among the small reservoirs, those less than 500 ha outnumber the rest (Fig. 4.2). Thus, Karnataka has 437 292 ha of water area under different categories of man-made impoundments.

Table 4.1. Riverine resources of Karnataka

Name of the river/ tributary	Annual mean discharge (million m³)	Main river	Flowing to (Neighbouring State)	Ultimate destination
East flowing				
Main river Krishna	37 000	-	Andhra Pradesh	Bay of Bengal
Bhima	12 690	Krishna	do	do
Ghataprabha	5 380	do	do	do
Malaprabha	1 980	do	do	do
Tungabhadra	14 700	do	do	do
Vedavati	1 410	do	do	do
Main river Cauvery	6 200	-	do	do
Herangi	1 130	Cauvery	Tamil Nadu	do
Hemavati	2 520	do	do	do
Shimsha	1 700	do	do	do
Arkavati	850	do	do	do
Lakshmanathirtha	425	do	do	do
Kabini	2 600	do	do	do
Suwarnavati	820	do	do	do
West flowing				
Sharavati	4 545	Sharavati	-	Arabian Sea
Kalinadi	6 537	Kalinadi	-	do
Netravati	4 615	Netravati	-	do
Gangavati	4 925	Gangavati	-	do

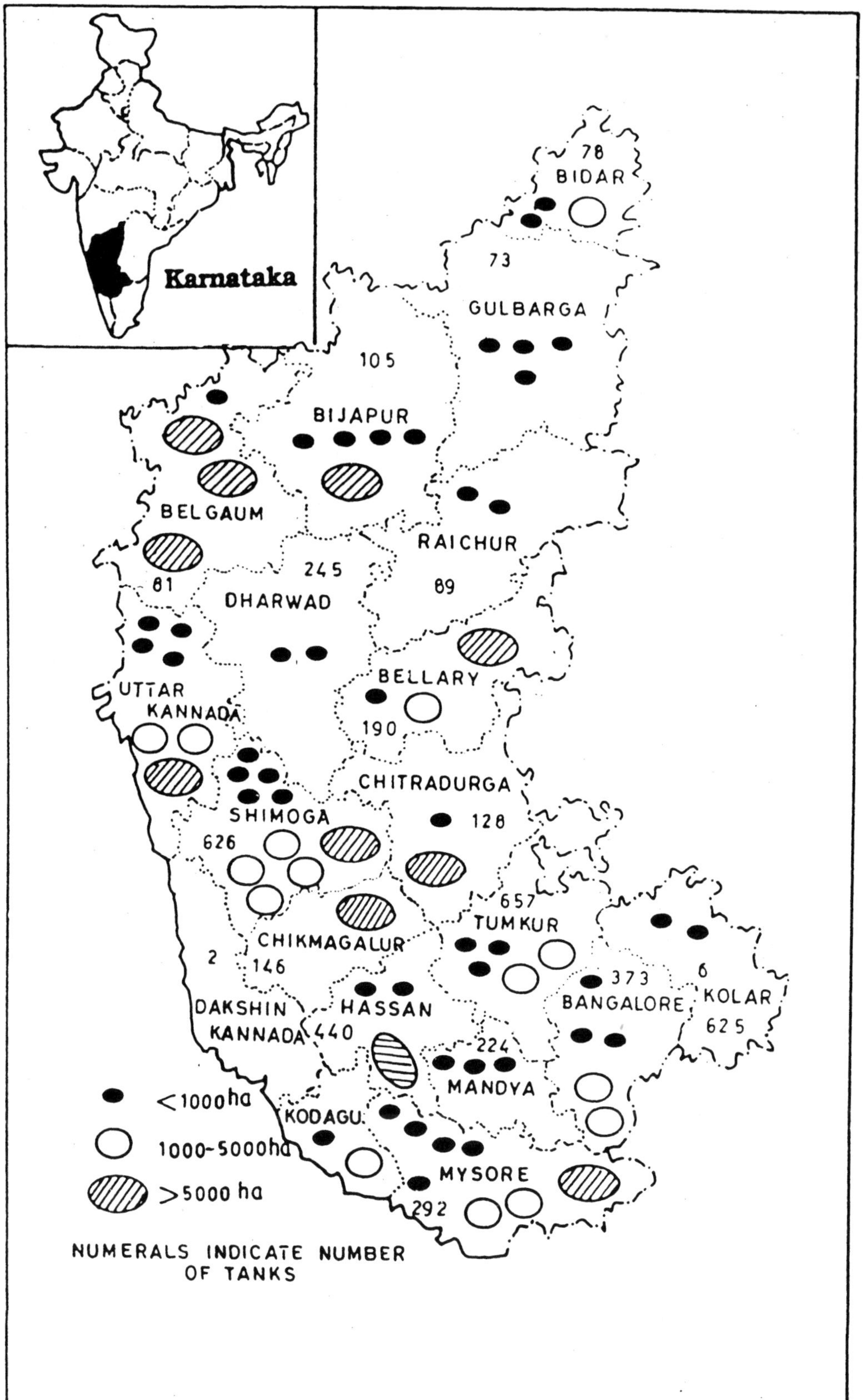

Figure 4.2. Distribution of reservoirs and tanks in Karnataka

Large reservoirs constitute 80% of the total area, followed by the medium (13%) and small (7%) ones. However, when the 4 605 minor irrigation reservoirs, designated as *tanks* are included in the small category, their share goes up to 51% (Table 4.2).

Table 4.2. Reservoir fisheries resources of Karnataka

Category	Area (ha)	Number of units
Large reservoirs		
>10 000 ha	*141 772*	*7*
5 000-10 000 ha	*37 784*	*5*
Total	**179 556**	**12**
Medium reservoirs		
1 000 - 5 000 ha	**29 078**	**16**
Small reservoirs		
500-1 000 ha	*7029*	*10*
< 500 ha	*8224*	*36*
Total	**15 253**	**46**
Tanks	**213 404**	**4 605**
TOTAL RESERVOIRS	**223 887**	**74**
GRAND TOTAL (Reservoirs and tanks)	**437 291**	**4679**

The largest reservoir Linganamakki is situated in Shimoga, a district having maximum number of reservoirs in the State. Bellary, Belgaum, Mandya, Bijapur, Uttar Kannada, Chitradurga, Hassan and Mysore also have large reservoirs (Table 4.3). Small reservoirs are mostly irrigation impoundments, distributed in all the districts, the richest being Kolar, with 625 tanks and 2 reservoirs. Shimoga (662 units; 13 762 ha) and Bellary (193 and 46 022 ha) are also rich in man- made lakes (Table 4.4). The list of reservoirs in Karnataka is given in Table 4.5.

Karnataka's total water area under man-made impoundments covering an area of 437 291 ha is undoubtedly one of the largest in the country, holding tremendous potential for fisheries development. Yet, very little scientific studies have been made on the reservoirs of the State. Of the seventy-four reservoirs, only the Tungabhadra has been studied on a long-term basis. The Central Inland Fisheries Research Institute (CIFRI) has investigated the lake from 1958 to 1966, covering a wide range of limnological parameters. Some information is also available on half a dozen other reservoirs in the State. The tanks have been studied, in detail, by the CIFRI in the 1960s. The information available on limnology and fisheries of reservoirs and tanks of Karnataka is summarised below.

Table 4.3. Reservoir area in Karnataka by category (in ha)

District	<1000 ha	1000-5000	>5000	Total
Bangalore (Rural)	(2) 877	-	-	877 (2)
Bangalore (Urban)	(2) 535	(2) 2 922		3 457 (4)
Belgaum	(1) 360	-	(3) 28 564	28 924 (4)
Bellary	(1) 280	(1) 1 209	(1) 37 814	39 303 (3)
Bidar	(2) 842	(1) 2 000		2 842 (3)
Bijapur	(4) 1 338	-	(1) 13 048	14 386 (5)
Chickmagalur		(1) 3 890	(1) 10 870	14 760 (2)
Chitradurga	(1) 780	-	(1) 8 640	9 420 (2)
Dharward	(2) 620	-		620 (2)
Gulbarga	(4) 1 852	-		1 852 (4)
Hassam	(2) 911	-	(1) 8 000	8 911 (3)
Kodagu	(1) 64	(1) 1 886		1 950 (2)
Kolar	(2) 462	-		462 (2)
Mandya	(3) 600	-	(1) 13 200	13 800 (4)
Mysore	(5) 853	(2) 2 494	(1) 6 020	9 366 (8)
Raichur	(2) 930	-		930 (2)
Shimoga	(5) 2 067	(4) 7 476	(1) 40 500	50 043 (10)
Tumkur	(3) 760	(2) 2 666	-	3 426 (5)
Uttar Kannada	(4) 1 122	(2) 4 536	(1) 12 900	18 558 (7)
Total	(46) 15 253	(16) 29 078	(12) 179 556	223 887 (74)

(Figures within paranthesis denote number of units)

Table 4.4. Distribution of reservoirs and tanks by districts

	Tanks		Reservoirs		Total	
	No.	Area (ha)	No.	Area (ha)	No.	Area (ha)
Bangalore (Rural)	373	14 579	2	877	375	15 456
Bangalore (Urban)	183	4 097	4	3 457	187	7 554
Belgaum	81	1 914	4	28 924	85	30 838
Bellary	190	6 719	3	39 303	193	46 022
Bider	78	2 508	3	2 842	81	5 350
Bijapur	105	5 964	5	14 386	110	20 350
Chickmagalur	146	5 580	2	14 760	148	20 340
Chitradurga	128	7 680	2	9 420	130	17 100
Dakshin Kannada	2	24	-	-	2	24
Dharward	245	15 181	2	620	247	15 801
Gulbarga	73	2 288	4	1 852	77	4 140
Hassam	440	19 042	3	8 911	443	27 953
Kodagu	-	-	2	1 950	2	1 950
Kolar	625	77 200	2	462	627	77 662
Mandya	224	9 778	4	13 800	228	23 578
Mysore	292	13 992	8	9 366	300	23 358
Raichur	89	4 219	2	930	91	5 149
Shimoga	626	11 287	10	50 043	636	61 330
Tumkur	657	10 336	5	3 426	662	13 762
Uttar Kannada	48	1 016	7	18 558	55	19 574
Total	4 605	213 404	74	223 887	4 679	437 291

Table 4.5. Reservoirs and their area by districts

District	Name of the reservoir	Waterspread area (ha)
Bangalore (Rural)	Kanva	440
	Byramangala	437
Bangalore (Urban)	Thippagondanahally	1 762
	Hessaragatta	1 160
	Manchanabele	365
	Iggalore	170
Belgaum	Rakasakoppa	360
	Gattaprabha (Hidkal)	7 800
	Malaprabha	13 440
	Dhupdal	7 324
Bellary	Tungabhadra	37 814
	Hegaribommanahalli	1 209
	Narihalli (Sandur)	280
Bidar	Mulamari (Upper Mulamari)	277
	Karanja	2 000
	Chulkinala	565
Bijapur	Narayanpura	13 048
	Ramanahalli	704
	Aresgankar	228
	Chitwadgi	206
	Kolaskop	200
Chickmagalur	Bhadra	10 870
	Jambadahalla	3 890
Chitradurga	Vanivilas Sagar	8 640
	Gayathri	780
Dakshin Kannada	--	
Dharward	Neerasagara	480
	Chickahandigola	140
Gulbarga	Hattikoni	120
	Chadrampalli	330
	Lower Mullamari	862
	Amarja	540
Hassan	Hemavathy	8 000
	Votehole	690
	Chakenahalli	221

Cont.

Table 4.5 (contd.). Reservoirs and their area by districts

District	Name of the reservoir	Waterspread area (ha)
Kodagu	Harangi	1 886
	Chiklihole	64
Kolar	Markendeya	145
	Bethamangala	317
Mandya	Nettkal	200
	Shivasamudra	320
	Bhima	80
	Krishnarajasagar	13 200
Mysore	Kabini	6 020
	Nugu	1 413
	Gundal	96
	Taraka	1 080
	Suvarnavathi	130
	Hebbehalla	312
	Chikkole	120
	Nallur Amanikere	195
Raichur	Kanakanala	360
	Rajolibunda	570
Shimoga	Anjanapura	698
	Ambligola	235
	Shanthisagar	2 488
	Sharavathi (Linganamakki)	40 500
	Sirur	144
	Aihole	50
	Tunga	1 200
	Chakra	1 228
	Saveakkalu	940
	Varahi	2 560
Tumkur	Markonahalli	1 336
	Boranakanive	1 330
	Mangala	320
	Theetha	200
	Mallaghatta	240
Uttar Kannada	Supa (Kali)	12 900
	Bommanahalli	1 836
	Tattihalla	2 700
	Chigalli	179
	Bachanki	203
	Sanavalli	60
	Malagi or Dharma	680
	Kadra	Under construction
	Kodasalli	-do-
	Kaneri	-do-
Total		**223 887**

4.2 TUNGABHADRA RESERVOIR (Fig. 4.3)

Tungabhadra reservoir was subjected to scientific study by the Central Inland Fisheries Research Institute from 1963 to 1965. Most of the information available on the reservoir stems from David *et al.* (1969a), Govind (1963 and 1969), Krishnamoorthy (1966) and Subba Rao and Govind (1964). A recent survey conducted by the CICFRI revealed the latest trends in limnology and fish productivity of Tungabhadra reservoir (Ramakrishniah, 1994).

Tungabhadra is the largest tributary of the river Krishna, contributing an annual discharge of 14 700 million m^3 at its confluence point to the main river. The reservoir was created by erecting a dam at Mallapuram, 5 km away from Hospet in the Bellary District. At the full level of 497.7 m above MSL, the reservoir extends over 37 814 ha; the lowest and average areas being 9 194 ha and 23 504 ha respectively. It has an extensive catchment of 28 168 km^2, chiefly fed by the southwest monsoon. The annual rainfall in the upper catchment of the river is 104 cm. Minor rivers that feed Tungabhadra are dammed at many places creating small to medium sized reservoirs such as Vanivilas Sagar and Anjanapur, and several large tanks such as Shantisagar and Madag. Orogenic and minerogenic materials, derived from both the main catchment and the submerged area influence the silt and nutrient content of the reservoir. Silt deposition within the reservoir is high, which has reduced the capacity of the lake by 13.5% in its first decade of existence. The reservoir receives rich humic and other organic and inorganic nutrients from the dense forests of the upper catchment and the cultivated areas around the tail end of the reservoir. Black cotton soils of Shimoga and Dharwar districts also bring in soil washings.

The climate at the reservoir site is mainly dry (humidity 80.7% to 93.7%); the average monthly maximum and minimum air temperatures ranging from 31.0 to 39.5 °C and 13.8 °C to 22.3 °C respectively. The basin soil is rich in calcium (200 to 300 mg 100 g^{-1}) and magnesium (50 to100 mg 100 g^{-1}). The high available nitrogen in soil (25.8 to 27.8 mg 100 g^{-1}) indicates intense biological activity at the soil phase. Phosphorus level is, however, very low.

Although many individual parameters of soil and water quality, coupled with the favourable thermal and solar regimes, portray a productive ecosystem, certain hydrographic, morphometric and meteorological features retard the growth and smooth succession of the standing crop of plankton, benthos and macrophytes. The water remains warm (23.1 to 29.5 °C) and rich in dissolved salts, throughout the year as evidenced by high values of total alkalinity and specific conductivity in the range of 30 to 100 mg l^{-1} and 240 to 359 µmhos respectively. The nutrient status of the impoundment is also moderately high, with level of nitrate at 0.2 to 0.5 mg l^{-1}, phosphate traces to 0.03 mg l^{-1} and silicate 9 to 16 mg l^{-1}. Dissolved oxygen in the range of 4.8 to 11.5 mg l^{-1} (80 to 90% saturation) suggests that the oxidation of organic matter is taking place easily at a reasonable rate. But the plankton community is not stable enough to utilize the nutrients. In the absence of a clear demarcation between the trophogenic and tropholitic zones, a steady decline of oxygen towards the bottom is not reported. Similarly, there is no vertical gradient in respect of pH, total alkalinity and carbon dioxide.

The sharp level fluctuations and abrupt drawdown of water have a destabilising effect on phytoplankton community, which fail to sustain themselves and thus utilize the congenial environment for primary productivity. Another retardant is the wind action. Apart from the monsoon turbidity, the wind-induced turbulence between May and September is caused by heavy particles and colloidal silt suspensoids, which severely restrict light penetration. Frequent water level fluctuations affect the substrata of benthic and periphytic communities. It is only for a brief period between October and December that the reservoir is stable and water

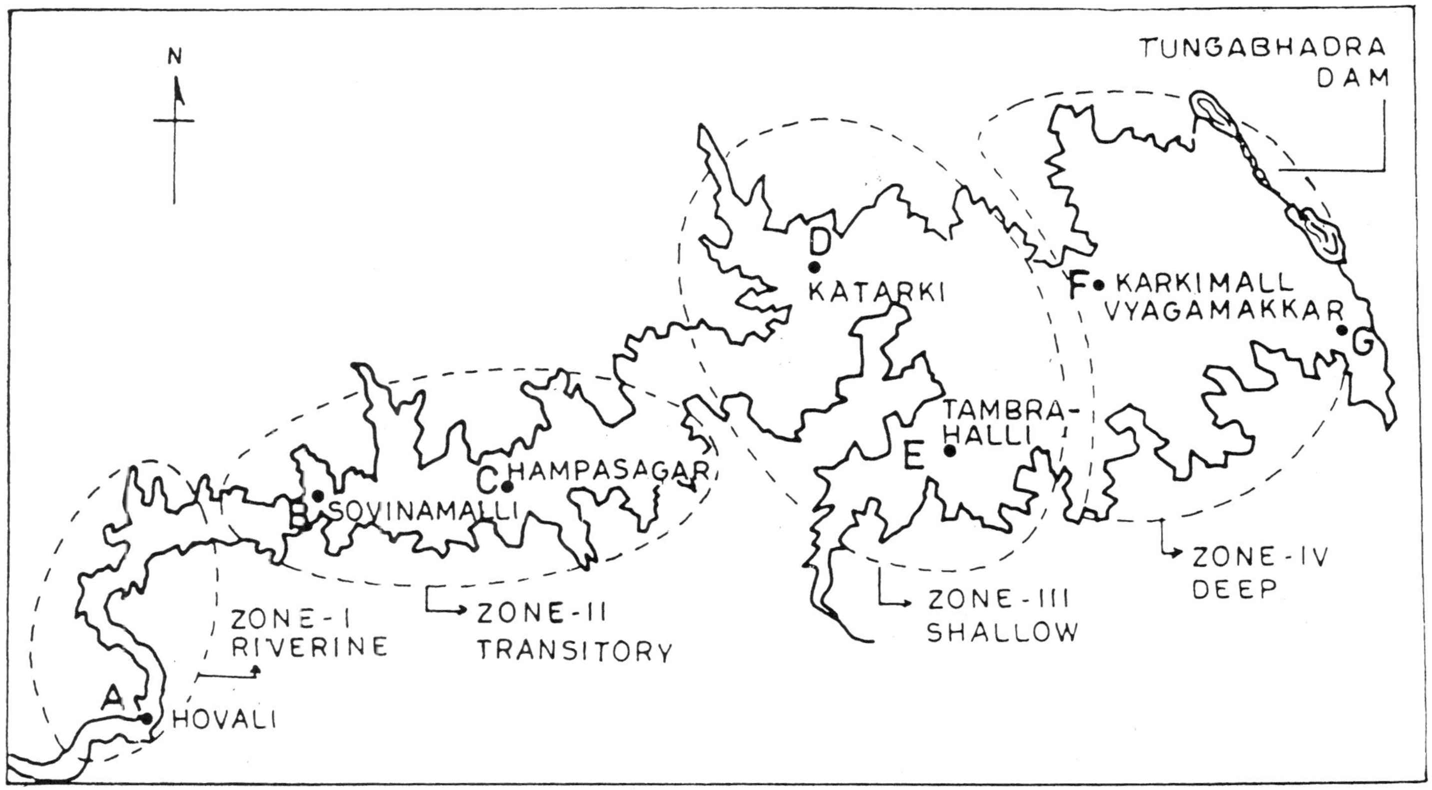

Figure 4.3. Tungabhadra reservoir, Karnataka

clear and undisturbed. The rate of primary production in the reservoir is reported to be low (Banerjee and Ray, 1979). Carbon production at various centres, at various depths, ranges within 16.2 mg C m^{-3} hr^{-1} to 189.3 mg C m^{-3} hr^{-1}. This is equivalent to annual gross production of 26.75 g C m^{-3} yr^{-1}. The total annual production of carbon is estimated at 57 000 t.

Govind (1963) demarcates three seasons, based on the plankton abundance, *viz.* the productive period (December to April, 85 to 233 units l^{-1}), retardation period (May to July, 14 to 27 units l^{-1}) and the recovery period (August to November, 49 to 74 individuals l^{-1}). Chlorophyceae and Cyanophyceae form the major components of phytoplankton, contributing 24.4% and 18.6% to the total plankton respectively. There is a rich spectrum of planktonic organisms in the reservoir. Blue-green and the green algae are represented by six genera each, while there are three genera of diatoms. *Ceratium* sp. is the lone member of Dinophyceae. Eighteen genera of zooplankton are recorded (Govind, 1969).

Benthic invertebrates are subjected to the vagaries of water level fluctuations, affecting their effective colonisation of the organically rich bottom (Krishnamoorthy, 1966). The admixture zone, where the riverine flow is neutralised by the backwash of reservoir, provides ideal habitat for molluscs and burrowing aquatic insects. The density of benthic organisms in the reservoir ranges from 147 to 421 unit l^{-1}. Insects (52%), molluscs (31%) and oligochaetes (16%) form the major constituents by number. Plankton and benthos communities did not change appreciably during the last few decades (Ramakrishniah, 1994).

Ichthyofauna of the reservoir comprises 81 species belonging to 8 orders and 14 families (David *et al.*, 1969a). Cyprinids with 37 species show maximum species diversity.

Fisheries

The fish catch of Tungabhadra reservoir during the mid- 1960s was dominated by predatory catfishes, an undesirable component of fisheries by any standards. The plankton, benthos and detritus resources were not directly utilized by any of the major commercial species. More than 75% of the total catch comprised *Aorichthys seenghala*, *Wallago attu*, *Silonia childreni* and *Pseudeutropius taakree*, all living on a long-food chain. This is clearly the result of management failure to induct fast-growing, short food chain fishes into the system, during the early years of the reservoir.

Tungabhadra reservoir, when it was impounded, had a population of *Puntius kolus*, which contributed up to a third of its total catch. Other species of *Puntius* (*P. dubius*, *P. sarana* and *P. pulchellus*), *Tor tor*, *Labeo fimbriatus*, *L. calbasu*, *L. porcellus*, *L. potail* and *L. pangusia* formed the other indigenous forms (Krishnamoorthy, 1979). Most of the native species found the changed environment after impoundment hard to cope with and started declining, the main reasons being destruction of breeding grounds, absence of fluviatile environment, and the changed trophic structure. Their share in the total fish catch has declined drastically from 74.89% in 1958 to 28.91% in 1965 (Table 4.6). Unlike the case of reservoirs in Tamil Nadu, no serious attempts were made in Tungabhadra to introduce Indo-Gangetic major carps to fill the vacant niches created by the receding population of *Puntius* and *Labeo* species. As a result, the carp minnows and minor weed fishes took advantage of the new spurt in plankton and benthic communities and these fishes, in turn, provided good forage to predatory catfishes.

Table 4.6 Changes in species composition of fish catches in Tungabhadra reservoir during 1958-1965

Year	Percentage				Total fish catch (t)
	Indo-Gangetic carps	Catfishes	Indigenous carps	Miscella-neous	
1958	0.18	22.15	74.89	2.78	15
1959	0.21	36.05	59.91	3.71	11
1960	0.23	47.00	48.93	3.88	29
1961	0.53	32.97	63.15	3.12	24
1962	0.38	33.14	64.47	1.95	24
1963	1.07	36.11	57.45	5.39	68
1964	0.18	67.90	31.46	1.57	133
1965	1.50	75.70	28.91	3.80	156

After Krishnamoorthy, 1979

Table 4.7. Fish production during 1980-81 to 86-87

Year	Catch	Yield	Percentage	
	(t)	*(kg ha^{-1})*	*Alivi*	*Other gear*
1980-81	3 529	93	92.17	7.83
1981-82	4 200	111	89.76	10.24
1982-83	3 301	87	83.50	16.50
1983-84	2 490	66	92.60	7.40
1984-85	3 255	86	90.90	9.10
1985-86	2 752	73	89.82	10.18
1986-87	2 068	55	88.36	11.16

Singit et al., (1987)

Tungabhadra reservoir seems to have come out of a long trophic depression as evidenced by the increase in catch from 15 to 156 t during 1950s and '60s to 2 068 to 4 200 t during 1980s. This increase in fish production is accompanied by matching increase of fishing effort, especially in the form of shore seines (Table 4.7). According to a latest survey, the fish production during 1993 was estimated at 1 500 to 1 600 t *i. e.*, 40 to 42 kg ha^{-1}. The higher production notwithstanding, the most disconcerting fact remains that 88 to 92% of the yield emanates from the destructive gear, *alivi* (a small-meshed giant shore seine which removes small fishes of all categories in large numbers) and 90% of the catch comprises trash fishes, which are sun-dried and sold. Only 10 to 12% of the catch, either from gill nets or from *alivi* is sold fresh. Moreover, the size of the economic catfishes and carps has gradually reduced and mostly 0+ and 1+ juveniles are harvested (Singit *et al.*, 1987). Breeding of *Puntius kolus, P. sarana, P. dobsonii* and *Labeo fimbriatus*, has been reported by

Bhatnagar (1963, 1979) who spotted their spent females and spawn. However, the spawning did not seem to have contributed to recruitment, as evidenced by the progressive decline in populations of these fishes. In reservoirs, it is often observed that even after successful spawning, the spawn drifts down to the main reservoir and perish due to unfavourable deep and lentic conditions. Occupying the apex of a grazing food chain, the commercial fishes of Tungabhadra poorly convert energy from the primary producer level to the fish flesh. Banerjee and Ray (1979) estimated the production potential of the reservoir at 33 333 t, on the basis of a conversion rate of 0.40% from primary carbon to fish. The actual fish yield during the period of study was considered as 0.8% of the actual yield potential. However, the higher fish yield reported during 1980s is 9.2% of the estimated potential.

Ills of the reservoirs are many. Firstly, the sharp decline in water levels reduces the fishing season and leads to over-exploitation (Singit *et al.*, 1987). Secondly, inadequate stocking results in vacant niches which encourage forage species and thirdly, destructive fishing practices like *alivi* and other small-meshed gear offset the advantages of any possible autostocking and artificial recruitment.

Stocking

Stocking, done so far, remains inadequate and ineffective. During 1958-59, 300 000 Gangetic carps, procured from Calcutta, (obviously undersized) were released into the reservoir. This practice was continued till 1963-64, when the farm facilities became available at the dam site. Later, the seed produced in the farm (40 to 60 mm in size) was stocked. However, these stocking attempts were grossly inadequate to make any dent on the undesirable fishery spectrum. Annual stocking figures ranged from 49 000 to 410 000 during a period of 17 years from 1963-64, at a per ha stocking rate of 1 to 11 fish ha^{-1}. Survival of the stocked fishes was very poor, as most of the stocking was done in wrong places (David *et al.*, 1969a). The spots selected for stocking did not have the essential pre-requisites to provide food and shelter to the growing fish seed. The stocked fishes were left to fend for themselves against all odds. The pervasive presence of predatory catfishes frustrated all efforts to build up a stock of desirable carps by stocking. Significantly, *Mystus punctatus*, the largest growing species of the genus commonly grows to a size of 25 to 35 kg in the reservoir and largest specimen recorded weighed an incredible 120 kg.

However, after the introduction of pen rearing in 1983, the annual stocking rate has been raised significantly touching more than 100 fingerlings per hectare in recent years. The stocking material consisted of rohu and mrigal of size 30-70 mm. In 1991 about 2.6 lakhs of catla (12 fingerlings ha^{-1}) have been stocked (Ramakrishniah, 1994). However, the impact of increased stocking of major carps to the catch has not been studied. As per the officials of Board fisheries the stocked carps of size 250 to 300 mm are exploited in large numbers during April-June when water levels go down. The size of catla in the catch during 1994 ranged 720 to 810 mm (7.0-9.0 kg). Stray specimens of over 900 mm (14 kg) are also observed. In contrast, *L. rohita* records three distinct groups viz., 300-350 mm, 500-575 mm and >800 mm. The 500-575 mm group forms the major segment of the catch.

The occurrence of major carps in considerable quantity reflects the success of stocking of these species. Catla is reported to have been stocked only during 1991 and the present catches could be traced to this stock. The size group 720-810 mm (7-9 kg) could be 2+ age indicating good growth of the species. This points towards the potential role of stocking catla in yield optimisation from the reservoir. The present fishery of *L. rohita* reflects the stocks introduced during different years. Though 500-575 mm group (2-3 kg) is dominant, specimens weighing 9 to 12 kg occur sporadically indicating the impressive size the fish attains in the reservoir (Ramakrishniah, 1994).

Craft and Gear

Coracle, the saucer-shaped country craft, made of split bamboo and covered with hide, is the most commonly used fishing craft. A normal unit has a diameter of 1.5 to 2 m. Fifty to sixty coracles are in operation in the reservoir at a given time; more than 70% of the licensed fishermen own one. Apart from the coracle, improvised rafts, made of empty cans, barrels and logs, are also in vogue, usually fabricated by fishermen themselves.

Surface gill nets (*Rangoon* nets), bottom set gill nets (*udu valai*), large shore seine (*alivi*), small shore drag net (*kunti valai*), cast nets, and hook and line are the commonly used fishing tackle. The gill nets, used throughout the year, catch all kinds of fish ranging from large catfishes to large and medium carps, depending on the mesh size. Small-sized gill nets (*bedisi valai*) bring in a host of small fishes which are normally processed by drying.

Alivi nets of Tungabhadra

The giant shore seine, *alivi* of Tungabhadra is characteristic to Tungabhadra and extremely rare in other reservoirs. This controversial fishing gear merits a separate discussion. Singit (1987) gives a graphic description of this net. Since the creation of reservoir in 1953, the fishermen of Tungabhadra have been using *konti valai, Rangoon nets* and other gill nets of various dimensions. Made of cotton, the above nets had a short life and the fish catch was too little for the fishermen to make both ends meet. As most of them were migrant fishermen from Andhra Pradesh, they brought in the *alivi* from their native place where it was known as *pedda alivi*. This experiment was not only a success, it gave them a windfall. Gradually, many fishermen took to *alivi* and soon most of them were operating more shore seines than gill nets. By 1970, there were 50 *alivis* in the reservoir.

The most disturbing feature of the gear is its capacity to remove indiscriminately fish of all hues in large numbers. Fishery scientists and conservationists have always viewed the net with concern. All attempts to ban the net have failed due to the passionate appeals made by the fishermen community and the lack of any enforcement machinery at the disposal of authorities. After considering all aspects of the subject, the Tungabhadra Board allowed *alivi* with a stipulation to release juveniles of desirable species back to the reservoir. This did not work and the net was finally banned by the Board in1975. The ban, however, was ineffective as the *alivis* were continued to be operated in a clandestine manner in the night, especially in the inaccessible areas of the reservoir. The ban was later withdrawn due to pressure from the fishermen community.

Today, licensed *alivis* are in operation contributing 88 to 92% of the total fish catch from the reservoir. *Alivi* brings in juveniles of economic carps, *Labeo calbasu, L. fimbriatus, Puntius kolus* and *P. pulchellus* (9%), important economic catfishes (11.5%), *Mastacembelus armatus, Channa* spp. and *Notopterus notopterus* (4.5%). Bulk of the catch (75%) comprises uneconomic weed and catfishes, *viz., Chela atpar, Oxygaster clupeoides, O. phulo, Barilius bendelensis, B. barila, B. barna, Danio aequipinnatus, Esomus dandricus, Rasbora daniconius, Amblypharyngedon mola, Puntius ticto, P. ambassis, P. sophore, Cirrhinus* spp., *Garra* spp., *Osteobrama vigorsii, Mystus cavasius, Pseudeutropius taakree, Ambassis ranga, A. nama* and *Glossogobius giuris*. All the above species are invariably sun-dried.

Alivi nets are operated on a cooperative basis with sixteen to eighteen people involved in operation. Four to six hauls are made in a day, each lasting 2 to 2.5 hours. Well-to-do fishermen pool their resources and fabricate the net which normally costs c. Rs. 30 000. The dry fish merchants often lend money for the venture. *Alivi* provides livelihood to at least 1 500 people who are involved in fishing and other ancilliary activities. Another 500 live on the dry fish trade. Other forms of fishing being grossly unremunerative, fishermen take to *alivi* on economic compulsions. This aspect needs consideration while banning the shore seining. Despite its drawbacks, *alivi* keeps the populations of predatory and weed fishes in check and, if operated with care, it can become an instrument to manoeuvre the fish stock. However, this cannot be achieved without the whole-hearted cooperation of the fishermen community. An awareness campaign is necessary to educate the fishermen about the need to conserve the fishes and increase fish production.

Pen culture

Rearing fry to fingerlings in the pen enclosures erected on the periphery of the reservoirs is an ideal way of raising stocking material. That the pen culture of fingerlings will cut down the production cost and obviate the necessity of cost-intensive nursery farms has long been known. Yet, Tungabhadra is the only reservoir in the country to try this as a management option. This practice has been followed in the reservoir for the last 11 years in a row. During 1992-93, pens were erected in Ladakanabhavi, 25 km from the dam site. There were 21 pens, covering a total enclosed area of 3.30 ha. The pen site was situated at 496 m above MSL and the erection job was completed during July, when the site was still exposed. Later, as the water level increased, the pen got inundated.

The pen area was pre-treated with organic manure to ensure a rich growth of plankton after the filling. A total of 15.0 million spawn were stocked in the pens which comprised 6.75 million *Labeo rohita* and 8.25 million *Cirrhinus mrigala*.. After three months of rearing, 2.41 million fingerlings were collected and stocked in the reservoir. This included 1.085 million rohu and 1.325 million mrigal, worth Rs.495 875. The present rate of survival (15 to 16%) and the size at harvest (40 to 70 mm) are rather low. Efforts are on to improve the survival rate and the growth of fingerlings. The break-up of cost involved during the pen culture operation of 1993 is as follows: recurring cost - Rs.30 100, non-recurring cost -of Rs. 16 180 and the cost of spawn- Rs.253 650. Thus, the net profit was estimated at Rs.165 945 for the whole operation. This is a commendable performance by the Tungabhadra board worth emulation in other reservoirs of the country.

4.3 MARKONAHALLI

Markonahalli is an old irrigation reservoir, created on the river Shimsha in the year 1939 to irrigate 6 070 ha of land in Tumkur district. The dam comprises a 139 m masonry structure and a pair of earth dams on either side extending to 1 470 m. The reservoir has a catchment area of 4 103 km^2 and a capacity of 68 million m^3 at the FRL of 731.57 m above the MSL. The catchment is, however, intercepted at several places to create at least 647 irrigation tanks, depriving the reservoir of its water and nutrient source. The reservoir area at FRL is 1 336 ha and at lowest storage level, it is reduced to 128 ha.

Morphometric features of the reservoir are very favourable to productivity. This includes a large catchment for the size of the reservoir and the rich allochthonous nutrient inputs derived from the catchment. A low mean depth, fairly good storage ratio (2.5), high temperature regime and the age of the reservoir are favourably disposed for organic productivity (Table 4.8).

High levels of methyl orange alkalinity and specific conductivity indicate ideal conditions for high photosynthetic rate. However, the phosphate and nitrate in water are low.

Table 4.8. Water quality of Markonahalli

Parameter	Range
Temperature (^{o}C)	22.0-26.2
Transparency (cm)	60-94
pH	7.2-8.4
Dissolved oxygen (mg l^{-1})	6.4-7.0
Methyl orange alkalinity (mg l^{-1})	104-140
Specific conductivity (μmhos)	167-204
Nitrate (mg l^{-1})	Tr.-0.028
Phosphate (mg l^{-1})	-tr.-
Silicate (mg l^{-1})	8.0-11.0

(CICFRI Barrackpore)

Rate of carbon fixation by phytoplankton is moderately high (gross primary productivity 125 mg C m^{-3} hr^{-1}), although the standing crop of net plankton is rather modest. The total count of plankton ranges from 7 to 35 units l^{-1}. *Ceratium hirudinella, Nitzschia, Navicula* and *Pinnularia* are the common phytoplanktonic forms generally encountered in the reservoir. Thirty-one genera of phytoplankton, belonging to Cyanophyceae, Chlorophyceae, Bacillariophyceae and Dinophyceae are recorded from the reservoir, though *Ceratium hirudinella* alone is present in appreciable quantity throughout the year. Zooplankton is rich in diversity as well as abundance. Probably, they live mainly on nannoplankton. The high rate of photosynthesis in the absence of rich phytoplankton suggests the possibility of nannoplankton playing the main role in carbon synthesis.

Benthic invertebrates of the reservoir comprise mayfly nymphs, damselfly nymphs, dragonfly nymphs and a rich molluscan population mainly represented by *Viviparus bengalensis, C. peninsularis, Lamellidens marginalis*. The common forms encountered in samples are *L. marginalis, L. corianus, V. bengalensis, Melania striatella tuberculata, Melania (Blotia) scabra, Bithynia stenothyroides, Lymnaea luteola, L. acuminata, Chironomus* spp. *Tubifex* spp. and *Ephemerella* spp. The thick colonisation of the bottom by the benthic macro-invertebrates indicates an organically rich bottom. The rich macrophyte community of the reservoir comprises *Hydrilla verticillata, Vallisneria appendicularis* and *Potamogeton* spp., among which *Vallisneria appendicularis* is the most dominant (4.4 kg m^{-2}).

Fisheries

Twenty-seven species of fishes have been recorded from the reservoir, among which 13 contribute to the commercial catch. *Puntius* has the maximum species diversity with five species. Even after 55 years of its existence, the reservoir sustains populations of indigenous fish species such as *Puntius sarana, P. dorsalis, Cirrhinus reba,* and *Labeo calbasu,* along with the transplanted carps. The fishery management should address the problems in enhancing the productivity of endemic fishes and the stocking should be limited to the extent of filling the vacant niches.

The stocking done so far, is arbitrary. In 1990, 91 000 fingerlings of rohu and catla were stocked in the reservoir (124 ha^{-1}). As a direct impact of this stocking, coupled with the increase in fishing effort, the yield has increased to 63 kg ha^{-1} in 1992-93 from the earlier level of 5.5 kg ha^{-1}. The catch of 46 t comprised catla (36.86%), rohu (23.74%), mrigal (1.6%), common carp (2.27%) *Wallago attu* (2.44%) miscellaneous fishes such as featherbacks, murrels etc. (32.7%) (Table 4.9).

Catch per unit effort was uniformly good, varying from 3.34 to 8.88 kg unit^{-1} (of gill net) from April 1992 to March 1993. Performance of introduced carps in the reservoir is good. *Catla catla* appears in the catch at a mean length of 378 mm (15 months old) in the range of 290 to 670 mm, the modes being at 390 mm 490 mm and 590 mm. *Labeo rohita* appears at a mean length of 272 mm in April (16 months) and at the end of 2 years attains a mean length of 474 mm. The length range of rohu in catches is 200 to 800 mm, with modes at 390, 490 and 590 mm.

Table 4.9. Fish catch (kg) of Markonahalli reservoir

	1990-91	1991-92	1992-93
C. catla	----	1 030 (19.8)	10 736 (23.34)
L. rohita	615 (15.1)	1 166 (22.4)	16 955 (36.86)
C. mrigala	1 190 (29.2)	69 (1.3)	735 (1.60)
L. calbasu	473 (11.6)	223 (4.3)	323 (0.70)
C. carpio	----	573 (11.0)	1 046 (2.27)
W. attu	265 (6.5)	272 (5.2)	1 124 (2.44)
Misc.	1 532 (37.6)	1 868 (36.0)	15 085 (32.79)
Total catch (kg)	4 074	5,201	46 004
Yield (kg ha^{-1})	5.56	7.01	62.77

(Figures in parenthesis are percentages)

(CICFRI Barrackpore)

During 1991-92, 600 000 fingerlings of catla (74%) and rohu (26%) have been stocked, which is equivalent to 922 fingerlings ha^{-1}. This was followed by putting in 450 000 more fingerlings of catla and rohu during 1992-93. Stocking during the last two years has a heavy bias in favour of catla and rohu at the cost of detritivorous species and therefore mrigal and *Cyprinus carpio* need stocking support.

The reservoir does not offer scope for the breeding of Indo-Gangetic major carps as the inflow from the southwest monsoon is negligible. Therefore, these species are to be managed on the basis of an annual stocking and harvesting schedule. Despite favourable conditions for its breeding, *Cyprinus carpio* failed to establish itself in the reservoir, probably due to predator pressure from *Gambusia* (predates on the larval stages), *Wallago attu* and *Channa marulius*.

Macrophytes remain as an unshared niche in the ecosystem which needs to be filled by stocking herbivorous species of fishes. The peninsular carp, *Puntius pulchellus*, an endemic species which can utilise the macrophytes, is an ideal candidate for stocking in the reservoir. The Central Inland Capture Fisheries Research Institute has suggested the following management measures for control of predator population and fish yield optimisation:

1. Introducing long line fishing, targeted against *Channa marulius* and *Wallago attu*,

2. extensive drag netting operations under Departmental supervision to remove weed fishes like *Gambusia*, using fine-meshed (3 mm bar) drag nets; fingerlings of the desirable species, collected in the nets, must be returned to the reservoir,

3. imposing restrictions on the use of small-meshed gill nets (<50 mm bar), encouraging fishermen to increase the mesh size to catch larger fishes, increase the total number of units only after careful monitoring of the catch per unit effort, and

4. initiation of programmes to train and educate the fishermen about the norms of conservation and sustainable use of resources.

4.4 HEMAVATHY (Fig. 4.4)

Hemavathy is an irrigation reservoir created in 1981, on the river Hemavathy at Gorur, Hassan district. The reservoir with an area of about 8 000 ha is situated at 12° 4' N and 76° 3' E at an elevation of 840 m above MSL. The river Hemavathy originates from the Western Ghats at Bellalarayanadurga in Chikmagalur district and joins the river Cauvery at Akkihebbal in Mandya district. River Yagachi and Aigoor Halla are the two main tributaries of the Hemavathy (Table 4.10).

Table 4.10. Salient features of the Hemavathy reservoir

Inflowing river	Hemavathy
Location	Gorur, Hassan district 12° 4 ' N 76° 3' E
Area at FRL	8 000 ha
Catchment	2 810 km^2
Gross storage capacity	1 050.63 million m^3
Live storage capacity	926.83 million m^3
Total length of dam	4 362 m
Height above river bed level	44.50 m
Crest level	881.482 m
Full reservoir level	854 m above MSL
Maximum water level	890.63 m
Minimum drawdown level	872.34 m
Maximum outflow	283 m^3 sec^{-1}
Minimum outflow	8 m^3 sec^{-1}
Maximum inflow	382 m^3 sec^{-1}
Minimum inflow	0.14 m^3 sec^{-1}
Volume development	0.754
Shoreline development	1.60

(After Devaraj et al.,1987)

Devaraj *et al.* (1987) made a comprehensive study on the reservoir, the salient findings of which are presented in the following account.

The climate at the reservoir site is mainly dry and the mean air temperature ranges from 13.5 to 20.0 °C (minimum) to 25.5 to 34.0 °C (maximum). Monthly rainfall varies from 1.33 to 122.60 mm. The reservoir is exposed to intense sunlight, skies being cloudless from October to May.

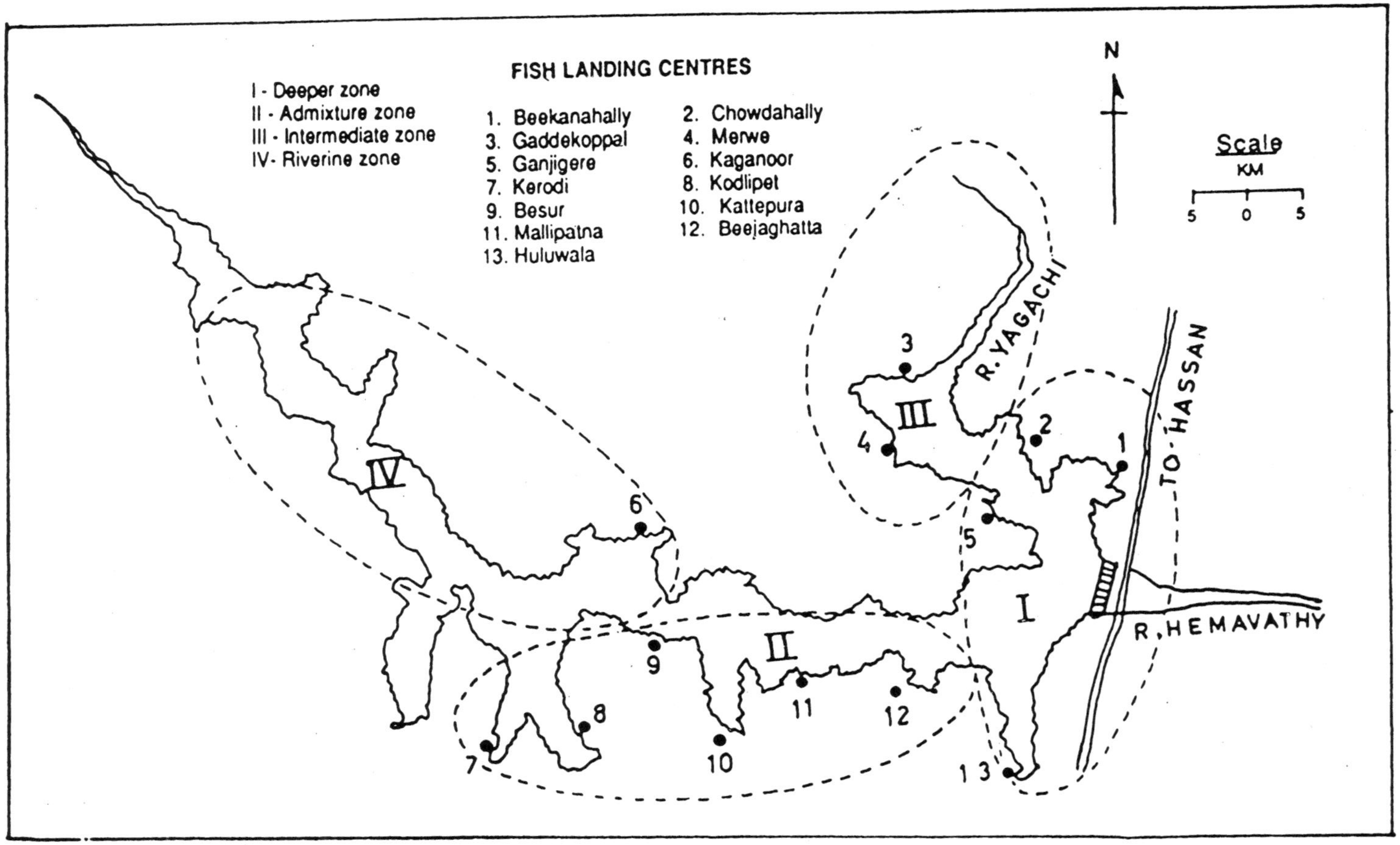

Figure 4.4. Hemavathy reservoir, Karnataka

The reservoir starts filling by late June or early July, due to early rains in the Hassan district. The water level reaches its peak of 37.19 m by the end of August or early September. During September and October or sometimes even in November, it becomes somewhat stable. From the month of November, the level declines steadily.

Soil and water quality

Basin soil is acidic (pH : 6.3 to 6.9) and generally coarse and sandy in texture. A mixture of silt, sand and clay, is however, found in the intermediate zone of the reservoir. The mean water temperature (25.85 ^{0}C) and transparency are high in the lentic zone, compared to the middle and riverine zones. Difference in temperature between the surface and bottom layers of water is not very significant. The level of dissolved oxygen at the surface is fairly high, with mean values for differenrt sectors ranging within 5.78 and 6.37 mg l^{-1}, and there is no oxygen deficit at the bottom. Incoming floodwaters chiefly determine the clarity of water . The Secchi disc transparency varies between 23.5 and 228 cm, the riverine zone being more turbid, compared to the rest of the reservoir. Hemavathy can be designated as a soft water reservoir, based on the total alkalinity values which vary within a range of 23 and 49 mg l^{-1}. The values are reported to be increasing progressively over the years, suggesting a possible improvement in terms of alkalinity in future. The pH is within the range of 7.05 to 8.00 (Table 4.11). There is decline in pH at the surface during summer months, leading to a weak vertical stratification.

Table 4.11. Physico-chemical characteristics of water in Hemavathy reservoir (surface) during 1984-1986

	DEEPER ZONE		MIDDLE ZONE		RIVERINE ZONE	
	Range	Average	Range	Average	Range	Average
Temperature (^{0}C)	23.25-29.75	25.85	22.75-29.75	25.69	22.40-29.50	24.96
Transparency (cm)	69.25-179.75	129.14	55.0-115.50	86.08	32.5-107.50	64.69
pH	7.2-8.2	7.65	7.2-8.0	7.52	7.05-7.80	7.39
Dissolved oxygen (mg l^{-1})	5.0-8.2	6.37	5.0-7.5	6.09	4.5-7.85	5.78
Total alkalinity (mg l^{-1})	24.5-43.0	32.72	26.0-39.00	33.17	23.0-39.00	30.79
Free CO_2 (mg l^{-1})	0.3-4.6	1.11	0.04-2.2	0.85	0.35-3.55	1.20
Inorganic PO_4 (mg l^{-1})	0.02-0.05	0.03	0.027-0.045	0.03	0.03-.04	0.03
Silicate (mg l^{-1})	0.567-0.71	0.643	0.757-0.81	0.78	0.62-0.75	0.68

After Devaraj et al., 1987

Average values of free carbon dioxide in water are below 3.0 mg l^{-1} and its concentration increases slightly towards the bottom. The young reservoir is rather poor in dissolved nutrients, the phosphate values mostly being in traces. On a few occasions, sudden rise in the phosphate level is reported, with values going up to 0.28 mg l^{-1}. Silicate is recorded at the rate of 0.51 to 6.72 mg l^{-1}, with an average of 1.59 mg l^{-1}.

Biotic communities

Plankton: The density and composition of the plankton do not reflect any eutrophic tendencies. The modest plankton community (68 to 112 organisms l^{-1}) comprises phytoplankton to the extent of 48.85 %. Characteristic feature of the phytoplankton is the domination of Chlorophyceae, followed by Cyanophyceae and diatoms (Table 4.12). Enrichment of the plankton community is influenced by the nutrients brought in by the rain washings (Devaraj *et al.*, 1987). Maximum plankton is during the post- monsoon period, extending from late August to February and a significant increase in phytoplankton is noticeable during April to August.

Littoral and bottom biota: Nutrient status of the soil is high, compared to the water and this is amply reflected in a rich community of littoral organisms comprising crustaceans, insects, molluscs, small fishes, tadpoles and oligochaetes, at a density ranging from 64 to 966 organisms m^{-2}. Insects, the major component of this community, are represented by chironomid larvae, waterbugs and the nymphs of dragonfly and damselfly. *Viviparus* and *Gyraulus* are the common genera of gastropods found among the littoral community, while the bivalves encountered are *Corbicula* and *Unio.* Insects, molluscs and worms constitute the benthic invertebrate community in that order of abundance, their density varying between 2 042 and 4 174 organisms m^{-2}. The riverine zone has a highpoper ulation density of benthic organisms than the rest of the reservoir.

Macrovegetation: The aquatic macrophytes of the reservoir comprise mainly submerged and rooted vegetation belonging to seven species *viz., Ceratophyllum demersum, Chara* spp., *Cyperus corymbosus, Hydrilla verticillata, Nymphaea odorata, Potamogeton nodosus* and *Utricularia vulgaris.*

Table 4.12. Composition of various groups of phytoplankton in Hemavathy reservoir

ORGANISMS/GROUPS	PERCENTAGE COMPOSITION			
	Deep zone	*Admixture zone*	*Intermediate zone*	*Riverine zone*
Microcystis	23.20	18.44	10.16	19.16
Anabaena	0.51	1.43	-	0.83
Oscillatoria	7.52	9.72	8.31	7.50
Nostoc	4.41	8.20	7.31	7.40
Cyancphyceae	**35.64**	**37.79**	**25.78**	**34.89**
Diatoma	13.20	17.68	21.00	21.20
Asterionella	4.54	4.90	2.60	5.00
Bacillariophyceae	**17.74**	**22.58**	**23.60**	**26.20**
Spirogyra	12.44	7.61	7.64	11.48
Pediastrum	11.23	12.52	11.93	10.18
Zygnema	18.34	14.46	26.30	12.96
Ulothrix	3.95	4.99	4.70	4.25
Chlorophyceae	**45.96**	**39.58**	**50.57**	**38.87**

Ichthyofauna and fisheries

Fish fauna of Hemavathy comprising 43 species includes a number of native fishes deserving attention from conservation point of view. This includes 11 species of *Puntius* and 3 species of *Labeo* and two species of *Ompok*. All the known predators *i. e. Wallago attu*, four species of *Channa*, 2 species of *Aorichthys* and 1 species of *Mystus* are present, apart from *Mastacembelus armatus* and the air breathing fish, *Heteropneustes fossilis*. Total estimated fish catch for 1984-85 is 115 t, which is equivalent to 12. 5 kg ha^{-1}, calculated on the basis of area at FRL. The catch and yield increased to 146 t and 14. 8 kg ha^{-1} during 1985-86. A disturbing feature of the catch structure is the disproportionately high numbers of catfishes and minnows. Eels, catfishes and murrels together constituted nearly 22% of the total catch during 1984-85 with another 43% contribution from minnows and other low value species. Thus, the share of Indian and exotic economic carps in the fish landing was only 35% (Table 4.13). The situation was more or less the same during the succeeding year, although a slight increase in the share of carps was noticed.

Table 4.13. Fish production in Hemavathy reservoir during 1984-85 and 85-86

Fishes	1984-85		1985-86	
	Catch (t)	(%)	Catch (t)	(%)
Major carps	13	*11.70*	18	*13.10*
Minor carps	12	*10.20*	19	*14.40*
Common carp	15	*13.20*	20	*15.10*
Murrels	5	*4.00*	12	*1.20*
Catfishes	17	*14.90*	23	*16.80*
Chela*	23	*19.80*	27	*19.70*
Gobies**	17	*15.20*	15	*10.80*
Eels	3	*3.00*	4	*2.80*
Misc.	9	*8.00*	8	*6.10*
Total	**114**	*100.00*	**146**	100.00

* Minnows, mainly *Salmostoma phulo* ** Small fishes, mainly *G. giuris*

Fishing in Hemavathy is yet to be organised. Barring a few migrants from the neighbouring State of Andhra Pradesh, most of the fishermen are the local farmers and farm labourers who changed their profession in favour of fishing. There are 5 main fishing camps in the reservoir namely, Kerodi, Beejaghatta, Haluvala, Beekanahally and Shettihally, each with 35 to 50 fishermen families. However, only 25 to 30 families among them have fisheries as a full time vocation. Average catch per individual fishermen on a normal day is 3 to 6 kg, which fetches them Rs. 10 to Rs.15, though during the peak fishing season the daily earnings can go up to Rs. 60.

Gill net, with mesh range of 2.5 to 25 cm, is the most popular gear in the reservoir, which are operated from coracles. Cast nets and traps are also used. The catch is sold to middlemen who cart the fish everyday by road to market. The womenfolk in the fishermen families often hawk fish in local markets and residential areas. A substantial part of the catch is sun-dried.

Fishery management

A study of the feeding habits of fish in the reservoir indicates that none of the the main fish food biotic communities are fully utilised by the major economic

Table 4.14. Food of some fish in Hemavathy reservoir

Fish	Major stomach contents
Catla catla	Copepods (56%), semi-digested organic matter (24%), diatoms (6%), mud (10%), miscellaneous items(4%)
Cirrhinus reba	Decayed plant and animal matter (29%), Cyanophyceae (15%), misc. (11%), Chlorophyceae (11%), semi-digested organic matter (13%), diatoms (12%)
Labeo calbasu	Decayed plant and animal matter (42%), diatoms (23%), sand and mud (21%)
L. fimbriatus	Semi-digested organic matter (37%), diatom (27%), sand and mud (24%) plant tissues (8.6%), miscellaneous items (4%)
Notopterus notopterus	Insects (42%), semi-digested organic matter (24%), fish remains (16%), prawns (7%), miscellaneous items (7%), plant tissues (4%)
Puntius sarana	Mud (29%), gastropods (24%), plant tissues (18%), misc. (15%), semi-digested organic matter (8%), insects (3%), ostracods (2%)
P. aurilius	Semi-digested organic matter (28%), decayed plant material (26%), Cyanophyceae (18%), copepods (13%), cladocerans (11%), miscellaneous items (3%)
Wallago attu	Fish (82%), insects (6.5%), miscellaneous items (5.4%), decayed matter (3%)
Mystus cavasius	Molluscs (36%), decayed matter (24%), insects (14%), animal matter (7%), mud (6%), crustaceans (6%), vegetable matter (2%)

fishes (Table 4.14). Although *Catla catla* feeds on copepods, given the negligible population of this fish in the reservoir, the niche can still be considered as underutilized. Presently, the carp minnows and other uneconomic species feed directly on plankton and littoral/bottom biota, and ultimately provide forage for predators. The energy loss involved in this rather long food chain is the main retardant to fish productivity. Interestingly, the detritivores, which are better converters of energy than the predators, are poorly represented in the catch. A qualitative change in the species spectrum, could be achieved through careful management.

Very little has been done to develop the fisheries of Hemavathy on scientific lines. The reservoir harbours a kaleidoscopic spectrum of the riverine fish fauna which need attention from conservation point of view. The reservoir is in its formative stage and unless effective measures are taken to induct some high value species that utilise the plankton and detritus, the present minnow-predator combination will get entrenched. Both peninsular and Gangetic species should be considered for stocking in the reservoir, after prior assessment of their potential impact on other fish.

4.5 VANIVILAS SAGAR

Vanivilas Sagar is situated on the river Vedavati in Chitradurga district, about 104 km northeast of the Babudan Hills, the source of the river. It is one of the oldest reservoirs in the State. Created in 1901, the impoundment has a water area of 8 640 ha at the full reservoir level of 621 m above MSL. The catchment area of the river is rich in iron ores, limestone and sodium salt deposits (Ray, 1969). The river, at its origin, receives rich rainfall to the tune of 375 cm a year. However, the precipitation in the local catchment is just 60 cm per year, as the reservoir is situated in the semi-arid plains.

The upper catchment of the river being dammed by numerous tanks for local irrigation, which leaves the inflow into the reservoir rather erratic. Low water renewal rate has an undesirable influence on the biotic communities. A substantial part of the reservoir is shallow and the annual level fluctuation of 0.2 to 0.6 m makes the environment ideal for rapid colonisation by aquatic macrophytes. Rich growth of *Hydrilla*, *Potamogeton*, *Aponogeton* and *Vallisneria* covers one third of the total reservoir bed, down to 2.4 m (Ray, 1969). In this way, the reservoir is rapidly changing itself into a wetland, an ecosystem akin to the floodplain lakes of Ganga and Brahmaputra, where the cut-off meanders become full of submerged vegetation. Excessive vegetation is undesirable from the fisheries point of view. The weeds remove nutrients rapidly from the system and utilise the solar radiation, leaving very little of the two inputs for the photosynthetic activities of plankton. Thus, the energy is diverted through a weed-detritus chain. The constant decay of vegetation at the bottom fouls the bottom bed, making anaerobic and toxic conditions. Moreover, they hasten the process of conversion of the reservoir to swamp and eventually dry land.

The reservoir water is uniformly warm throughout the year (22.3 to 26.3 °C) and no thermal stratification develops. With minimum disturbances from the level fluctuations, and due to the presence of submerged plants, which arrest and settle the suspended load brought in by the inflowing river, the water is clear most of the time. Although the pH (8.4 - 8.5), total alkalinity (194-228 mg l^{-1}) and hardness (108-128 mg l^{-1}) are high, the macrophytes take advantage of such environment, leaving little room for the plankton community to flourish. Consequently, the phytoplankton productivity is a negligible 0.39 to 3.25 mg C m^{-3} day^{-1}.

The bottom, having been colonised by the macrophytes, does not offer much substrate for benthic invertebrates. Gastropods and bivalves are abundant among the plants, forming an important food of fish. The ubiquitous presence of macrophytes and their constant photosynthetic activities enrich the dissolved oxygen content of water (6.56 to 9.8 mg l^{-1}). There is no decline of oxygen towards the bottom, indicating a slow pace of organic decomposition at the bottom. Lack of organic acids at the bottom is confirmed by the high pH of soil (8.0 to 8.5). This alkaline nature is characterised by high soil calcium (200 to 300 mg 100 g^{-1}). Phosphate is low both in the water and soil phase. No published account is available on the fish and fisheries of the reservoir.

4.6 SUPA RESERVOIR (Fig. 4.5)

Supa reservoir is one of the three impoundments of the Kalinadi hydro-electric project, the other two being Bommanahalli (1 836 ha) and Tattihalla (2 700 ha). The project is aimed at tapping the hydel potential of the west-flowing river, Kalinadi at the confluence of its three tributaries *viz.*, Pandri, Naginalla and Kaneri (diverted). Supa is one of 10 dams to be constructed on Kalinadi to tap its full hydel power potential of 1 298 MW. The reservoir has a total area of 12 900 ha and a catchment of 1 067 km^2, draining mainly the Western Ghats. The 161 km long Kalinadi has its catchment rich in iron and manganese ores, limestones, quartz, bauxite and silica, and the mean annual rainfall is 274 cm, largely from the southwest monsoon.

Despite draining a mineral-rich, heavily wooded catchment area, the reservoir remains oligotrophic. Birasal *et al.* (1991) points out the poor ionic composition of the lotic region of the reservoir. The total dissolved solids increase substantially from the lotic to the lentic zone, as reflected by the specific conductivity (68.8 and 99.2 μmhos respectively). It is reported that during the initial phase of the impoundment, decomposition of submerged vegetation contributed to the ionic build-up. The decomposing vegetation also increased the acidity of water.

Vertical profiles of oxygen and other chemical parameters are not available. Being a hydel reservoir with a substantial part of its water locked up in a deep basin, the bottom accumulation of H_2S is natural. Heavy accumulation of the offending gas is reported in the lentic sector of the reservoir (0.24 to 0.68 mg l^{-1}) and in the outflowing water (0.68 to 1.82 mg l^{-1}). Massive fish kills are reported from the reservoir and the dead fishes usually pass through the turbines. While metalic ions are present (Fe^{3+} 1.08 mg l^{-1} ; Mn^{2+}, 1.35 mg l^{-1}) in the water phase, nitrate is in low concentration (0.08 mg l^{-1}). Phosphate is high (0.2 mg l^{-1}), in comparison with many other South Indian reservoirs. On account of the high concentration of iron and manganese, the water of Supa is not potable. The available information on physico-chemical parameters is inadequate to draw conclusions about the productivity of the reservoir. Birasal *et al.* (1989) reported the presence of 25 species of zooplankton in the reservoir, *i.e.*, rotifers (7), copepods (9) and cladocerans (9), but gave no information on their quantitative abundance. The reservoir is oligotrophic, but this has to be further verified through biological parameters.

There is no organised fisheries in the reservoir. The Karnataka Power Corporation (KPC), which owns the reservoir, has plans to develop the fisheries of Supa and three other reservoirs in the Uttar Kannada district. In 1990, 265 000 fry/fingerlings of *Catla calta*, *Labeo rohita*, and *Cirrhinus mrigala*, were stocked, along with a few silver carp, *Hypophthalmichthys molitrix* (Rahman, 1993). During 1992-93, 724 000 seed of Indo-Gangetic carps, and milkfish *Chanos chanos* were stocked. The KPC intends to stock 1.07 million fish seed during 1993-94. They have also plans to establish mahseers for attracting tourists, to stock the non-predatory catfish *Pangasius pangasius*, peninsular herbivorous carp *Puntius pulchellus*, and the giant freshwater prawn *Macrobrachium rosenbergii*.

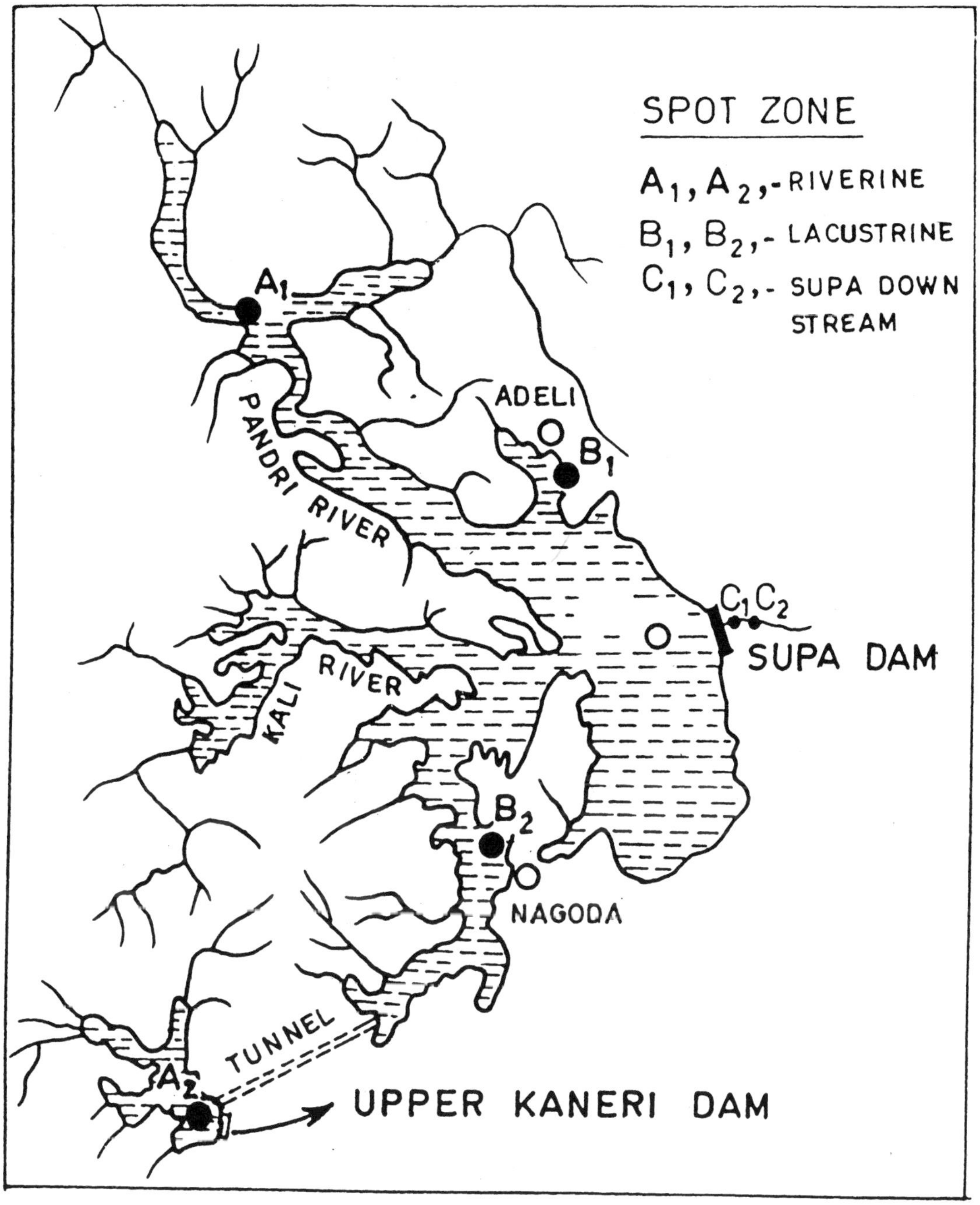

Figure 4.5. Supa reservoir, Karnataka

4.7 KABINI RESERVOIR

Kabini reservoir is situated on the river Kabini, a major tributary of the Cauvery at Sargur in Mysore district. The 2 732 m long dam was constructed in 1974 for the twin purpose of irrigation and hydel power generation. At full reservoir level, the lake has a waterspread of 6 020 ha and storage capacity of 553 million m^3. No ecological studies have so far been made on the reservoir. Neither have any records been maintained about the fish production, as the fishermen themselves dispose off the catch to the merchants. The following information on Kabini reservoir is gathered from Murthy *et al.* (1986) and Srivastava *et al.* (1985).

There are eleven important fish landing centres around the reservoir. Fishermen of Antharasanthe, K. G. Hundi, Ponapura and Gundathalu centres use gill nets while *sosalu bale*, a small drag net, is used by fishermen of Hosur, Nerale, Ramenhally, Hosamala, Jogihally, Bidarhally and Thoramante. Rod and hook, and long lines are also used for fishing. The licence for the former is issued to those who fish for pleasure. The catch being negligible, they often consume the catch themselves. Long lines often comprise 50 hooks, licence for which is owned by part-time fishermen who catch fish as an extra source of income.

Table 4.15. Fish production trends in Kabini

	1980-81	1981-82	1982-83	1983-84
Total catch (t)	**450**	**151**	**131**	**75**
Yield (kg ha^{-1})	**74**	**25**	**21**	**20**
Percentage composition				
Cirrhinus reba	70	60	30	20
Wallago attu	8	8	10	5
Tilapia	8	12	40	60
Carps and others	14	20	20	15

(After Murthy et al., 1986; Srivastava et al., 1985)

The fish production from Kabini reservoir is decreasing over the last four years (Table 4.15), total fish catch plummeting from 450 t to 75 t (74 kg ha^{-1} to 20 kg ha^{-1}). *Cirrhinus reba* which used to contribute 70% of the catch in 1980-81 has declined to 20% with a similar fall in the percentage of carps and other fishes. Tilapia has improved its position from 8% to 60%, while the predatory catfishes maintained its position of 8 to 10%. The reservoir has been stocked regularly with the fish seed of Indian and exotic carps (Table 4.16).

The stocking in the reservoir has become effective only after 1979. All the earlier efforts in this direction were nullified due to the complete draining of the reservoir in 1976 and 1978. Over the years, nearly 2.3 million Indian major carps (mainly *Labeo rohita* and *Cirrhinus mrigala*), and 2.9 million common carp were released. In 1980-81, mrigal contributed 90% of the seed stocked and during 1981-82 there were only 100 specimens of catla in the seed. The species-mix of stocking material has always been arbitrary. The stocked fish neither appeared in the catch

Table 4.16. Details of stocking in Kabini reservoir during 1980-81 to 1984-85 (x 1000)

Fish	Years				
	1980-81	1981-82	1982-83	1983-84	1984-85
Major carp	900	692	289	400	-
Common carp	1 073	457	371	825	200
Silver carp	5	-	-	-	-
Grass carp	5	-	-	-	-
Total	1 983	1 149	660	1 225	200

(Murthy *et al.*, 1986)

nor was there any improvement in the yield. On the contrary, there was a steady decline and the indigenous *C. reba* is fast giving way to tilapia which competes with the desirable carps for food. Major factors contributing to the negative trends have been identified as:

1. *Ad-hoc and arbitrary approach towards stocking*: species selection and stocking density are governed more by expediency than any ecological considerations.

2. *Lack of conservation measures:* No closed seasons are observed to protect the brooders; no attempts have been made towards mesh size control and effective monitoring of fishing effort.

3. *Draining of the reservoir:* Sometimes the reservoir is totally drained for repair of the dam, causing destruction of the brood stock and stocked fingerlings.

4. *Social problems:* Lack of credit for the migratory fishermen and lack of social support for the fishermen are the disincentives for fisheries development.

Srivastava *et al.* (1985) pointed out the need to remove the bottlenecks in the transport of fish from fishing sites to the assembly centres. It is also recognized that cooperative societies need to play a more meaningful role in market intervention and smooth flow of credit to the fishermen.

4. 8 KRISHNARAJASAGAR (Fig. 4.6)

Krishnarajasagar, a multi-purpose reservoir on the main river Cauvery below the confluence of the tributaries, Hemavati and Lakshmanathirtha was constructed during 1911 to 1932. The dam is a brainchild of the celebrated Indian engineer and visionary, Mokshakundam Visweswaraya. The reservoir was created originally for irrigating 50 585 ha of semi-arid land of Mandy district and to generate hydro-electric power at the Shivasamudram Power Station. Now, as the water requirement for power generation is being met from the Kabini river, Krishnarajasagar water is utilised for irrigation, water supply to Mysore city and for meeting the demands for the industries nearby.

At the full level of 752.23 m above MSL, the reservoir has an area of 13 200 ha (average 8 156 ha). The catchment area of 10 619 km^2 comprises rocky hills,, wooded forests and agricultural land, bringing in different types of water. The reservoir has a relatively deep basin (mean depth 30.17 m) and a long and irregular shoreline (shoreline development index: 7.25)

A characteristic feature of the lake is its rapid level fluctuation, the water level dropping by 15 m from the FRL every year during May-June. Apart from influencing the physio-chemical quality of water and soil, this has a direct effect on the fishery, by allowing intensive fishing activity during the low level phase. The mainstream Cauvery and its tributaries, Hemavathy and Lakshmanathirtha are the main source of inflow and all the catchment areas are fed by the southwest monsoon.

Water quality and plankton of the Krishnarajasagar have much in common with the low productive Western Ghat reservoirs of Kerala. Water is mostly clear, primarly because of the pick-up tanks in the upper catchment that act as silt arresters. The pH value rises during summer when water stagnates, with very little inflow from the river. Bicarbonate alkalinity ranges from 36 to 140 mg l^{-1} and CO_2 is usually present. Low values of phosphate and nitrate are recorded during the flood season.

The water is thermally stratified and a weak oxycline (oxygen deficit of 1.3 mg l^{-1}) develops, but it is unaccompanied by any increase in bicarbonates towards depth. Since CO_2 is present in dissolved form at the top layer all the time, there is no extraction of the gas from the bicarbonates for photosynthetic purpose. Thus, there is no decline in bicarbonate at the top all the and the top layer does not derive CO_2 from bicarbonates, resulting in the absence of any stratification. At the same time, not much carbonate is left at the bottom in the presence of CO_2. This is confirmed by the lesser increase in H^+ and thereby no appreciable decrease in pH values (deficit 0.3) towards the bottom.

Water quality parameters point towards a very slow rate of organic production which is further confirmed by the subdued rate of carbon fixation by phytoplankton (gross production nil to 63 mg C m^{-3} hr^{-1}) and a lower plankton density (13 to 121 units l^{-1}). There is a conspicuous absence of phytoplankton blooms. Benthic and littoral biota, on the other hand, is rich and varied, playing a vital role in the trophic events of the reservoirs. The shallow, irregular margin of the reservoir provides a conducive environment and suitable substratum for the benthic invertebrates.

A variety of insects, their nymphs and larvae form 50% of the benthic invertebrates, which give mean density of 1 342 organisms m^{-2}. A number of gastropods and bivalves, mainly *Bithynia stenothyroides*, *Viviparus bengalensis*, *Melania striatella*, *M. scabra*, *Lymnaea acuminata*, *Lamellidens marginalis*, and *Corbicula peninsularis*, insects such as *Caenis*, *Notonecta*, *Ranatra*, *Diplonychus*, *Corixa*, *Cybister* and the nymphs of dragonfly and damselfly, and oligochaetes are the

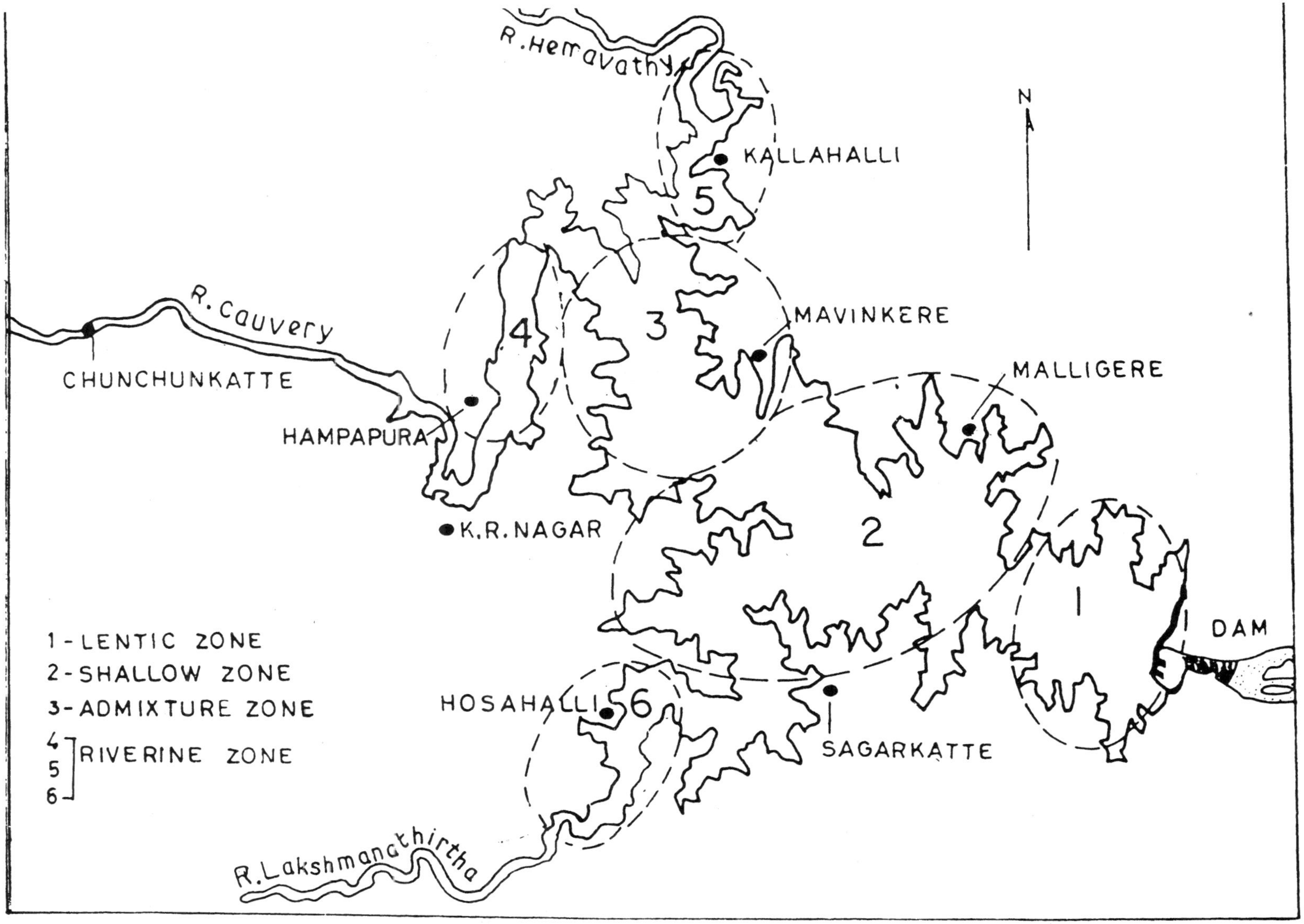

Figure 4.6. **Krishnarajasagar reservoir, Karnataka**

common members of the benthic littoral community. Food and feeding habits of 7 species of fish (*Labeo calbasu, L. rohita, Cyprinus carpio, Cirrhinus reba, Ompok bimaculatus, Glossogobius giuris* and *Oreochromis mossambicus*) were studied and all of them except *C. reba* have the bottom detritus and bentic invertebrates as the main or a major component of diet (Table 4.17).

Table 4.17. Food of some important species of fish in Krishnarajasagar

Species	Length (mm)	Food
Labeo calbasu	165-495	Mainly detritus, phyto- and zooplankton
Labeo rohita	195-419	Detritus, phyto- and zooplankton
Cyprinus carpio	---	Detritus
Cirrhinus reba	142-305	Phytoplankton, algae, and diatoms
Ompok bimaculatus	168-410	Insects and insect larvae
Glossogobius giuris	146-247	Insects, zooplankton
Oreochromis mossambicus	105-269	Diatoms, insects.

CICFRI Barrackpore

Thirty-three fishes belonging to 11 families are recorded from the reservoir. These include the mahseer, *Tor tor* and the chocolate mahseer *Acrossocheilus hexagonolepis*. The indigenous *Puntius* has the maximum diversity (6 species), followed by *Labeo* (5 species). Predators belong to catfishes of Bagridae and Siluridae and the murrels, *Channa* spp. Significantly, the minnows and weed fishes thriving on plankton are not many. The carp minnows are mainly *Rasbora daniconius, Puntius ticto, P. sophore and Chanda ranga.* Plankton-feeding minnows being fewer, the predatory catfishes and murrels do not exceed 12% of the total catch.

A perusal of fish production trends for 25 years from 1956-57 (Srivastava *et al.*, 1985) reveals that the total catch was erratic (45.83 to 189.0 t) at an average of 96.2 t. This is equivalent to a yield rate of 7.28 kg ha^{-1} (at full area). Between 1980-81 and 1987-88 the yield has been gradually increasing (Table 4.18).

Fish catch in Krishnarajasagar is related to the water level, the catch being the highest in the months of low water levels (Fig. 4.7). During 1987-88, common carp, contributing 36% to the catch, was the single largest component of the fisheries, the indigenous economic carps (*Labeo* spp. and *Puntius* spp.) representing only 21% and the transplanted Indo-Gangetic carps a negligible 1%. The predator catfishes and murrels contributed nearly 12% and the forage fish share was 30%.

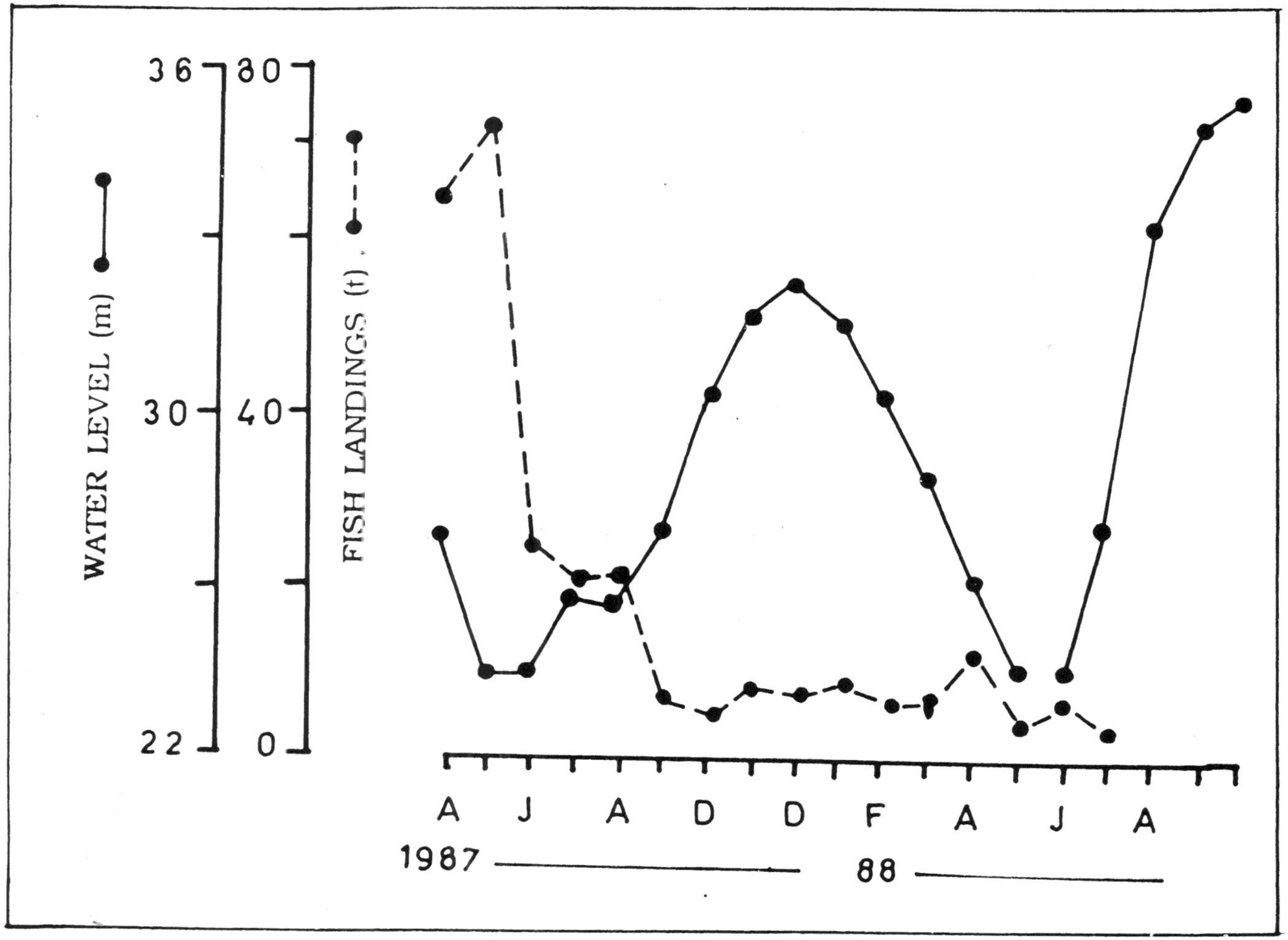

Figure 4.7. Fish landings in relation to water level in Krishnarajasagar reservoir

Table 4.18. Fish catch trends in Krishnarajasagar

Year	Total landings (t)	Yield (kg ha^{-1} $year^{-1}$)
1956 to 1980	*96.20*	*7.28*
1980-81	75.65	5.73
1981-82	57.77	4.38
1982-83	115.72	8.77
1983-84	244.40	18.52
1984-85	185.95	14.09
1985-86	170.00	12.88
1986-87	175.00	13.26
1987-88	271.00	20.53
1980-81 to 1987-88	*161.93*	*12.27*

There has been a qualitative change in the fish species compositon over the years. The resident carps of the Cauvery such as *Puntius dubius* and *P. carnaticus* were reported to be the dominant species in the late 1950s contributing 39% and 25% of the catch respectively. Obviously, these native species have suffered a setback due to the changed conditions, especially in the fish food biotic communities. As at present, most of the energy transfer is channelled through the detritus/benthic chains, it is not surprising that the common carp gets an edge over others. This fish is a prolific breeder and a competitor to *Cirrhinus* sp. for food. Failure of Indian major carps to appear in the fishery is due to the poor plankton community, inadequate stocking and competition with common carp (in case of *C. mrigala*).

Stocking

Stocking has been arbitrary both in quantitative and qualitative terms. The total number of fish seed stocked during 1956-57 to 1980-81 varied from 2 000 in the year 1964-65 to 888 000 in 1980-81. The selection of species was equally erratic. A large number of common carp was stocked during certain years, obviously because of their easy availability, and not based on any scientific reasoning. For instance, in the year 1978-79, 100 000 Indian major carps were stocked along with 352 000 common carp. Considering that the fish breeds readily in the reservoir, there is no necessity for the continued stocking of common carp. Moreover, the fish, especially if it is very big in size, has a low consumer preference, compared to *Cirrhinus mrigala* and the indigenous competitors like the other *Cirrhinus* species. Bulk of the gill nets used in the reservoir are surface gill nets and they do not effectively catch the common carp. Stocking, done so far, has little impact on the fisheries.

Fishing activities

There are 25 fishing villages around the reservoir and a local fishermen population of 2 000. Fishing activities continue round the year without any closed season or fishing holidays.

The main fishing gear employed are gill nets, drag nets, cast nets, and lines. Locally fabricated coracle is the main craft used in the reservoir. Coracles of Krishnarajasagar are prepared with HDPP (high density polypropylene), and coal tar as an external covering, in place of the normal hide. This improvised coracle is cheaper and more durable. About 90% of the fishermen use gill nets, licence for which is issued by the local fisheries authorities without any limit, on payment of fee. Each fisherman normally uses 7 pieces of gill nets. There are different rates for different kinds of gear; the charge varying from Rs. 1 to Rs. 129. Total number of licences has gone up from 646 to 1 150 during the 25 years from 1956-57 to 1980-81, with a corresponding increase in the revenue from Rs.4 333 to Rs. 18 075.

Long lines, hook and lines, traps and a few giant shore seines (*alivi*) are also in operation.

Production problems

Krishnarajasagar reservoir produces much less fish than its potential. The main problem is an undesirable species spectrum, where neither the indigenous carps nor the Gangetic carps could carve out a niche for themselves. The stocking done, so far, being totally ineffective, a scientific stocking policy needs to be evolved, based on ecosystem considerations, with a complementary conservation effort aimed at the indigenous ichthyofauna. The absence of restrictions on fishing during the breeding season and use of small meshed nets result in destruction of broodstock and juveniles with far-reaching ecological consequences. This situation, coupled with, the predator pressure from catfishes and murrels is the main factor responsible for the inadequate population of desirable carps.

4.9 OTHER RESERVOIRS

Nalligudda

The 112 ha Nalligudda reservoir (listed as tank in Karnataka Fisheries Department records) is situated in the Bangalore Urban district, 40 km away from the city. Though free from pollution, it shows strong eutrophic tendencies. Water quality parameters indicate a rapid photosynthetic activity at the surface with high dissolved oxygen rate and a matching decomposition process at the bottom, as evidenced by heavy accumulation of CO_2 and oxygen consumption. Nalligudda is rich in nitrate and phosphate. There is a rich standing crop of plankton characterised by blooms of *Microcystis* sp. This high natural productivity could be used beneficially for aquaculture.

Byramangala

Byramangala reservoir is created on the river Vrishabhavati, 38 km away from the city of Bangalore. The 437 ha man-made lake has been receiving a steady inflow of treated sewage from the city and wastes from many industrial establishments for the last two decades. The water is slightly alkaline (pH 7.5) with high concentartion of CO_2 that ranges from11 to 20 mg l^{-1}. Total Kjeldhal nitrogen (27 mg l^{-1}) and ammonia (16.0 mg l^{-1}) show heavy nitrogen loading due to pollution. A very low dissolved oxygen (1.0 mg l^{-1}) and high levels of chlorides (164 mg l^{-1}), BOD (7.0 mg l^{-1}) and COD (58 mg l^{-1}) indicate the eutrophic nature of the water body.

The heavy organic load results in luxuriant growth of water hyacinth, *Eichhornia crassipes*, which chokes the reservoir almost completely. Fish mortality has become a recurring feature in the reservoir. The problem is more acute during the initial incursion of rain washings and during the summer peak. Raghavan *et al.* (1977), reported a case of a very heavy mortality in May 1977, when more than 1 500 kg of fish were killed. The dead fish comprised *Catla catla, Labeo rohita, Cyprinus carpio, Puntius dorsalis, P. ticto, Mystus cavasius, M. vittatus* and *Ompok bimaculatus*. Among them, catla was in the size range of 1.25 to 9.75 kg and rohu 1.25 to 1.75 kg. During the fish kill, the water was acidic, with no dissolved oxygen and the CO_2 level was 292 mg l^{-1}. No hydrogen sulphide, or any other poisonous substances were reported. The high carbon dioxide and gill coating by suspensoids and the absence of dissolved oxygen were the chief reasons attributed to the fish kill. Significant accumulation of heavy metals such as zinc, chromium and copper in the soil and water phase and their uptake by water hyacinth has been reported in Byramangala reservoir (Joshi, 1990).

4.10 STATUS OF RESERVOIR FISHERIES IN KARNATAKA

Of the 74 reservoirs in the State, 10 have at least some ecological data. A proper scientific study covering all the major environmental parameters, the biotic communities and the production processes has been conducted only in Tungabhadra, Krishnarajasagar, Markonahalli and Hemavathy. Data available on the others are too insufficient to evaluate the ecodynamics of the system. Moreover, almost all the reservoirs subjected to the scientific study belonged to the large category, leaving the medium and small ones, which together constitute 84% of the total number. In the absence of a reliable scientific database on the essential ecological attributes, it is not yet possible to make any generalisations, based only on their morphometry, location, or the nature of drainage.

Deep reservoirs created in the uplands of Malnad tend to be oligotrophic and organically less productive. Supa and Krishnarajasagar are very deep and the benefits from the rich nutrient input from the heavily wooded catchment is almost nullified as the allochthonous nutrients get locked up in the deep water. Nevertheless, the conducive temperature regime and the ionic build-up over the years due to ageing have started showing their impact on the production process. After a long depression, the aquaticc productivity in Krishnarajasagar has started improving. However, the management measures have not been adequate to take advantage of this increasing productivity.

Supa reservoir, after the initial trophic burst, has just entered a phase of trophic depression and may take a few years to regain its initial level of production. Tungabhadra is a typical Deccan Plateau reservoir, having all the ingredients for a productive water body such as warm, well-illuminated and ion-rich water. However, the fish production in the earlier years has been very low due to the undesirable species-mix. Having failed to induct the fast -growing carps at an early stage, a forage fish-predator chain has been established at the cost of fish yield. Due to accumulation of nutrients over the years, the productivity is on the rise and effective management can enhance the production substantially. Careful manoeuvering of gill nets, shore-seines and long lines can check the growth of predators and carp minnows. At the same time production of the desirable carps can be enhanced through stocking and proper conservation measures.

A third category of reservoirs represents the small impoundments of the plains which are shallow and highly productive. Nalligudda has a rich natural build-up of nutrients which makes it very ideal for culture-based fisheries. Experience elsewhere in the country indicates enormous scope for enhancing yield from such water bodies. A large number of them are situated in the northern plains, where the temperature is high and by virtue of their small size, light penetrates up to the bottom. As most of them do not offer scope for autostocking, annual stocking and harvesting has to be adopted. Markonahalli in Tumkur district is a testimony to the effectiveness of this *stock and harvest* system.

The reservoirs of Karnataka are the sanctuaries of a rich ichthyofauna, especially the species of *Puntius* and *Cirrhinus*. Some of them grow better than the transplanted fishes. Despite being riverine species, they have adapted themselves well to the reservoir conditions. The main factors that can lead to their decline are the breeding failure and predator pressure. Small reservoirs leaving little room for autostocking and conservation, focus should be on the large and medium reservoirs in our efforts for conserving fish biodiversity.

4.11 TANKS OF KARNATAKA

Small irrigation impoundments listed as tanks (Please see the chapter on resources) constitute an important inland fisheries resource of the State. David (1974) made a comprehensive survey of such water bodies in Karnataka, covering a number of morpho-edaphic limno-chemical and biotic parameters. The status of these water bodies is given in the foregoing account.

The soil and water qualities of tanks in Karnataka depend, to a large extend, on their location. Therefore, David *et al.* (*op. cited*) classified them as:

1. *Coastal and Malnad tanks.*

2. *Transitional zone tanks of Shimoga, Chickmagalur and stretches of Hassan, Mandya and Mysore.*

3. *The black soil zone tanks of Bidar, Gulbarga and Bijapur.*

4. *Red soil tanks of Bellary, Tumkur, Kolar, Bangalore, parts of Mysore, Mandya, fringes of Hassan, Chitradurga and Dharwar.*

Coastal and Malnad tanks:

The lateritic soils of Malnad and coastal plains contain only traces of lime, magnesium and potash and are poor in silica and finer fractions of soil like clay and silt. They are acidic and low in nutrients except in the densely forested areas, where laterites are covered by forest litter, and very rich in organic matter and nitrogen. Soil quality plays a vital role in determining the soluble salts, nutrients and hydrogen-ion concentration of water. Disintegration of surface humid matter release elements like calcium, magnesium and other salts. Exchangeable salts in the presence of humic matter are more readily released, which help the growth of aquatic plants. Water of Malnad tanks, as a whole, has pH, alkalinity hardness and specific conductivity in the ranges of 6.8 to 8.4, 16 to 118, 21 to 158 mg l^{-1} and 213 to 42 µmhos respectively. Majority of them is full of aquatic vegetation. Apart from the marginal vegetation comprising mainly *Pseudoraphis* and *Scirpus*, all the 25 species of floating, submerged and emergent vegetation recorded from the tanks of Karnataka are found in this region. Floating weeds are restricted to *Pistia* and *Lymnanthemum* in most of the districts, but *Lemna* is more common in Dharwar. *Hydrilla*, *Chara* and *Ceratophyllum* constitute the submerged vegetation. Most tanks in Malnad are swampy with various rooted vegetation including *Typha*, *Scirpus* and *Nelumbium*. The vegetation offers good substrate for a number of associated fauna, mainly tubificids, oligochaetes, *Stylaria* and *Chaetogaster* spp. But the typical forms associated with disintegrating vegetation, such as trichopteran and neuropteran insects, are conspicuous by their absence.

The main channel of energy transformation taking place through the macrophytic chain-plankton community has a subordinate role in the trophic structure and functions. Of the plankton count of 567 l^{-1}, phytoplankton constitutes 409 units. Among the blue-greens, *Phormidium* and *Rivularia* are common in the North Canara and Shimoga areas, whereas *Microcystis, Anabaena, Chathrocystis, Osillatoria* and *Cydindrosperium* are found in Chackmagalur and Hassan districts. Diatoms, as a group, are not well-represented in the tank ecosystems of Malnad and coastal belt. Desmids are abundant and include the genera *Cosmarium, Selenastrum, Xanthidium, Microasterias, Closterium, Euastrum, Desmidium* and *Gonatozygon*. A rich variety of Chlorophyceae is reported, the common forms being *Spirogyra, Mougeotia Pediastrum simplex, P. duplex, Zygnema, Oedogonium, Crucigenia, Tribonema, Chaetophora* and *Microspora*.

Transitional zone tanks:

These tanks are situated in the transitional zones between the Malnad (hills) and plateau, characterised by high soil and water fertility. Soil vary mostly from laterite to red and water quality parameters such as pH (8.7), alkalinity (223 mg l^{-1}) hardness (150 mg l^{-1}) and specific conductivity (737 μmhos) are higher than in the Malnad and coastal tanks. The improved ionic build-up is due to indirect fertilization from disintegrating organic debris and humus from the forests, fertilizer leachings from the coffee plantations of Chickmagalur and the fields in the surrounding Hassan, Mandya and Mysore districts.

In most of the tanks, aquatic plants, though present, are not choking the water body. Emergent weeds are confined to the margins. The composition of littoral and bottom biota is the same as that of Malnad-coastal tanks. Macrophytes being relatively sparse, the plankton count reaches to 1 708 units l^{-1} (phytoplankton 1 616 units). This region has the maximum concentration of Cyanophyceae and Bacillariophyceae. Blooms of *Microcystis aeruginosa* are very common in tanks especially in Chickmagalur, Hassan and Mysore districts.

Black soil zone:

Tanks of the black soil zone are characterised by either black or mixed, black and red soil. Black soil is derived from rocks containing soda lime felspars, produced under impeded drainage conditions. The soil has a rich clay content, a high proportion of alumina, lime, magnesia, and potash. Humus content is 1 to10% and the soil is very fertile and high in base status and base exchange capacity. The area being warmer, the rate of decomposition of organic matter is high and water quality, in general, is quite conducive for organic productivity (pH 7.8-9.0, alkalinity 108 to 263 mg l^{-1}, hardness 52 to 97 mg l^{-1}, and specific conductivity 297 to 1197 μmhos).

Aquatic plants are mostly submerged and emergent (*Hydrilla, Chara* and *Nitella*) or filamentous rather than spatulate type. They do not pose any major impediment for fishing as the swampy condition prevails only in very few tanks in the area. With the vegetation cover considerably reduced, the associated fauna is less in variety and density, compared to the two preceding zones. Invertebrates, are mostly chironomids and tubificids. Culicidae are also found (mostly in Chitradurga district). Plankton population is more or less similar to that of transitional zone with a preponderance of Cyanophyceae, with *Anabaena, Microcystis* and *Phormidium* manifesting into blooms in many tanks.

Some characteristic features of the tanks in the black soil region are their high productivity, good soil and water quality. They are very conducive for fish production.

Red Soil zone:

Tanks in the red soil zone are characterised by the red loam or tropical red earth which are derived mainly from granites and gneisses with low water holding capacity. These soils are usually neutral, but tend to become acidic. Poor in nitrogen, phosphates and humus, manuring is required to make them productive. However, potash and iron oxide are present. Compared to the other zones, water portrays a poorer quality in terms of total alkalinity (61-125 mg l^{-1}), hardness (58 to 88 mg l^{-1}) and specific conductivity (185 to 474 μmhos). The low concentrations of phosphorus and calcium in the slightly acidic water is considered unfavourable for plankton.

Most tanks are seasonal and the littoral vegetation renews itself every year, after a period of quiescence in summer. Weed-choking, though rare, is reported when influx of sewage into the system promotes a luxuriant growth of *Eichhornia*. The plankton density is 847 individuals l^{-1} and the phytoplankton count of 734 unit l^{-1} is more comparable with the black soil zone.

In general, where tanks are full of aquatic plants, they are deficient in plankton, as rooted weeds utilise all nutrients, depriving plankton of its share. In the absence of many herbivores in the tank ecosystem, the weeds are not directly utilised by fish. Even the *Microcystis* blooms are not utilized by fish. Both macrophytes and *Microcystis* contribute towards detritus and indirectly form food of detritivores. As most of the tank fish, such as *Cyprinus carpio, Cirrhinus, Labeo*, and *Puntius*, live mainly on detritus, energy channelled through plankton and macrophytes is utilised at the detritus phase. Invertebrates associated with the plants, especially the molluscs, are an important link in the food chain. *Pangasius pangasius*, the non-predatory catfish and the herbivorous peninsular carp, *Puntius pulchellus* are very important fish in the tak the tank ecosystem by virtue of their short food chains..

Fish fauna of tanks

Ichthyofauna of a tank reflects the faunistic composition of the river system to which it belongs and the transplanted Indo-Gangetic carps (Table 4.19). More than 70 species of fish have been recorded from various tanks in the State, out of which 20 are considered to be economically important. They include the large fishes *viz.*, *Catla catla, Cirrhinus mrigala, Cyprinus carpio, Labeo rohita, L. calbasu, L. fimbriatus, Puntius kolus, P. carnaticus, Tor* spp; and *Wallago attu* and the medium sized fishes such as *Notopterus, Puntius sarana, Ompok* spp., *Clarias batrachus, Channa* spp. *Mastacembelus* spp., *Labeo ariza, L. boggut, C. reba* and *C. fulungee*. The majority (60-70%) of the catch, however, is represented by minor uneconomic species such as *P. dorsalis Amblypharyngodon* spp. The growth of some important fish species in tanks of Karnataka is given in Table 4.20.

Fisheries management

Fishing rights, including exploitation, stocking and disposal by leasing or licencing of inland waters (rivers, reservoirs and tanks), as a rule, are vested with the Department of Fisheries. In some areas, the Revenue Department has either the exclusive right or a conditional one with a stipulation to pay earnings from fisheries to the Fisheries Department. In Coorg, Department of Fisheries pays 50% of the total catch from tanks to the local village panchayat (local self government). Minor tanks are managed by the village panchyats. Generally, licences are issued by the Fisheries Department, according to the fishing gear.

A variety of fishing gear is employed for fishing. Gill nets, both surface and bottom set, are the most common. The fishing tackle is used mainly in large water bodies. Drag nets, cast nets and an improvised triangular scoop net are the other favourite fishing implements. A variety of traps are employed for catching prawn, air breathing catfishes and murrels, while rod and line are sometimes employed to hook *Wallago attu, Ompok* spp. and *Mastacembelus* spp. Poisoning, shooting and use of dynamite, though reported, are not very common.

According to David *et al.* (1974), the fish yield of perennial tanks was 148 kg ha^{-1} and that of seasonal tanks 10 kg ha^{-1}. Total fish production from tanks in 1974 was estimated at 30 000 t.

Table 4.19. List of fishes recorded from tanks of Karnataka

	Species	River catchment	District/Taluk/Tank
1.	*Notopterus notopterus* (Pillas)	All river catchments	Almost all districts and in all large perennial tanks
2.	*Chela atpar* (Ham.)	-do-	-do-
3.	*Chela laubuca* (Ham.)	-do-	-do-
4.	*Oxygaster argentea* (Day)	-do-	-do-
5.	*Oxygaster clupeoides* (Bl.)	-do-	-do-
6.	*Oxygaster phulo* (Ham.)	-do-	-do-
7.	*Barilius barila* (Ham.)	All river catchments except Palar	Tanks connected to rivers by canals
8.	*Barilius bendelensis* (Ham.)	-do-	-do-
9.	*Barilius gatensis*	-do-	-do-
10.	*Danio* (Brachydanio) *rerio* (Ham.)	-do-	-do-
11.	*Danio* (Danio) *aeguipinnatus*	-do-	-do-
12.	*Esomus barbatus* (Jesion)	All catchments	Tanks throughout all the districts
13.	*Rasbora daniconius* (Ham.)	-do-	-do-
14.	*Rasbora rasbora* (Ham.)	-do-	-do-
15.	*Amblypharyngodon melittinus* (Cuv. & Val.)	-do-	In most tanks in all districts
16.	*Amblypharyngodon mola* (Ham.)	-do-	-do-

Cont.

Table 4.19 (cont.). List of fishes recorded from tanks of Karnataka

No.	Species	River catchment	District/Taluk/Tank
17.	*Aspidoparia morar* (Ham.)	Restricted to the Krishna	Most northern and central district tanks
18.	*Catla catla* (Ham.)	Propagated	Selected tanks
19.	*Cirrhinus fulungee* (Sykes)	Krishna	Tanks in northern and central districts
20.	*Cirrhinus mrigala (Ham.)*	Propagated	Selected tanks
21.	*Cirrhinus reba* (Ham.) Cauvery	Krishna, Godavary, districts	Tanks in most
22.	*Garra spp.*	All catchments except Palar	Canal connected tanks or large reservoir like tanks
23.	*Labeo ariza* (Ham.)	Godavary, Krishna, Cauvery	Occurs variously in one or the other tank in all districts
24.	*Labeo bata* (Ham.)	-do-	-do-
25.	*Labeo boggut* (Sykes)	-do-	-do-
26.	*Labeo calbasu* (Ham.)	All catchments	-do-
27.	*Labeo fimbriatus* (Bl.)	Krishna, Godavary	
28.	*Labeo rohita* (Ham.)	Propagated	Selected tanks in almost all districts
29.	*Labeo boga* (Ham.)	-	-
30	*Labeo porcellus* (Sykes)	-	-
31.	*Oreichthys cosuatus* (Ham.)	Krishna	Malnad tanks
32.	*Puntius amphibius* (Cuv. & Pakke Val.)	All catchments	Almost in every district in one or other tank
33.	*Puntius carnaticus* (Jerdon)	Cauvery	Recorded in Muguna-halli tank in Mysore district and tanks in Mandya
34.	*Puntius chola (Ham.)*	All catchments	Tanks in every district

Cont.

Table 4.19 (cont.). List of fishes recorded from tanks of Karnataka

No.	Species	River catchment	District/Taluk/Tank
35.	*Puntius conchonius* (Ham.)	-do-	-do-
36.	*Puntius dorsalis* (Jerdon) *P. puckelli* (Day)	Cauvery, Krishna -	Tanks in Bangalore, Tumkur, Chitradurga and Raichur district
37.	*Puntius filamentosus* (Cuv. & Val.)	Krishna, Cauvery	Tanks in Western Ghat
38.	*Puntius pulchellus (Day)*	Tungabhadra	Shimoga district
39.	*Puntius kolus* (Sykes)	Godavary, Krishna	Bijapur (Bhuthnal tank)
40.	*Puntius sarana* (Ham.)	All catchments	In almost every tank of all districts
41.	*Puntius stigma* (Day)	-do-	-do-
42.	*Puntius ticto puntius* (Ham.)	-do-	-do-
43.	*Puntius dobsonii* (Day)	-	-
44.	*Tor spp. (Mahseers)*	Krishna & Cauvery	Tanks connected by canals
45.	*Lepidocephalus quntea*	Godavary, Krishna, Sharavathy,	In all tanks
46.	*Lepidocephalus thermalis* (Cuv. & Val.)	Cauvery, Palar	
47.	*Nemacheilus sp.*	All catchments	In most tanks
48.	*Ompok bimaculatus* (Bl.)	-do-	In most tanks in all districts
49.	*Ompok pabo* (Ham.)	-do-	
50.	*Wallago attu* (Bl. & Schn.)	-do-	In most large tanks
51.	*Mystus cavasius* (Ham.)	-do-	In most tanks
52.	*Mystus seenghala* (Sykes)	Krishna	Bhuthnal tank
53.	*Mystus vittatus* (Bl.)	All catchments tank	In almost every
54.	*Heteropneustes fossilis* (Bl.)	-do-	All tanks
55.	*Clarias batrachus* (Linn.)	-do-	-do-

Cont.

Table 4.19 (cont.). List of fishes recorded from tanks of Karnataka

No.	Species	River catchment	District/Taluk/Tank
56.	*Clarius dussumeiri dussumeiri Val.*	Krishna	Belgaum, Dharwar, & Shimoga
57.	*Gambusia affinis* (Baird & Girard)	Southern Karnataka	Bangalore, Mysore & Shimoga
58.	*Oryzias melanostigma* (Mc. Cl.)	All catchments	Mostly in perennial tanks
59.	*Panchax lineatus* (Val.)	-do-	-do-
60.	*Channa gachua* (Ham.)	-do-	In all perennial and seasonal tanks
61.	*Channa leucopunctatus* (Sykes)		
62.	*Channa marulius* (Ham.)	All catchments	In all perennial and seasonal tanks
63.	*Channa striatus* (Bl.)	-do-	-do-
64.	*Channa punctatus* (Bl.)	-do-	-do-
65.	*Ambassis nama* (Ham.)	-do-	In many tanks of various districts
66.	*Ambassis ranga* (Ham.)	-do-	
67.	*Macropodus cupanus* (Cuv. & Val.)	Tungabhadra	Shimoga Dist.
68.	*Glossogobius giuris* (Ham.)	All catchments	In all districts
69.	*Macrognathus aculeatus* (Bl.)	-do-	-do-
70.	*Mastacembelus armatus* (Lac.)		
71.	*Mastasembelus pancalus* (Ham.)		
Transplanted spp.			
	1. *Cyprinus carpio* (Lin.)	Exotic form	Stocked in several tanks
	2. *Mugil corsula* (Ham.)		
	3. *Oreochromus mossambicus*		
	4. *Etroplus suratensis* (Bl.)		
	5. *Osphronemus gourami* (Lacep)		
	6. *Chanos chanos* (Forskal)	Collected from the sea	
	7. *Ctenopharyngodon idella*		
	8. *Hypophthalmichthys molitrix*		

After David et al., 1974

Table 4.20. Growth rate of fishes in the tanks of Karnataka

FISH	Growth cm	Year
Major Carps		
Catla catla (Ham.)	45-60	I-II year
Labeo rohita (Ham.)	23-38	I Year
	35-40	II Year
Cirrhinus mrigala (Ham.)	38-60	I Year
Cyprinus carpio (Linn)	23-30	I Year
C. carpio specularis (Linn.)	50	II Year
Cirrhinus cirrhosa (Bl.)	40	I Year
	61	II Year
Labeo fimbriatus (Bl.)	23-30	I Year
Labeo calbasu (Ham.)	25-30	I Year
	30-40	II year
Labeo kontius (Jerdon)	25-30	I year
Puntius dobsonii (Day)	13	I Year
P. pulchellus (Day)	31	II Year
	47	III Year
Puntius carnaticus (Jerdon)	23-25	I Year
Puntius dubius (Day)	-	-
Medium sized carps		
Puntius sarana (Ham.)	14-16	I Year
	21-24	II Year
Puntius kolus (Sykes)	10-15	I Year
Labeo bata (Ham.)	18-23	I Year
Labeo boga (Ham.)	30-38	I Year
Labeo ariza (Ham.)	40-46	I Year
Labeo boggut (Sykes)	12-15	I Year
Cirrhinus reba (Ham.)	22-28	I Year
Cirrhinus fulungee (Sykes)	15-20	I Year
Labeo porcellus (Sykes)	-	-

Cont.

Table 4.20 (cont.). Growth rate of fishes in the tanks of Karnataka

FISH	Growth cm	Year
Backwater & other forms		
Chanos chanos (Forskal)	40-60	I-II Years
Mugil cephalus (Forskcl)	30-60	I-II Years
Etroplus suratensis (Bl.)	20-26	I-II Years
Osphronemus gorami (Lace.)	15-20	I Year
Oreochromis mossambicus	21-33	I Year
Predaceous species		
Channa marulius (Ham.)	40-60	I-II Years
Channa striatus (Bl.)	30-36	I-II Years
Channa punctatus (Bl.)	20-26	I-II Years
Ompok spp. (2)	25-33	I Year
Wallago attu (Bl. & Schn.)	30-34	I Year
	35-60	II Year
Mastacembelus armatus (Lace)	10-16	I Year
	20-26	II Year
	37-43	III Year
Mastacembelus pancalus (Ham.)	15-20	I Year
Notopterus notopterus (Pallae)		I Year
Lates calcarifer (Bl.)	30-46	I Year
	50-60	II Year
Others		
Ctenopharyngodon idella (V)	43-49	I Year
Hypophthalmichthys molitrix (V)	38-57	I-II Year

(After David et al., 1974)

Table 4.21. Salient features of some reservoirs in Karnataka

Parameters	Tunga-bhadra	Krishnarajasagar	Vanivilas-Sagar	Markona-halli	Hemavathy	Supa
Year of sealing	1953	1932	1901	1936	1981	-
Area at FRL (ha)	37 814	13 200	8 640	1336	9 162	12 900
Maximum length (km)	80	40	-	-	-	-
Maximum width (km)	-	8	-	-	-	-
Volume (m m^3)	3 709	1 348	-	68	1 050	-
Maximum depth (m)	41	-	-	-	-	-
Mean depth (m)	9.8	30.17	-	5.08	-	-
Catchment area (km^2)	28 168	10 619	-	4 103	2 810	1 067
Elevation (m above MSL)	498	752.23	621	731.6	854	
Length of shoreline (km)	337	292	-	-	-	-
Shore dev. index	5.251	7.25	-	-	1.6	-
Volume dev. index	0.737	0.78	-	-	0.754	-
Annual level fluctuations (m)	16.15	15	-	-	18.3	-
Inflowing rivers	Tungabhadra	Cauvery Hemavathy Lakhshmana-thirtha	Vedavati	Shimsha	Hemavathy Yagachi	Kalinadi Pandri Naginalla Kaneri
Outflowing rivers	Tungabhadra	Cauvery	Vedavati		Hemavathy	Kalinadi
Maximum outflow ($m^3 sec^{-1}$)	4199	6 035	-	-	283	-
Latitude (N)	15^0 15'	12^0 33'	13^0 21'	12^0 55'	12^0 4'	-
Longitude (E)	76^0 20'	76^0 37'	76^0 46'	76^0 55'	76^0 3'	-
Soil						
Organic matter (%)	6.6-26.7	1.25-1.86	-	0.55-0.63	0.55-0.63	-
Available P (mg 100 g^{-1})	1.07-1.25	0.6-1.6	2.25	-	-	-
Available N (mg 100 g^{-1})	25.8-27.8	-	-	-	-	-
pH	7-8.5	6-7.2	8.4-8.7	6.3-6.9	6.3-6.9	-
Spec. cond. (μ mhos)	35-261	-	-	-	-	-
Calcium (mg 100 g^{-1})	200-300		200-300			
Water						
Water temp. (°C)	23.1-29.5	20.1-29.8	22.3-26.3	22.0-26.2	22.4-29.75	
pH	7.9-8.5	7.2-8.4	8.4-8.5	7.2-8.4	7.2-8.2	6.57-7.57
Transparency (cm)	100-825	45-225	-	60-94	32.5-179.5	-
DO (mg l^{-1})	4.8-11.5	6.0-8.0	6.56-9.8	6.4 - 7.0	4.5-8.2	4.19-7.6
CO_2 (mg l^{-1})	-	0-3.0	-	trace	0.04-4.6	0.47-2.02
Total alkalinity (mg l^{-1})	30-100	36-140	194-228	104-140	23-46	6.84-15.6
Spec. cond. (μmhos)	240-359	-	474-682	167-204	-	68.8-99.2
Total hardness (mg l^{-1})	31-75	-	108-128	-	-	29.28-35.37
Calcium (mg l^{-1})	-	-	-	-	-	-
Nitrate (mg l^{-1})	0.2-0.5	tr-0.011	0.11-0.55	tr-0.028	-	0.004-0.46
Phosphate (mg l^{-1})	tr-3.3	tr-0.05	tr-0.03	tr	0.02-0.05	0.12-0.91
Silicate (mg l^{-1})	9-16	5.6-11.0	8-12	8.0-11.0	0.05-0.08	12.84-25.1
Chlorides (mg l^{-1})	11.9-42.1	-	46-77	-		15.2-32.44
GPP (mg C $m^{-3} d^{-1}$)	194.4-2271.6	-	0.39-3.25	-	-	-

5. ANDHRA PRADESH

Andhra Pradesh is the fifth largest State in India, both demographically as well as geographically. The State consists of the eastern coastal belt, the Telengana area of the north and the Rayalseema area of the Southwest. The northern part of Telengana is mountainous receiving an annual rainfall of the order of 102 cm. In the south, the terrain is undulating with isolated hills and an average precipitation of only 50 cm a year. The State is a hot, semi-arid dryland, except the coastal districts which are very fertile and have a highly productive agriculture.

Andhra Pradesh is drained , by three large rivers (catchment > 20 000 km^2), including the Godavari and Krishna, which have an annual discharge of 105 000 and 67 675 million m^3 respectively. Pennar, the third major river carries 3 238 million m^3 annually. The medium rivers (catchment 2001- 20 000 km^2) comprising Nagavali, Sarda, Eluru, Gumdlakamma, Musi, Paleru, Muneru and Kunleru have an annual discharge of 6 430 million m^3. More than 33 minor rivers (catchment < 2000 km^2) of the State are small coastal streams flowing into the Bay of Bengal, carrying a total of 6 764 million m^3 of water. All the rivers in the State are east-flowing and are harnessed for irrigation and power generation.

People living in the dry regions of Telengana and Rayalseema realised the need for irrigation from time immemorial and learned to create reservoirs on small streams, rivulets and creeks to store water for irrigation. Like their counterparts in Karnataka and Tamil Nadu, these improvised irrigation reservoirs, locally called *tanks* are often overlooked while compiling the reservoir resources data.

5.1 RESERVOIR FISHERIES RESOURCES

Andhra Pradesh has a long tradition of constructing big dams and reservoirs. Some of the oldest man-made lakes in the country are situated in the State. Hussainsagar in the heart of Hyderabad city is 500 years old, followed by Saroornagar and Mir Alam, which had been in existence for the last 275 and 170 years respectively. Banjara, Osmansagar and Himayatsagar are the other old reservoirs, created between 50 to 65 years ago. Srivastava *et al.* (1985) have listed 117 reservoirs, along with their areas at FRL and the fish production rate. However, a recent communication from the State Fisheries Department mentions only 102 reservoirs. After reconciling the discrepancies by verification with other sources such as maps and other documents, a list of 137 reservoirs has been prepared (Table 5.1 ; Fig. 5.1). More than 46 reservoirs in Andhra Pradesh are said to be under dispute, where no fishery activities exists and Sreenivasan (1993) reports that hardly a dozen man made lakes in the State have been developed for fisheries.

Andhra Pradesh has 98 small reservoirs, 2 800 tanks, 32 medium reservoir and 7 large reservoirs with a total surface water area of 458 507 ha (Table 5.2). Vishakhapatnam district has the maximum number of reservoirs (20), followed by Rangareddy, Prakasam, Nellore and Karimnagar districts with 10 reservoirs each. While Vizianagaram and the two Godavari (east and west) districts have only small reservoirs, Ranga Reddy, Nellore, Nalgonda, Karimnagar and Chitoor have large concentration of small reservoirs. Medium reservoirs are well-dispersed among the districts and the seven large reservoirs are distributed among Guntur, Karimnagar, Kurnool, Nellore and Nizamabad districts (Table 5.3). Majority of the reservoirs in the State are small (98), followed by those in the medium (32) and large (7)

Table 5.1. Reservoirs of Andhra Pradesh

	Name of the reservoir	District	Area at FRL	Area. average	Annual fish production (t)	Yield kg ha-1
1.	Kadam	Adilabad	2 445	2 160	25.0	10.22
2.	Swarna Jonlivellivagu	do	683	519	NA	
3.	C. R. Gudi	Anatapur	158	120	1.2	7.6
4.	Upper Pennar	do	776	677	10.0	12.9
5.	Mid Pennar	do	1 536	883	11.1	7.2
6.	B. T. Project	do	1 529	829	15.0	9.8
7.	Pennahobilam	do	2 049	1 557	NA	-
8.	Araniyar	Chittoor	1 378	761	47.0	34.10
9.	Kolangi	do	960	141	60.0	62.5
10.	Mallemadugu	do	268	121	1.0	3.7
11.	Siddalagandi	do	68	55	0.5	7.3
12.	Bahuda	do	18	142	NA	-
13.	Krishnapuram	do	66	50	NA	-
14.	Podderau	do	138	105	NA	-
15.	Pincha	Guddapah	160	110	0.5	3.1
16.	Lower Sagiluru	do	280	160	0.5	1.8
17.	Mylavaram	do	2 000	1 520	2.2	1.1
18.	Gangareru	Guddapah	57	43	NA	
19.	Pampa	Godavari (E)	495	286	20.0	40.4
20.	Maddigadda	do	80	60	NA	
21.	Vottigada	do	75	57	15.0	200.0
22.	Kolagummigadda	do	16	12	3.0	187.0
23.	Mangaligadda	do	20	8	NA	
24.	Subbareddy Sagar	do	71	54	15.0	211.0
25.	Lingavaram	do	16	12	6.0	375.0
26.	Krishna Barrage	Krishna	800	608	150	187.0
27	Edurumondi	do	105	80	NA	
28.	Tammileru	Godavari (W)	820	623	NA	
29.	Nagarjunasagar	Guntur	28 474	25 844	70.00	2.5
30.	Buggavagu	do	1 228	933	15.0	12.2
31.	Hussainsagar	Hyderabad	4 016	1 942	-	-
32.	Mir-Alam	do	1 280	1 165	5.0	0.39
33.	Himayatsagar	do	3 584	2 724	40.0	11.16
34.	Osmansagar	do	4 016	3 052	50.0	12.45
35.	Foxsagar	do	1 024	778	15.0	14.6
36.	Lower Mannair	Karimnagar	8 817	6 701	-	-
37.	Upper Mannair	do	1 400	842	20.0	14.3
38.	Shanigaram	do	659	104	10.0	15.2
39.	Tegumfoor	do	140	106	1.0	7.1
40.	Rallavagu	do	236	180	1.5	6.3
41.	Cheggaon	do	200	150	2.0	10.0
42.	Yellamcherru	do	200	150	1.0	5.0
43.	Tangalmana	do	200	150	2.0	10.0
44.	Kalwala	do	294	223	3.0	10.2
45.	Kamalapur	do	160	122	1.5	9.4
46.	Wyra	Khammam	1 894	1 670	60.0	31.7
47.	Paleru	do	1 707	1 511	60.0	35.1
48.	Kennerassani	do	2 048	1 584	10.0	4.9
49.	Prisithipalli	do	358	272	15.0	41.9
50.	Gangavaram Lankasagar	do	516	298	20.0	38.7
51.	Srisailam	Kurnool	51 200	46 080	NA	-
52.	Gajuladinne	do	2 053	1 560	1.0	0.48
53.	Thommeru	do	140	111	NA	-

Cont.

Table 5.1(cont.). Reservoirs of Andhra Pradesh

No.	Name of the reservoir	District	Area at FRL	Area average	Annual fish production (t)	Yield kg ha-1
54.	Palair	do	96	73	12.0	125.0
55.	Pocharam	Medak	1 664	1 238	5.0	3.0
56.	Mallavagu	do	856	650	NA	-
57.	Royampally	do	139	106	1.0	7.2
58.	Koilsagar	Mehaboobnagar	1 024	366	NA	-
59.	Sarlasagar	do	837	636	NA	-
60.	Chandrasagar	do	186	107	NA	-
61.	Laknavaram	Warrangal	3 600	1 821	9.0	2.5
62.	Ramappa	do	1 428	1 085	12.5	8.7
63.	Ghanapur	do	1 066	539	9.0	8.4
64.	Salivogu	do	772	390	4.5	5.8
65.	Pakhal	do	2 072	1 575	3.5	1.7
66.	Dindi	Nalgonda	4 342	3 300	25.0	5.7
67.	Moosi	do	2 304	1 294	50.0	21.7
68.	Vemulum	do	500	380	NA	-
69.	Pendlipakala	do	760	396	NA	-
70.	Bhemapalli	do	225	171	NA	-
71.	Gandhinagar	do	122	93	NA	-
72.	Edavalbi	do	237	180	NA	-
73.	Yathavakkal	do	768	584	NA	-
74.	Sochijorram	do	400	304	NA	-
75.	Kanigiri	Nellore	6 230	4 610	580.0	93.1
76.	Nellore tank	do	2 110	1 096	300.0	142.2
77.	Suryapalli	do	3 287	1 074	400.0	121.7
78.	Gamdipalem	do	-	701	NA	-
79.	Edukodur	do	355	270	280.0	788.7
80.	Kollur	do	270	205	260.0	963
81.	Vidavalu	do	147	112	175.0	1 140
82.	Anantasagaram	do	710	540	450.0	634
83.	Atmakur	do	439	334	380.0	866
84.	Kalboy	do	212	161	270.0	1273
85.	Nizamsagar	Nizamabad	12 800	8 875	150.0	11.7
86.	Sriramsagar	do	37 830	28 750	NA	-
87.	Ramadugu	do	140	106	NA	-
88.	Alisagar	do	44	33	NA	-
89.	Maulalisagar	do	160	122	NA	-
90.	Pochampalli	do	44 800	34 050	NA	-
91.	Cumbum	Prakasam	2 304	1 467	40.0	17.4
92.	Mopadu	do	2 048	897	40.0	19.5
93.	Panduramgadi	do	84	64	NA	-
94.	Rallapadu	do	1 280	375	20.0	15.6
95.	Gollapally	do	100	75	NA	-
96	Morasipally	do	150	113	NA	-
97.	Peddabommalepuram	do	100	75	NA	-
98.	Veerabhadrapuram	do	28	21	NA	-
99.	Pengodu	do	142	108	NA	-
100.	Nakkalangadi	do	1 024	778	15.0	14.6
101.	Tandava	Vishakhapatnam	1 689	438	0.5	0.3
102.	Tajangi	do	108	82	12.0	111.1
103.	Gambheeramgadda	do	400	53	NA	-
104.	Kalyanapulova	do	250	190	NA	-
105.	Kottali	do	120	91	NA	-
106.	Ravanipalli	do	151	115	5.0	33.1
107.	Konam	do	553	420	NA	-
108.	Raywada	do	165	125	NA	-

Cont.

Table 5.1(cont.). Reservoirs of Andhra Pradesh

No.	Name of the reservoir	District	Area at FRL	Average area	Annual fish production (t)	Yield kg ha-1
109.	Girigarthi	do	46	35	NA	-
110.	Mampa	do	66	50	NA	-
111.	Limbarthi	do	41	31	NA	-
112.	Pottipittavagu	do	46	35	NA	-
113.	Timiragadda	do	22	17	NA	-
114.	Thunkalagadda	do	59	45	NA	-
115.	Mallavaram	do	10	7	NA	-
116.	Kongasingi	do	10	6	NA	-
117.	Uppusileru	do	576	438	NA	-
118.	KR Project	do	154	117	NA	-
119.	Tatipudi	do	532	404	NA	-
120.	Mudasalona	do	60	46	NA	-
121.	Gadigadda	Vizianagaram	338	257	NA	-
122.	Vattigadda	do	134	102	NA	-
123.	Kondalaveedi	do	45	34	NA	-
124.	Dulikuppa	do	21	16	NA	-
125.	Gajjuvanigadda	do	30	23	NA	-
126.	Frradigadda	do	18	14	NA	-
127.	Kiran dole	do	20	15	NA	-
128.	Kothapally	Ranga Reddy	560	118	15.0	26.8
129.	Lakshmanpur	do	182	138	5.0	27.5
130.	Juntipalli	do	600	87	15.0	25.0
131.	Salarnagar	do	105	80	NA	-
132.	Sarampally	do	131	100	NA	-
133.	Nadivagu	do	112	85	NA	-
134.	G. Sriramulivari	do	21	16	NA	-
135.	Indirammasagar	do	24	18	NA	-
136.	Chinnavagu	do	20	16	NA	-
137.	Allapur	do	67	51	NA	-
	Total area	**137**	**280 785**	-	-	-

N.A.= Not available

Table 5.2. Reservoirs of Andhra Pradesh (by category)

	Number	Area (at FRL)
Small reservoirs	*98*	*24 178*
Tanks	*2 800*	*177 749*
Small reservoirs and tanks (*<1 000 ha*)	**2 898**	**201 927**
Medium reservoirs (*1 000-5 000 ha*)	**32**	**66 429**
Large reservoirs (*> 5 000 ha*)	**7**	**190 151**
TOTAL	**2 937**	**458 507**

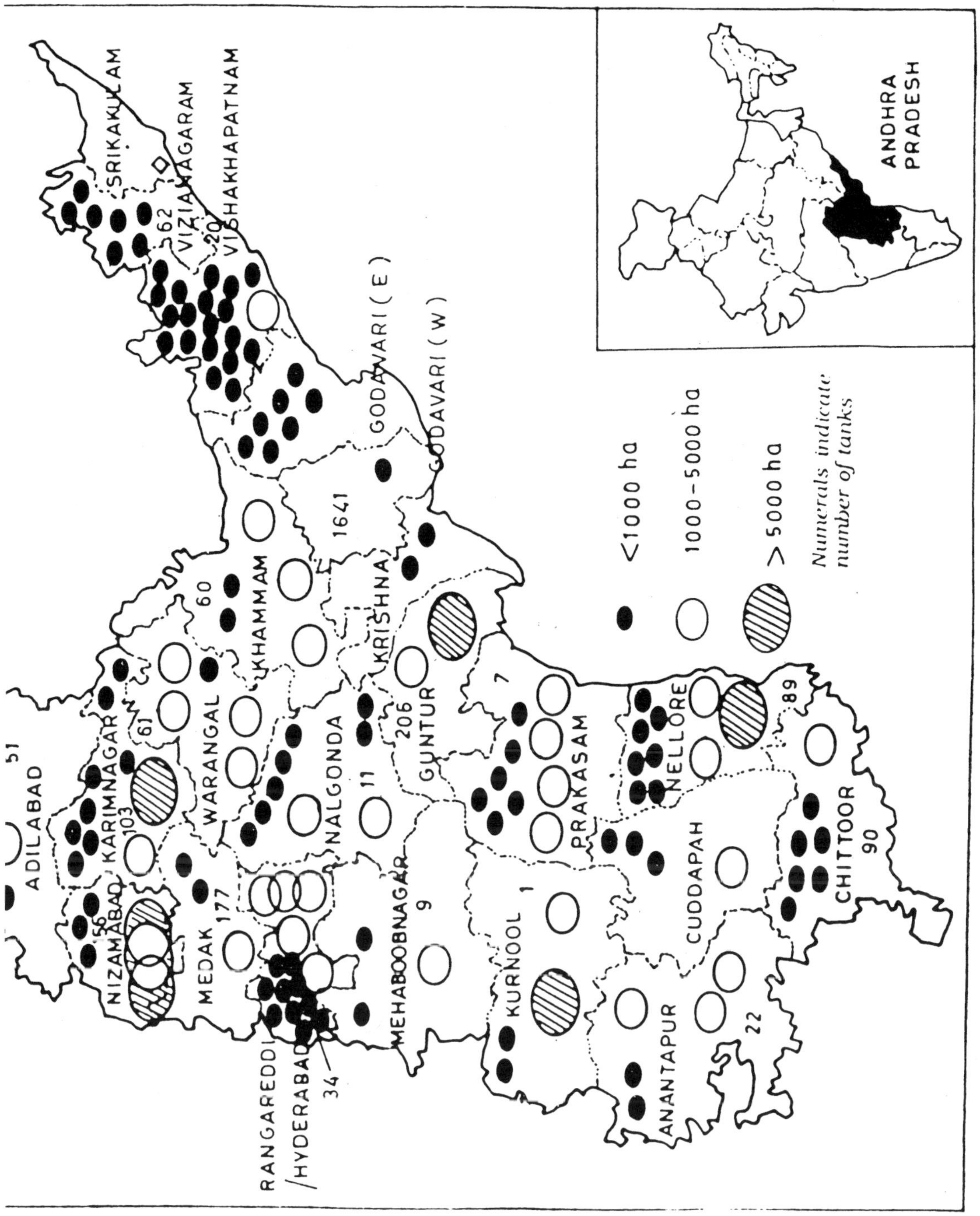

Figure 5.1. Distribution of reservoirs in Andhra Pradesh

categories. The seven large reservoirs have a total surface area of 190 151 ha, while the medium and small reservoirs cover 66 429 ha and 24 178 ha respectively. Taking the irrigation tanks into the account the total number and area is much higher.

Irrigation tanks of Andhra Pradesh are classified into two categories, *viz.*, *perennial* and *long seasonal* (Anon., 1994a). Although an inventory of all perennial tanks in the State is not available, their number and area by districts are known. All the perennial tanks, except those of Srikakulam, East Godavary and Krishna districts (which have an average size less than 10 ha), have been considered as man-made lakes. More than half of the perennial tanks are situated in East Godavari district, both numerically and in terms of waterspread. The 2 800 perennial tanks cover a total area of 177 749 ha in 18 districts.

Reservoirs and tanks together constitute 458 507 ha of water area in Andhra Pradesh, (Table 5.4) which is the second largest in India.

Considering the size of the resource and its potential for fisheries of the State, scientific attention received by the reservoirs of Andhra Pradesh is inadequate. A systematic study was done only in Nagarjunasagar as part of the All India Coordinated Project on Reservoir Fisheries from 1971 to 1981. Hussainsagar in Hyderabad, the oldest reservoir in the country, has attracted attention mainly on account of pollution. A cursory appraisal of eight man-made impoundments in the vicinity of Hyderabad city was done by Zafar (1986).

Table 5.3. Distribution of reservoirs in Andhra Pradesh (by districts)

District	<1 000 ha		1 000-5 000 ha		>5 000 ha		Total	
	No.	Area (ha)	No.	Area (ha)	No.	Area (ha)	No.	Area (ha)
Adilabad	1	683	1	2 445	-	-	2	3 128
Anantapur	2	934	3	5 114	-	-	5	6 048
Chittoor	6	1 687	1	1 378	-	-	7	3 065
Cuddappah	3	497	1	2 000	-	-	4	2497
Godavari (E)	7	773	-	-	-	-	7	773
Godavari (W)	1	820	-	-	-	-	1	820
Guntur	-	-	1	1 228	1	28 474	2	29 702
Hyderabad	-	-	5	13 920	-	-	5	13 920
Karimnagar	8	2 089	1	1 400	1	8 817	10	12 306
Khammam	2	874	3	5 649	-		5	6 523
Krishna	2	905	-	-	-	-	2	905
Kurnool	2	236	1	2 053	1	51 200	4	53 489
Medak	2	995	1	1 664	-	-	3	2 659
Mehaboobnagar	2	1 023	1	1 024	-	-	3	2 047
Nalgonda	7	3 012	2	6 646	-	-	9	9 658
Nellore	7	2 133	2	5 397	1	6 230	10	13 760
Nizamabad	3	344	-	-	3	95 430	6	95 774
Prakasam	6	604	4	6 656	-	-	10	7 260
Rangareddy	10	1 822	-	-	-	-	10	1 822
Visakhapatnam	19	3 369	1	1 689	-	-	20	5 058
Warrangal	1	772	4	8 166	-	-	5	8 938
Vizianagaram	7	606	-	-	-	-	7	606
Total	98	24 178	32	66 429	7	190 151	137	280 758

Table 5.4. Distribution of tanks and reservoirs in Andhra Pradesh

District	Tanks Number	Tanks Area (ha)	Reservoirs Number	Reservoirs Area (ha)	Total Number	Total Area (ha)
Adilabad	51	3 478	2	3 128	53	6 606
Anantapur	22	1 490	5	6 048	27	7 538
Chittoor	90	3 170	7	3 065	97	6 235
Cuddappah	-	-	4	2 497	4	2 497
Godavari (E)		-	7	773	7	773
Godavari (W)	1 641	99 011	1	820	1 642	99 831
Guntur	206	15 578	2	29 702	208	45 280
Hyderabad	-	-	5	13 920	5	13 920
Karimnagar	103	12 508	10	12 306	113	24 814
Khammam	60	5 143	5	6 523	65	11 666
Krishna	-	-	2	905	2	905
Kurnool	1	204	4	53 489	5	53 693
Medak	177	4 028	3	2 659	180	6 687
Mehaboobnagar	9	380	3	2 047	12	2 427
Nalgonda	11	1 638	9	9 658	20	11 296
Nellore	89	5 893	10	13 760	99	19 653
Nizamabad	156	8 407	6	95 774	162	104 181
Prakasam	7	374	10	7 260	17	7 634
Rangareddy	34	1 300	10	1 822	44	3 122
Visakhapatnam	20	283	20	5 058	40	5 341
Warrangal	61	13 925	5	8 938	66	22 863
Vizianagaram	62	939	7	606	69	1 545
Total	2 800	177 749	137	280 758	2 937	458 507

5.2 NAGARJUNASAGAR (Fig. 5.2)

Nagarjunasagar reservoir is situated on the main river Krishna within the geographical coordinates, 16^{0} 34' N and 79^{0} 19' E. The 28 475 ha reservoir extends to Nalgonda and Guntur districts, which are divided by the Krishna. Predominantly an irrigation reservoir serving 1.37 million ha of paddy land, the reservoir also generates 905 MW of hydro-electric power. Being on the mainstream, the reservoir has a vast catchment area of 215 194 km^2. Most of the drainage originates from the Western Ghats, receiving an annual precipitation of 90 cm from the southwest monsoon.

The reservoir, at FRL, holds an enormous 11 560 million m^3 water and shrinks to 1 410 million m^3 at dead storage level, resulting in a water level fluctuation of 58 m. The mean depth at the average level is 29.9 m. Shoreline of Nagarjunasagar is very irregular with a shoreline development index of 7.89. Inflowing water readily gets mixed with the whole column. The high flushing rate (3.0 to 6.2) point to the possibility of nutrient loss. Even during the months of summer stagnation, the deeper outlets drain away nutrients. Nevertheless, the productivity remains high due to the input of nutrients through the inflow and the conducive environment of the bays.

Prevailing climatic conditions play a key role in the production of the reservoir. Nagarjunasagar is situated in one of the warmest regions of India, the summer air temperature often touching 45° C. With no winter season, minimum temperature never drops below 16° C, thus preventing the formation of any thermal stratification in water, while the warm bottom accelerates bacterial decomposition. The wind induced turbulance churns the water to facilitate nutrient mixing.

The high productivity of the reservoir is manifested through many physico-chemical attributes of water. It is a typical hard water lake with a high nutrient and mineral content. The hardness as $CaCO_3$ varies between 50 to 200 mg l^{-1} and the pH between 7.6 to 8.6. The inundated area is rich in limestone and the free CO_2 present in the inflowing water causes dissolution of CO_3 into bicarbonates. Thus, the water retains a very high ionic concentration in terms of total alkalinity and specific conductivity. Carbondioxide is present in the surface, in free state, only during the flood season. During the other seasons, it is used up at a faster rate than the supply. Carbon dioxide for photosynthesis is derived from bicarbonates. Though the nutrient build-up in terms of phosphates is not high, the levels of nitrate and silicate are high (Pathak, 1979).

There is a sharp discontinuity between the oxygen-rich trophogenic layer and the oxygen-deficient tropholytic zone at the bottom during summer, accompanied by the decrease in pH and increase in specific conductivity and total alkalinity towards the bottom.

The productivity manifests itself by a flourishing plankton community. The annual water renewal plays an important role in its seasonal succession. The massive annual floods bring, in minerals and nutrients, but they also dislodge and sweep away the plankton. As soon as the destabilising effects of the flood discharges are over, the plankton community recovers. The phytoplankton cycle more or less conforms to a general pattern of aestival (summer) and autumnal (post-monsoon) seres, punctuated by the hiemal (monsoon) minima (Sugunan, 1980). The protected bays which are shallow, undisturbed and nutrient-rich bear blooms of *Microcystis aeruginosa*. The deep lentic sector has diatom blooms in summer and winter. Sectoral variations in the composition and diversity of plankton are clearly discernible. The fluviatile lotic sector retains a rich variety of algae, dominated by Chlorophyceae, blue-greens and diatoms. The lentic waters, especially the bays, have

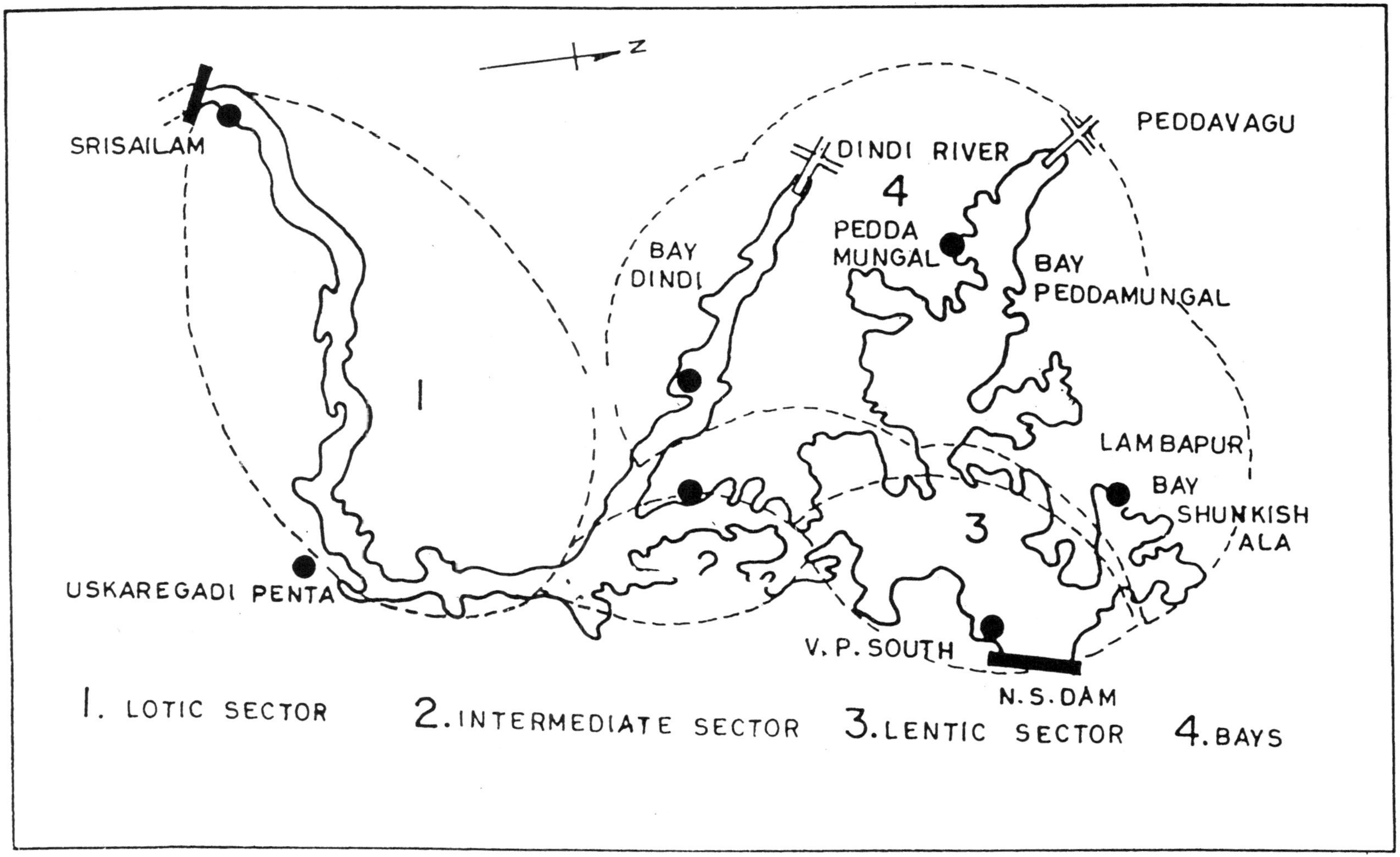

Figure 5.2. Nagarjunasagar reservoir, Andhra Pradesh

blue-green algae in blooms and a very low species diversity index (Sugunan, 1991). Though the stress factors are responsible for keeping the diversity index low, they do not lower the productivity in the main sectors of the reservoir, as stability is more directly related to diversity than does the productivity.

Barring the rich populations of molluscs in the bays, the benthic invertebrate fauna in the reservoir is rather poor due to the predominantly rocky nature of the substratum and the frequent and abrupt level fluctuations which prevent the effective colonisation of the available substrata (Sugunan and Das, 1983). That the vicissitudes of level fluctuations retard the growth of periphyton on natural substrate has been corroborated by the rapid and dense deposition of these organism on artificial substrata (Sugunan and Pathak, 1986). Steadily charged with nutrients and salts from the allochthonous sources, the well-mixed and well-illuminated reservoir has all the prerequisites for natural eutrophication, despite its deep basin. The rate of primary production is very high, up to 590 mg C m^{-2} day^{-1} during the summer. The energy fixed by the producers range from 19 577 x 10^3 to 24 674 x 10^3 k cal ha^{-1} yr^{-1} (as oxygen), which is 0.33% of the light energy. In term of carbohydrates, the conversion is 0.34%.

David *et al.* (1969b) described 38 species of fish from the reservoir to which Bhatnagar and Sugunan (1978) added another 31. There are more than 20 species of carp minnows, 25 small catfishes and other uneconomic species which compete with the commercial species resulting in low yield of commercially important fish.

The commercial catch comprises *Labeo fimbriatus, L. calbasu, Tor khudree, Catla catla, Cirrhinus mrigala, L. rohita, Pangasius pangasius, Aorichthys aor, A. seenghala, Silonia childreni,* and *Wallago attu*. Among them *L. fimbriatus* and *L. calbasu* formed the indigenous carps, which together contributed more than 40% of the catch in the initial years. Later, their percentage declined to 13% in 1975-76. During 1979-80, the two species together formed 16% of the catch (Table 5.5).

Food habits of most of the fishes are known. *L. fimbriatus* and *L. calbasu* feed mainly on detritus (Anon., 1982; Vinci and Sugunan, 1981), while *A. aor* (Ramakrishniah, 1984) *A. seenghala, Silonia childreni* (Vinci, 1986 and 1984) and *W. attu* (Anon., 1982) feed on carp minnows and other small fishes. *Pangasius pangasius* lives almost exclusively on molluscs. Even among the less important and the weed fishes, plankton does not figure as a major item of food except for *Ambassis* spp. and *Rohtee* sp. (Table 5.6 to 5.8).

Organic detritus forms the major food of *Rhinomugil corsula* and *Osteobrama vigorsii* (86% and 60% of the gut contents respectively), the other food items of the mullet being diatoms, blue- green algae, green algae, zooplankton and benthic fauna. *O. vigorsii* consumes insects, diatoms and molluscs, in addition to organic detritus. *Pseudeutropius taakree* is the only fish having insects as the predominant food (74.2%), the other components of the stomach contents being fish (14.8%), prawn, organic detritus (30%) and plankton. *Channa marulius* is essentially a predatory fish living on trash fishes (76.50%), prawn and insect.

Salmostoma phulo phulo which form 86.45% of the trash fishes is detritophagus; the detritus forming 80% of its gut contents, followed by fish remains (10.75%), crustaceans (5.12%), insect parts (3.13%), Chlorophyceae (0.50%), diatoms (0.25%) and Cyanophyceae (0.25%). *Barbus sp., Barilus spp., Chela atpar, Dania aequipinnatus, Osteobrama cotio, Oryzias melastigmus* and *Puntius sophore* also had their stomach contents dominated by organic detritus. *Ambassis* spp., and *Rohtee ogilbii* lived mainly on zooplankton. *Gagata itchkeea* and *Glossogobius giuris* were predatory on other fishes.

Table 5.5. Fish yield (kg) and species composition from 1971-72 to 1979-80

	1971-72		1972-73*		1973-74		1974-75		1975-76		1976-77		1977-78	
	Catch	%	Catch	%	Catch	%	Catch	%	Catch	%	Catch	%	Catch	%
L. fimbriatus	56 300	35.49	9 906	30.65	22 436	29.28	5 262	7.61	8 631	7.71	10 240	9.27	26 009	15.03
L. calbasu	8 880	5.60	1 588	4.81	8 265	10.83	11 808	15.92	7 086	6.33	7 571	6.85	11 872	6.86
L. rohita	640	0.40	50	0.15	-	-	-	-	1 332	1.19	1 414	1.28	1 179	0.68
C. catla	6 530	4.12	6 610	20.25	1 408	1.95	1 938	2.80	2 160	1.93	2 200	1.99	3 038	1.76
C. mrigala	890	0.56	729	2.26	-	-	1 761	2.55	660	0.59	542	0.49	1 155	0.67
T. khudree	2 120	1.34	500	1.56	3 888	5.09	1 074	1.58	2 015	1.80	2 631	2.38	3 780	2.18
A. aor	8 550	5.39	4 142	12.51	8 356	10.95	9 457	13.67	18 112	16.18	12 854	11.63	28 990	16.76
A. seenghala	11 590	7.31	727	2.21	7 267	9.52	6 233	9.01	11 295	10.09	5 515	4.99	7 100	4.10
P. pangasius	30 630	19.31	2 290	7.09	8 493	11.13	13 043	18.66	23 654	21.13	37 525	33.95	63 432	36.66
S. childreni	23 590	14.37	1 727	5.34	5 535	7.25	8 492	12.88	22 089	19.74	12 048	10.90	13 957	8 07
W. attu	1 970	1.24	934	2.89	1 679	2.20	2 416	3.49	4 164	3.72	2 457	2.2	2 966	1.71
Misc.	4 270	4.37	3 134	10.28	9 009	11.80	8 451	11.83	10 736	9.57	15 534	14.05	9 534	5.51
Total catch	158 650		32 318		76 328		69 155		111 934		110 531		173 012	
Kg/ha	8.6		1.75		4.14		3.75		6.04		6.00		9.4	

*Fishing effort was low due to civil strife in the State

Table 5.6. Food habits of important commercial fishes of the reservoir

Name of fish	Food in the order of importance	Source of information
Labeo calbasu	Organic detritus (70.15%); Bacillariophyceae, mainly *Fragilaria, Cymbella, Amphora, Asterionella, Navicula, Tabellaria, Frustulia* (10.57); Chlorophyceae, mainly *Spirogyra, Pediastrum* (5.04%); Cyanophyceae, mainly *Oscillatoria, Microcystis* (0.059%); Zooplankton, mainly nauplii, *Cyclops, Diaptomus, Keratella* (0.98%); miscellaneous items (0.37%); mud (11.66%)	Vinci & Sugunan, 1981
L. fimbriatus	Organic detritus (65.0%); diatoms (25.0%), Chlorophyceae, mainly *Spirogyra* , Cyanophyceae (5.50%); *Merismopedia, Microcystis* (1.53%) mud (1.36%); Rotifera, mainly *Keratella* (0.90%), crustacean larvae (0.71%)	Anon., 1982
L. rohita	Organic detritus (97.33%), *Fragilaria* (1.66%); *Merismopedia* (0.67%); *Amphora* (0.17%); *Penium* (0.07%); *Oscillatoria* (0.07%); *Syneccocus* (0.03%)	Anon., 1982
Tor khudree	Organic detritus (32.1%); gastropod shells (22.8%), bivalve shells (15.6%); mud (23.5%); macrovegetation (6%)	Anon., 1982
Cirrhinus mrigala	Organic detritus (85%), mud (12.0%); diatoms (2.5%); Chlorophyceae , mainly *Oedogonium* (0.5%)	Anon., 1982
Aorichthys aor	Fish, mainly, *O. phulo, O. cotio, G. giuris* (42.2%); *Macrobrachium lamarrei* (20.46%); gastropods, mainly *Viviparus* and *Melanoides* (18.05%); Insects, mainly nymphs and larvae (13.20%); detritus (2.40%); miscellaneous items (3.7%)	Rama-krishniah, 1984

Cont.

Table 5.6 (cont.). Food habits of important commercial fishes of the reservoir

Name of fish	Food in the order of importance	Source of information
A. seenghala	Fish, mainly *G. giuris, Ambassis* sp. *Etroplus, S. phulo phulo, O. cotio* (76.12%); semidigested organic matter (11.14%); prawn, mainly *M. lamarrei* (9.23%); insects (3.51%)	Vinci, 1984
Silonia childrenii	Fish, mainly *Chanda* sp., *O. phulo, O. vigorsii, O. clupeioides, P. sophore* (42.31%); *M. lamarrei* (7.31%); insects, mainly larvae and nymphs (7.87%); amphipods (4.25%); semidigested matter (38.26%)	Vinci, 1986
Pangasius pangasius	Gastropods (91%); bivalves (7%); semidigested matter (2%)	Anon., 1982
***Mystus* sp.**	Fish 1(00%)	Anon., 1982
Wallago attu	Fish (100%)	Anon., 1982

The reservoir has a rich plankton community, comprising Cyanophyceae, (mainly *Microcystis aeruginosa*), Bacillariophycea and Chlorophyceae, apart from a dense population of zooplankton, dominated by copepods and rotifers. The plankton production in Nagarjunasagar is one of the richest, among Indian reservoirs. A perusal of Tables 5.6, 5.7 and 5.8 clearly shows that the plankton is hardly utilized as a major food ingredient by any of the commercial fishes of the reservoir.

M. aeruginosa, Merismopedia sp. and *Oscillatoria* sp., which are the mainstay in the plankton community, occur in the stomachs of *Labeo calbasu, L. fimbriatus, L. rohita, Rhinomugil corsula* in very low percentages (never exceeding 0.59% of food, by volume). They were not traced in the stomach contents of trash fishes either.

Among Chlorophyceae, *Pediastrum* sp. and *Spirogyra* sp. form 5% of stomach contents in *L. calbasu*. Green algae contribute insignificantly to the food of *L. fimbriatus* (5.5%) and *Cirrhinus mrigala* (0.5%). Among trash fishes, *Barilius* spp. have 20% of their gut contents contributed by Chlorophyceae. *Spirogyra* sp., *Pediastrum* spp. and *Cosmarium* sp. are reported from the gut contents of *Chela atpar, Glossogobius giuris* and *Salmostoma phulo phulo.*

Table 5.7. Food habits of less important commercial fishes

Name of fish	Food	Source of information
Rhinomugil corsula	Detritus (86.04%); Diatoms, mainly *Navicula, Cymbella, Nitzschia, Amphora Synedra, Tabellaria, Gyrosigma, Fragilaria, Melosira* (10.6%); Cyanophyceae, mainly *Oscillatoria, Merismopedia, Microcystis* (0.59%); Chlorophyceae, mainly, *Spirogyra, Ulothrix, Pediastrum, Cosmarium* (1.16%); zooplankton, mainly *Cyclops*, nauplii, *Diaptomus, Keratella,, Lecane, Brachionus, Chydorus Daphnia, Ceriodaphnia* (1.24%); benthos, mainly insect larvae, mayfly nymphs, insect parts (.0.81%)	Sugunan & Vinci, 1981
Pseudeutropius taakree	Insects, mainly coleopterans, gyrinids, ditiscids, dipteran larvae, chironomid pupae (74.2%); fish, mainly *O. phulo, Osteobrama cotio* (14.8%); molluscs, mainly gastropods (6.7%), prawns (1.9%); miscellaneous items (2.46%)	Ramakrishniah, *1984*
Puntius dobsonii	Gastropod shell pieces (60%); mud (15%); macrovegetation (25%);	Anon., 1982
P. kolus	Molluscs, mainly gastropods, bivalves (60%); Chlorophyceae, mainly *Spirogyra, Pediastrum* (3%); diatoms, mainly *Fragilaria, Melosira, Navicula* (5%); organic detritus (30%), zooplankton, mainly copepods (2%)	Anon., 1982
Osteobrama vigorsii	Detritus (60%); insects, mainly chironomid larvae, beetles (5%), diatoms (15%), molluscs , mainly *gastropods, bivalves* (20%)	Anon, 1982
Channa marulius	Fish, mainly *Oryzias melastigmus, Esomus danricus, Ambassis* spp. (76.50%); Prawns, mainly *M. lamarrei* (14.85%); semidigested organic matter (6.30%)	Anon, 1982

Table 5.8. Food habits of trash fishes of Nagarjunasagar Reservoir

Name of fish	Food
Salmostoma phulo phulo	Organic detritus (80%); fish remains (10.75%); crustacean remains, mainly prawns, *Cyclops, Keratella, Diaptomus, Daphnia*, nauplii (5.12%); insect parts (3.13%); Chlorophyceae, mainly *Spirogyra, Pediastrum*, (0.50%); diatoms, mainly *Fragilaria, Melosira, Navicula* (0.25%); Myxophyceae, mainly *Microcystis, Merismopedia, Oscillatoria* (0.25%).
Ambassis baculis	Copepods, mainly *Diaptomus, Cyclops*, nauplii (70%); insects, mainly dipteran larvae, mayfly nymphs (15%); organic detritus (15%)
A. nama	Copepods, mainly nauplii, *Cyclops, Diaptomus* (65%); insects, mainly dipteran larvae, chironomids, mayfly nymphs (17.50%); semidigested organic matter (15%); fish remains (2.50%)
***Barbus* sp.**	Organic detritus (90.50%); insect parts (2.15%); mud (7.35%)
***Barilius* spp.**	Organic detritus (60%); Chlorophyceae, mainly *Spirogyra, Pediastrum, Ulothrix* (20%); insect parts (18%); diatoms, mainly *Navicula, Fragilaria, Melosira, Amphora* (2%)
Chela atpar	Organic detritus (80%); prawn remains (10%); insects, mainly dipteran larvae, nymphs (5%); Chlorophyceae, mainly *Spirogyra, Pediastrum Cosmarium* (3%); Cyanophyceae, mainly *Merismopedia, Microcystis* (2%)
Dannia aequipinnatus	Organic detritus (100%)
Gagata itchkeeu	Semidigested organic matter (65%); fish remains (20.25%); prawn remains (14.75%)
Glossogibius giuris	Fish, mainly *P. sophore, Ambassis* spp. (79.67%); prawn, mainly *M. lamarrei* (15.33%), insects (3%); (mayfly nymphs); amphipods (1.50%); Chlorophyceae, mainly *Spirogyra* (0.50%)
Osteobrama cotio	Organic detritus (70%), zooplankton, mainly *Diaptomus, Cyclops*, nauplii, *Chydorus, Keratella, Lecane, Daphnia* (20%), phytoplankton, mainly *Navicula* , *Fragilaria, Cymbella, Spirogyra, Pediastrum* (10%)

Cont.

Table 5.8 (cont.). Food habits of trash fishes of Nagarjunasagar Reservoir

Name of fish	Food
Oryzias melastigmus	Organic detritus (69%); phytoplankton, mainly *Navicula, Synedra, Cymbella, Spirogyra Pediastrum, Fragilaria, Tabellaria, Microcystis* (20%)
P. sophore	Organic detritus (71.48%); diatoms, mainly *Fragilaria, Navicula, Amphora, Synedra, Melosira* (22.52%); Cyanophyceae, mainly *Merismopedia, Oscillatoria, Microcystis* (22.52%); Chlorophyceae, mainly *Spirogyra, Pediastrum* (2.28%)
Rohtee ogilbii	Zooplankton, mainly nauplii, *Keratella, Daphnia, Ceriodaphnia, Diaptomus, Lecane, Cyclops* (62%); organic detritus (25%), insects, mainly dipteran larvae, mayfly nymphs (13%)

(Anon., 1982)

Plankton and benthos as fish food organisms

Share of diatoms in the gut contents of fishes is better, compared to green and blue-green algae, though it could not be established whether the diatom shells encountered in the stomach were engulfed from the plankton phase or devoured along with organic detritus. *Fragilaria, Cymbella, Amphora, Asterionella, Navicula, Tabellaria* and *Frustulia* form 10.57% of stomach contents in *L. calbasu*. Diatoms form 25% of the gut contents of *L. fimbriatus*, 2.5% in *C. mrigala*, 1.66% in *L. rohita*, 10.16% in *Rh. corsula* and 5% in *P. kolus*. Among trash fishes, *P. sophore* consumes *Fragilaria, Navicula, Amphora, Synedra* and *Melosira* (22.52%) while *Oryzias melastigmus* feeds on *Navicula, Synedra* and *Cymbella*.

Among zooplankton, copepods like, *Cyclops, Diaptomus* sp. and nauplii were very rare in the stomachs of commercial fishes. They form 0.98% in *L. calbasu* along with rotifers and 2% in *P. kolus*. Copepods form the major food of *Ambassis baculis* (70%) and *A. nama* (65%). Rotifers and cladocerans are hardly consumed either by commercial or trash fishes excepting *Rohtee ogilbii* (62%). Gastropods are largely consumed by *Pangasius pangasius* almost as a sole food item. Molluscs are taken by *P. dobsoni* (60%), *P. kolus* (60%), *T. khudree* (22.8%) and *Osteobrama vigorsii* (20%). But considering the size of their populations, the latter species does not possibly make any impact on the benthic community by way of niche utilization.

Niche utilization

Plankton as niche remains largely unutilised. The carps *viz.*, *L. fimbriatus* and *L. calbasu*, which together constitute 16.2% of the catch live mainly on detritus. Catfishes other than *Pangasius pangasius* are predatory on carp minnows and other trash fishes. This group comprising *Aorichthys aor, A. seenghala, Silonia childreni, Wallago attu, Bagarius bagarius*, etc., are living on a long food chain and hence considered uneconomic species. *P. pangasius*, on the other hand, feeds on molluscs and not considered to be on a long food chain.It is compatible with carps and utilizes a major niche namely bottom macrofauna (molluscs being the most dominant component of bottom fauna).

L. calbasu, *L. fimbriatus* and *P. pangasius* account for more than 50% of the total fish produced in the reservoir. While all the above mentioned fishes depend on detritus for food, either directly or indirectly, none of them feeds on plankton. The rich plankton crop of the reservoir remains more or less unutilised and enriches the bottom detritus on death and sinking. In all probability, the diatom shells encountered in the guts of fishes are a part of detritus. Thus, there are three basic routes of energy transformation. *L. fimbriatus*, *L. calbasu* and *C. mrigala* (20.38%) have the shortest food (detritus) chain followed by that of *P. pangasius* (27-37%) which is on a detritus-molluscs chain. The forage feeders (40%) such as *A. aor*, *A. seenghala*, *Si. childreni* and *W. attu* are on a long chain *i. e.*, either a primary energy - herbivore - predator (grazing) chain or primary energy - dead organic matter - detritivores-predators (detritus chain), both involving considerable dissipation of energy at each trophic level (Natarajan and Pathak, 1983).

Fish production potential

Natarajan *et al.* (1979) estimated that the reservoir can produce 3408 t of fish at secondary consumer level. Thus, the actual catch falls short by more than 3 200 t. The shortfall seems to be due to the imbalance in the species spectrum. Nagarjunasagar is an example of a situation where management missed the opportunity of stocking desirable fish species at the time of trophic burst. The initial bloom of plankton remained unutilised as direct food and contributed to the energy at detritus phase. Though the indigenous detritiphagous fish such as *L. fimbriatus* and *L. calbasu* could take advantage, they suffered setback due to pressure from predators. The non-predatory *P. pangasius* seems to have found a favourable breeding and feeding environment. Table 5.9 shows the change in favour of catfishes over the years. The poor performance of detritiphagous carps and large proportion of non-*Pangasius* catfishes in the catch are a cause of concern.

Table 5.9. Percentage of carps and catfish from 1971-72 to 1978-79.

Year	1971-72	72-73	73-74	74-75	75-76	76-77	77-78	78-79
Carps	47.50	60.19	47.15	30.46	19.55	22.26	27.18	23.66
Catfishes	48.11	38.98	41.05	57.71	70.86	63.69	67.30	72.80
Catfishes (without P. pangasius)	34.81	30.89	28.92	37.05	49.73	29.74	30.64	38.66

Stocking

Stocking has been erratic, both in qualitative and quantitative terms (Table 5.10), the total number stocked in a year ranging from nil to 833 700 during the period from 1964 to 1978-79. Species composition of the stock was also inconsistent. The stocking density was only 0.8 ha^{-1} yr^{-1} up to 1969, but increased to 3.49 ha^{-1} yr^{-1} during the four years from1970. Although it reached 19.2 ha^{-1} in 1978, it made no difference in the fish catch composition, which continued to be dominated by catfishes (Table 5.9). Considering the vacant niche of zooplankton feeder in the reservoir, an annual stocking of 1.8 million fingerlings of *Catla catla* was suggested by the All India Coordinated Project on Reservoir Fisheries (Anon., 1977). *L. calbasu* (600 000), *C. mrigala* and *Cyprinus carpio* (500 000 each) were also recommended for better utilisation of detritus. This required a stocking rate of 3.4 million yr^{-1}, but only 64 000 to 353 000 fingerlings could be stocked per annum (Table 5.11).

Table 5.10. Details of stocking in Nagarjunasagar reservoir

Year	*C. catla*	*L. rohita*	*C. mrigala*	*C. carpio*	*L. fimbriatus*	Total
1964-65	-	-	-	-	690	690
1965-66	-	-	-	-	-	nil
1966-67	150	-	-	2 075	730	2 955
1967-68	41 630	-	-	-	-	41 630
1968-69	-	-	26 770	17 100	-	43 870
1969-70	-	-	9 326	-	-	9 326
1970-71	6 900	12 700	26 350	101 535	1 000	148 485
1971-72	16 750	8 250	129 244	47 950	-	202 194
1972-73	-	-	-	80 254	-	80 254
1973-74	2 000	23 950	500	21 800	1 850	50 100
1974-75	-	-	93 000	-	-	93 000
1975-76	67 000	185 000	407 200	124 500	-	833 700*
1976-77	4 000	103 000	75 000	100 000	4 000	422 000*
1977-78	112 000	121 700	24 200	41 700	-	415 600*
1978-79	9 500	148 250	154 250	-	-	312 000

* Total includes 50 000 in 1975-76, 100 000 in 1976-77 and 116 000 in 1977-78 of *M. malcolmsonii.*

Other constraints

Apart from the undesirable species-mix and understocking, there are other logistic problems coming in the way of yield optimisation. The reservoir lies in a reserve forest area with very little inhabitation along its margin. All the productive areas, lying in remote places not connected by road, remain unexploited, since there is no means to market the fish without spoilage. The craft used are obsolete and tedious to operate. A state-run motor boat ensuring quick transport of fish would be an incentive for fishermen to fish untried fishing grounds of intermediate and bay sectors, which are known to be very productive. The present *free for all* licensing system is not conducive to development. There is no check on the number of units operated or their quality and dimensions. Neither is the catch monitored regularly by the authorities. Indiscriminate capture of desirable species at the breeding grounds of Peddamunigala bay continue unchecked and this single factor may be chiefly responsible for the decline of indigenous carps. Although there is a closed season, it is not enforced. The fishermen live in abject poverty and are exploited by private money lenders, who pay them a very low price for their catch. Formation of efficient cooperatives, provision of credit to the fishermen, and assistance in marketing their catch are options which should be seriously considered as a way to improve the socio-economic condtions of the fisherfolk.

Table 5.11. Stocking of major carps in Nagarjunasagar

Period	Fish stocked	Rate of stocking No/ha/yr	*C. catla*	*L. rohita*	*C. mrigala*	*L. fimbriatus*	*C. carpio*	*L. calbasu*
Actual stocking								
Upto 1969	79 296	0.8	41 780	-	36 096	1 420	-	-
1970-74	321 594	3.49	25 650	44 900	249 094	1 950	-	-
1975-78	1 414 200	19.2	192 500	557 250	660 450	4 000	-	-
Stocking requirement (Anon, 1977)	3 400 000		1800 000	-	500 000	-	500 000	600 000

5.3 HUSSAINSAGAR

Hussainsagar reservoir was constructed in1562 in the city of Hyderabad. Because of its age and the anthropogenic environmental stress it is subjected to, Hussainsagar is in an advanced stage of eutrophication. On account of its proximity to centres of learning, it attracted the attention of many scientific workers. Srinivasan *et al.* (1965) and Zafar (1966) made the first comprehensive ecological investigations in Hussainsagar. At full level, it covers 4 016 ha, which shrinks to 1 942 ha by the end of the dry season. The 'V' shaped basin holds 9.25 million m^3 of water and has a very irregular shoreline. Hussainsagar receives water from three channels, *viz.*, Kukatpally *nala*, Bownpalli *nala* and Balkur channel (Fig. 5.3). The outlets comprise the flood gates in the centre of the dam and the sluice gates on either side. Floodgates are opened only during the flood season.

The twin cities of Hyderabad and Secunderabad are 2 040.6 km^2 in area with a population of 1.8 million. Average rainfall in the city is 76 cm, mostly from the southwest monsoon. Catchment of the lake, which includes the stormwater from the city as cover 241 km^2.

Hussainsagar is one of the highly polluted reservoirs in India. Kukatpally *nala*, the main feeding channel of the reservoir, is flanked by 261 major and minor industries discharging untreated or inadequately treated effluents. Bownpalli and Balkur channels are comparatively small and discharge mainly the stormwater combined with domestic and city wastes. The pollutants include strong acids, alkalis, burnt diesel and heavy oils, pesticides, asbestos, cement and biochemical wastes (Hingorani *et al.*, 1977). Although there is no direct sewage discharge into the lake, the city sewage channel, known as *K-main*, which runs parallel to Kukatpalli *nala* has developed leaks and sewage gets mixed with the *nala* water, finding its way into the reservoir (Hingorani, 1980).

The nitrogen and phosphorus loading in the reservoir is considerable. The maximum level of inorganic nitrogen was 3.2 mg l^{-1} and that of PO_4- P 0.43 mg l^{-1} (Zafar, 1986). The pollutional effects, become acute during the summer. The pH (which is normally around 8), falls rapidly to 7 with a heavy build up of CO_2 to the level of 95 mg l^{-1} (Ghosh and George,1989). The bicarbonate alkalinity of 92 to 385 mg l^{-1}, high chlorides, little or very low level of dissolved oxygen and accumulation of H_2S and heavy metals are the other impacts of pollution.

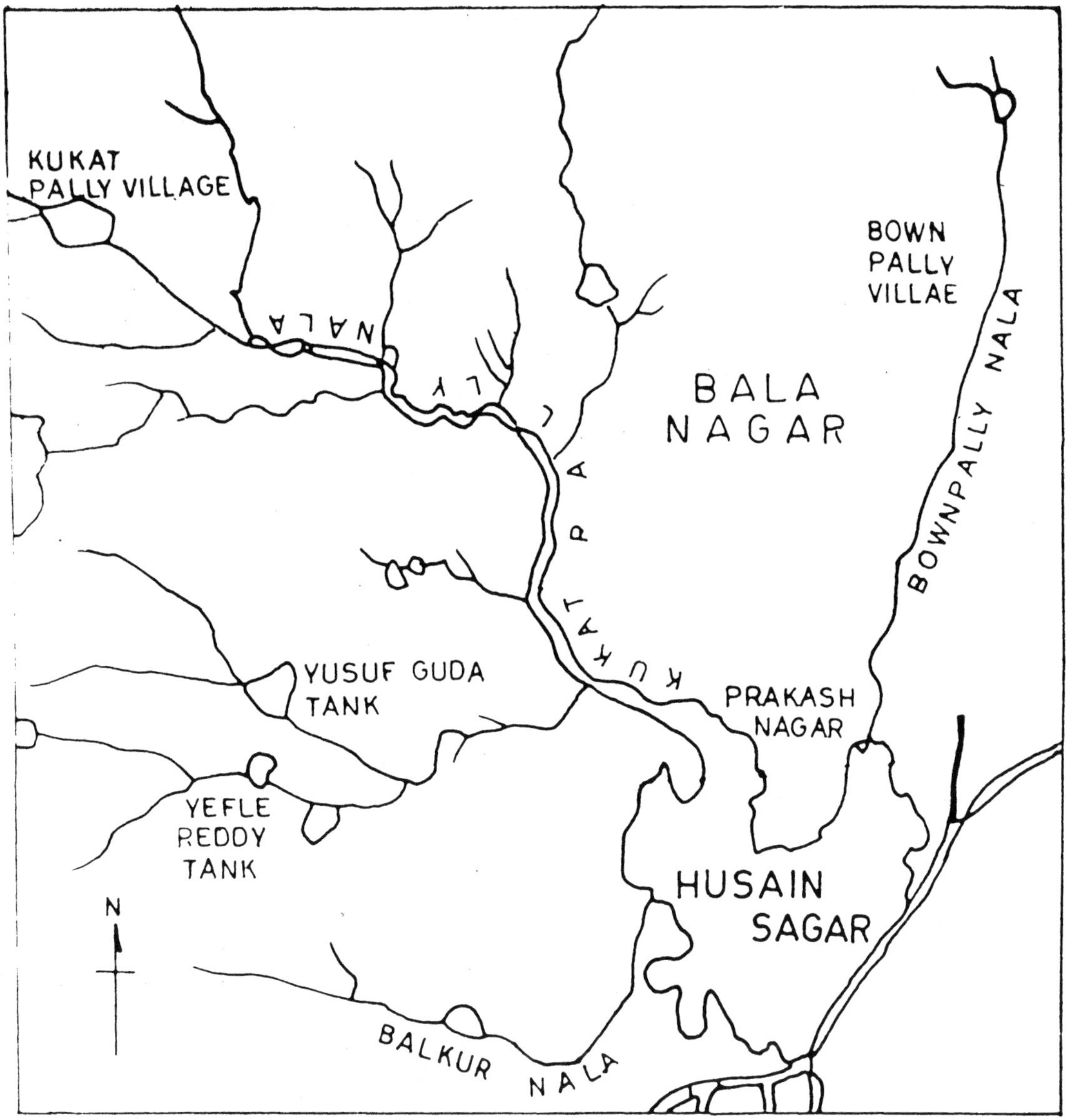

Figure 5.3. Hussainsagar reservoir, Andhra Pradesh

The effluent discharge point near Indian Drugs and Pharmaceuticals Limited (IDPL) at Kukatpally *nala* has dissolved free CO_2 up to 200 mg l^{-1} and total alkalinity of 280 to 560 mg l^{-1}. The effluent has a pH of 6.1, suspended solids 4 540 mg l^{-1}, BOD level of 4 080 mg l^{-1}, and COD 10 470 mg l^{-1}. Rao (1990), after comparing the CO_2, dissolved oxygen alkalinity, chloride content and the nutrient flux during the mid '60s and late '80s, concluded that the pollution of the reservoir has considerably increased .

Bioaccumulation of chromium, manganese and mercury is in excess of the permissible limits in water and sediment phase (Seenayya and Prahalad, 1987 and Prahalad and Seenayya, 1988; Table 5.12).

Table 5.12. Bioaccumulation of heavy metals

Profile	Cr	Mn	Hg
Water ($\mu g\ l^{-1}$)	72.33	1800.00	10.00
Nannoplankton ($\mu g\ g^{-1}$ ww)	2.09	4.18	10.46
Phytoplankton ($\mu g\ g^{-1}$ ww)	6.00	13.50	43.00
Zooplankton ($\mu g\ g^{-1}$ ww)	6.00	3.50	68.00
Particulate matter ($\mu g\ g^{-1}$ ww)	4.70	7.06	40.48
Fish ($\mu g\ g^{-1}$ ww)	6.05	1.50	23.07
Surface sediments (mg kg^{-1} dw)	120.60	912.00	12,400.00

ww= wet weight, dw=dry wt. *After Prasad, 1993*

Frequent fish mortality due to pollution has been reported from the reservoir, the most vulnerable sites being Nallagadda, Patigadda, Railway bridge, and Road bridge near IDPL. The fishes affected are mainly *Etroplus suratensis, Ompok pabda, Puntius* spp., and *Mystus* spp. The cause of death is attributed to asphyxiation (Hingorani, 1977).

Since the mid 1970s floating weeds, especially the water hyacinth, choke the lake, covering one third to half of the surface. The problem is acute during the summer, when the floating plants attain a height of 1 m. Municipal authorities incur enormous expenditure in cleaning the weeds manually. But the plants soon grow back to their original density. Clear portions of the lake have thick blooms of *Microcystis aeruginosa* and other blue-green algae. Ghosh and George (1989) report a poor species diversity in terms of phyto and zooplankton.

The fish fauna comprises *Wallago attu, Channa marulius, Mystus* spp., *Chela* sp., *Ompok* sp., *Heteropneustes fossilis, Puntius* spp., *Etroplus* spp., and *Labeo* spp. The transplanted carps, *L. rohita, Cirrhinus mrigala* and *Cyprinus carpio* appear in the catches very rarely. In the absence of any regular stocking, the wild populations are being exploited by the fishermen. A change in population structure of fishes has been reported due to rapid eutrophication, carp species giving way to air-beathing fishes and catfishes.

Twelve common species of fish including *Mastacembelus* spp., *Chela clupeoides*, *Rasbora buchanani*, *Puntius kolus*, *Brachydango reria*, *Labeo calbasu*, *L. fimbriatus*, *Danio aequipinnatus*, earlier reported by Rahimullah (1943) are now found only in low numbers. But hardy species like *Heteropneustes fossilis*, *Clarias batrachus*, *Mystus bleekeri*, *Garra mullya*, *Gambusia affinis*, *Poecilia reticulata* and *Notopterus notopterus* have become common (Rao, 1990). There is no system to collect the catch statistics or to monitor the fishing effort. An estimated 10 to 18 t of fish catch was reported in 1975-76

There are no fishermen regularly operating in this resservoir and the fishing is carried out by part-time fishermen who employ gill nets, cast nets and rod and line for raising extra income. A gill net measures 10 x 2 m with a mesh size varying from 10 to 30 mm. On account of the unremunerativeness of fishing operations, fishermen cannot afford to own a boat. They depend on improvised crafts such as floats, inflated tyres and rafts for laying and lifting of nets.

Ecosystem recovery is an uphill task. Problems posed by the man-induced eutrophication, and industrial pollution are enormous. Pollution abatement can be achieved only by:

1) diverting the city sewage to treatment plants without polluting the channels,

2) effective treatment of industrial effluents, and

3) control of macrophytes

Thereafter, the problems arising from the pollution coming from the adjoining slums will have to be addressed on a social plane. In the present situation, not only the environment is unhealthy for fish populations, but the existing fishes are a potential health hazard, in view of bioaccumulation of heavy metals and chemicals.

5.4 YERRAKALAVA RESERVOIR

Yerrakalava, an important rivulet originating from the Eastern Ghats of Khammam district, joins Upputeru, the outlet channel of the Kolleru lake. Yerrakalava is dammed 54 km away from Eluru, creating a waterspread of 1 550 ha at FRL of 81.05 m above MSL. Situated between the geographical ordinates 81° 15' 22" E and 17° 5' 44' N, the dam was originally conceived to provide relief to 8 100 ha of downstream land from flood. However, subsequently, the water from the reservoir was also diverted to irrigate 10 000 ha of agricultural land. The Yerrakalava reservoir also improves water circulation in the Kolleru wetlands. The reservoir has a catchment of 1 073 km^2 and holds 125.3 million m^3 water at FRL. The mean depth at FRL is 8 m.

Limnological studies were conducted in the reservoir during October-December 1993. The reservoir is productive, with high values of total alklinity (90 to 110 mg l^{-1}) and total hardness (80 to 100 mg l^{-1}). The water is warm (22.0 to 32.5 °C) and clear (transparency 43 to 178 cm), with a pH range of 7.4 to 8.4. Dissolved oxygen at the surface ranges from 2.0 to 8.8 mg l^{-1} and the carbon dioxide, when present, 4 to 6 mg l^{-1}. The rate of carbon fixation was high with values recorded at the surface and in sub-surface ranging between 41.66 mg C m^{-3} $hr^{-1}$and 104.16 mg C m^{-3} hr^{-1} as observed in the months October to December, 1993 (Table 5.13).

The flushing rate of Yerrakalava is very low, which causes the growth of a variety of submerged vegetation. *Hydrilla* and *Chara* dominate the the margin, while the deeper zones have patches of *Vallisneria*. Mats of *Spirogyra* are also reported from some places.

Table 5.13. Water quality parameters of Yerrakalava reservoir

Parameter	Range
Air temperature (°C)	15.0 - 30.5 °C
Water temperature (°C)	22.0 - 32.5 °C
Transparency (cm)	43-178
Dissolved oxygen (mg l^{-1})	2.0-8.8
Carbon dioxide (mg l^{-1})	nil - 6.0
Phenophthalene alkalinity (mg l^{-1})	9 - 80
Methyl orange alkalinity (mg l^{-1})	90 - 110
Total hardness (mg l^{-1})	80 - 100
Net primary productivity (mg C m^{-3} hr^{-1})	41.66-104-16

(CICFRI Barrackpore)

The untreated effluents from Adivasi Paper Mill, located at Aswaopeta in Khammam district find their way into this reservoir through Jalleru, a rivulet tributary to Yerrakalava. Dowstream, the river Yerrakalava receives waste discharge from a distillery and a paper mill. This harms the recruitment of fish in the estuarine zone.

There are about 175 active fishermen residing in 6 villages contiguous to the reservoir, who make a living out of fishing,(Table 5.14). All the 175 permanently settled active fishermen exploit the fish of the reservoir throughout the year and there is no restriction on their fishing operations. They are free to operate their gear anywhere in the reservoir. In addition to them, about 25 fishermen of A. Polavaram, a distant village, also exploit the fishery wealth of the reservoir during seasons of god catch. About 200 fishermen migrate to Yerrakalava reservoir area in February-June. They come from villages around Kolleru lake and return with the commencement of monsoon.

Fishing operations on Yerrakalava reservoir mainly use set gill nets, cast nets, long lines and basket traps. Since the reservoir is not cleared of tree stumps, other types of nets cannot be operated. Set gill nets and cast nets are employed for capturing carps and catfishes, while *pandiri mavulu*, the basket traps are used for prawns and murrels. Among the traps, *gari* is utilised for eels, and *sanna mavulu* are used for minor carps, perches, spiny eels and barbels.

Fishermen who are permanently settled and conduct fishing year round in the reservoir, usually possess the following gear for use in different seasons and for catching different species.

.	Set gill nets -	10 to 25 pieces, measuring 30 to 40 m each in length and 4 to 6.5 height, mesh sizes 3.5, 5.0, 7.0, 8.5, 12.0,16.5, 21.5 and 25.0 cm, made of synthetic yarn.
.	Cast nets	One to 3 nets each of different mesh sizes usually 2, 5 and 7 cm, 7 to 8 m diameter, made of synthetic yarn.

Table 5.14. Fishermen population at Yerrakalava reservoir

Name of villages	Permanently settled		Migratory	
	Fisherfolk	Active fishermen	Fisherfolk	Active fishermen
Villages on northern side:				
Chakradevarapalli (or Konguvarigudem)	70	15	230	50
Vegavaram	200	40	250	50
Taduvai	250	55	250	50
Villages on southern side:				
Devulapalli	40	10	80	20
Vallampatla (or Singarajupalem)	80	15	80	20
Borrampalem	180	40	110	30
Total	**820**	**175**	**1000**	**220**

(CICFRI Barrackpore)

c. Traps

1. *Pandiri mavu* - 100 to 150, usually in three sizes (big, medium and small), rectangular shape.

2. *Sanna mavu* - 50 to 100, with or withoutflaps (*gandava mavu*), rectangular in shape.

3. *Gari* - The cone shaped traps numbering 10-25, the location of traps set are indicated with the help of a float.

d. Long lines - Usually 2, each with 50 hooks.

e. Utchulu - About 50 scare traps

Gill nets are usually set for about 14 hours.The fish catch from these units ranges between nil to 9.75 kg and the catch per man hour 0.29 kg. The cast net is operated for 6 hours with a yield of one kg of *Macrobrachium malcolmsonii* and the catch per man hour is 0.16 kg.

Almost all of the active fishermen possess a flat-bottom dug out canoe made of teak. Others employ a round bottom dug out canoe made from palm trees.

Ichthyofauna

Thirteen species of fish have been recorded from the reservoir, as well as a freshwater prawn, *Macrobrachium malcolmsonii* (Table 5.15).

Table 5.15. Fish fauna of Yerrakalava reservoir

Species	% in the catch	Size range in mm
Cirrhinus reba	68.5	160-270
Labeo rohita	9.5	575-670
Puntius sarana	6.7	126-270
Notopterus notopterus	5.2	220-320
L. calbasu	4.6	200-521
Macrognathus aculeatus	2.6	265-510
Wallago attu	0.8	510
Nandus nandus	0.7	105-130
Ompck bimaculatus	0.6	175-210
Heteropneustes fossilis	0.5	200
Labeo pangusia	0.1	215
Glossogobius giuris	0.1	210
Anguilla nebulosa	0.1	400
M. malcolmsonii	-	130-196

Catla, rohu, mrigal and *Wallgo attu* are caught during monsoon months in substantial quantities. Their size ranges from 3 to 6 kg. The summer fishery is dominated by minor carps, prawns and murrels. Middlemen purchase the fish from the fishermen. There is no sale of fish by the fishermen directly to consumers. Demand for fish is high on Thursday, the weekly market day at Jangareddygudem, the nearby town.

5.5 RESERVOIR FISHERIES IN ANDHRA PRADESH

Fish production in the man-made lakes of Andhra Pradesh shows a great variety. Reservoirs of the Nellore district are particularly productive, the yield rate ranging from 93 to 142 kg ha^{-1} among the medium category, and 643 to 1 273 kg ha^{-1} among the small ones. All types of reservoirs yield less fish in Karimnagar, Hyderabad, Kurnool, Medak and Warrangal districts (Table 5.1). Production figures are available for 38 small, 28 medium and three large reservoirs (Srivastava *et al.*, 1985; Table 5.16).

Very little information is available on the reservoir fisheries of Andhra Pradesh. Production figures, stocking details, craft, gear and manpower deployed in respect of individual reservoirs are not kept by any agency, mainly due to the free fishing system followed in the State.

Two types of fishing practices are followed in the State *i. e.*, free fishing and auctioning. The large water bodies are managed on the basis of common property norms where fishermen are free to fish after obtaining a license from the local Assistant Director of Fisheries, issued free of cost. The number of units or quantum of fishing effort to be employed by a licencee is restricted. However, the local authorities are not provided with sufficient manpower, resources or executive power to enforce such restrictions. Similarly, the rules regarding closed season meant for conserving and protecting the broodstock are not strictly enforced. Free fishing is not conducive to the development of reservoir fisheries. Licence fee or royalty of some sort need to be levied from the fishermen to infuse a sense of accountability and to provide the State with the much- needed resources to provide pre- and post-harvest infrastructure facilities for the fishermen. In many reservoirs, the fishermen do not operate their gear in remote places due to lack of quick transport of their catch. State's intervention is necessary for providing facilities for transport of catch and marketing channels.

Table 5.16. Yields of reservoirs in Andhra Pradesh

	Smal <1 000 ha	Medium 1 000-5000	Large >5 000	Pooled
No. of reservoirs, production figures from which are available	38	28	3	69
Area (ha)	13 463	59 885	13 450	86 798
Production (t)	2 560	1 297	800	4 657
Range in yield (kg ha^{-1})	2-1 273	0.3-142	2.5-93	0.3-1 273
Mean yield (kg ha^{-1})	190	22	59	54

(Srivastava et al., 1985)

Stocking in reservoirs is done arbitrarily without any considerations on stocking density or species-mix. Largely it is determined by the seed availability.

Fishing rights in the small reservoirs in the State designated as *tanks* are auctioned off to private contractors, usually for a period of 12 months. The contractor hires fishermen for netting, on a daily wage or royalty basis. This short-term lease is not in conformity with norms of conservation and development since the contractors capture the entire stock, leaving nothing for the succeeding year. A large part of the profit is kept by the contractors, leaving very little for the fishermen. A long-term leasing policy or direct management of the reservoirs by the cooperatives is more desirable from conservation point of view and to ensure a sustainable yield.

Reservoir fishermen constitute one of the weakest sections of the society. They are exploited by the middlemen and money lenders, and they receive a very low price for their catch. The reservoir fisheries need to be brought under the fold of cooperative societies, which can take care of their credit needs, market their catch effectively and provide all necessary infrastructure. This is possible only with a strong help and participation from the State.

5.17 Salient features of some reservoirs in Andhra Pradesh

	Yerrakalava	Hussainsagar	Nagarajunasagar	Mir Alam
Date of closure	1988	1562	1967	1806
Area at FRL (ha)	1550	4 016	28 475	160.58
Area (minimum)	506	1 942	8383	-
Area (average)	1028	2 979	18 429	-
Maximum length (km)	-	3.3	96	-
Maximum width (km)	-	2.8	-	-
Volume (million m^3)	125.3	9.25	11 560	8.1
Maximum depth (m)	15	12.5	-	-
Mean depth (m)	8	2.5	40.5	5
Catchment Area	1 073	-	215 194	16.49
Elevation (m above MSL)	81.05	510	180	-
Length of shoreline (km)	18	-	471	11.2
Shore dev. index	1.3	1.9	7.89	8.68
Volume dev. index	1.5	0.525	1.13	1.19
Annual level fluctuations (m)	-	-	58	-
Latitude (N)	17 ° 5'	15 °	16 ° 34'	-
Longitude (E)	81 ° 15	81 °	79 ° 19	-
Inflowing rivers	Yerrakalava Jalleru	Kukatpally	Krishna	-
Outflowing river	Yerrakalava	Musi	Krishna	-
Maximum outflow (m^3 $sec.^{-1}$)	38.8	-	869	-
Soil				
pH	-	-	7.5-8.1	-
Sand (%)	-	-	39	-
Silt (%)	-	-	21	-
Clay (%)	-	-	40	-
Organic carbon (%)	-	-	1.25	-
Available N (mg 100 g^{-1})	-	-	18	-
Available P (mg 100 g^{-1})	-	-	0.45	-
$CaCO_3$ (%)	-	-	0.79-14.93	-
Water				
Water temperature (°C)	22.0 - 32.5	20.5-30.5	22.7-31.0	-
pH	-	7.0-8.1	7.6-8.6	8.6-8.7
Transparency (cm)	43 - 178	26.0-90.5	13.0-510.0	-
DO (mg l^{-1})	2.0 - 8.8	-	5.12-8.9	-
CO_2 (mg l^{-1})	nil - 6.0	nil-95	nil-7.0	-
Total alkalinity (mg l^{-1})	90 - 110	92-385	69.98-169.27	-
Spec. cond. (μ mhos)	-	-	219.6-1114.7	-
Total hardness (mg l^{-1})	80 - 100	-	50-200	-
Calcium (mg l^{-1})	-	-	20.40-38.0	-
Nitrate (mg l^{-1})	-	-	0.2-3.2	0.0165-0.0195
Phosphate (mg l^{-1})	-	0.43	0.001-0.003	0.140-0.252
Silicate (mg l^{-1})	-	-	19.0-40.0	-
GPP (mg C m^{-3} d^{-1})	41.66 - 104.16	-	90-1 205	-

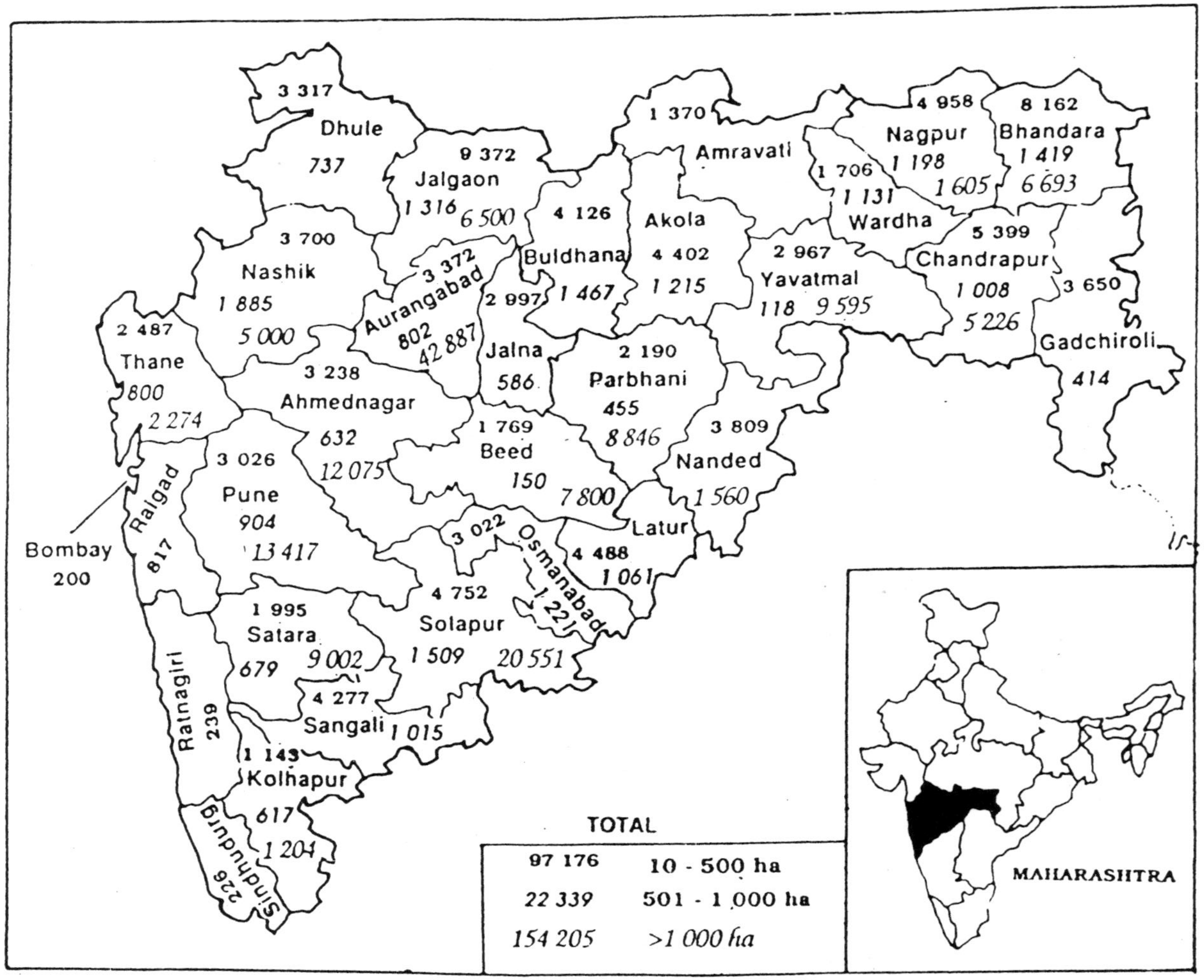

Figure 8.1. District-wise distribution of reservoirs area (ha) in Madhya Pradesh

6. MAHARASHTRA

Maharashtra, the third largest State of the Indian Union, both in population and geographical area, is surrounded by the Arabian Sea in the West, Andhra Pradesh in the southeast, Karnataka in the south, Gujarat in the northwest and Madhya Pradesh in the north. The State has three distinct physiographical regions, *viz.*, the *coastal belt (Konkan)*, the *Western Ghats* and the *eastern plateau*.The tall hills of the Western Ghats run parallel to the coastline of 720 km. The 80 km wide (average) strip of land between the Western Ghats and the sea known as Konkan, is one of the richest alluvial plains, receiving an average annual rainfall of 200 cm. The region east of the hills is a vast plateau land drained by the Godavari, Bhima, and Krishna which originate in the Western Ghats and flow eastward across the Indian peninsula into the Bay of Bengal. The rainshadow areas of the eastern Maharashtra receive rainfall of 60 to 75 cm a year or even < 50 cm in certain areas. A characteristic feature of these low rainfall districts such as Nasik, Dhule, Jalgaon, Sangali and Solapur is the traditional small irrigation impoundments. The landscape of the plateau is dotted with thousands of such small reservoirs created to tap the surface drainage (Fig. 6.1).

6.1 RESERVOIRS OF MAHARASHTRA

The State Fisheries Department of Maharashtra does not keep a list of reservoirs. Nevertheless, it has information on the total area under four different categories of water bodies, *viz.*, <10 ha, 11 to 500 ha, 501 to 1 000 ha and >1 000 ha (Table 6.1). The water bodies below 10 ha in size have been considered as of aquaculture systems and have not been included under man-made lakes. Details of the three other categories have been presented after reclassifying them as per the format followed in this report. The small reservoirs (<1 000 ha) in Maharashtra occupy 119 515 ha (44%), out of the total 273 750 ha. Among the remaining 152 205 ha, the large reservoirs (>5 000 ha) constitute 115 054 ha, the other 39 181 ha falling under the medium (1 000 to 5 000 ha) category (Table 6.2).

Information on 33 reservoirs has been gathered from Srivastava *et al.* (1985), Lonkar (1992) and a number of publications cited elsewhere. The eight large reservoirs in the State are Jayakwadi (39 777 ha) in Aurangabad district, Ujjaini (29 000 ha) in Solapur district, Shivsagar (11 216 ha) in Satara district, Yeldari (9 472 ha) and Darna (7 500 ha) in Parbhani district, Itiadoh (7 345 ha) in Bhandara district, Girna (5 420 ha) in Nasik district and Mula (5 324 ha) in Ahmednagar district (Table 6.3). Out of the 39 181 ha of medium reservoirs, the 17 listed in Table 6.3 cover an area of 38 860 ha. Details of only eight small reservoirs with a total waterspread of 4 302 ha are available.

Production figures in respect of 23 reservoirs are available (Tables 6.3 and 6.4), on the basis of which, the yield is estimated at 11.11 kg ha^{-1}. Minor fishes form a substantial component of the catch in most of the reservoirs, irrespective of the size of the water bodies or the river systems to which they belong. This indicates some lapse in the management of fish populations. In Pawana, Pus, Saikheda, Panset and Shivsagar reservoirs, the minor carps contribute 75 to 95% of the total catch.

The small reservoirs have the highest average yield (28.68 kg ha^{-1}), followed by the medium (14.44 kg ha^{-1}) and large (10.21 kg ha^{-1}) reservoirs (Table 6.4).

Table 6.1. Districtwise area estimates of different categories of water bodies in Maharashtra

Name of the district	Area under reservoirs and lakes			
	10-500 ha	500-1 000 ha.	>1 000 ha	Total
Greater Bombay	200	-	-	200
Bombay Suburban	-			
Thane	2 487	800	2 274	5 561
Raigad	817	-	-	817
Ratnagiri	239	-	-	239
Sindhudurg	226	-	-	226
Nashik	3 700	1 885	5 000	10 585
Dhule	3 317	737	-	4 054
Jalgaon	9 372	1 316	6 500	17 188
Ahmednagar	3 238	632	12 075	15 945
Pune	3 026	904	13 417	17 347
Satara	1 995	679	9 002	11 676
Sangali	4 277	1 015	-	5 292
Solapur	4 752	1 509	20 551	26 812
Kolhapur	1 143	617	1 204	2 964
Aurangabad	3 372	802	42 887	47 061
Jalna	2 997	586	-	3 583
Parbhani	2 190	455	8 846	11 491
Beed	1 769	150	7 800	9 719
Nanded	3 809	-	1 560	5 369
Osmanabad	3 022	1 221	-	4 243
Latur	4 488	1 061	-	5 549
Buldhana	4 126	1 467	-	5 593
Akola	4 402	1 215	-	5 617
Yavatmal	2 967	118	9 595	12 680
Amravati	1 370	-	-	1 370
Wardha	1 706	1 131	-	2 837
Nagpur	4 958	1 198	1 605	7 761
Bhandara	8 162	1 419	6 693	16 274
Chandrapur	5 399	1 008	5 226	11 633
Gadchiroli	3 650	414	-	4 064
Total	**97 176**	**22 339**	**154 205**	**273 750**

Table 6.2. Area under different categories of reservoirs in Maharashtra

Category	Area at FRL (ha)
<1000 ha	119 515
1 000-5 000 ha	39 181
>5 000 ha	115 054
Total	**273 750**

Table 6.3. Details on selected (investigated) reservoirs of Maharashtra

Name of the reservoir	Area at FRL (ha)	Name of the river	Production ($t\ yr^{-1}$)	Catch composition (%)		
				Major carps	Local major	Local minor
Jayakwadi	39 777	Godavari	20	5	45	50
Ujjaini	29 000	Bhima	705	19	-	-
Shivsagar	11 216	Bhagwati	-	-	-	-
Yeldari	9 472	Purna	29	80	-	20
Darna	7 500	Darna	-	-	-	-
Itiadoh	7 345	Gadhvi	40	80	20	60
Girna	5 420	Girna	-	-	-	-
Mula	5 324	Mula	-	-	-	-
Erai	4 250	Erai	45	3	5	92
Sidheswar	3 360	Purna	10	20	30	50
Ghod	3 200	Ghod	-	-	-	-
Pench	2 791	Pench	43	35	5	60
Bhatghar	2 800	Yelwandi	13.5	4	37	59
Manar	2 640	Manar	25	40	50	10
Chankapur	2 635	Girna	-	-	-	-
Gangapur	2 400	Godavari	-	-	-	-
Pawana	2 360	Pawana	25	5		95
Dhom	2 306	Krishna	-	-	-	-
Bhandardara	1 822	Pravara	21	12	22	66
Laxmi	1 813	Bhagawati	-	-	-	-
Khadakwasla	1 472	Muthai	25	15	65	20
Katepurna	1 350	Katepurna	3	-	-	-
Bor	1 334	Bor	53	23	25	52
Panset	1 252	Mishi	10	10	5	85
Pus	1 075	Pus	40	20	5	75
Saikheda	960	Khani	30	20	5	75
Chandani	815	Chandani	9	-	-	-
Wana	677	Wana	1.6			
Waghada	454	Kolwar	-	-	-	-
Morna	430	Morna	n	-	-	-
Harani	402	Harani	6	-	-	-
Galhati	343	Galhati	5.5	10	45	45
Mekhari	221	Mekhari	20	40	25	35

n - Negligible

(After Srivastava *et al.*, 1985; Lonkar, 1992)

Table 6.4. Yields of some reservoirs of Maharashtra

Name	District	Area (ha)	Year of construction	Yield (kg ha^{-1})
Jayakwadi	Aurangabad	39 777	1975	0.50
Ujjaini	Solapur	29 000	1978	24.3
Shivsagar	Satara	11 216	1963	-
Yeldari	Parbhani	9 472	1962	3.06
Darna	-	7 500	1915	
Itiadoh	Bhandara	7 345	1970	13.00
Girna	Nasik	5 420	-	-
Mula	Ahmednagar	5 324	1910	11.52
Average yield of large reservoirs				*10.21*
Erai	Chandrapur	4 250	1983	11.00
Sidheswar	Parbhani	3 360	1962	2.98
Ghod	Pune	3 200	-	-
Pench	Nagpur	2 791	1976	28.00
Bhatghar	Pune	2 800	1928	4.82
Manar	Nanded	2 640	1969	9.47
Chankapur	-	2 635	1973	
Gangapur	Aurangabad	2 400	1957	-
Pawana	Pune	2 360	1971	10.56
Dhom	Satara	2 300	-	-
Bhandardara	Ahmednagar	1 822	1910	11.52
Laxmi	Kolhapur	1 813	-	-
Khadekwasla	Pune	1 472	1938	16.98
Katepurna	-	1 350	1970	2.22
Bor	Wardha	1 334	1965	30.62
Panset	Pune	1 252	1959	7.99
Pus	Yavatmal	1 075	1970	37.20
Average yield of medium reservoirs				*14.44*
Saikheda	-	960	1969	37.21
Chandani	-	815	1964	11.04
Wana	Nagpur	677	1965	2.38
Waghada	Nashik	454	1901	-
Morna	-	430	1968	-n-
Harani	Osmanabad	402	1964	14.93
Galhati	Aurangabad	343	1964	16.03
Mekhari	-	221	1966	90.50
Average yield of Small reservoirs				*28.68*

n= negligible (After Srivastava *et al.*, 1985; Lonkar, 1992)

6.2 DHOM RESERVOIR

Dhom is primarily an irrigation reservoir created by damming the mainstream of the river Krishna at the village Dhom in Satara district. Situated at 17^0 5' N and 73^0 33' E, the 241 m long masonry dam impounds 382 million m^3 of water at the FRL of 741.7 m above MSL. The reservoir has a catchment area of 217.55 km^2 and the area at FRL is 2 306 ha. The two irrigation outlets of the dam have a discharge capacity of 26 m^3 sec^{-1}.

Trivedy (1993) places the reservoir under the oligotrophic category, based on the values of specific conductivity (usually within 50 to 90 μmhos), and COD (<20 mg l^{-1}). The level of dissolved oxygen is 5.6 to 12 mg l^{-1}, as against 12-15 mg l^{-1}, observed in organically enriched waters in the same area. Ammonia, calcium, sodium and potassium concentrations suggest oligotrophy but the high values of nitrogen and phosphorus point towards pollution. This is further confirmed by high values of organic matter and nutrients in the sediment phase. A mild faecal contamination is evident from the count of coliforms (4 to 63 100 ml^{-1}).

Phytoplankton density is a low 416 to 8 128 units l^{-1}. Other water bodies in the area have counts of several thousands. Trivedy (*op. cit.*) suggests that the water body is slowly getting mesotrophic and heading for eutrophy. *Chlorella, Navicula, Nitzschia, Synedra* and *Phormidium* figuring in the Palmer's list of 60 pollution-tolerant genera in the world form a substantial part of the planktonic algae. However, the most important genus in the reservoir is *Raphidiopsis*, a blue-green alga, usually recorded from unpolluted water bodies. Important zooplankters are *Keratella* sp., *Keratella serrulata*, and *Notholca* sp.

6.3 BHATGHAR RESERVOIR

Situated near Bhor village in Pune district, the Bhatghar dam, erected in 1928 across the river Yelwandi at 18^0 10' N and $73^0$52' E. The 1 625 m long dam created a water body that covers up to 2 800 ha at FRL and shrinks to 1 500 ha at the lowest level. The full sill level is 626.28 m and the storage capacity, 672.64 million m^3.

The river at the dam site has a catchment area of 336 km^2. The maximum length of the reservoir and the mean depth are 45 km and 24.02 m respectively. The rate of inflow varies from 5.45 to 21.94 million m^3 at an average of 12.86 million m^3, the maximum yield being in the third quarter. All the outflowing channels together discharge 2.07 to 7.85 million m^3 of water (mean 5.32 milion m^3). The Central Inland Capture Fisheries Research Institute has conducted detailed studies on the reservoir during 1987 to 1992.

The soil of Bhatghar reservoir is sandy and neutral to alkaline in reaction. The soil quality is poor in terms of organic carbon (0.31 to 0.46%), available phosphorus (1.5 to 2.3 mg 100 g^{-1}) and available nitrogen (15.5 to 22.2 mg 100 g^{-1}). The low values of specific conductivity (38.5 to 79.1 μ mhos) indicate poor ionic concentration. Nevertheless, the calcium and magnesium concentartions seem to be good. The values of total hardness (mean) at the intermediate and lotic sectors are 130.42, 101.45 and 97.00 mg l^{-1} respectively. Phosphate level at lentic sector is a high 0.36 mg l^{-1} and the total alkalinity ranges from 10 to 36 mg l^{-1}.

The soil and water quality parameters do not indicate a high productivity, except for the high values of phosphate and total hardness. The reservoir is usually weakly stratified, with an oxyclione (8.4 mg l^{-1} to 7.1 mg l^{-1}) at the bottom and increase in specific conductivity (58.2 to 63.5 μmhos) and total alkalinity (16 to 18 mg l^{-1}) towards the bottom.

Gross primary production rate varies from 20.83 mg C m^{-3} day^{-1} during July to 145 mg C m^{-3} day^{-1} in November, an average of 71.20. The plankton community of Bhatghar reservoir confirms the oligotrophic nature of the water body. Total plankton count varies from 8 to 6 601 units l^{-1}, with Chlorophyceae, represented by *Gonatozygon*, *Hormidium*, *Eudorina*, *Closteridium* and *Kirchneriella* forming the major group. The blue-green algae, *Microcystis aeruginosa* and other taxa indicative of organic enrichment are present in small numbers. Diatoms (*Navicula*, *Nitzchia*, *Syndera*, and *Surirella*), Dinophyceae (*Ceratium*) and desmids (*Staurastrum*) are the other forms recorded.

The periphyton community is represented by 20 genera of diatoms, 14 green algae, 5 blue-green algae, 3 desmids and one genus of Dinophyceae. The reservoir harbours very little benthic invertebrate and macrophyte communities.

Ichthyofauna

A total of 49 species of fishes have been recorded from the lake, belonging to 30 genera, representing 12 families.

Fisheries management

The State Fisheries Department has leased the fishing rights of the reservoir to the Gajanan Fisheries Cooperative Society of Bhor village. The Society has a total membership of 120. Nylon gill net is the chief fishing gear employed in the reservoir, which has a mesh bar range of 20 to 150 mm. A fishing unit comprises two fishermen and 10 to 15 units of gill nets of assorted mesh sizes. No boats are used.

Total fish landings from the reservoir during 1987 were estimated at 24.7 t, which declined to 12.7 t in 1988, 9 t in 1989 and 1990 and increased to 12.1 t during 1991 (Table 6.5). The mean fish production was 13.5 t and the yield 4.8 kg ha^{-1}.

Table 6.5. Fish landings (kg) of Bhatghar reservoir

	Species	1987	1988	1989	1990	1991	Σ	$\bar{X}$	%
1.	*Catla catla*	140	43	26	441	420	1 070	214	1.50
2.	*Labeo rohita*	.9	17	-	-	-	26	5	0.04
3.	*Cirrhinus mrigala*	374	727	155	184	285	1 725	345	2.54
4.	*Tor khudree*	361	85	1 024	16	88	1 574	315	2.32
5.	*Puntius dobsonii*	143	427	442	-	-	1 012	202	1.49
6.	*Wallago attu*	368	1 126	1 440	194	318	3 446	689	5.08
7.	*P. kolus*	9 574	4 667	1 751	1 019	2 278	19 289	3 858	28.45
8.	*Chela fulungee*	10 726	5 576	4 254	1 729	2 386	24 671	4 934	36.39
9.	*Chela* spp.	2 930	-	-	3 095	3 383	9 408	1 842	13.87
10.	Eels	41	-	140	-	59	240	48	0.35
11.	Murrels	17	40	-	121	485	663	132	0.98
12.	*Mugil* sp.	7	10	-	-	-	17	3	0.03
13.	Miscellaneous	15	4	48	2 200	2 398	4 665	933	6.88
	Total	24 705	12 722	9 280	8 999	12 100	67 806	13 520	100.00

1 to 3 major carps, 4 to 6 local major, 7, 11, 12 local minor *CICFRI Barrackpore*

Based on the rate of primary energy fixation, a fish yield potential of 65 kg ha^{-1} has been estimated for Bhatghar reservoir. Accordingly, a stocking rate of 250 fingerlings ha^{-1} has been proposed, thereby necessitating a stocking of 700 000 fingerlings a year. However, the actual stocking over the years varied from nil to 482 000 at an average rate of 27 fry ha^{-1} (Table 6.6). In the absence of recruitment of fish stock, sustained stocking of adequate number of fingerlings holds the key for efficient management of the reservoir.

Table 6.6. Stocking of Bhatghar reservoir

Year	No. of fry stocked (in thousands)
1979-80	100
1980-81	482
1981-82	nil
1982-83	nil
1983-84	1
1984-85	5
1985-86	10
1986-87	25
Total	**623**
$\bar{x}$	**77**
Stocking rate ($\bar{x}$)	***27.5 ha^{-1}***

6.4 OTHER RESERVOIRS

Yeldari

Yeldari reservoir is on Purna, a tributary of the Godavari. Draining a catchment area of 2 830 km^2, the reservoir covers 9 472 ha at FRL and 600 ha during the lowest level. The maximum depth is 40.6 m. Fish catch details of the reservoir from 1973-74 to 1977-78 are given in Table 6.7.

Table 6.7. Fish production trend in Yeldari reservoir

Years	Major carps	Major carps weighing < 5 kg and other fishes	Total
1973-74	11	3	14
1974-75	19	12	31
1975-76	25	14	39
1976-77	70	36	106
1977-78	72	72	144
$\bar{x}$	39	27	67

Values in t *(After Valsankar, 1980)*

Stocking is done jointly by the local cooperative society and the State Fisheries Department. Fry of Indian major carps (10 mm to 25 mm size) were stocked during 1959-70 at an average rate of 0.3 million year^{-1}.

Jayakwadi reservoir

The 39 777 ha Jayakwadi reservoir is the largest man-made lake of Maharashtra, formed by constructing a 10 200 m dam across river Godavari at Paithan in the district of Aurangabad. The river at the reservoir site has a fertile catchment area of 21 750 km^2 (Desai, 1980). The reservoir produces 20 t fish per year, equivalent to 0.5 kg ha^{-1}.

Bhandardara reservoir

One of the oldest reservoirs in Maharashtra, Bhandardara was impounded in 1910 across the river Pravara near Bhandardara village, a holiday resort in Ahmednagar district. The 1 822 ha reservoir shrinks to 259 ha during the dry season. The catchment area comprises of 121.37 km^2 of hilly forested areas of the Western Ghats, receiving a rainfall of 544 cm annually. The climate is cool, being located at an altitude of 741.51 m above MSL. Velsangkar (1993) estimated an annual fish production of 21.23 t (11.5 kg ha^{-1}) from the reservoir.

Major component (65.48%) of the landings is trash fish (13.9 t), represented by *Chela* sp. Indian and exotic major carps, *viz.*, *Catla catla, Labeo rohita, Cirrhinus mrigala and Cyprinus carpio* form 22.60% of the catch (4.8 t), followed by the mahseer, *Tor khudree* which contributes 11.92% (2.53 t).

Girna reservoir

Girna reservoir is located in the arid zone of the Malegaon Tahsil of Nashik district. The 5 420 ha reservoir has a storage capacity of 609 million m^3, with maximum and minimum depths of 43 m and 10 m respectively. The catchment area of 4 279 km^2 is hilly and it receives high rainfall (200 cm). On its way to the reservoir, water is stored in irrigation tanks, such as the Chankapur and Kelzar tanks. The reservoir was understocked in the past (Valsangkar, 1987). Arbitrary stocking was done during 1974-75, and 1977-78 by the State Government. Although regular stocking at the rate of 100 000 fish fry had been carried out by the local cooperative society since 1983-84, the stocking rate is inadequate. One significant feature of the Girna reservoir is the consistency of common carp fishery. Total fish production from the reservoir registered an increase from 73.35 t in 1980-81 to 117.58 t in 1985-86. The peak production was recorded in 1982-83 (222.5 t). Common carp has gained a steady foothold in the reservoir, perhaps at the cost of local major carps (Table 6.8). Valsangkar (*op. cit.*) attributes its success to the favourable conditions during the water renewal stage.

Table 6.8. Fish landing (t) in Girna reservoir

Year	Common carp	Local major	Local minor	Total
1980-81	12.52	35.25	26.58	74.35
1981-82	15.51	40.00	37.75	93.26
1982-83	35.50	15.00	172.00	222.50
1983-84	32.22	21.51	120.00	173.73
1984-85	37.82	17.81	130.05	185.68
1985-86	34.81	28.56	117.58	180.47
Mean	28.06	26.36	100.66	155.08

(After Valsangkar, 1987)

Table 6.9. Salient features of some reservoirs of Maharashtra

	Bhatghar	Dhom
Date of closure	1928	-
Area at FRL (ha)	2800	2306
Mean depth (m)	24.02	16.6
Volume (million m^3)	672.64	382.32
Catchment area	336	217.56
Elevation (m above MSL)	626.28	-
Annual level fluctuations (m)	31	-
Inflowing rivers	Yelvandi	Krishna
Total outflow (m^3 $sec.^{-1}$)	45	-
Latitude (N)	18^0 10'	17^0 5'
Longitude (E)	73^0 52'	73^0 33'
Water temperature (^{0}C)	17.0-31.0	21-29.5
pH	6.2-8.2	6.5-8.17
Transparency (cm)	5-266	-
DO (mg l^{-1})	2.0-16.0	5.6-12.2
CO_2 (mg l^{-1})	1.6-16.0	-
Total alkalinity (mg l^{-1})	10.0-36.0	-
Spec. cond. (μmhos)	38.5-84.25	50-90
Total hardness (mg l^{-1})	44.0-300.0	-
Calcium (mg l^{-1})	-	2.4-20.04
Phosphate (mg l^{-1})	tr.-0.36	-
Silicate (mg l^{-1})	2.5-35.0	-
Chlorides (mg l^{-1})	4.0-17.5	3.52-71
Gross primary production (mg C m^{-3} d^{-1})	20.83-145.80	-

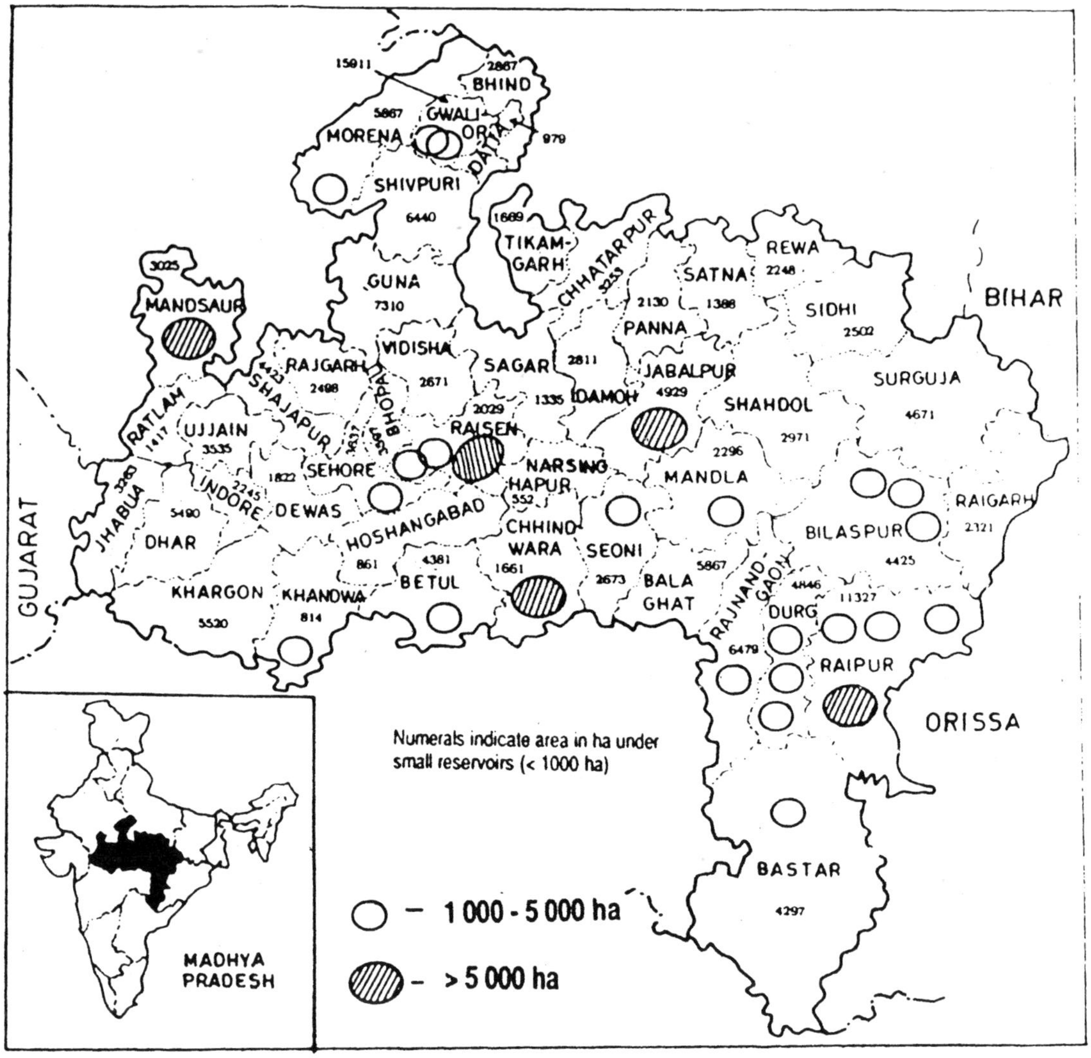

Figure 7.1. Distribution of reservoirs in Madhya Pradesh

7. MADHYA PRADESH

Situated in the centre of India, Madhya Pradesh is the largest geographic unit of the country. Except for the valleys of the Narmada and Tapti, Madhya Pradesh is mainly plateau land, with a mean elevation of 488 m above sea level, interspersed with mountains of the Vindhya and the Satpura ranges. With an average rainfall of 76 to 150 cm, the State gives birth to four great rivers of the country, *viz.*, Narmada, Tapti, Mahanadi, and Mahi and it also contains sub-catchments of the Ganga, Yamuna, Godavari and Brahmani rivers.

Rivers originating from the highlands of Madhya Pradesh drain out in three different directions. The great Narmada basin with its mainstream runs across the State from east to west, receiving tributaries on either side. The rivers Chambal, Betwa and Son descend from the northern slopes of the Vindhyas, feeding the Ganga river system. The Mahanadi flows eastwards and the Tapti cuts across the State diagonally in a southwesterly direction to enter the Bay of Bengal. The rivers, tributaries, rivulets and streams in the State have a combined length of 12 000 km.

During the monsoons, all the rivers in Madhya Pradesh turn into raging torrents, carrying enormous quantities of water. During the dry season, they shrink to rivulets. Thus, harnessing of water for irrigation and power generation is more relevant in the State, than anywhere else in India and its track record in tapping the water resources has been impressive.

7.1 RESERVOIR FISHERIES RESOURCES OF MADHYA PRADESH

Madhya Pradesh, with its 0.46 million ha of reservoirs, has the maximum waterspread under man-made lakes of all Indian States. Gandhisagar reservoir is the second largest reservoir (in area) in the country, next only to the Hirakud of Orissa. However, the State is all set to top the table when the proposed Narmadasagar reservoir (91 348 ha) becomes a reality.

The State Fisheries Department of Madhya Pradesh does not have an account of all man-made water bodies. The enumeration is especially poor in respect of the small reservoirs. However, the Statistics Wing of the State Fisheries Department maintains records on total area under different categories of inland water bodies by districts. The Department has classified the inland water bodies into four size groups, *viz.*, <10 ha, 10 to 500 ha, 501 to 1 000 ha and > 1 000 ha. The water bodies less than 10 ha in size are ponds and small tanks which are excluded from the purview of this study. The remaining three categories, by and large, comprise man-made impoundments, with very few exceptions. Among them, the second and third groups together represent small reservoirs (<1 000 ha), as defined in this study. Reservoirs of the fourth group include both medium and large reservoirs.

Within limitations of the available data, an attempt has been made to separate various categories of impoundments and place them under the present classification. The small, medium and large reservoirs in the State are estimated at 172 575, 169 502 and 118 307 ha respectively with a total of 460 384 ha (Table 7.1). Raipur district with 11 327 ha and Guna with 7 310 ha have the largest areas under small reservoirs, while Narasingpur, (552 ha), Hoshangabad (861 ha) and Khandwa (814 ha) have the least. The five large reservoirs of the State are in Raisen, Mandsaur, Jabalpur, Chindwara and Raipur districts, while 19 districts have medium reservoirs.

Description of 32 reservoirs comprising 6 small, 21 medium and 5 large (Fig. 7.1), covering a total area of 173 901 ha, is available. The average size of these small, medium and large reservoirs is 350, 2 527 and 23 661 ha respectively (Table 7.2).

Fish production trends are available for 25 reservoirs, 20 of which belong to the medium category. The Sarni reservoir (1 012 ha) in the district of Betul, producing 61.95 t of fish (61.21 kg ha^{-1}) is the most productive medium reservoir, followed by Kolar (40.0 kg ha^{-1}) and Maniyari (25.75 kg ha^{-1}). Bhimgarh (0.51 kg ha^{-1}) and Manoharsagar (1.13 kg ha^{-1}) have the poorest yield. The three large reservoirs, *viz.*, the Gandhisagar, Barna and Totladoh for which catch data are available, produce 9.21, 13.40 and 60.80 kg ha^{-1} respectively. Yields of only two small reservoirs, *viz.*, the Govindgarh (59.64 kg ha^{-1}) and Loni (28.99 kg ha^{-1}) are known (Table 7.3).

Commensurate with the size of the resource and its pivotal role in inland fisheries development, reservoirs of Madhya Pradesh received considerable research attention, compared to other States. Excepting Tamil Nadu, no other State has subjected as many reservoirs to scientific research as Madhya Pradesh did. At least 16 reservoirs find a place in the literature dealing with some aspects or another, be it ecology or fisheries management. Some reservoirs like Gandhisagar, and Ravishankarsagar are well researched. Efforts have been made in the State to develop the reservoir fisheries on scientific lines.

Table 7.1. Waterspread areas under different categories of reservoirs in Madhya Pradesh

Districts	Area of reservoirs under		
	<1 000 ha	1 000 - 5 000 ha	> 5 000 ha
Bhopal	3 597	4 000	-
Raisen	2 029	8 252	7 690
Sehore	3 637	2 268	-
Rajgarh	2 498	-	-
Vidisha	2 671	-	-
Hoshangabad	0 861	20 243	-
Betul	4 381	-	-
Jhabua	3 263	-	-
Dhar	5 490	-	-
Khargone	5 520	-	-
Khandwa	0 814	1 206	-
Indore	2 245	-	-
Ujjain	3 535	-	-
Dewas	1 822	-	-
Mandsaur	3 025	-	66 000
Shajapur	4 423	-	-
Ratlam	1 417	-	
Jabalpur	4 929	-	27 296
Narasinghpur	0 552	-	-
Mandla	2 296	2 129	-
Seoni	2 673	4 952	-
Chindwara	1 661	-	7 781

Cont.

Table 7.1 (cont.). Waterspread areas under different categories of reservoirs in Madhya Pradesh

Districts	Area of reservoirs under		
	<1 000 ha	1 000 - 5 000 ha	> 5 000 ha
Balaghat	5 867	-	-
Bilaspur	4 425	26 238	-
Raigarh	2 321	-	-
Surguja	4 671	-	-
Raipur	11 327	12 615	9 540
Rajnandgaon	6 479	3 257	-
Durg	4 846	9 462	-
Bastar	4 297	5 626	-
Gwalior	15 911	8 177	-
Datia	0 979	-	-
Shivpuri	6 440	-	-
Guna	7 310	2 500	-
Morena	5 867	3 109	-
Bhind	2 867	1 679	-
Rewa	2 248	-	-
Shahdol	2 971	51 684	-
Sidhi	2 502	-	-
Satna	1 388	-	-
Sagar	1 335	1 074	-
Damoh	2 811	-	-
Tikamgarh	6 991	1 031	-
Chatarpur	3 253	-	-
Panna	2 130	-	-
Total	172 575	169 502	118 307
Grand total	*460 384*		

Table 7.2. Average size of different categories of reservoirs

Category	Total Area (ha)	Number on units, areas of which are known	Area (ha)	Average of a unit (ha)
Small	172 575	6	2 102	350
Medium	169 502	21	53 492	2 527
Large	118 307	5	118 307	2 366
TOTAL	*460 384*	*32*	*173 901*	*5 434 (pooled)*

Table 7.3. Fish production and yields in some reservoirs of Madhya Pradesh

District	Reservoir	Area in ha (at FRL)	Production (t)	Yield rate kg ha^{-1}
Raipur	Kodar	3 583	33.58	9.37
"	Sikasar	1 862	35.32	18.97
"	Murumsilli	2 508	23.85	9.51
Durg	Kharkhara	2 975	31.51	1059
"	Gondli	1 862	12.03	6.46
"	Tandula	4 410	46.28	10.49
Rajnandgaon	Manohar Sagar	3 257	3.68	1.13
Bilaspur	Maniyari	2 460	63.35	25.75
"	Kharang	2 760	12.52	4.53
"	Hasdeo Bango	2 510	5.68	2.26
Bastar	Dudhawa	1 868	14.02	7.50
Morena	Kotwal	3 332	23.38	7.02
Gwalior	Harsi	2 523	38.38	15.21
"	Tighra	1 952	5.50	2.82
Khandwa	Sukta	1 206	23.50	19.48
Mandsour	Gandhisagar	66 000	607.80	9.21
Betul	Sarni	1 012	61.95	61.21
Sehore	Kolar	2 500	100.00	40.00
Raisen	Halali	2 528	73.47	29.17
"	Barna	7 690	103.00	13.40
Rewa	Govindgarh	0 307	18.31	59.64
"	Loni	0 208	6.03	28.99
Mandla	Thawar	2 219	14.54	6.55
Chindwara	Totladoh	7 781	473.07	60.80
Seoni	Bhimgarh	4 656	2.37	0.51
Jablapur	Bergi	27 296	-	-
Bhopal	Mansarovar	0 78	-	-
Indore	Yashwantnagar	-	-	-
Raipur	Ravishankarsagar	9 540	-	-
Raisen	Dahod	1 509	-	-
Betul	Sampua	0 520	-	-
Satna	Kulghari	0 193	-	-
Bilaspur	Dulhara	0 077	-	-

7.2 GANDHISAGAR RESERVOIR

Gandhisagar reservoir is impounded on the river Chambal, which is the largest tributary of Yamuna, a component of the Ganga river system. The river, at the dam site, receives catchment from the Vindhya ranges in the south and Aravalli in the northeast, covering a drainage area of 23 025 km^2. Important tributaries that discharge water into the Chambal are Shipra, Chhoti, Kalisindh, Ansar, and Rupniya on the eastern side and Tilsoi, Edar, Retum and Shivna in the west. The 513.6 m long, 64 m high dam has a gross storage capacity of 7 744 m m^3, covering 66 000 ha at FRL (Fig. 7.2). The mean depth is 11.73 m, shore development index 4.78, and volume development index 0.601 at the full level. Situated between the geographical ordinates, 24^0 44' N and 75^0 33' E, at an elevation of 399 m above MSL, the maximum length and width of the reservoir are 67.8 and 26.1 km respectively. Gandhisagar is one among the four hydraulic structures conceived as a part of the *Integrated Chambal Development Programme*, the other three being Ranapratapsagar, Jawaharsagar and the Kota barrage.

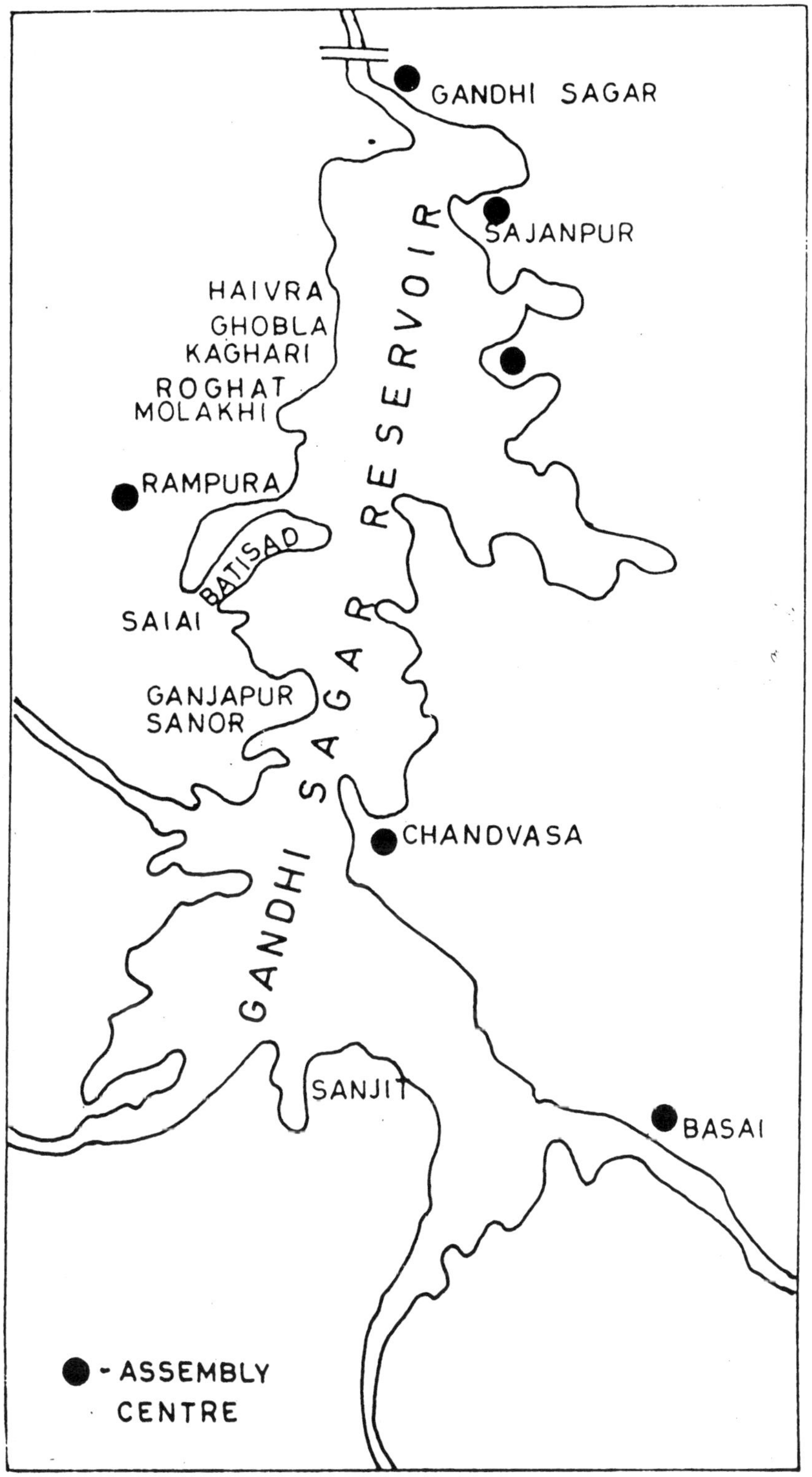

Figure 7.2. Gandhisagar reservoir, Madhya Pradesh

Gandhisagar has been the subject of scientific studies for at least a dozen research papers, the notable among them being the pre-impoundment study by Dubey and Mehra (1959) and the post- impoundment studies by Dubey and Chattterjee (1976), Rao *et al.* (1988, 1990) and Kartha and Rao (1990 and 1993). Physico-chemical attributes of water have been described by Dubey and Chatterjee (1976). In the absence of the details on seasonality, the vertical profile and the diurnal behaviour of these parameters, it is difficult to characterise the productive propensities of the water body. Nevertheless, the available information indicates that the reservoir is productive. For instance, the incoming water from the extensive catchment areas of the reservoir has nitrate nitrogen (0.9 mg l^{-1}) and phosphate (0.3 mg l^{-1}) which is high for Indian reservoirs. Total alkalinity (82 to 124 mg l^{-1}) and total dissolved solids (170 to 200 mg l^{-1}) are relatives high as well. Free carbon dioxide is not present, at least at the top of the water column, the photosynthetic requirement being met from the bicarbonates as in the case of other productive reservoirs of South India. The pH is always alkaline of 8.2 to 8.9. Unni (1993) throws light on some additional attributes like total hardness (> 100 mg l^{-1}) and calcium (36 mg l^{-1}), which are conducive for productivity.

The reservoir water has a moderate to high rate of primary productivity up to 102 mg C m^{-3} hr^{-1}, at an annual average of 40.83 and 36.75 mg C m^{-3} hr^{-1}, at the surface and sub-surface water column respectively, corroborates the positive attributes of water. This matches with a rich standing crop of phytoplankton which attains densities up to 3 211 units l^{-1} during the peaks (Rao and Choube, 1990). Diversity of phytoplankton species is notably rich, comprising 12 species of Cyanophyceae, 22 of Bacillariophyceae, 3 of Chlorophyceae and one species of Dinophyceae. In harmony with the other productive, warm, hardwater reservoirs of the country, there is one plankton pulse each during the summer and post-monsoon seasons. The planktonic organisms reach their highest density during the summer months when, due to the low water levels, productive areas of the lake are exposed to sunlight. The high temperature and a relatively stable water level aid the production process during this period. The flourishing standing crop of summer is disturbed and dislodged due to the torrential inflow and outflow during the monsoons. Nevertheless, the influx of nutrients results in another pulse of plankton, soon after the monsoon inflow.

Macrobenthic community of the reservoir is dominated by gastropods (70%), followed by bivalves, insects and oligochaetes. The molluscs colonise the littoral region. Kartha and Rao (1993), after observing 11 main fish collection centres, reported the presence of 41 species of fish, caught mainly by bottom and mid-water trawling and gill netting. This ichthyofauna comprises, among others, the commercially important carps, *Catla catla, Cirrhinus mrigala, C. reba, Labeo rohita, L. calbasu* and *Tor tor*. The predatory species comprise the murrels (*Channa marulius, C. striatus* and *C. punctatus*) and catfishes such as *Wallago attu, Aorichthys aor, A. seenghala, Bagarius bagarius* and *Silonia silondia*. The fish fauna also includes the less important commercial fishes, *Labeo boggut* sp., *L. bata, L. gonius, Rhinomugil corsula, Glossogobius giuris, Xenentedon cancila*, and *Notopterus notopoterus* along with the uneconomic species, *Osteobrama cotio, Barilius bendelisis* and *Chanda* spp. The transformation of fluviatile environment into a lentic one during the impoundment has had an impact on the fish species of Gandhisagar. Dubey and Mehra (1959) recorded 54 species of fishes in the river stretch during the pre-impoundment survey and soon after impoundment, only 39 native species were seen by Dubey and Chatterjee (1976). This included the major carps, *C. mrigala* and *L. calbasu*, and the mahseer, *T. tor*.

Fisheries

Gandhisagar is the best-managed reservoir in the State. Stocking, though inadequate has been regular ever since the programme was iniated in 1959-60 (Table 7.4). The impact of stocking has been phenomenal. Catch particulars are available for the period from 1963-64 to 1980-81 and 1984-85 to 85-86.

Table 7.4. Stocking in Gandhisagar reservoir (Number of fish stocked)

Year	*C. catla*	*L. rohita*	*C. mrigala*	*C. carpio*	Others	Total
1959-60	17 560	25 930	-	-	15 090	58 580
1960-61	-	-	-	-	171 900	171 900
1961-62	342 911	166 545	60 392	100	229 275	788 223
1962-63	450 000	148 440	-	-	125 775	724 215
1963-64	148 125	29 137	24 533	500	2 817	205 112
1964-65	36 958	55 435	92 392	-	-	184 785
1965-66	291 700	102 010	3 200	140	-	397 050
1966-67	18 180	17 715	720	-	-	36 615
1967-68	186 211	290 027	122 816	-	98 651	697 705
1968-69	47 880	55 688	107 654	-	334	211 556
1969-70	27 642	59 067	66 944	53	7 424	161 130
1970-71	-	91	85	8 142	-	8 318
1971-72	266 705	172 738	122 148	23 815	-	585 403
1972-73	-	50 565	229 146	-	40 329	320 040
1973-74	95 931	33 200	71 419	185 500	-	386 050
1974-75	68 971	60 538	203 474	103 348	2 192	438 523
Total	1 998 774	1 267 126	1 104 923	321 598	693 787	5 386 208

(After Dubey and Chatterjee, 1976)

The catch structure could be brought to a reasonably desirable level over the years by inducing a shift in favour of *Catla catla* (Table 7.5). However, the improvement in the percentage composition of rohu and mrigal was not commensurate with the stocking rate. Catla seems to be breeding in the lake, and this autostocking, catalysed by the stocking support, has resulted in the steady increase of its stock. The indigenous *C. mrigala*, is on the decline, in all probability due to the deep nature of the reservoir which retards the utilization of its favourite food, the detritus. *Labeo rohita* has staged a recovery in recent years.

The fish yield from the reservoir fluctuated from 0.51 kg ha^{-1} during 1962-63 to the level of 12 to13 kg ha^{-1} during 1978-79 and 1979-80 (Table 7.6). During the two years when the percentage of catla has been the highest (76 to 77%), the yield was also high. A close examination of Tables 7.4 to 7.6 bears out that the fluctuation in the catch is a function of stocking, more precisely the stocking of catla. It has been reported that the production of 1 342 t (20 kg ha^{-1}), was achieved at a stocking rate of 4.6 million fingerlings (70 ha^{-1}). The State Government plans to increase the stocking rate to 200 ha^{-1} to achieve a target of 2 640 to 3 960 t (40 to 60 kg ha^{-1}). September, the best month for fishing, brings in the maximum catch. This is attributed to the turbidity and the post-breeding sluggishness of fish (Dubey and Chatterjee, 1976). Fish catch during January and February is also high. From March onwards, the catch declines due to the heavy wind action, which renders gill nets less effective.

Table 7.5. Percentage of major carps in catch (Gandhisagar)

Year	Percentage by weight						
	Catla	*Rohu*	*Mrigal*	*Tor tor*	*Calbasu*	LMJ	LMN
1963-64	-	-	55.00	15.00	25.00	5.00	-
1964-65	2.30	0.26	29.30	9.35	14.55	16.22	28.02
1965-66	3.92	1.43	35.05	11.17	21.79	19.66	6.98
1966-67	17.00	2.00	17.00	15.20	27.20	7.90	13.70
1967-68	49.90	1.90	3.60	7.40	10.90	9.70	16.60
1968-69	37.15	0.37	1.31	2.95	15.82	4.46	37.94
1969-70	65.90	0.75	1.48	2.90	12.65	8.39	9.55
1970-71	63.05	1.50	1.00	1.65	16.50	8.35	7.95
1971-72	71.11	0.97	2.23	3.01	12.06	6.50	4.12
1972-73	77.55	0.98	2.87	3.90	7.58	4.68	2.44
1973-74	76.07	0.74	3.64	4.96	3.77	8.32	2.50
1974-75	71.81	0.88	2.37	5.01	6.06	10.88	2.99
1984-85	58.67	17.14	22.18	-	-	1.93	0.08
1985-86	77.12	20.83	2.01	-	-	0.03	0.01

LMJ: Local major species LMN ; Local minor species

(After Dubey and Chatterjee, 1976; Kartha and Rao, 1993)

Table 7.6. Catch and yield (Gandhisagar)

Year	Catch (t)	Yield rate (kg ha^{-1})
1962-63	34	0.51
1963-64	64	0.97
1964-65	312	4.73
1965-66	310	4.69
1966-67	214	3.24
1967-68	128	1.95
1968-69	118	1.79
1969-70	157	2.38
1970-71	172	2.60
1971-72	267	4.05
1972-73	586	8.88
1973-74	697	10.56
1974-75	359	5.44
1975-76	487	7.38
1976-77	458	6.94
1977-78	630	9.54
1978-79	778	11.79
1979-80	879	13.32
1980-81	584	8.85
1984-85	2 474	37.48
1985-86	1 342	20.33

(After Dubey and Chatterjee, 1976; Kartha and Rao, 1993; Srivastava et al., 1985)

Size groups represented in the catch are worth-noticing. Catla in the size groups 75 to 80 and 80 to 85 cm dominated the catch during 1972-73 and 73-74. This has shifted to 85-90 and 90-95 cm class during the next year. Catla, in this size group, represents IV and V year class and thus has a comfortable position vis-a-vis population. There seems to be scope for increasing the fishing effort for this species (Table 7.7). A similar shift in the size group of *Labeo rohita* from 45-50 cm to 50-55 cm has been reported. Continued domination of larger *C. mrigala* suggests that their population in the reservoir is larger than what is reflected in the catch composition. If so, they are probably not caught in adequate numbers in the gill nets, which entangle more catla due to the latter's presence in the water column for feeding on plankton.

Mode of exploitation

The fishing rights for the reservoirs were handed over to a public sector corporation in 1980. Prior to that, fishing was done either by the cooperative societies or a group of fishermen from whom a royalty ranging from Rs. 0.80 to Rs. 3.50 was charged by the Government. The royalty rate varied with fish groups. While Rs.2 to 3.50, was levied on a kg of major carps, Rs. 1.50 to 2.00 was the rate for *local major species* (such as catfishes and mahseers) and rohu. The *local minor species*, comprising small-sized miscellaneous fishes attracted a royalty of Re.0.80 to Rs. 1.00 per kg. Fish merchants, through middlemen and local agents, purchased the fish from the cooperative societies or the groups. The fishermen generally got wages for their daily labour.

There were two types of cooperative societies, *viz*., the primary cooperatives and apex society, the former engaged in fishing and the latter taking care of the marketing and providing marketing infrastructure and logistics. However, the apex society often failed to discharge the market functions and the private merchants succeeded in making inroads into the system. The royalty system had many drawbacks, such as:

a) the fishermen did not get adequate income,

b) middlemen and merchants made considerable profit,

c) the cooperatives could not effectively undertake the market functions, and

d) non-payment by the merchants/societies deprived the Government of its income in the form of royalty. It is estimated that while the Government earned Rs. 2.87 million on royalty, more than one million was spent on its collection.

Since 1980, the Madhya Pradesh State Fisheries Development Corporation (MPSFDC), a public sector undertaking has taken over the functions of the apex cooperatives. The primary societies are now actively involved in fishing, their members fishing on a catch-sharing basis.

Craft and gear

Submerged obstructions in the reservoir restricts the use of many of fishing gear and the gill net remains the mainstay of the fishing tackle employed in the reservoir. On an average, 20 gill nets, each 3.0 m in length are operated from one boat. Nylon twines of 210 d/1/3, 210 d/2/3 and 210 d/3/3 are used for the webbing, the hanging coefficient being 0.5. The mesh size ranges from 25 to 225 mm bar.

Generally, the large-meshed nets of 100 to 125 mm bar are used for the capture of large catla and mahseer during winter months and the smaller ones during summer for the other fish species. In addition to gill nets, long lines with live bait are employed to catch murrels and catfishes. Cast nets are also used along the margin.

Table 7.7. Size groups of major carps in the catch (Gandhisagar)

Size group (cm)	Percentage (number) 1972-73	1973-74	1974-75
(A) *C. catla*			
50-60	-	-	0.15
60-65	0.67	0.37	2.46
65-70	5.55	3.36	1.56
70-75	11.12	9.51	2.39
75-80	46.00	26.87	2.76
80-85	24.00	31.16	16.52
85-90	6.66	15.86	36.12
90-95	4.00	6.34	23.11
95-100	1.34	4.48	9.05
100-105	0.44	0.93	3.96
105-110	0.22	0.93	1.47
110-115	-	0.19	0.45
(B) *L. rohita*			
30-35	3.03	-	-
35-40	-	0.32	0.68
40-45	12.12	4.53	0.34
45-50	27.27	30.10	21.02
50-55	15.15	18.76	26.57
55-60	-	9.07	6.46
60-65	6.06	9.39	8.76
65-70	9.09	11.33	9.00
70-75	6.06	7.44	9.52
75-80	9.09	4.21	10.30
80-85	9.09	4.21	5.44
85-90	3.04	0.65	1.91
(C) *C. mrigala*			
35-40	-	0.11	-
40-45	-	0.47	0.44
45-50	2.02	4.19	0.76
50-55	7.58	10.30	16.33
55-60	31.82	30.03	27.81
60-65	32.32	32.36	18.98
65-70	18.18	12.22	17.65
70-75	5.05	6.75	11.41
75-80	3.03	2.68	2.65
80-85	-	0.70	3.09
85-90	-	0.36	0.88

(After Dubey and Chatterjee, 1976)

A plank-built, flat bottomed canoe, 2 to 3 m in length and 0.6 to 0.7 m in width, is the most popular fishing craft. The other, less common variety, is the Bengal type *dinghy*, which is 5 to 7 m in length, introduced by the repatriates from the erstwhile East Pakistan. The dinghy has the additonal facility for setting sails for wind propulsion. There are about 300 boats in the reservoir, each operated by 2 to 3 fishermen. The cooperative societies arrange loans for procurement of boats to those fishermen who could afford to own one and have the capacity to repay. Cooperative societies also use mechanised boats, provided by the Department of Fisheries on rental basis, for navigation and fish transport.

Kartha and Rao (1990) report their interesting findings on the efficacy of various types of trawling in increasing the productivity of commercial carps and checking the predator and weed fish populations. After experimental trawling with single-boat bottom trawling, two-boat bottom trawling and two-boat mid-water trawling under different speeds, it has been found that more than 92% of the total catch consisted of economic fish such as catla, rohu, murrels, mullets and featherbacks in two-boat mid-water trawling. This was in sharp contrast to the bottom trawling, of both single boat and two boat variety, which yielded mostly (64 to 91%) the non-commercial species of fish. The authors have recommended two-boat mid-water trawling at a speed of 3 to 4 knots for exploitation of commercial species and single and two-boat bottom trawling at 2 to 3 knots for eradication of uneconomic species of fishes.

Conservation

The State Government prohibits fish capture from rivers and reservoirs including Gandhisagar, during 16 June to 15 August. The upper stretches of the river Chambal and its tributaries such as Chhoti Kali Sindh and Shipra falls on the inter-State border with Rajasthan and lack of jurisdiction and diverse conservation rules of the two States often pose difficulties in the strict enforcement of the closed season. A 10 km stretch of the river Shipra, where the fish congregrate for breeding is declared as a sanctuary. Exploitation of under-sized fishes has been checked by the mesh and fish size regulations. Use of nets below 25 mm bar is not allowed, and this is being enforced without any difficulty. Capture of rohu, mrigal, mahseer and common carp below 1 kg is prohibited under the law. The upper size limit for catla is 4 kg.

7.3 RAVISHANKARSAGAR

Ravishankarsagar reservoir was formed in 1978, on the river Mahanadi, at 20^0 38' N, 81^0 34' E, 92 km south of the city of Raipur. A multi-purpose reservoir, serving irrigation, hydro-electric power-generation and the industrial requirements of the Bhilai Steel Plant, it has an area of 9 540 ha at the FRL of 348.7 m above MSL. The lowest waterspread during the lean season is 3 220 ha. Total catchment area is estimated at 3 620 km^2, of which 625 km^2 is intercepted by the upstream dam, Dudhawa and 486 km^2 by Murumsilli reservoir (Fig. 7.3). At full level, the reservoir holds 909 million m^3 of water (mean depth of 9.53 m). Maximum depth is 32 m.

The Central Inland Capture Fisheries Research Institute has worked on the ecology and fisheries of this reservoir for five years from 1988. The details presented here are collected from various reports of the Institute with the permission from its Director. The reservoir has a shoreline of 102.4 m and a low shoreline development index of 2.0 which suggests a low productivity. One salient hydrographic feature of this reservoir is the continuous drawdown for industrial, irrigation and hydel purposes. The outflowing channels let out water in the order of 11 000 to 30 000 m^3 per second in the peak season, the rate coming down to 230 m^3 during the summer. However, at no time of the year, the outlets are closed completely preventing the formation of stagnant water. Water level fluctuates by 3 to 5 m in a year. The

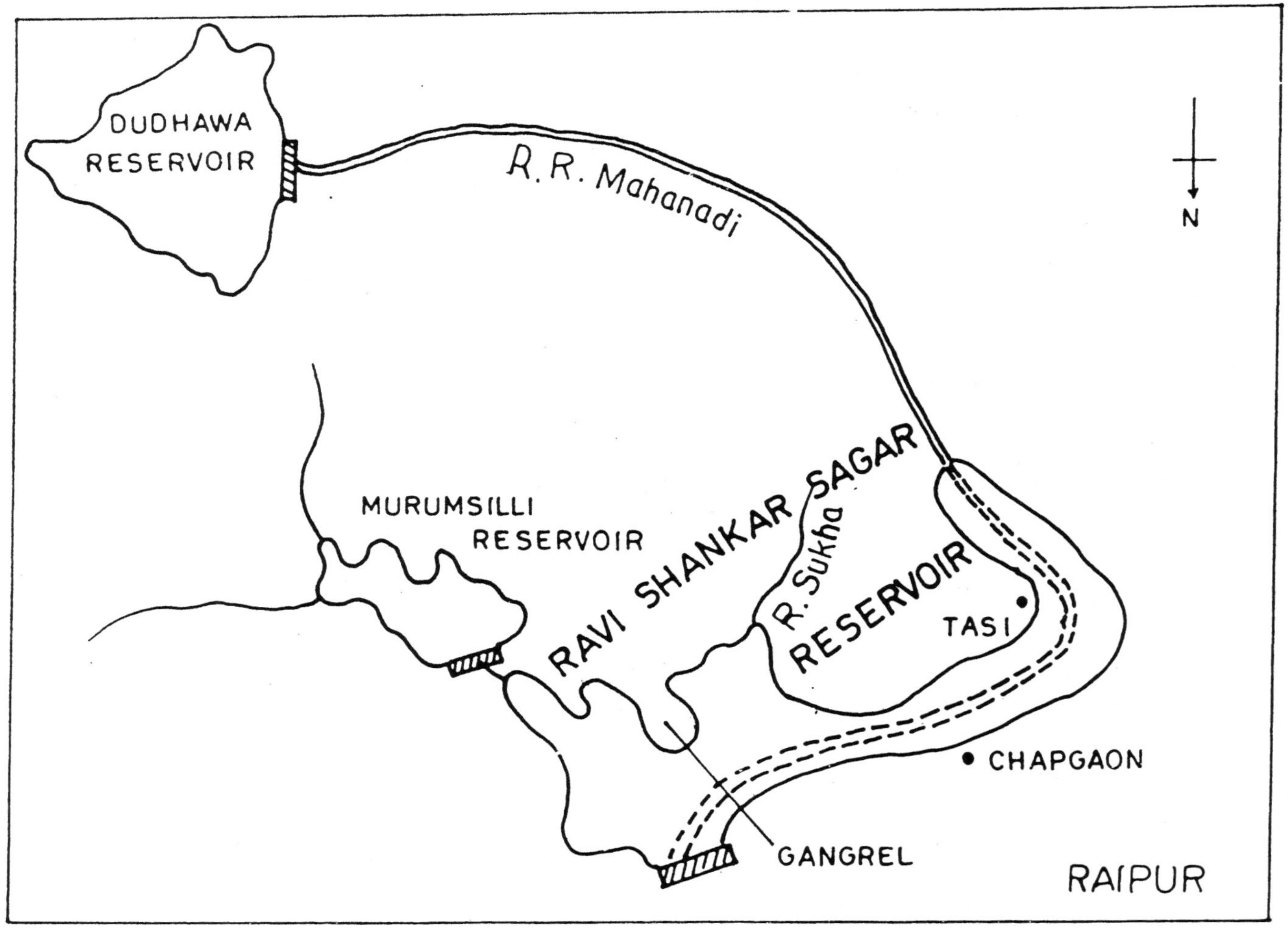

Figure 7.3. Ravishankarsagar, Dudhawa and Murumsilli reservoirs, Madhya Pradesh

sediments of Ravishankarsagar is poor in nutrients and organic matter. With respect to the available nitrogen (12 to 30 mg 100 g^{-1}) and organic carbon (0.2-1.2%), the reservoir has a low productivity, but the level of phosphorus (2.7 to 8.0 mg 100 g^{-1}) suggests a high productivity .

Unlike the thickly forested catchment of Gandhisagar, the Ravishankarsagar reservoir derives water from rocky terrain or deforested areas. There are several impoundments above Ravishankarsagar which deprives it of suspended matter. enrichment of the reservoir by allochthonous nutrients is limited. The physico-chemical attributes of water bear out this fact well. Free CO_2 is present in the surface during most part of the year, indicating that it is not utilized for photosynthesis. Only during April-July the CO_2 is depleted with a corresponding increase in the values of dissolved oxygen, suggesting a higher rate of photosynthesis. Overall ionic build-up is rather moderate, as expressed by the values of total alkalinity (38 to 94 mg l^{-1}; mean 66.5), specific conductivity (44 to100 μmhos; mean 67.2) and total dissolved solids (22.0 to 50.0 mg l^{-1}; mean 33.5). Total hardness ($CaCO_3$) of 48.7 mg l^{-1}, calcium of 37.9 mg l^{-1} and magnesium of 10.8 mg l^{-1} place the lake under medium hard water category. Phosphate and nitrate are either below detectable levels or very low (mean 0.05 mg l^{-1} nitrate, 0.08 mg l^{-1} phosphate) (Table 7.8).

Thermal stratification is very weak, with the temperature dropping 2 to 3 °C in a column of 16 m. During a period of three years, except for one month (April 1992) the oxygen depletion in the tropholytic zone was negligible. Similarly, there is no appreciable decrease in pH or increase in total alkalinity and specific conductivity with depth, as is common in the eutrophic warm water reservoirs. All the limno-chemical attributes suggest a low to moderate rate of organic productivity, compared to other reservoirs in the similar geo-climatic regimes.

The low productivity of the reservoir can also be seen from the low rates of carbon synthesis and a low standing crop of plankton and benthic communities. Gross primary production rate is estimated at 16 to 62 mg C m^3 hr^{-1}, which is equivalent to 197 to 750 mg C m^{-3} day^{-1}, the yearly mean being 285 mg C m^{-3} day^{-1}. Total plankton count ranges from 812 to 3 455 units l^{-1}, the higher values normally recorded for a brief period in summer months, when the reservoir is relatively stable. The main component of the net plankton being zooplankton (88%), the photosynthesis is carried out mainly by the nannoplankton. Phytoplankton is dominated by *Microcystis, Coelosphaerium, Pediastrum, Microspora, Navicula, Fragilaria* and *Synedra* and there are 25 genera of zooplankton mainly *Diaptomus, Cyclops, Daphnia, Diaphnosoma Bosmina, Keratella, Brachionus, Difflugia* and *Arcella*. Benthos is rich, when compared to the plankton communities. Dipteran larvae (52.6%), gastropods (30.6%) caddis worms (13.7%), bivalves (2.6%) and oligochaetes (0.5%) are the major groups. The bottom biota is particularly rich during the months with minimum water level fluctuation. The annual average density of benthic macrofauna is 254 individuals m^{-2}, but the figure doubles during the years of less water movements (550 individuals m^{-2} in 1991-92).

The fish yield potential estimates, vary from 160 to 200 t or 25 to 33 kg ha^{-1} yr^{-1}. Against this, the present yield is round 8 kg ha^{-1}. Fish production has registered a four-fold increase during the last five years (Table 7.9). Catch structure is biased in favour of predatory catfishes, mainly *Aorichthys aor, A. seenghala* and *Wallago attu*. Their higher percentage, along with the other predators like *Channa* spp. and *Notopterus* spp. lowers the fish production by diverting the energy through a longer food chain. Predatory fishes along with the uneconomic species always outnumber the carps and the stocking done so far has had only a limited impact on the catch structure.

Table 7.8. Water quality of Ravishankarsagar

	Sector				
	Lentic	*Inter*	*Lotic*	Reservoir mean	Remarks
Temperature (°C)	21.0-29.0 (26.23)	23.0-30.5 (27.5)	23-30.5 (27.3)	26.8 ±2.9	Vertical stratification absent or weak in summer,
DO ($mg\ l^{-1}$)	6.76-9 (8.17)	7.0-9.8 (8.57)	0.8-10.2 (8.69)	8.47±1.0	Orthograde
CO_2 ($mg\ l^{-1}$)	Nil-4.0 (1.09)	nil-10.0 (1.80)	nil-10.0 (1.75)	1.58±2.68	Present only during August-February
Carbonate ($mg\ l^{-1}$)	nil-12.0 (2.01)	nil-12.0 (2.8)	nil-12.0 (3.6)	3.35 ± 4.5	Absent from August to February
Bicarbonate ($mg\ l^{-1}$)	46-72 (60.54)	38-94 (64.8)	40-90 (64.4)	63.16±14.2	Present throughout the year
pH	8-8.4 (8.13)	7.6-8.4 (8.08)	7.4-8.4 (8.05)	8.08±0.30	Low values (7.4 to 7.7) during monsoon, otherwise above 8.0. No stratification
Total alkalinity ($mg\ l^{-1}$)	46-80 (64.18)	40-94 (67.6)	40-90 (68)	66.5±14.1	No stratification
Ca ($mg\ l^{-1}$)	32-39 (34)	37-44 (39.33)	38-43 (40.33)	37.88±4.1	Based on observation during December to March
Mg ($mg\ l^{-1}$)	8-14 (11)	6-12 (9)	10-15 (12.33)	10.77±2.7	-do-
Hardness ($mg\ l^{-1}$)	43-47 (45)	46-50 (48.33)	50.58 (52.66)	48.66±4.0	-do-
Phosphate ($mg\ l^{-1}$)	0.04-0.08 (0.06)	0.04-0.16 (0.07)	0.04-0.16 (0.10)	0.08±0.05	-do-
Nitrate ($mg\ l^{-1}$)	0.05 (0.05)	0.05 (0.05)	- (0.05)	0.05±0	-do-
Spec. cond. (µmhos)	50-94 (67.09)	45-98 (67.5)	44-100 (67.2)	67.25±14.9	No stratification
TDS ($mg\ l^{-1}$)	25-47 (33.54)	22.5- 49 (33.75)	22-50 (33.6)	33.62± 7.4	No stratification

(CICFRI Barrackpore)

Table 7.9. Fish catch (Ravishankarsagar)

Year	Total catch (t)	Yield (kg ha^{-1})	Percentage composition
1987-88	13.0	2.04	MC 9%, CF 44%, Misc. 32%
1988-89	17.7	2.77	MC 17%, CF 47%, Misc, 21%
1989-90	41.2	6.46	MC 32%, CF 24%, Misc. 44%
1990-91	50.4	7.90	CF 40%, CF 28%, Misc. 31%
1991-92	53.0	8.30	MC 21%, CF 52%, Misc. 24%
1992-93*	12.8	2.01	MC 11%, CF 52%, Misc. 36%

MC=Major carps, CF=Catfishes, Misc.=Miscellaneous (*low yield due to non-fishing; only 89 days fishing during Nov.-Jan.)

The stocking done so far, though erratic both qualitatively and quantitatively, has been on a large scale compared to many other Indian reservoirs. Annual stocking figures during 1983-84 to 1990-91 range from 1.2 to 3.3 million, which is equivalent to 198 to 516 ha^{-1} (Table 7.10).

Table 7.10. Stocking (Ravishankarsagar)

Year	Total seed stocked *(in millions)*	Stocking rate *(per ha)*	Percentage *Catla*	*Rohu*	*Mrigal*
1983-84	1.3	198	12.5	79.0	8.5
1984-85	1.5	237	11.0	19.0	70.0
1985-86	3.3	516	3.0	16.0	31.0
1986-87	2.1	335	8.0	50.0	42.0
1987-88	1.2	193	56.0	32.0	11.0
1988-89	1.5	230	25.0	41.5	25.5
1989-90	1.3	206	33.0	16.0	51.0
1990-91	2.0	312	48.0	11.0	41.0
Total	**14.2**	-	-	-	-
Mean	***1.77***	***278***	***19.0***	***44.5***	***36.5***

The species-mix of stocking material changes during different years, but the number of fingerlings stocked has been considerable (14.2 million in eight years, *i. e.*, @ 278 ha^{-1} yr^{-1}). Poor impact of stocking on the catch structure is attributed to the presence of predatory catfishes which take a heavy toll of the stocked fingerlings. Breeding of economic carps such as *Catla catla*, *Labeo rohita*, *L. calbasu*, *L. fimbriatus* and *Cirrhinus mrigala*, has been reported to be hampered due to want of submergence of breeding grounds in the upper stretches. The breeding failure, coupled with the predatory pressure, results in the undesirable species ratio of the commercial catches.

Management

Fishing rights of Ravishankarsagar vest with the Madhya Pradesh State Fisheries Development Corporation (MPSFDC) since 1984-85. There are two modes of fishing operations *viz.*, direct fishing and contract fishing. Under the direct fishing system, the corporation provides all fishing implements to the fishermen and procures the fish brought ashore by them on payment of remuneration, fixed on the basis of the size and category of fish they catch. These rates are periodically revised by the Corporation. The rates prevalent during 1991-92 are as follows:

	Category of fish	Remuneration
1.	Catla above 5 kg	Rs. 8.00 kg^{-1}
2.	Major carps	Rs. 5.00 kg^{-1}
3.	Local major (catfishes, murrels, etc.)	Rs. 3.00 kg^{-1}
4.	Local minor (other species below 0.5 kg, *Puntius* spp., *L. bata*, etc.)	Rs. 2.00 kg^{-1}
5.	Minnows	Rs. 0.50 kg^{-1}

Under the contract fishing system, the full reservoir or a part of it is let out to contractors who, in turn, hire fishermen on daily wages for fishing. The contractors are free to sell the fish in the private market. A royalty is levied on the contractor at the rate fixed by the Corporation, the current rate (1991-92) being Rs. 14.00, 12.00 and 8.50 per kg for the major carps, local major and the local minor categories respectively.

Craft and gear

The gear applied are mostly gill nets, although drag nets, traps and a triangular dip nets (*pelna*) are used occasionally. Everyday, 50 to 60 country boats operate in the reservoir. They are either owned by the fishermen or rented from the MPSFDC. There are six cooperative societies operating in the reservoir, to which the loyalties of 225 fishermen are divided. The cooperatives get their revenue either from the fishing contract deals with the MPSFDC or from the members in the form of commission charged on their earnings. The societies also advance loans, procure fishing gear and arrange bank loans for the member fishermen. Despite a relatively low production capacity, the reservoir's fish yield could be substantially enhanced by adopting the following measures as perceived by the Central Inland Capture Fisheries Research Institute, Barrackpore:

1) A stocking rate of 400 to 500 ha^{-1}, with an allowance of 60%, to make good the loss due to the predators and the escapement through the outflowing channels.

2) Rationalisation of the species-mix in stocking to build up good fish stocks of desirable species. Catla and rohu should be around 30 and

20% respectively. With the subdued presence of plankton communities, emphasis needs to be laid on the detritivores, such as, *Cyprinus carpio* and *Cirrhinus mrigala*.

3) Selective fishing is to be adopted employing suitable gear to control the population of predatory and weed fishes. Drag netting with a provision to release the carps of smaller sizes back to the reservoir would be a welcome step.

4) Stricter enforcement of fishing restrictions during monsoon is required as a conservation measure.

5) Under the contract system, the reservoir should be leased out only to cooperatives, lest the private contractors exploit the fishermen.

6) The MPSFDC should create and extend post-harvest and market infrastructure facilities to the fishermen/ cooperatives as an incentive.

7.4 GOVINDGARH

Govindgarh is one of the oldest reservoirs of Madhya Pradesh, construction of which was started in 1856 and completed in 1916. It is a typical small irrigation reservoir of the State, created by blocking the small streams in the upper catchment of a tributary of the Ganga river system called the river Bichia. This is the type of water body, usually designated as *tank* in South India. The 307 ha Govindgarh reservoir has a small local catchment of 25.12 km^3 from the adjoining hills of Kaimur ranges, receiving 111 cm yr^{-1} of rain. Situated 22 km south of Rewa, (geographical ordinates, 24° 24' N and 81° 15' E), the reservoir has a live storage capacity of 8.69 million m^3. Water quality, primary productivity and plankton are described by Mathew (1975, 1977 and 1985). Not much is known about its ichthyofauna and fisheries management.

Physico-chemical profile of the reservoir presents a rather enigmatic picture, with certain parameters indicating an oligotrophic regime while others pointing towards eutrophy. The water is uniformly warm throughout the year with an upper temperature limit of 30.5 °C and the lower values never sliding below 18.6 °C. Higher pH values, transparency and dissolved oxygen also indicate the reservoir's propensity towards productivity. However, low total alkalinity and nutrients suggest an unfavourable environment for a high rate of primary productivity. Considering the age of the reservoir, this poor build-up of nutrients is intriguing (Table 7.11). The heavy rainfall in the catchment and the high flushing rate result in washdown of nutrients and standing crop of plankton and benthos over the sluice gates. This not only disturbs but also arrests the process of primary community succession.

The low rates of dissolved solids and nutrients do not seem to have effect on the production. The rate of primary productivity, though not very high, does not portray a picture as bad as reflected by some of the chemical indicators. Annual average gross production of carbon for the years 1968 and '69 was estimated by Mathew (1977) at 1.104 and 1.482 g C m^{-3} day^{-1}, respectively, assuming a photosynthetic activity for 12 hours a day. This is equal to 404 and 541 g C m^{-3} for the two years respectively. The primary production also exhibits a seasonality, with a spurt in this activity during July-August in consonance with the water renewal. Total alkalinity during this period is the highest, thus suggesting the importance of allochthonous input of salts. Rapid utilization of phosphate and nitrate is also indicated during this period. This seasonality of carbon fixation rate confirms the role of water renewal pattern in retarding eutrophication in a shallow, and warm tropical lake.

Table 7.11. Water quality of Govindgarh reservoir

Parameter	Range	Mean
Water temperature ^{0}C	18.67-30.55	26.71
Transparancy (cm)	55.80-91.60	75.07
Dissolved oxygen (mg l^{-1})	7.58-10.64	8.67
Free Co_2 (mg l^{-1})	1.95-4.56	3.07
pH	8.00-8.20	8.02
Total alkalinity (mg l^{-1})	30.74-53.90	39.01
Phosphate (mg l^{-1})	neg-0.033	0.013
Nitrate (mg l^{-1})	neg-0.130	0.031
Silicate (mg l^{-1})	neg-12.93	7.75

Figures for 1969 neg- negligible *(After Mathew, 1975)*

Given the level of certain parameters indicating oligotrophy, the community structure and total count of the plankton are rather good. Mean density of plankton ranged from 1 030 to 1 185 units l^{-1} during 1968 and 1969; the share of phytoplankton being 898 and 1 048 unit l^{-1} respectively (Table 7.12). Mathew (1975) suggests the presence of *Microcystis, Anabaena, Melosira* and *Ceratium* among the plankton as an indication that the reservoir is on its road to eutrophication. The contention is fortified by the absence of desmids and pennales, and a high rate of photosynthesis. A good standing crop of plankton, despite a low ionic concentration and available nutrients, is not uncommon in tropical lakes.

Table 7.12. Composition of plankton (Govindgarh)

Groups	Count per litre		Percentage
	Range	Mean	
Chlorophyceae	6-281	130	10.97
Bacillariophyceae	171-823	425	35.87
Cyanophyceae	5-750	115	9.70
Dinophyceae	49-840	378	31.90
Phytoplankton	380-1637	1 048	88.44
Protozoa	1-31	9	0.76
Rotifera	13-89	70	5.90
Copepoda	3-26	13	1.27
Nauplii	8-63	27	2.28
Cladocera	5-29	16	1.35
Zooplankton	49-348	137	11.56
TOTAL	568-1 728	1 185	100.00

(After Mathew, 1975)

Table 7.13. Fish landings (Govindgarh)

Year	Total catch (kg)	Yield (kg ha^{-1})
1961-62	7 215	23.5
1962-63	707	2.3
1963-64	2 361	7.7
1964-65	132	0.4
1965-66	2 957	9.6
1966-67	9 634	31.4
1967-68	3 918	12.8
1968-69	4 504	14.7
1974-75	5, 350	17.5
1975-76	7, 450	24.3
1976-77	8, 080	26.4
1977-78	4, 880	15.9
1978-79	5, 270	17.2
1979-80	5, 760	18.8
Average	4 872	15.92

A perusal of fish catch statistics of Govindgarh reservoir from 1961-62 to 1968-69 and 1974-75 to 79-80 (Table 7.13) reveals sharp fluctuations in yield rates, varying from 0.4 to 26.4 kg ha^{-1} at an average of 15.92 kg ha^{-1}. Indian major carps, *Catla catla, Labeo rohita, Cirrhinus mrigala* and *Tor tor* constitute 90% of the catch (Mathew, 1975) and the predators are very few. It is not known whether Indian major carps constitute a part of the indigenous ichthyofauna of the reservoir by virtue of its connection with Ganga river system, since they are also stocked regularly. They are reported to be breeding in the reservoir (Bhatnagar *et al.*, 1977). *T. tor* is one of the local species which contributes substantialy to the catch. Carps, especially catla, registers impressive growth, specimens weighing 12 kg being a very common sight in the reservoir. A giant catla weighing 40 kg is reported have been caught. Average sizes of rohu and mrigal in the catch are 3 and 5 kg respectively, while the mahseer, *T. tor* grows to a smaller size but outnumbers the other fishes in the catch. According to various estimates, the fish production potential of the reservoir varies from 67.79 to 128.38 t and thus only 4 to 7% of the potential is presently harvested.

Mathew (1975) emphasised the need to increase the stocking rate to enhance productivity. However, details of stocking during that period are not known. Srivastava *et al.* (1985) reported stocking of 5 704, 6 000 and 121 000 fingerlings during the three years from 1972-73 to 1974-75. This is equivalent to 19, 20 and 394 fingerlings ha^{-1}, a very low rate, considering the size and category of the reservoir.

A waste weir limits the water level of the reservoir to 30.05 m allowing the excess water to flow down to the *Jhiria nala*, a stream which ultimately joins the Ganga river system through the Bichia river. This open overflow not only results in loss of fish eggs, but also may be letting out a considerable number of young fish, depriving the reservoir of its stock strength. Necessary allowances have to be made while estimating the stocking density. Moreover, a stocking and harvesting schedule needs to be evolved, allowing the stocked fish to grow for a maximum period of time during the non-overflow months, even at the cost of some reduction in the size at harvest. In such a system, larger stock-density can be maintained to make good the production loss due to early harvest.

7.5 MANSAROVAR (Fig. 7.4)

Mansarovar, situated in the heart of the city of Bhopal, is a typical example of a reservoir subjected to man-induced eutrophication. The 79.7 ha man-made lake is created on a tributary of Kaliasole river and a local stream near a thickly populated southern part of the city. Although the main source of water is rain-fed catchment comprising 8.29 km^2, the storm water channels and domestic wastes of a very high order from the city find their way into the lake. This heavy organic load and the shallow basin (mean depth: 2.87 m) set the ideal conditions for hyper-eutrophication (Kulshrestha *et. al.*, 1992).

Suspended particles inhibit the light penetration which is limited to 11 to 41 cm and the pH exhibits extreme temporal and spatial variations. During the heavy inflow of waste water in the rainy season the water turns acidic. Similarly, low pH values are recorded at the sewage discharge points, compared to other areas of the reservoir. On account of accumulation of oxidisable organic matter and the consequent demand on oxygen, the O_2 level drops to 1.2 mg l^{-1} with a corresponding build-up of CO_2 (up to 150 mg l^{-1}). Specific conductivity ranges from 560 to 9 900 µmhos cm^{-1}, against total alkalinity which goes up to 760 mg l^{-1}. This clearly illustrates the presence of various ionisable salts in water, other than the carbonates and bicarbonates. Similarly, the hardness of 30 to 350 mg l^{-1} (calcium 28 to 195 mg l^{-1}; magnesium 1.14 to 71.15 mg l^{-1}) reflects the chlorides and other anions. Phosphate loading of 4.0 to 13.5 mg l^{-1} is phenomenal. High BOD and COD values have also been recorded (Adholia, 1991, 1992).

Eutrophication, though conducive to fish production within certain limits, can hamper productivity if the organic loading continues unchecked to the level of hyper-eutrophication. Carbon fixation through phytoplankton gets retarded on account of poor light penetration, thick infestation with water hyacinth and low pH.

Although the phytoplankton community is rich (2 080 to 30 240 units l^{-1}) and diverse (63 species), an equally luxuriant macrophytic growth including littoral, emergent, and floating types competes for nutrients and the sunlight. On many occasions, the reservoir gets completely choked with weeds with a consequent collapse of plankton communities which gets restored only on weed clearance. Weed choking, especially with the obnoxious water hyacinth, if left unchecked, can lead to swamp formation and destruction of the aquatic ecosystem.

The 19 species of fishes recorded from the reservoir include 10 from the family of Cyprinidae and representatives from the following families: Clupeidae, Notopteridae, Siluridae, Bagridae, Clariidae, Channidae, Belonidae and Mastacembelidae. On account of the organically polluted nature of the water body, air-breathing fishes (*Clarias batrachus*, *Notopterus notopterus*) and catfishes (*Aorichthys seenghala*) dominate the ichthyofauna, along with the detritophagous *Labeo calbasu*, *L. fimbriatus* and *L. bata*.

Ecosystem recovery programme through pollution abatement measures should precede any attempt to develop fisheries in the lake.

7.6 KULGARHI RESERVOIR

Commissioned in 1966 across a seasonal stream *Durha nalla* of the Ganga basin, in Satna district, the 193 ha Kulgarhi is much akin to Govindgarh in its size, catchment particulars and water quality. The catchment, covers only 27.69 km^2 and water is impounded with the help of a waste weir. Despite its small size and low mean depth (4.5 m), the water has a low productivity (Dwivedi *et al.*, 1986). The catchment with acidic soil (pH 6.6), with low calcium carbonate (2.28%) and organic carbon (0.61%), is very poor in nutrients such as phosphorus (0.20 mg 100 g^{-1}) and

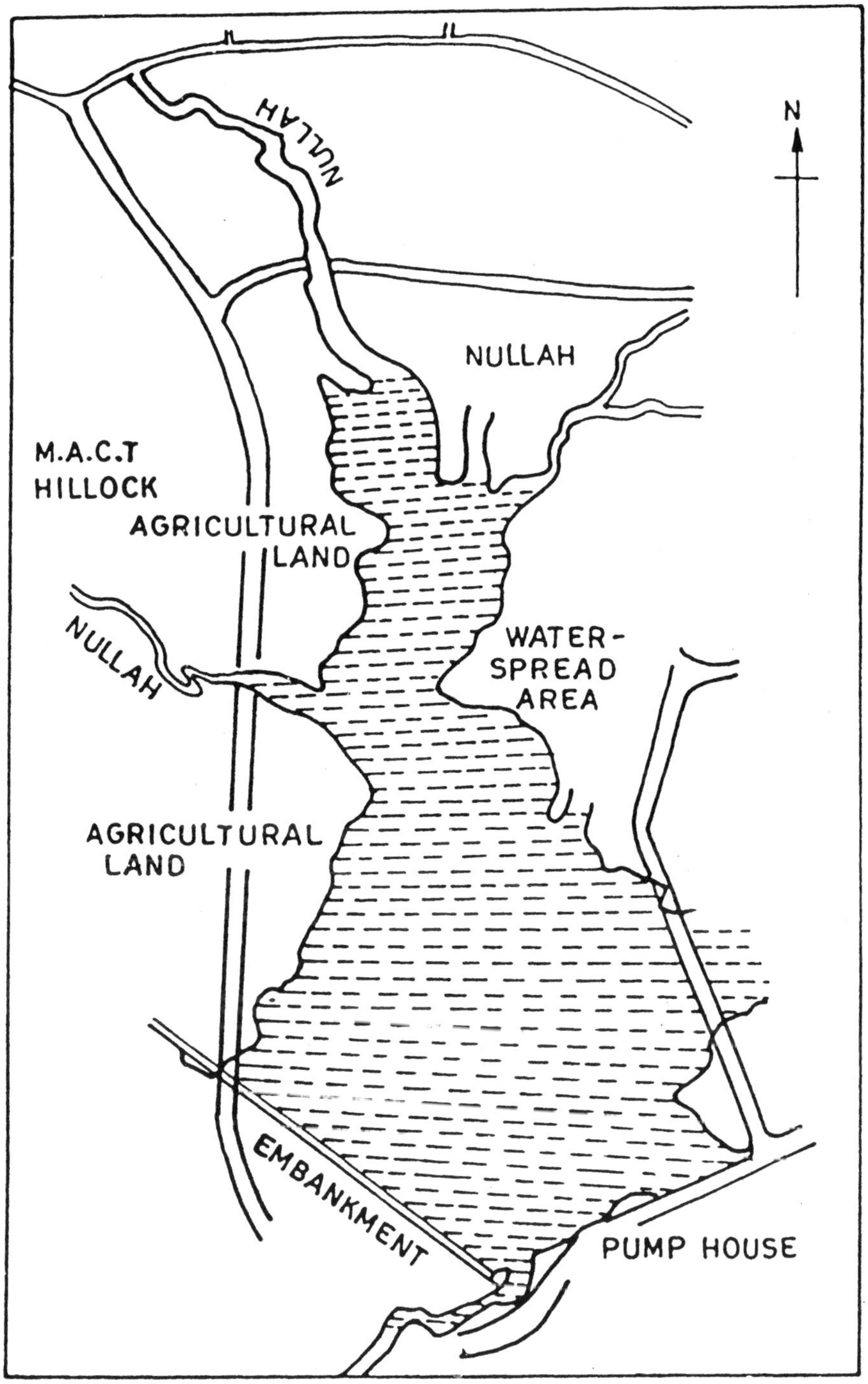

Figure 7.4. Mansarovar reservoir, Bhopal, Madhya Pradesh

nitrogen (0.081 mg 100 g^{-1}). The water quality can be categorised, at best, as medium productive, depending on total alkalinity (47.7 to 140 mg l^{-1}) and total hardness (42.5 to 125.3 mg l^{-1}) . The water pH (8.0 to 8.3) and dissolved oxygen (5.9 to 11.5 mg l^{-1}) are high while the nutrients like phosphate (tr-0.08 mg l^{-1}), and nitrate (tr-0.05) in water are of low order (Table 7.14).

The reservoir has a fairly good standing crop of plankton, comprising up to 75% (by number) of phytoplankton and a high rate of primary productivity. Plankton consists of *Peridinium, Melosira, Synedra, Fragilaria, Microcystis, Anabaena* and other eutrophic species. During the period between 1968 and 1972, the plankton density ranged from 91 to 210 units l^{-1} and the average gross primary productivity was estimated at 11.4 g C m^{-3} day^{-1}. Both the plankton count and carbon fixation rate go up substantially during the summer months. The reservoir fish production ranged from 363 kg to 812 kg, the average being 517 kg (7.4 kg ha^{-1}). The fish catch represented by *Catla catla* (75.6%), *Cirrhinus mrigala* (11.5%), *Labeo rohita* (8.3%) and hybrids of carps (2.5%). Dwivedi *et al.* (1986) stressed the need for proper management measures to be taken to improve the fish yield rate.

Table 7.14. Water and soil quality in Kulgarhi reservoir

Parameter	Range in values
Water	
Water temperature (^{0}C)	15.03-29.70
pH	8.0-8.3
Transparency (cm)	94.0-128.4
Dissolved oxygen (mg l^{-1})	5.9-11.5
Carbon dioxide (mg l^{-1})	nil- 6.3
Total alkalinity (mg l^{-1})	47.7 - 140.1
Total hardness (mg l^{-1})	42.5-125.3
Nitrates (mg l^{-1})	tr.-0.05
Phosphates (mg l^{-1})	tr.-0.08
Silicate (mg l^{-1})	2.5-20.2
Soil	
Sand (%)	30.0 - 43.7
Silt (%)	15.5 - 28.7
Clay (%)	28.8 - 43.7
Available phosphorus (mg 100^{-1})	0.15 - 0.25
pH	6.8 - 6.3
Organic carbon (%)	0.61- 1.11
Total nitrogen (%)	0.072 - 0.085

tr= traces

Kulgarhi is the first reservoir in India, where the exotic silver carp, *Hypophthalmichthys molitrix* was first introduced on an experimental basis. A small consignment of 229 fingerlings in the size range 121 to 216 mm was released in the reservoir on 11-2-1969 and eight specimens were recovered during the period between 2-12-1969 and 4-6-1971. The growth rate registered by recovered silver carp was remarkable, the fastest being 404 mm in 293 days or 1.4 mm day^{-1}. Even the slowest growth was 597 mm in 783 days, *i. e.*, 0.76 mm day^{-1} (Rao and Dwivedi, 1972). Karamchandani and Mishra (1980), while evaluating the impact of the coexistence of silver carp with the indigenous *Catla catla*, observed that the two fishes shared a common niche and competed with each other for food. The percentage composition of phytoplankton in the guts of both silver carp and catla caught during the same period, was more or less the same (18.24 and 15.24% respectively). Zooplankton, the favourite menu of catla, formed 21% of the gut contents of silver carp. Comparing the growth performance of the two carps, the authors concluded that the silver carp negatively affected the growth of catla in the reservoir and advocated caution before undertaking large-scale stocking in Indian reservoirs. A detailed discussion on the impact of silver carp population on catla is given elsewhere in this publication.

7.7 OTHER RESERVOIRS

At least a dozen more reservoirs in the State of Madhya Pradesh have been subjected to investigations by scientists. Loni reservoir (350 ha) has medium-soft, nutrient-poor water with moderate primary productivity and plankton count (Anon., 1968). Macrophytes, weed associated fauna and the benthic invertebrates form the mainstays of biotic communities through which the energy transformation takes place. *Hydrilla, Vallisneria, Najas*, and *Potamogeton* are particularly abundant among the aquatic plants which colonise the littoral region up to 3.15 m depth during post-monsoon months (Gupta, 1976). The nutrients brought in along with the monsoon inflow are better utilised by the aquatic plants, than the plankton. There is an equally rich community of benthic macrofauna, chiefly represented by molluscs, annelids and insect larvae. The mean density of macrobenthos in Loni varies from 1 829 (in 1969) to 2 386 individuals m^{-2} (in 1968). Benthic community exhibits a strong seasonal and spatial rhythm. Commercially important fish of the reservoir comprise small clupeids, detritophagous and predatory catfishes. Among them, biological traits of *Puntius sarana* (Sinha, 1972; 1976) and *Labeo calbasu* (Pathak, 1975; Pathak and Jhingran, 1977) are known.

Bergi (27 296 ha), Tawa (20 055) and Barna (7 705 ha) are three large and Sarni (1 012 ha) and Sukta (1 350 ha) are two medium reservoirs in the Narmada river basin. Some limnological attributes of these lakes are reported by Unni (1993). Sarni reservoir has organic carbon, available phosphorus and available nitrogen of 0.2-1.2%, 2.7 to 8 mg 100^{-1} and 12 to 30 mg 100^{-1} respectively. In Sukta reservoir, Desai (1992) reported a phenomenal increase in fish yield from 0.2 kg ha^{-1} in 1980-81 to 32 kg ha^{-1} in 1985-86 along with a shift in population in favour of major carps. Percentage composition of catla, rohu, mrigal, mahseer, *Labeo calbasu* and common carp increased from 3% to 20% during the above period.

Sampna, Undasa and Yeshwantnagar are the three small reservoirs on which cursory studies have been made. The 520 ha Sampna reservoir is built on the river Sampna which receives local catchment from the cultivated fields. Specific conductivity (114 to 220 μmhos), phenophthalein alkalinity (2 to 10 mg l^{-1}), methyl organge alkalinity (70 to 168 mg l^{-1}) and pH (7.9-9.4) of the reservoir water indicate medium productivity, but the phosphate content is very low (10 to 20 μg l^{-1}). The algal flora is very rich with regular blooms of *Ankistrodesmus, Anabaena* or *Ceratium hirudinella* (Unni, 1982). Undasa reservoir is a 176.5 ha small irrigation impoundment, created by erecting embankments to gather the local catchment. The nutrient content of the reservoir is reported to be very high (Singh, 1986).

Yeshwantnagar, another minor irrigation impoundment, created to gather the local catchment for irrigation is a hard water (total hardness 106 to 130 mg l^{-1}) reservoir, rich in phosphate (0.14 to 0.22 mg l^{-1}) and nitrate (1.3 to 2.5 mg l^{-1}) (Sharma and Diwan, 1989).

Table 7.19 gives water quality and associated data in respect of Halali, Kolar and Dahod reservoirs of Madhya Pradesh (Unni, 1993).

7.8 STATUS OF RESERVOIRS IN MADHYA PRADESH

In total, 5 large, 6 medium and 8 small impoundments have been studied. Irrespective of the size of the water bodies and the elevation at which they are situated, the reservoirs have medium to low productivity. Even smaller shallow lakes usually are not eutrophic. This is basically due to geo-climatic reasons. Madhya Pradesh, in a higher latitude, receives less incident solar radiation and the reservoirs, in general, are situated at a higher altitude, than their counterparts in Andhra Pradesh and Tamil Nadu. The average altitude of the reservoir in Madhya Pradesh is 488 m above MSL. Sharp fluctuations between the summer and winter air temperature have the impact on the thermal and nutrient regime of the water bodies, affecting their biogenic potential. Above all, the geo- chemical make up of catchments play and important role in determining the water quality. Soils of Madhya Pradesh are normally deep black, medium black, shallow black, mixed red and skeletal, low in nitrogen and low to medium in phosphorus (Dubey, 1994).

While the reservoirs in the plains of Andhra Pradesh and Tamil Nadu, spreading over the limestones and other calcareous rocks, have a predominantly hard-water regime, those of Madhya Pradesh are soft to medium soft in nature. However, this does not mean that they are actually poor in biotic communities. Under the Indian conditions, even the lakes categorised as less productive on the basis of morphometric and limno-chemical characteristics are often found to be rich in terms of primary, secondary and fish productivity due to the relatively high turnover rate. The small reservoirs of the State, except those receiving organic enrichment through effluents, are as a rule, less eutrophic than the larger ones.

Gandhisagar, with a relatively low nutrient status, harbours a rich standing crop of plankton with a moderate to high rate of primary production, which is reflected in the fish yield. This reservoir receives its drainage from a vast, forest covered drainage and its large; deep basin facilitates storage of allochthonous nutrient and mineral inputs. Large deep reservoirs have a good percentage of their water as dead storage and their flushing rate is low. The low level of available nutrients, measured in the water phase could be due to the immediate utilisation, on their release into water.

The small reservoirs with small catchments and low volume water have a high water renewal rate, especially during the monsoon. Thus, they rapidly flush out the incoming nutrients rather than storing them. Reach blooms of blue-green algae are much less common in Madhya Pradesh, than in the reservoirs further south where most of the small, shallow reservoirs have summer blooms and many of them have a permanent bloom of *Microcystis*.

The oligotrophic tendencies character of the reservoirs in Madhya Pradesh are compensated by the sheer magnitude of the resource. Madhya Pradesh will further increase its water surface area on completion of all the Narmada basin projects. The State has realised the importance of reservoir fisheries way back in 1962 and initiated several steps to increase their fish production. During the Vth Five Year Plan (1975-76 to 1979-80), the reservoirs were categorised into five groups according to size and their rate of stocking was specified (Table 7.15).

Table 7.15. Stocking rates proposed for the reservoirs in Madhya Pradesh

Reservoir size (ha)	Stocking rate no. ha^{-1}
40-200	1 250
201-1 000	1 250
1 001-4 000	750
4 001-12 000	500
above 12 000	250

The stocking targets could not be achieved due to constraints in seed production. In order to fulfil the stocking requirements, the State required 180 million fish seed, but could produce only 10 million (Srivastava *et al.*, 1985).

In some reservoirs like Gandhisagar, Ravishankarsagar and Sukta, there has been a considerable improvement in productivity through good management. Gandhisagar registered a four-fold increase in the yield rate from 0.5 kg to 20 kg ha^{-1} during the last three decades and Ravishankarsagar increased its fish production from 13 to 50.3 t (1990-91) in four years. Sukta a small reservoir increased its yield rate from 0.2 kg to 32 kg ha^{-1} within a period of five years. The Government, realising the importance of the resource, has entrusted the management of many reservoirs to the Madhya Pradesh State Fisheries Development Corporation. Total fish production from the reservoirs in the state during 1979-80 was 1 251 t, which increased to 2 000 t during 1989-90. This production is still low. The current yield is just 8.9 kg ha^{-1}, indicating enough room for further increase through effective management.

7.9 IMPOUNDMENTS IN THE NARMADA BASIN

The Narmada river basin is well-poised for water resources development of an unprecedented magnitude. The large number of multi-purpose reservoirs to be created in the basin will soon earn Madhya Pradesh the most prominent place on the reservoir map of the country.

Narmada basin, from Omkareswar to Harinphal receives an average rainfall of 1 000 mm with the maximum of 2 030 mm at Panchmarhi. The Narmada cascades down from east to west with an average slope upto 1:1 000 at the origin to 1:10 000 in the lower reaches (Table 7.16). Most of the rainfall occurs during 2 to 3 months of the year. Narmada basin comprises 95 726 km^2, out of which 81% is situated in Madhya Pradesh. Twenty of the 45 districts of the State are in the Narmada basin, receiving 22 of the 41 principal tributaries of the Narmada. Thus, a substantial part of the reservoirs will be situated in Madhya Pradesh.

Table 7.16. Average slope of the Narmada river at different reaches

Stretch	Slope (Av.)
Origin to Bargi	1:1 000
Bargi to Hoshangabad	1:3 700
Hoshangabad to Omkareshwar	1:1 700
Omkareshwar to Harinpal	1:2 300
Harinpal to Gujarat border	1:1 600
In Gujarat	1:10 000

(After Dubey, 1994)

Table 7.17 Projects proposed in the Narmada basin in Madhya Pradesh

Major Project	
Irrigation	21
Hydel	5
Multipurpose	3
Medium irrigation projects	450
Minor irrigation projects	3 500
Total	***3 979***

(After Dubey, 1994)

The 29 major projects planned include 21 irrigation, 5 hydroelectric and 3 multi-purpose schemes. This is in addition to the 450 medium and 3 500 minor irrigation schemes (Table 7.17) also planned. All the 3 979 structures when completed, will create an additional waterspread of 260 000 ha in Madhya Pradesh alone. Three large reservoirs *viz.*, Tawa (20 055 ha), Barna (7 705 ha) and Bargi (27 296 ha), 20 medium and 975 small reservoirs have already been created and many others are in various stages of completion. One of the dams that awaits completion is the 91 348 ha Narmadasagar, which will be the largest reservoir of India (Figs. 7.5 and 7.6). All dams on the main river are in Madhya Pradesh, except the Sardar Sarovar which is just across the border with Gujarat. More than 55% of the huge 37 030 ha Sardar Sarovar reservoir, however, will lie in Madhya Pradesh. A list of 30 projects with their waterspread is given in Table 7.18.

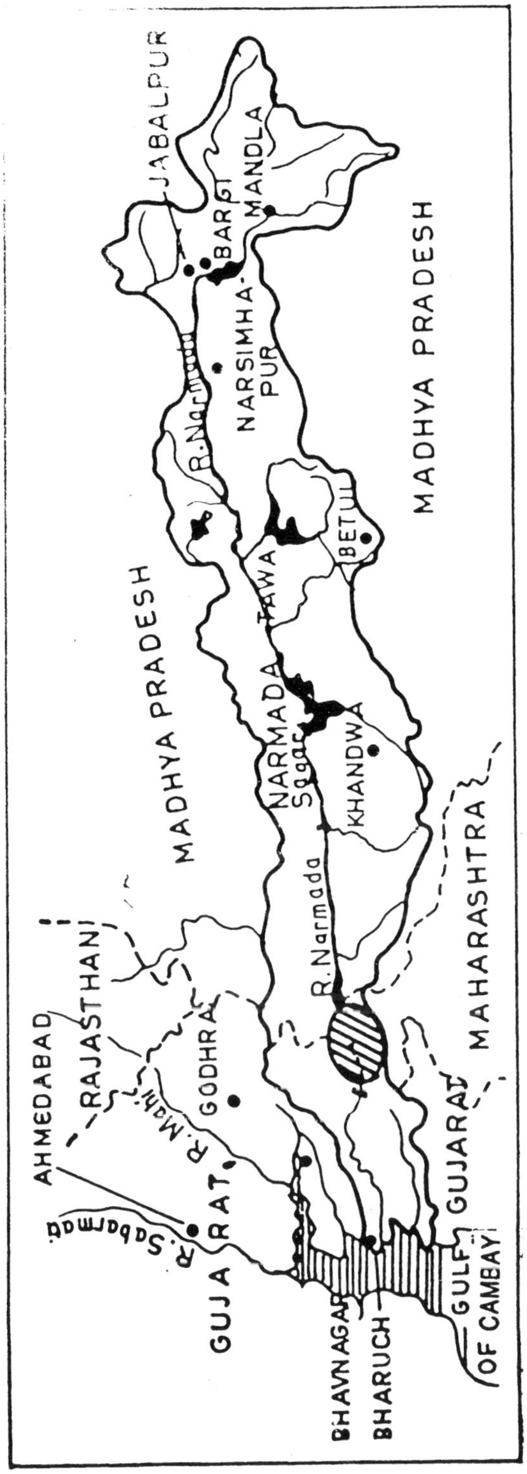

Figure 7.5. Narmada basin and the main impoundments

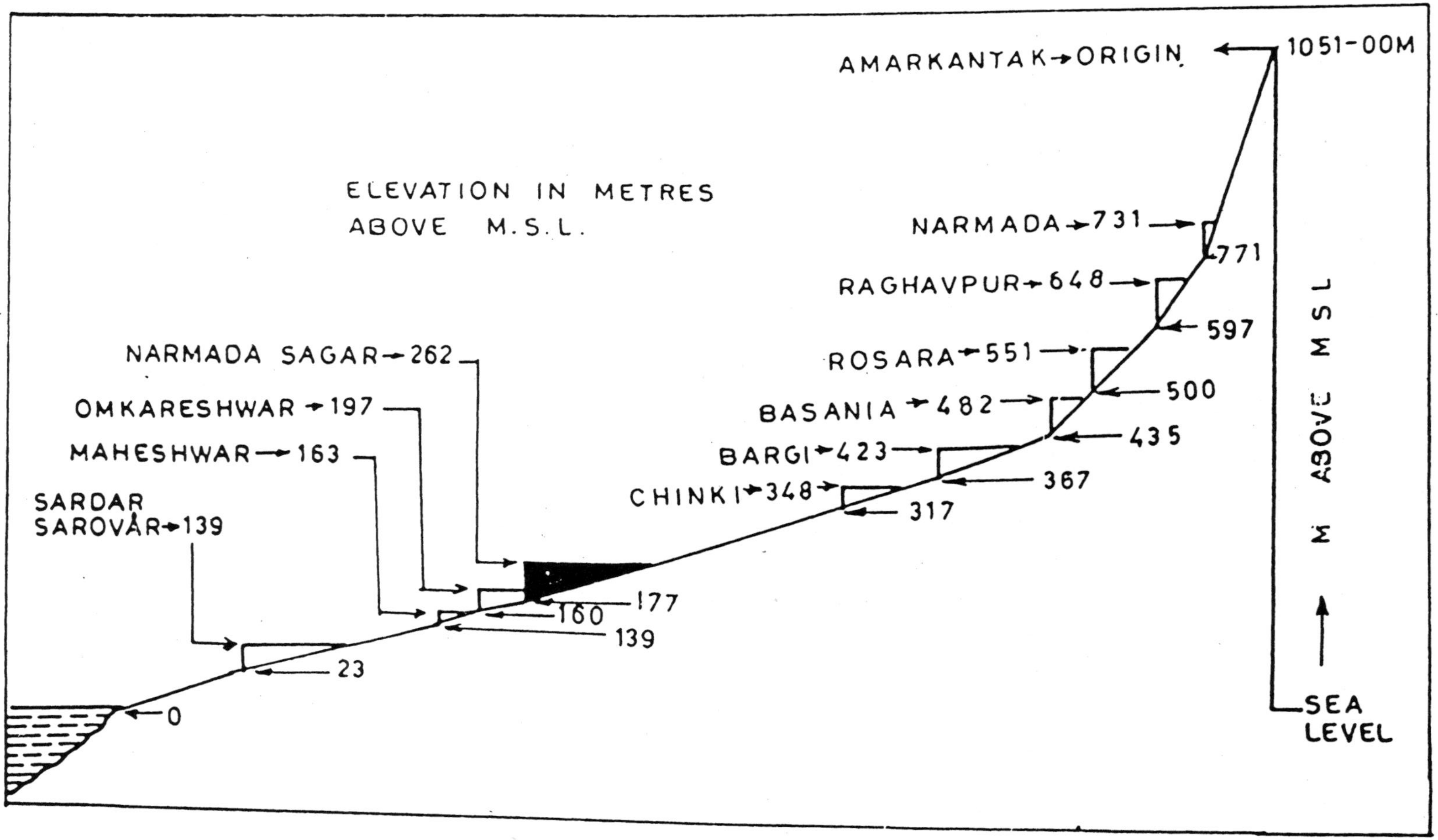

Figure 7.6. Diagrammatic representation of the cascade of reservoirs on the river Narmada

Table 7.18. Location and area of dams and reservoirs of proposed projects in Narmada basin

	Name of project	Kms from the source	FRL m above MSL	Area (ha)
On main river				
Upper zone				
	Upper Narmada	77	731.52	3 043.23
	Raghavpur (H)	135	648.00	2 370.00
	Rosra (H)	206	551.08	3 200.00
	Basania (H)	252	481.80	14 201.00
Middle Zone				
	Bergi (M)	378	422.76	27 296.00
	Chinki	489	348.08	11 007.00
Lower zone				
	Narmadasagar (M)	845	262.13	91 348.00
	Omkareshwar (M)	899	196.60	9 393.00
	Maheshwar (H)	947	162.76	4 856.00
	Sardar Sarovar (M)	1163	138.68	37 030.00
On tributaries				
Upper zone				
	Upper Burhner	111	616.61	6 154.84
	Halon	109	623.01	2 070.00
	Dhobatoria	55	479.17	1 801.9
Middle zone				
	Ataria	148	409.04	2 161.10
	Sher	53	412.70	2 256.00
	Machrawa	55	420.93	506.87
	Shakkar	74	445.31	2 121.00
	Sitarewa (H)	80	690.67	947.00
	Dudhi	71	388.10	3 849.00
	Barna	29	348.55	7 705.00
	Tawa	32	355.39	20 055.00
	Kolar	76	462.20	2 380.00
	Morand	63	362.71	1 940.00
	Ganjal	61	373.68	1 092.00
	Chhota Tawn	45	299.00	2 914.00
	Sukta	80	410.7	1 350.00
Lower zone				
	Beda Upper	72	310.28	1 300.00
	Han	37	297.65	1 077.00
	Lower Goi	53	295.81	1 020.00
	Jobat	58	260.17	970.37
			Total	267 412

Source: Narmada Planning Agency

7.19 Salient features of some reservoir in Madhya Pradesh

Parameters	Gandhisagar	Tawa	Barna	Ravishankarsagar	Halali
Year of sealing	1960	-	-	-	-
Area at FRL (ha)	66 000	20,055	7705	9540	2528
Average area (ha)	41 317	-	-	-	-
Maximum length (km)	112	-	-	-	-
Maximum width (km)	16	-	-	-	-
Maximum depth (m)	62	46	-	-	-
Mean depth (m)	11.73	-	-	9.53	-
Volume (million m^3)	7 744	-	-	909.3	-
Catchment area (km^2)	23 025	-	-	3,620	-
Elevation (m above MSL)	399	355.39	348.55	348.7	-
Shore dev. index	4.78	-	-	2.96	-
Volume dev. index	0.601	-	-	2	-
Inflowing rivers	Chambal, Shipra, Shivna, Chottikali Sindh, Ansar, Rupnia, Tilosi Ratum, Edar	Tawa	-	-	-
Outflowing river	Chambal	Tawa	-	-	-
Latitude (N)	24^0 44'	-	23^0 05'	20^0 38'	23^0 40'
Longitude (E)	75^0 33'	-	78^0 07'	81^0 34'	76^0 33'
Water temperature (^{0}C)	2.7-32.5	22-34	20-34	21-30.5	19-32
pH	8.2-8.9	7.4-8.4	7.5-8.4	7.4-8.4	7.0-7.5
Transparency (cm)	11.5-200	32-80	70-161	61-166	-
DO (mg l^{-1})	6.4-11.8	6-8.8	6.4-9.6	6.76-10.2	5.3-6.8
CO_2 (mg l^{-1})	Nil	Nil-4	Nil-8	Nil-10	0.42
Total alkalinity (mg l^{-1})	82-124	54-128	54-94	40-94	-
Spec. cond. (μmhos)	121	121		44-100	-
Total hardness (mg l^{-1})	100	48	110-138	43-58	102-218
Calcium (mg l^{-1})	36	32	84-89	32-44	74-118
Nitrate (mg l^{-1})	Tr-0.9	-	-	0.05	-
Phosphate (mg l^{-1})	Tr-0.3	-	-	0.04-0.16	-
Chlorides (mg l^{-1})	-	5.0	28-48	-	26-72
Organic carbon (mg l^{-1})	441-490	240-1440	240-1200	-	-

tr= traces

7.19 (cont.). Salient features of some reservoir in Madhya Pradesh

Parameters	Kolar	Yeshwant-nagar	Dahod	Sarni	Bergi
Area at FRL (ha)	2268		1509	1012	27,296
Maximum depth (m)	-	-	-	29	
Elevation (m above MSL)	462.2	-	-	440	422.76
Latitude (N)	22° 51'	-	23° 3'	22° 8'	-
Longitude (E)	77° 24'	-	77° 29'	77° 11'	-
Water temperature (°C)	24-30	18-33	18-32	22-32	27-35
pH	7.5-8.0	8.0-9.0	7.0	7-8.0	7.5-7.9
Transparency (cm)	-	60-72	-	-	-
DO (mg l^{-1})	4.8-8.0	6.6-8.4	5.1-6.8	7.4-10.8	7.14-9.24
CO_2 (mg l^{-1})	0-0	0	0	0	0-2.0
Total alkalinity (mg l^{-1})	-	80-120	-	-	-
Spec. cond. (μmhos)	-	-	-	-	192-221
Total hardness (mg l^{-1})	174-186	106-130	94-118	152-184	96-108
Calcium (mg l^{-1})	115-126	-	48-78	64-84	44-64
Nitrate (mg l^{-1})	-	1.3-2.5	-	-	-
Phosphate (mg l^{-1})	-	0.14-0.23	-	-	-
Chlorides (mg l^{-1})	10-14	22-26	26-38	18-36	4.2-6.4
GPP (mg C m^{-2} d^{-1})	441-490	240-1 440	240-1 200	-	-

7.19 (cont.). Salient features of some reservoir in Madhya Pradesh

Parameters	Mansarovar	Sukta
Area at FRL (ha)	79.7	1 350
Volume (million m^3)	2.29	-
Mean depth (m)	2.87	-
Catchment area (km^2)	8.29	468.79
Elevation (m above MSL)	488.30	410.7
Volume dev. index	1.07	-
Latitude (N)	23° 16'	21° 36'
Longitude (E)	77° 26'	76° 16'
Water temp. (°C)	17-39	-
pH	5.5-9.5	-
Transparency (cm)	11.41	-
DO (mg l^{-1})	1.2-18.0	-
CO_2 (mg l^{-1})	0-150	-
Total alkalinity (mg l^{-1})	2-760	-
Spec. cond. (μmhos)	560-9 900	-
Total hardness (mg l^{-1})	30-350	-
Calcium (mg l^{-1})	28-195	-
Nitrate (mg l^{-1})	0.11-3.0	-
Phosphate (mg l^{-1})	4.0-13.5	-
Chlorides (mg l^{-1})	9.8-95.0	-
GPP (mg C m^{-2} d^{-1})	542-6 000	-

7.19 (cont.). Salient features of some reservoir in Madhya Pradesh

Parameters	Kulgarhi	Govindgarh	Sampna
Area at FRL (ha)	193	307	520
Volume (million m^3)	8.66	8.69	0.44
Maximum depth (m)	-	-	6
Mean depth (m)	4.5	6.12	-
Catchment area (km^2)	27.69	25.12	3 180.5
Elevation (m above MSL)	-	-	597
Volume dev. index	-	1.54	-
Latitude (N)	24° 28'	24° 24'	21° 22'
Longitude (E)	80° 44'	81° 15'	78° 3'
Water temperature (^oC)	15.03-29.7	18.67-30.55	16-32
pH	8.0-8.3	8.0-8.2	7.9-9.
Transparency (cm)	9.4-128.4	55.80-91.6	10-9
DO (mg l^{-1})	5.9-11.5	7.58-10.64	5.1-10.8
CO_2 (mg l^{-1})	nil-6.3	1.95-4.56	4-30
Total alkalinity (mg l^{-1})	47.7-140.1	30.7-53.90	72-178
Spec. cond. (μmhos)	-	-	140
Total hardness (mg l^{-1})	42.5-125.3	-	65-76
Calcium (mg l^{-1})	-	-	38-46
Nitrate (mg l^{-1})	tr-0.05	tr.-0.130	0.04-0.06
Phosphate (mg l^{-1})	tr.-0.08	tr.-0.033	0.010-0.020
Silicate (mg l^{-1})	2.5-20.2	tr.-12.93	-
Chlorides (mg l^{-1})	-	-	2.4-10.0
GPP (mg C m^{-2} d^{-1})	195.0-1 162.0 (per 6 hrs)	-	-

Table 7.20. Profiles of some reservoirs in Madhya Pradesh

***The Reservoir:* Kodar**

Location

District: Raipur

Altitude : 295 m above MSL

Latitude: 21° 11' 50" N

Longitude: 82° 10' 40" E

Year of Commissioning : 1981

Depth: 9.3 m

Volume: 149.02 million m^3

Surface Area At Full Res. Level: 3 583 ha

At Dead Res. Level: 567 ha

Length of Shoreline: 2 087 m

Annual Fluctuation in Water Level: 9.296 m

Discharge: 623 m^3 sec^{-1}

Major inflowing river: Kodar Nala

Outflowing river: Mahanadi

Physical and chemical features

Surface temperature: 27 °C

Conductivity: -

Dissolved solids: 47.9 mg l^{-1}

pH: 7.2

Fisheries

Number of species: 16

Major species: C. catla L. rohita, L. calbasu L.bata, C. mrigal, C. reba, L. fimbriatus, W. attu, A. seenghala, A. aor, M. armatus, C. striatus, C. marulius, C. batrachus, H. fossilis, O. pabda.

Total catch : 33. 58 t
(year 1992-93)

Number of fishermen: 217

Number of boats : 49

Fishing gear: Gill nets

Fish stocked: Catla, rohu, mrigal, grass carp

***The Reservoir:* Sikasar**

Location

District: Raipur

Altitude in : 406 m above MSL

Latitude: 20° 31' 15" N

Longitude: 82° 19' E

Year of Commissioning : 1977

Depth: 31.69 m

Volume: 118.1 million m^3

Surface Area At Full Res. Level: 1 862 ha

At Dead Res. Level: 1 174 ha

Length of Shoreline: 10 464 m

Annual Fluctuation in Water Level: 6.59 m

Discharge: 33.97 m^3 Sec^{-1}

Major inflowing rivers: Pairi

Outflowing river: Pairi

Physical and chemical features

Surface temperature: -

Conductivity: 45.5 µmhos

Dissolved solids: 23.7 mg l^{-1}

pH: 6.8

Fisheries

Number of species: 23

Major species: *C. catla L. rohita, L. calbasu, L. bata, L. fimbriatus, L. gonius, C. mrigal, C. carpio, T. tor, M. armatus, C. marulius, O. bimaculatus, N. notopterus, C. bacaila, P. sophore, B. barna, M. cavasius*

Total catch : 3 532 t
(year 1992-93)

Number of fishermen: 78

Number of boats : 26

Fishing gear: Gill nets

Fish stocked : Catla, rohu, mrigal.

The Reservoir: Murumsilly

Location

District: Raipur

Latitude: 20° 32' 15" N

Altitude : 377 m above MSL

Longitude: 81° 40' E

Year of Commissioning : 1923

Depth: 32.1 m

Volume: 3.43 million m^3

Surface Area *At Full Res. Level:* 2 508 ha
At Dead Res. Level: 1 377 ha

Length of Shoreline: -

Annual Fluctuation in Water Level: -

Discharge : 1000 m^3 sec^{-1}

Major inflowing river: Silyari

Outflowing river: Mahanadi

Fisheries

Number of species:

Major species: *C. catla L. rohita, L calbasu C. mrigal, O. pabda*

Total catch : 23.8 t
(year 1992-93)

Number of fishermen: 45

Number of bcats : 18

Fishing gear: -

Fish stocked : Catla, rohu, mrigal.

***The Reservoir:* Kharkora**

Location

District: Durg

Altitude : 334 m above MSL

Latitude: 20° 50' N

Longitude: 80° 59' E

Year of Commissioning : 1966

Depth: 24.7 m

Volume: - million m^3

Surface Area At Full Res. Level: 2 975 ha

At Dead Res. Level: 778 ha

Length of Shoreline: -

Annual Fluctuation in Water Level: 4.57 m

Discharge: 14.0 m^3 sec^{-1}

Major inflowing rivers: Kharakhara nala

Outflowing river : Kharakhara nala

Physical and chemical features

Surface temperature: -

Conductivity: 96.2 μmhos

Dissolved solids: 48.1 mg l^{-1}

pH: 6.8

Fisheries

Number of species: 20

Major species: N. notopterus, N. chitala, W. attu, M. armatus, P. ticto, O. pabda, A. seenghala, L.calbasu, Channa spp.

Total catch : 3 .151 t
(year 1992-93)

Number of fishermen: -

Number of boats : -

Fishing gear: -

Fish stocked : Catla, rohu, mrigal.

***The Reservoir:* Gondali**

Location

District: Durg

Altitude : 343 m above MSL

Latitude: 20° 44' 10" N

Longitude: 81° 07' 55" E

Year of Commissioning : 1956

Depth: 25.38 m

Volume: 96.73 million m^3

Surface Area At Full Res. Level: 1 862 ha

At Dead Res. Level: 375 ha

Length of Shoreline: -

Annual Fluctuation in Water Level: 10.36 m

Discharge: 28.0 m^3 sec^{-1}

Major inflowing river: Jujhara nala

Outflowing river: Jujhara *nala*

Physical and chemical features

Surface temperature: -

Conductivity: -

Dissolved solids: 0.112 mg l^{-1}

pH: 7.2

Fisheries

Number of species: 7

Major species: *C. catla, C. mrigala , L. rohita, W:attu, N. notopterus, P. sarana*

Total catch : 120.36 t
(year 1992-93)

Number of fishermen: - 733

Number of boats : 6 987

Fishing gear: -

Fish stocked : Catla, rohu, mrigal.

The Reservoir: Tandula

Location

District: Durg

Latitude: 20° 42' N

Altitude : 332 m above MSL

Longitude: 81° 17' E

Year of Commissioning : 1920

Depth: 25 m

Volume: 302 million m^3 -

Surface Area At Full Res. Level: 4 410 ha
At Dead Res. Level: 141 ha

Length of Shoreline: -

Annual Fluctuation in Water Level: 5 to 10 m

Discharge: 54.40 $m^3 sec^{-1}$

Major inflowing river: Tandula & Sukha

Outflowing river: Tandula

Physical and chemical features

Surface temperature: -

Conductivity: 136.2 μmhos

Dissolved solids: 68.1 mg l^{-1}

pH: 7.2

Fisheries

Number of species: 5

Major species: *C. rohu, mrigal, W. attu, Mystus spp., N. chitala*

Total catch (average) : 46.28 t

Number of fishermen: 13 355

Number of boats : 8 862

Fishing gear: -

Fish stocked: *Catla, rohu, mrigal*

***The Reservoir:* Manohar Sagar**

Location

District: Rajnandgaon

Altitude : 350 m above MSL

Latitude: 20° 3' N

Longitude: 80° 27' E

Year of Commissioning : 1969-70

Depth: 11-37 m

Volume: 193 million m^3

Surface Area At Full Res. Level: 3 257 ha

At Dead Res. Level: 1 779.5 ha

Length of Shoreline: -

Annual Fluctuation in Water Level: 11.37 m

Discharge: 31.83 m^3 sec^{-1}

Major inflowing river: Bagh river

Outflowing river: Bagh

Physical and chemical features

Surface temperature: 22 °C

Conductivity: 61.6 µmhos

Dissolved solids: 30.8 mg l^{-1}

pH: 7.4

Fisheries

Number of species: 13

Major species: C. catla, L. rohita, L. bata, L. calbasu, C. mrigala, W. attu, P. sarana, C. punctatus, M. bleekeri, Chela bacaila, Barbus ticto, Chanda sanga, C. nama

Total catch (average) : 3.68 t

Number of fishermen: 1 417

Number of boats : 369

Fishing gear: -

Fish stocked : Catla, rohu, mrigal

***The Reservoir:* Maniyari**

Location

District: Bilaspur

Altitude : 358 m above MSL

Latitude: 22° 23' N

Longitude: 81° 35' E

Year of Commissioning : 11 930

Depth: 22.86 m

Volume: 147.7 million m^3

Surface Area At Full Res. Level: 2 460 ha

At Dead Res. Level: 900 ha

Length of Shoreline: 2 040 m.

Annual Fluctuation in Water Level: 13.7 m

Discharge: 38.5 m^3 Sec^{-1}

Major inflowing river: Maniyari

Outflowing river: Maniyari

Physical and chemical features

Surface temperature: -

Conductivity: -

Dissolved solids: -

pH: -

Fisheries

Number of species: 8

Major species: C. catla, L. rohita, C. mrigala, O. pabda, L. calbasu, Mystus spp.,

Total catch (average) : 63 to 3525 t

Number of fishermen: 317

Number of boats : 89

Fishing gear:

Fish stocked : Catla, rohu, mrigal

The Reservoir: Kharung

Location

District: Bilaspur

Latitude: 22° 7' N

Altitude : 298 m above MSL

Longitude: 82° 13' E

Year of Commissioning : 1930

Depth: 21.31 m

Volume: 192.32 million m^3

Surface Area At Full Res. Level: 2 760 ha

At Dead Res. Level: 1 670 ha

Length of Shoreline: -

Annual Fluctuation in Water Level: 2.83 m

Discharge:

Major inflowing river: Kharang river

Outflowing river: -

Physical and chemical features

Surface temperature: -

Conductivity: -

Dissolved solids: -

pH: -

Fisheries

Number of species: 8

Major species: C. catla, L. rohita, C. mrigala, O. pabda, L. calbasu, Mystus spp.,

Total catch (average) : 12.520 t

Number of fishermen: 23

Number of boats : 20

Fishing gear: -

Fish stocked : Catla, rohu, mrigal

The Reservoir: **Hasdeo Barrage (Darri)**

Location

District: Bilaspur

Altitude : 286 m above MSL

Latitude: 22^{0} 24' N

Longitude: 88^{0} 42' E

Year of Commissioning : 1967

Depth: 8.76 m

Volume: 75.50 million m^3

Surface Area At Full Res. Level: 2 510 ha

At Dead Res. Level: 1 000 ha

Length of Shoreline: -

Annual Fluctuation in Water Level:

Discharge: 19 820 m^3 sec^{-1}

Major inflowing river: Hasdeo

Outflowing river: Hasdeo

Physical and chemical features

Surface temperature: -

Conductivity: -

Dissolved solids: -

pH: -

Fisheries

Number of species: -

Major species: *C. catla, L. rohita, T. tor, L. calbasu, C. mrigala W. attu, Channa* spp., *Mystus* spp., *Notopterus* spp.

Total catch (average) : 5.686 t

Number of fishermen: 98

Number of boats : 45

Fishing gear: Gill nets, and drag nets

Fish stocked : *Catla, rohu, mrigal, common carp*

The Reservoir: **Dudhawa**

Location

District: Bastar

Altitude : 427 m above MSL

Latitude: 20^{0} 19' N

Longitude: 81^{0} 45' E

Year of Commissioning : 1963-64

Depth: 24.53 m

Volume: 283.5 million m^3

Surface Area At Full Res. Level: 1 867.68 ha

At Dead Res. Level: 872.06 ha

Length of Shoreline : -

Annual Fluctuation in Water Level: 1.98 m

Discharge: 1 400 m^3 sec^{-1}

Major inflowing river: Siiyari

Outflowing river: Mahanadi Basui

Physical and chemical features

Surface temperature: -

Conductivity: -

Dissolved solids: -

pH: -

Fisheries

Number of species:

Major species: Catla catla, L. rohita mrigala, Mystus spp., *L. calbasu*

Total catch (·average) : 140 t

Number of fishermen: 188

Number of boats : 70

Fishing gear: -

Fish stocked : Catla, rohu, mrigal

The Reservoir: Kotwal

Location

District: Morena

Latitude: 26° 29' N

Altitude : 168 m above MSL

Longitude: 78° 10' E

Year of Commissioning : -

Depth: 10.95 m

Volume: 84.289 million m^3

Surface Area At Full Res. Level:

At Dead Res. Level:

Length of Shoreline: 294.13 m.

Annual Fluctuation in Water Level:

Discharge: 1 490.37 $m^3 sec^{-1}$

Major inflowing river: Asan

Outflowing river: Asan

Physical and chemical features

Surface temperature: -

Conductivity: -

Dissolved solids: 102 to 200 mg l^{-1}

pH: 7.4 to 8.1

Fisheries

Number of species : 45

Major species: C. catla, L. rohita, C. mrigal, W. attu, Channa spp. *L. calbasu*

Total catch (average) : 23.4 t

Number of fishermen: 55 to 60

Number of boats : 25

Fishing gear: -

Fish stocked : Major carps

The Reservoir: Harsi

Location

District: Gur

Altitude : 266 m above MSL

Latitude: 25° 58 ' N

Longitude: 77° 67' E

Year of Commissioning : 1935

Depth: 15.24 m

Volume: 230.447 million m^3

Surface Area *At Full Res. Level:* 2 523 ha
At Dead Res. Level: 484 ha

Length of Shoreline: -

Annual Fluctuation in Water Level: -

Discharge: -

Major inflowing river: Parvati

Outflowing river: Parvati

Physical and chemical features

Surface temperature: -

Conductivity: -

Dissolved solids: -

pH: -

Fisheries

Number of species : 45

Major species: C. catla, L. rohita, C. mrigala, Mystus spp., *L. calbasu, Mahsheers*

Total catch (average) : 38.38 t

Number of fishermen: 55 to 70

Number of boats : 27

Fishing gear:

Fish stocked: *Catla, rohu mrigal*

The Reservoir: Tighra

Location

District: Gur

Altitude : 225 m above MSL

Latitude: 26° 13 ' N

Longitude: 78° 30' E

Year of Commissioning : 1929

Depth: 21 m

Volume: -

Surface Area *At Full Res. Level:* 1 952 ha
At Dead Res. Level: 198 ha

Length of Shoreline: -

Annual Fluctuation in Water Level: -

Discharge:

Major inflowing river: Sank

Outflowing river: Sank

Physical and chemical features

Surface temperature: -

Conductivity: -

Dissolved solids: -

pH: -

Fisheries

Number of species : 34

Major species: L. rohita, C. mrigala, Cirrhinus spp., *L. calbasu, Mystus,* spp.,

Total catch (average) : 5.5 t

Number of fishermen: 70 to 80

Number of boats : 10

Fishing gear: -

Fish stocked : Catla, rohu, mrigal

The Reservoir: **Bhagwantsagar (Sukta)**

Location

District: Khandwa

Latitude: 21° 36 ' N

Altitude : 411 m above MSL

Longitude: 76° 16' E

Year of Commissioning : 1974

Depth: 32 m

Volume: 0.121 million m^3

Surface Area *At Full Res. Level:* 1 350 ha
At Dead Res. Level: 330 ha

Length of Shoreline: 6 553 m.

Annual Fluctuation in Water Level: 2 m

Discharge: 2431.04 m^3 sec^{-1}

Major inflowing river: Sukta

Outflowing river: Sukta

Physical and chemical features

Surface temperature: -

Conductivity: -

Dissolved solids: -

pH: -

Fisheries

Number of species : 6

Major species: C. catla, L. rohita, C. mrigala, C. carpio, L. calbasu

Total catch (average) : 23.5 t

Number of fishermen: 92

Number of boats : -

Fishing gear: -

Fish stocked: Catla, rohu, mrigal, common carp

***The Reservoir:* Sarni**

Location

District: Betul

Altitude : 440 m above MSL

Latitude: 22° 8 ' N

Longitude: 78° 11' E

Year of Commissioning : 1967

Depth: 28.95 m

Volume: 92.52 million m^3

Surface Area *At Full Res. Level:* 1 012 ha

At Dead Res. Level: 800 ha

Length of Shoreline: -

Annual Fluctuation in Water Level: 3.65 m

Discharge: 135 $m^3\ sec^{-1}$

Major inflowing river: Tawa

Outflowing river: Tawa

Physical and chemical features

Surface temperature: 22-32 °C

Conductivity: 125 μmhos

Dissolved solids: 70 mg l^{-1}

pH: 7.0-8.0

Fisheries

Number of species : 23

Major species: C. catla, L. rohita, C. mrigal, L. calbasu

Total catch (average) : 61.96 t

Number of fishermen: 45

Number of boats : 25

Fishing gear: -

Fish stocked: Grass carp, common carp, Indian major carps

***The Reservoir:* Kolar**

Location

District: Schore

Altitude : 462 m above MSL

Latitude: 22° 51 ' N

Longitude: 77° 24' E

Year of Commissioning : 1989

Depth: 42 m

Volume: 0.27 million m^3

Surface Area *At Full Res. Level:* 2 268 ha

At Dead Res. Level: 125.7 ha

Length of Shoreline: 29.50 km m

Annual Fluctuation in Water Level: 15 m

Discharge: 16.40 $m^3\ sec^{-1}$

Major inflowing river: Kolar

Outflowing river: Narmada

Physical and chemical features

Surface temperature: 24-30 (°C)

Conductivity: 220 μmhos

Dissolved solids: 225 mg l^{-1}

pH: 7.5-8.0

Fisheries

Number of species : 10

Major species: *C. catla, L. rohita C. mrigala, W. attu*

Total catch (average) : 100 t

Number of fishermen: 110

Number of boats : 60

Fishing gear: Gill nets

Fish stocked: *Catla, rohu, mrigal*

***The Reservoir:* Halali**

Location

District: Raisen

Latitude: 23° 30 ' N

Altitude : 458 m above MSL

Longitude: 77° 30' E

Year of Commissioning : 1973

Depth: 29.57 m

Volume: 227 million m^3

Surface Area *At Full Res. Level:* 2 528 ha
At Dead Res. Level: 2 590 ha

Length of Shoreline:

Annual Fluctuation in Water Level: 15.24 m

Discharge: 5 665 m^3 sec^{-1}

Major inflowing river: Halali, Chamari, Firozi

Outflowing river: Halali, Betawa

Physical and chemical features

Surface temperature: -

Conductivity: -

Dissolved solids: -

pH: -

Fisheries

Number of species : 16

Major species: *C. catla L. rohita, C. mrigala, W. attu Mystus* spp., *N. chitala*

Total catch (average) : 73.478 t

Number of fishermen: 50

Number of boats : 35

Fishing gear: Gill nets, drag nets

Fish stocked: *Catla, rohu, mrigal*

The Reservoir: Barna
Location

District: Raisen
Latitude: 23° 5 ' N
Altitude : 349 m above MSL
Longitude: 78° 7' E

Year of Commissioning : 1975
Depth: 47.70 m
Volume: 539 million m^3

Surface Area *At Full Res. Level:* 7 705 ha
At Dead Res. Level: 337.40 ha

Length of Shoreline: -

Annual Fluctuation in Water Level: 11.15 m
Discharge: 13 557 $m^3\ sec^{-1}$

Major inflowing river: Barna
Outflowing river: Barna

Physical and chemical features

Surface temperature: 20-34 °C
Conductivity: -

Dissolved solids: -
pH: 7.5-8.4

Fisheries

Number of species : 16
Major species: *C. catla, L. rohita C. mrigala, L. calbasu, W. attu, Mystus,* spp.,

Total catch (average) : 103 t
Number of fishermen: 100

Number of boats : 50
Fishing gear: Gill nets, drag nets

Fish stocked: *Catla, rohu, mrigal*

The Reservoir: Loni
Location

District: Rewa
Latitude: 25° 8 ' N
Altitude : 83 m above MSL
Longitude: 81° 31' E

Year of Commissioning : 1954
Depth: 9.4 m
Volume: 1.13 million m^3

Surface Area *At Full Res. Level:* 350 ha
At Dead Res. Level: 50.000 ha

Length of Shoreline: -

Annual Fluctuation in Water Level: 2.8 m
Discharge: -

Major inflowing river: Loni
Outflowing river: Waste weir spill

Physical and chemical features

Surface temperature: 18-40 °C

Conductivity: 250 µmhos

Dissolved solids: 0.2 mg l^{-1}

pH:

Fisheries

Number of species : 8

Major species: C. catla, L. rohita, L. bata, C. mrigala, N. notopterus

Total catch (average) : 6.039 t

Number of fishermen:

Number of boats : 205

Fishingf gear: -

Fish stocked: Catla, rohu, mrigal

***The Reservoir:* Thawar**

Location

District: Mandla

Latitude: 22° 34 ' N

Altitude : 27 m above MSL

Longitude: 80° 04' E

Year of Commissioning : 1979

Depth: 27.10 m

Volume: 139.25 million m^3

Surface Area At Full Res. Level: 2 219.1 ha
At Dead Res. Level: 463.9 ha

Length of Shoreline: 85 km.

Annual Fluctuation in Water Level: 11.10 m

Discharge: 3 993 $m^3\ sec^{-1}$

Major inflowing river: Thawar

Outflowing river: -

Physical and chemical features

Surface temperature: 33 °C

Conductivity: -

Dissolved solids: -

pH: -

Fisheries

Number of species : 20

Major species: Catla

Total catch (average) : 14.537 t

Number of fishermen: 2 618

Number of boats : -

Fishing gear: -

Fish stocked: Catla, rohu, mrigal, common carp

***The Reservoir:* Meghdoot (Totladoh)**

Location

District: Chindwara

Altitude: 490 m above MSL

Latitude: 21° 30' N

Longitude: 79° 14' E

Year of Commissioning : 1990

Depth: 70 m

Volume: 1241.109 million m^3

Surface Area At Full Res. Level: 7781 ha

At Dead Res. Level: 1 494 ha

Length of Shoreline: 190 km.

Annual Fluctuation in Water Level: -

Discharge: -

Major inflowing river: Pench

Outflowing river: Pench

Physical and chemical features

Surface temperature: -

Conductivity: -

Dissolved solids: 40 mg l^{-1}

pH: 7.70

Fisheries

Number of species : 15

Major species: 5

Total catch (average) : 473 t

Number of fishermen: 227

Number of boats : 103

Fishing gear: -

Fish stocked: *Catla, rohu, mrigal*

Courtesy: Mr. A. R. Qureshi, Asst. Director of Fisheries, Raipur;
Mr. Aslam Khan, Asst. Fisheries Officer, Sikasar reservoir;
The Regional Manager, MPSFDC, Raipur;
Mr. R. P. Varma, Fisheries Inspector, Kharkora, Durg;
The Executive Engineer, Itiadoh Irrigation Division;
Mr. K. S. Sengar, Asst. Fisheries Officer, Rajnandgaon;
The Manager, MPSFDC, Bilaspur;
The Asst. Fisheries Officer Bilaspur;
Mr.S. K. Jain, Manager, MPSFDC, Morena;
Mr. S. K. Sinha, Manager, MPSFDC, Gwalior;
Mr. G. S. Sinha, Asst. Director of Fisheries, Khandwa;
The Regional Manager, MPSFDC, Gandhisagar Dam;
Mr. P. K. Singh, Asst. Director of Fisheries, Betul;
Mr. R. R. Siddiqui, Asst. Director of Fisheries, Sehore;
Mr. M. E. Khan, Asst. Director of Fisheries, Bhopal;
Mr. Laxman Singh, Investigator, Rewa ;
Mr. P. S. Rao, Manager, MPSFDC, Totladoh.

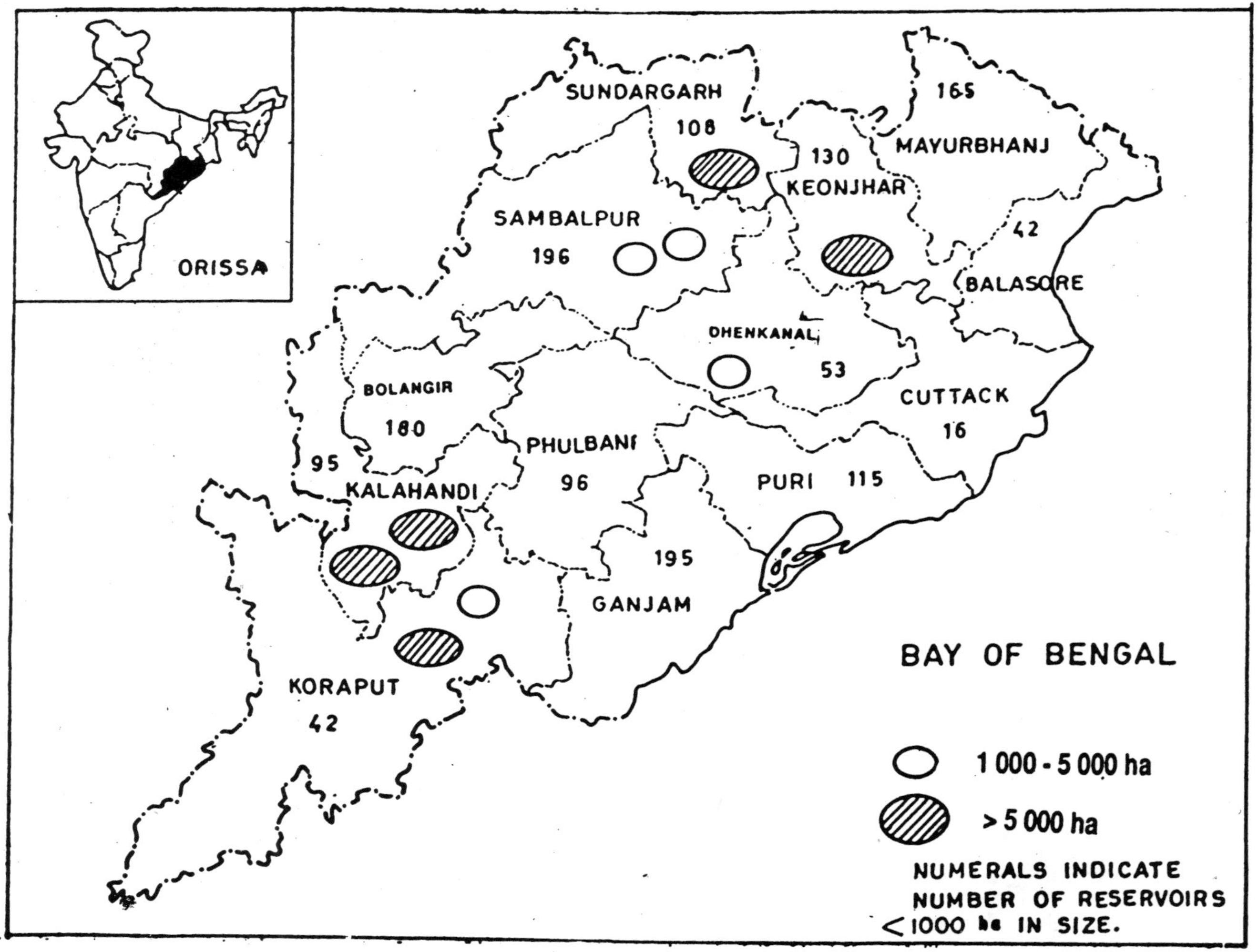

Figure 8.1. Distribution of reservoirs in Orissa

8. ORISSA

Orissa, a coastal State with a coastline of 480 km, is situated in eastern India, between West Bengal in the north and Andhra Pradesh in the south. The entire State lies in the tropical zone and is divided into four distinct parts, *viz.*, the northern plateau, the Eastern Ghats, the central part and the coastal plains. Three great rivers of the subcontinent *viz.*, Mahanadi, Brahmani and Baitarani drain the State, along with a number of east-flowing smaller rivers.

The Mahanadi discharges annually 66 640 million m^3 of water into the Bay of Bengal. The river basin comprises 7 million ha of agriculture area, 22% of which is irrigated. India's largest (in area) man-made lake, the Hirakud, is situated on Mahanadi to irrigate 264 228 ha of land and to generate 270 MW of power. The river has a total catchment of 141 600 km^2, of which 46.5% is in Orissa. The Brahmani has a total length of 800 km, of which a 540 km stretch and 57% of its catchment are in Orissa. The Brahmani carries 18 310 million m^3 of water and the basin comprises 1.7 million ha of cultivated area, 17% of which is irrigated. The State's second largest reservoir, the Rangali is on the Brahmani. The Baitarani and its tributary, the Salandi have a drainage area of 1 793 km^2. Dams and barrages on these rivers irrigate 61 920 ha. Matai is another tributary of Baitarani. The proposed Bhimkund and upper Baitarani multi-purpose projects envisage many more dams across this river and its tributaries to provide irrigation to more than 100 000 ha. Orissa has a surface water potential of the order of 9.9 million m^2, which is considered to be sufficient for meeting all irrigation needs of the State. However, only a negligible percentage of the potential is currently utilised, leaving enough room for creating more water storage reservoirs.

8.1 RESERVOIR FISHERIES RESOURCES

According to the State Fisheries Department, Orissa has a total waterspread of 256 000 ha in the form of major, medium and minor reservoirs. Unlike Tamil Nadu, Andhra Pradesh and Karnataka, the State does not designate man-made lakes as *tanks*. All the small irrigation impoundments above 10 ha are probably listed as reservoirs and hence the higher hectarage in this category. However, information on the number of reservoirs in each category is not available. The Department of Fisheries surveyed the State in 1982-83 and identified 1 442 reservoirs, covering an area of 197 198 ha, which are suitable for fisheries development activities (Table 8.1).

There are three large reservoirs, Balimela in Koraput district (19 440 ha), Hirakud in Sambalpur (71 963 ha) and Rangali in Dhenkanal district (28 000 ha). Of the total six medium reservoirs, two are in Kalahandi (Sunder and Batrajore), and one each in Koraput (Upper Kolab), Sambalpur (Gohira), Sundargarh (Mandira), and Keonjhar (Salandi) (Fig. 8.1). All the remaining 1 433 reservoirs are small (<1 000 ha) with a combined waterspread of 65 047 ha. So far, the Department of Fisheries has initiated stocking and other management measures only in 62 reservoirs (Table 8.2).

Very little scientific information is available on the ecological attributes and production dynamics of the reservoir ecosystems in the State. The only man-made lake studied in detail is the Hirakud.

Table 8.1. Reservoirs with fishery potential (Orissa)

Name of the district	Small		Medium		Large		Total	
	No.	Area	No.	Area	No	Area	No.	Area
Cuttack	16	2 126	-	16	-	-	2 126	
Puri	115	5 857	-	-	-	-	115	5 857
Balasore	42	75	-	-	-	-	42	75
Mayurbhanj	165	1 070	-	-	-	-	165	1 070
Ganjam	195	18 510	-	-	-	-	195	18 510
Koraput	42	864	1	3 500	1	19 440	54	23 804
Kalahandi	95	895	2	2 880	-	-	97	3 775
Phulbani	96	25 000	-	-	-	-	96	25 000
Sambalpur	196	267	1	1 094	1	71 963	198	73 324
Bolangir	180	1 982	-	-	-	-	180	1 982
Sundargarh	108	1 230	1	3 328	-	-	109	4 558
Dhenkanal	53	3 651	-	-	1	28 000	54	31 651
Keonjhar	130	3 520	1	1 946	-	-	131	5 466
Total	*1 433*	*65 047*	*6*	*12 748*	*3*	*119 403*	*1442*	*197 198*
Others				*Data not available*				*58 802*
Grand total								**256 000**

*Handbook on Fisheries Statistics, Orissa. (1992-93)

Table 8.2. Reservoirs with fisheries management (Orissa)

District / Reservoir	Average area (ha)	Average annual yield (kg ha^{-1})
Cuttack		
1. Kalkala	270	0.77
Puri		
2. Budhabudhani	717	2.84
3. Kuanria	200	8.38
Mayurbhanj		
4. Sunei	907	1.55
5. Kalo	546	5.11
6. Khadkei	620	3.72
7. Nesa	163	7.62
8. Bankabal	770	8.87
9. Arikul	50	7.72
10. Kukudajodi	60	5.23
11. Paunsia	60	6.45
12. Haldia	40	21.22
13. Badjore	80	17.76
14. Kattapal	20	96.35
Sambalpur		
15. Jharbandh	427	NA
16. Gohira	1 094	0.51
17. Padmapumala	152	8.28
18. Duari	56	NA
19. Kamboo	98	NA

Table 8.2 (cont.). Reservoirs with fisheries management (Orissa)

District Reservoir	Average area (ha)	Average annual yield (kg ha^{-1})
20. Khandijharana	115	NA
21. Jambunala	83	NA
22. Bankasal	216	5.83
23. Hirakud	74 000	6.60
Dhenkanal		
24. Derjang	530	21.07
25. Pamial	648	3.83
26. Dadaraghati	480	3.27
27. Rangali	28 000	0.52
Sundargarh		
28. Pitamohal	230	26.08
29. Talasara	210	12.66
30. Sarafgarh	140	16.06
31. Mandira	3 330	53.13
32. Kanjhari	260	35.50
33. Garh	50	2.00
34. Khajuria	46	25.89
35. Ordei	65	19.78
36. Ramal	456	0.70
37. Sanamachhakandana	44	3.20
38. Salandi	1 946	8.70
Ganjam		
39. Bhanjanagar	660	6.49
40. Sorada	400	18.75
41. Daha	700	6.19
42. Dhanei	100	7.42
43. Salia	175	20.94
44. Marood	70	14.47
45. Ghodahad	150	24.32
Phulbani		
46. Palisalki	120	152.75
Koraput		
47. Bhaskal	670	15.74
48. Jaganathsagar	72	56.80
49. Balimela	19 440	1.95
50. Upper Kolab	3 500	10.55
Kalahandi		
51. Sundar	1 280	2.70
52. Bhatrajore	1 646	0.97
53. Dumarbahal	500	2.59
54. Saipala	460	5.37
55. Behera	270	NA
56. Kentesir	50	4.14
57. Asurgam	39	NA
58. Tangarkana	50	7.00
59. Kiding	50	12.18
60. Kanjharkate	50	NA
61. Chhurigarh	80	19.45
62. Pipinala	44	3.22

8.2 HIRAKUD

Hirakud is one of the earliest major multi-purpose river valley projects in India. The reservoir was commissioned in 1957, on the river Mahanadi, a little below the confluence with its tributary, Ib (Fig. 8.2). Situated within the geographical ordinates of 21° 30' and 21° 50' N, and 83° 30' and 84° 05' E, the reservoir has a waterspread area of 71 963 ha at FRL. The 1 248 m long masonry dam has a height of 61 m and this, along with the earthen dams, has a combined length of 25.8 km (Table 8.3). Besides having an installed capacity of 270 MW power generation, the reservoir serves the irrigation needs of 264 038 ha of land and produces an yearly fish crop of 350 t.

The river Ib originating in Raigarh, Madhya Pradesh, has a total length of 251 km and drains a catchment of 12 447 km^2. It is rain-fed and hence nearly 80% of the runoff occurs during the monsoon months of June to October. Geologically, granite and gneisses underlie the catchment's soil mantle. A mixed red and black soil dominates the region surrounding the reservoir. With a long shoreline of 643 km, the reservoir has a high shore development index of 13.5. More than 65% of the vast catchment area (83 400 km^2), stretching over the central Indian plateau is fertile land area. The total annual inflow (1981 figures) has been estimated at 2.22×10^4 million m^3 against an outflow of 2.31×10^4 million m^3. The average annual rainfall in the region is 152 cm.

Nutrient status and production functions

The 38 year old Hirakud reservoir has emerged out of the initial phases of trophic burst and trophic depression and has now entered into a stage of stabilised productivity. Morphometrically, the massive waterspread and high shore development index portray a positive influence on the productivity levels of the reservoir.

The water regime of the reservoir can be compartmentalised into three zones. The shallow upper zone (18 621 ha) retains, to certain extent, the lotic characteristics and serves as breeding and nursery grounds of commercial carps. The lower zone, next to the dam (40 338 ha), is lacustrine in nature and supports a fishery round the year. The middle zone (13 004 ha) also sustains an year round commercial fishery.

High transparency of water is a remarkable feature of the reservoir, especially during October to May. It decreases during the floods in June to September. Hirakud, by virtue of its location within the tropics, is subjected to intense sunlight. Usually cloudless sky prevails from October to May. The reservoir, during this period, experiences algal blooms, chiefly of *Microcystis*. The high primary production rates (gross: 900-2 250 mg C m^{-3} d^{-1} ; net: 765-2 025 mg C m^{-3} d^{-1}) suggest the rich density of primary producers in the reservoir. Physico-chemical variables of water (Table 8.4) also reflect the moderate to high productivity status of the reservoir (Yadava and Sugunan, 1992). Hirakud has a rich species diversity in terms of phytoplankton. Thirty nine species of phytoplankton at a density range of 2 590 to 42 474 units l^{-1} have been reported (Dash *et al.*, 1993). There is a preponderance of blue-green algae represented by 12 species including *Microcystis* aeruginosa, *Oscillatoria* spp. and *Merismopedia* sp. There is a serious concern about the discharge of alkaline effluents into the reservoir from three paper mills (Dash *et al.*, 1993).

The reservoir experiences a steep decline in water level during April to June. Based on satellite pictures, the reservoir area is estimated at 46 381 ha during pre-monsoon and 55 243 ha during post-monsoon seasons. Alternate drying and wetting of the margins help mineralization and adds to the nutrient content of the water. The high surface runoff also adds considerable quantity of allochthonous nutrients.

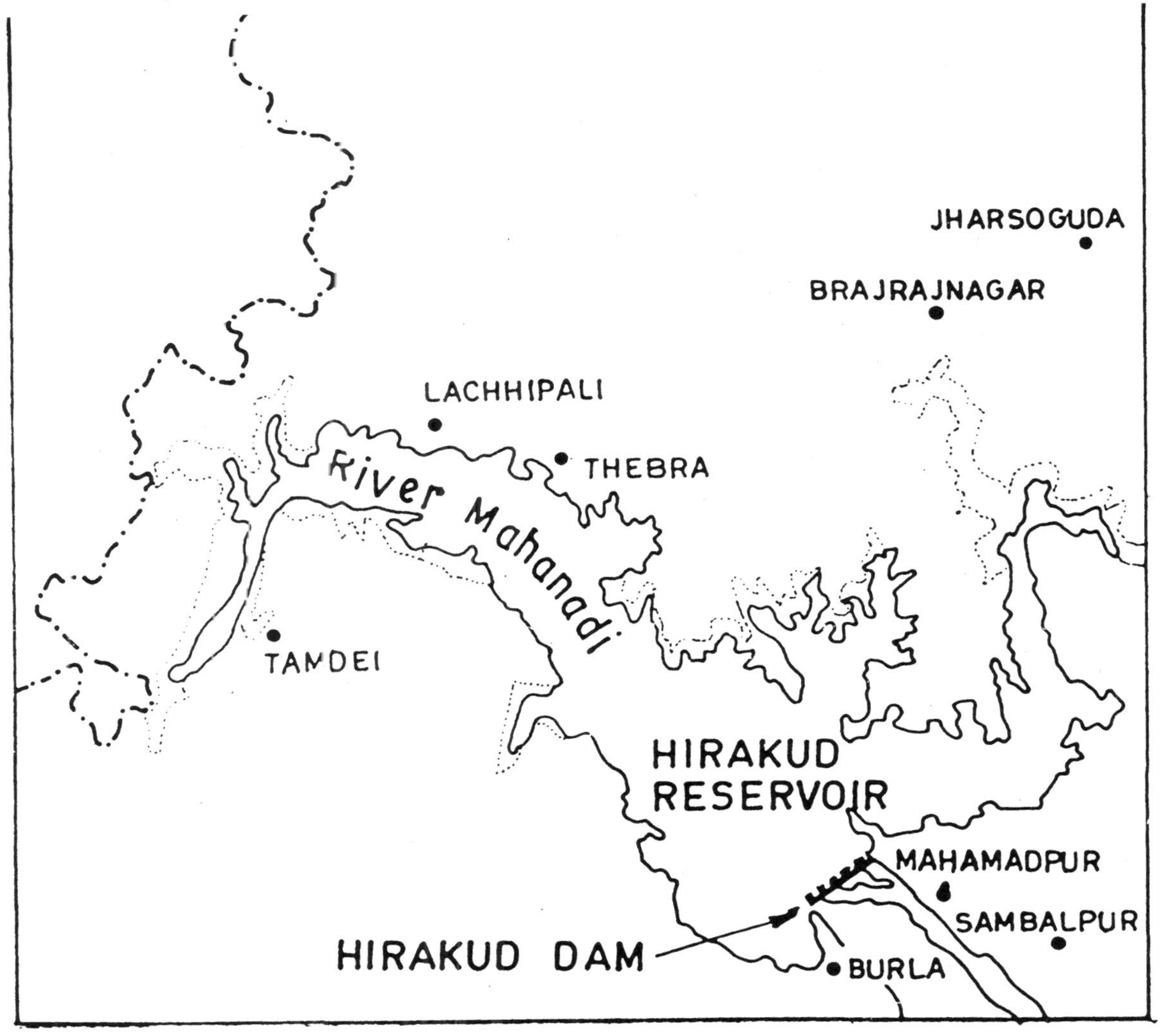

Figure 8.2. Hirakud reservoir, Orissa

Table 8.3. Morphometric features of Hirakud reservoir

General		
Location	:	District Sambalpur (Orissa)
Name of the river	:	Mahanadi
Year of commissioning	:	1956
Type	:	Multi-purpose
The reservoir		
Geographical ordinates	:	21° 30'-21° 50' N & 83° 30'-84° 05'E
Catchment area (km^2)	:	83 400
Elevation at FRL (m above MSL)	:	192.024
Elevation at DSL (m above MSL)	:	179.830
Gross storage capacity (million m^3)	:	8 316
Live storage capacity (million m^3)	:	5 818
Dead storage capacity (million m^3)	:	2 318
Waterspread area at FRL (ha)	:	71 963
Waterspread area at DSL (ha)	:	36 190
Mean depth (m)	:	11.3
Length of shoreline (km)	:	643
Shoreline development index	:	13.5
Volume development index	:	0.52
The dam		
Length of the masonry dam (m)	:	1 248
Number of sluices	:	64
Number of crest gates	:	34
Crest level of spillway dam (m above MSL)	:	185.928
Spillway capacity (m^3 sec^{-1}.)	:	42.450
Maximum annual runoff (million m^3)	:	9.19×10^4
Minimum annual runoff (million m^3)	:	1.24×10^4

(After Yadava and Sugunan. 1992)

Table 8.4. Water quality in Hirakud reservoir

Parameters	Lentic sector	Bay sector
Transparency (cm)	110	
pH	8.2	8.2
Dissolved oxygen (mg l^{-1})	9.6	8.6
Chloride (mg l^{-1})	7.4	5.5
Total alkalinity (mg l^{-1})	39.6	39.6
Total hardness (mg l^{-1})	60.00	48.0
Nitrate (mg l^{-1})	0.05	0.06
Phosphate (mg l^{-1})	0.04	0.025
Silicate (mg l^{-1})	7.0	10.0
Calcium (mg l^{-1})	52.0	40.0
Magnesium (mg l^{-1})	8.0	8.0
Spec. Cond. (µmhos)	151.0	151.0

(After Yadava and Sugunan. 1992)

Heavy deforestation and poor soil conservation measures in the catchment have resulted in siltation (at the rate of 23.14 million m^{-3} yr^{-1}) of the reservoir bed, affecting the basin topography. The washed off soil input is also responsible for reducing transparency during the monsoon months.

Fish and fisheries

The fish fauna of the Hirakud reservoir, including that of the parent rivers comprises both plain and hill river species with sizeable representation of carps and catfishes. A survey conducted in the 1950s recorded 86 species belonging to 20 families, of which 24 were of economic significance. Migratory fishes like *Tor mosal* and *Rhinomugil corsula* and prawns, (*Macrobrachium* spp.) have been affected adversely by the dam, while Indian major carps and bagrid catfishes have adapted well to the lentic conditions. The euryhaline sciaenids form a lucrative fisheries in the reservoir now. Presently, the commercial fishery comprises 40 species:

Carps:	*Catla catla, Labeo rohita, Cirrhinus mrigala, L. calbasu, L. gonius, L. fimbriatus, L. bata, C. reba, Tor mosal,*
Catfishes:	*Wallago attu, Silonia silondia, Ompok bimaculatus, O. pabda, Aorichthys aor, A. seenghala, Mystus tengra, M. cavasius, Pangasius pangasius. Rita chrysea, Bagarius bagarius, Eutropiichthys vacha,*
Featherbacks:	*Notopterus notopterus, N. chitala,*
Live fishes:	*Heteropneustes fossilis, Clarias batrachus, Channa gachua, C. punctatus, C. striatus,*
Prawns:	*Macrobrachium malcolmsonii,*
Miscellaneous species:	*Mastacembalus armatus, Nandus nandus, Glossogobius giuris, Gudusia chapra, Esomus danrica, Rasbora daniconius, Amblypharyngodon mola, Puntius sarana, P. ticto, Rohtee cotio* and sciaenids.

Before the impoundment, a comprehensive fisheries survey was conducted in a stretch of river Mahanadi by Job *et al.* (1955). More than 103 species were recorded, of which 24 were of commercial importance. *Puntius sarana, Tor mosal, Labeo fimbriatus, L. calbasu* and Gangetic major carps were the most common carps. Interestingly, *Wallago attu*, the major predator today, was very rarely encountered before the reservoir formation. The two indigenous species, viz., *T. tor mahanadicus* and *T. mosal* were protected in temple tanks which later on submerged.

An evaluation of the commercial fisheries of the reservoir since inception indicates progressive increase in the fish catch (Table 8.5). From 51.9 t in 1961-62, the present yield has increased to 489.5 t (19990-91), with an annual average of 350 t for the period 1985-91. The gradual increase in effort (number of fishermen) from 184 numbers in 1968 to 2 044 in 1987-88 is primarily responsible for the enhanced catch during the latter phase.

The yield from the reservoir has been estimated at 6.95 kg ha^{-1} (based on the area allotted to cooperative societies and the government sector). This is very low, against the State average of 12 kg ha^{-1} yr^{-1} and the national average of 20 kg ha^{-1} yr^{-1}. However, the catch accounted for the estimation of the present yield level excludes a sizeable quantity harvested through illegal fishing.

Table 8.5. Trends in fish landings of Hirakud reservoir

Year	Catch (t)
1961-62	51.9
1962-63	32.4
1963-64	14.4
1964-65	15.1
1965-66	12.4
1985-86	332.8
1986-87	483.0
1987-88	192.1
1988-89	263.5
1989-90	337.4
1990-91	489.5

Although a sound database for proper stock assessment of Hirakud is lacking, the available information indicates a decreasing trend in the populations of carnivores (catfishes and featherbacks) and *Catla catla*. The omnivores among the carps have, on the contrary, registered an increase (Table 8.6). *Aorichthys seenghala* and *Silonia silondia* among catfishes and *C. catla* among carps dominated the catch during 1978 and 1979 (Table 8.7).

Table 8.6. Fluctuation of major commercial species in the annual landings

	Percentage of abundance				
	1967-86	1977-78	1978-79	1980-81	1987-88
Carnivores (catfishes, featherbacks)	56.2	-	54.44	52.1	45.2
Omnivores					
Labeo spp., *C. mrigala* & smaller cyprinids	27.68	34.44	-	-	44.6
Catla catla	16.3	12.3	-	13.8	10.8

Data on catch per unit of effort are not available in respect of all fishing units. Based on the available information, as depicted in Table 8.8, with the progressive increase in effort, the maximum production of 843.1 t was achieved against deployment of 1 609 individuals. However, with the increase in number of fishermen, the catch per unit of effort (CPUE) decreased. In 1990-91, with the total catch of 489.5 t, the CPUE was only 239 kg. This shows the over-exploitation in the reservoir.

The low yield and the undesirable species-mix is the result of poor management. The State Fisheries Department has now initiated a stocking programme under which a total of 4.26 million fingerlings of Indian major carps were stocked during the period 1988-89 to 1990-91. This low stocking density of 10 to 47 fingerlings ha^{-1}. has had little impact on improving the catch structure. Better stock management and adequate conservation measures are necessary to reach the MSY, keeping the present effort level static.

Table 8.7. Fish catch (t) from Hirakud reservoir

Species	1978	1979
Aorichthys seenghala	95.3	125.1
Silonia silondia	94.2	97.8
Catla catla	91.8	62.5
Wallago attu	65.3	44.5
Rohtee cotio	56.3	55.0
Labeo rohita	49.6	27.6
Labeo calbasu	45.8	60.4
L. fimbriatus	37.7	43.8
Notopterus chitala	36.7	30.6
Gadusia chapra	31.7	52.1
Eutropiichthys vacha	30.6	26.2
Cirrhinus mrigala	21.1	12.9
Rita chrysea	20.2	15.1
L. bata	18.7	28.2
Puntius sarana	15.6	15.5
Aorichthys aor	7.4	6.6
Rhinomugil corsula	6.3	4.9
Glossogobius giuris	6.0	10.3
Sciaeneid spp.	5.3	4.1
Barbus tor	4.7	1.9
Chela bacaila	4.6	5.3
Channa sp.	3.5	3.9
Notopterus notopterus	1.9	8.7
Mystus tengra	1.3	1.4
Bagarius bagarius	1.1	3.1
Others	1.0	2.3
Total	**753.7**	**749.8**

Source: *Nair et al. (1981)*

The fishing rights of Hirakud were transferred to the State Fisheries Department in 1960. The reservoir is divided into six sectors, of which five are leased to the cooperative societies at the rate of Rs. 100 $mile^{-2}$ yr^{-1}. The sector adjacent to the dam (sector III) is not fished for security reasons. However, experimental fishing and licensed non-commercial fishing are permitted in this sector. More than 94% of the approximately 3 000 fishermen dispersed over 120 peripheral villages have fisheries as the major means of sustenance.

Locally built canoe, (*dingi;* 5.5 to 6 m size)) is main fishing craft. The Central Institute of Fisheries Technology (CIFT) has introduced a wooden mechanised boat of 9.1 m length and a fibre glass, flat bottomed craft into the reservoir (Khan *et al.*, 1992).

Table 8.8. Fish yield in relation to effort in Hirakud reservoir

Year	Effort (nos)	Catch (t)	CPUE	Remarks
1968	184	63.9	347	-
1969-70	262	65.7	251	-
1976-77	341	220.2	646	Marginal increase in efforts
1977-78	464	576.6	1243	Introduction of shore seine and surrounding nets increased the yield.
1980-81	1609	843.1	524	Maximum catch recorded in the reservoir.
1981-82 to 1987-88	2044	185.5	91	Declining trend
1990-91	2044	489.5	239	Improvement in the catch

Effort = number of fishermen

Fishing and post-harvest operations

Gill nets, drag nets, cast nets, stake nets and long lines are the major fishing gear operated in Hirakud. Gill nets, mainly of 3 types, are operated within 3 to 10 m depth. Drag nets of 4 to 5 types, shore seines of different sizes and hooks and lines are used round the year. Varying mesh sizes (130 mm to 203 mm) are regularly used. Use of gill nets of < 100mm mesh size is prohibited.

The presence of numerous underwater obstructions limits the use of active gear like trawls, seines and drift nets in most parts of the reservoir. The choice is well neigh limited to gill nets. Extensive siltation has, however, eased the operation of dragged gear in some areas of the reservoir. Recent trials on trawling by the CIFT have given results.

George (1979) reports increased availability of *S. silondia* in the middle reaches during summer, and *Labeo fimbriatus* during June and July in the same fishing grounds. *Catla catla* and *C. mrigala* are more evenly distributed. Fish population in the reservoir being sparse, in relation to the volume of water stored, the chances of fish being caught in gill nets are reduced. Thus, the entangling nets such as frame, vertical line and trammel nets are much more effective. Activity periods of fishes such as *Gudusia chapra, Osteobrama cotio cotio* and *E. vacha* suggest that they are best caught in gill nets during 00.00 to 06.00 hrs. (Varghese *et al.*, 1993).

The monsoon months of August and September witness a maximum and the period April to July records minimum catch in the reservoir and more than 70% of the fish are caught in entangling nets (Sulochanan *et al.*, 1960). Commercial fishermen set the gill nets at night and hauling them next morning. During the period September to November, a large number of juveniles of Indian major carps in the size range 5 to 25 mm are caught by shore seines. It is thus necessary to impose seasonal restrictions on the use of shore seines. Larger boats are used for transportation. Mechanisation of boats may make the fishing operations more remunerative.

The fish catch of the individual fisherman/party is handed over to the cooperative society, which is sold to traders. Barring a small percentage that is sold locally, the entire catch is despatched to Howrah and urban agglomerates in Orissa as well as neighbouring States. A flourishing trade of dried and smoked fish exists in the region, where the involvement of women fisherfolk is very high. Unmarketed surplus is usually dried and the dry fish is sold in 1 kg cans, the cost ranging from Rs. 20 to Rs. 40 per can. Post-harvest channels are largely influenced by the traders or the middlemen. Their influence also promotes direct disposal of catch to the agents by the fishermen, leading to unaccountability. This also leads to erosion of the functions of cooperatives, ultimately resulting in their collapse.

Cooperatives and their management

Fishing activities in Hirakud are under the cooperative sector. The five sectors of the reservoir, where commercial fishing is permitted, have been leased out to the respective local primary fishery cooperative society (PFCS). Their share capital is supplemented by a government share and subsidies (Table 8.9).

Table 8.9. Status of primary fishery co-operative societies of Hirakud reservoir

COOPERATIVES	FISHING AREA (km^2)	MEMBERSHIP	TOTAL SHARES OF (Rs.) Members	Govtment
TAMDEI	67.34	555	3 333.00	2 000.00
LACHHIPALI	41.44	379	21 830.00	23 200.00
THEBRA	111.37	657	48 010.00	207 600.00
MAHAMADPUR	116.55	822	12 870.00	42 200.00
IB	98.42	617	6 170.00	9 500.00
TOTAL	435.12*	3 030	92 213.00	284 500.00

*Excluding 68 sq. km area reserved for dam security

The Mahamadpur PFCS has the maximum operational area (11 655 ha), and Lachhipali has the minimum (4 144 ha.). Based on the average catch for 1989-90 and 1990-91, Thebra attained maximum yield (11.02 kg ha^{-1}) from its area (Table 8.10). The societies appoint merchants to lift the catch from the landing centres. Price of fish is fixed according to its size class *i. e.*, carps big, carps small, non carps big, non-carps small, Channa etc. The societies normally reimburse the sale price of fish to the fishermen after deducting a commission ranging from Rs. 0.5 (*Channa*) to Rs. 3.00 (carps big). The procurement price of fish and the commission vary from society to society.

Established in 1979, the Lachhipali primary fishermen's cooperative society has 379 members. The scheduled caste and scheduled tribes constitute 51.45 and 31.33% of the total membership respectively. The society fishes 4 144 ha of the reservoir, most of which dries up during summer months due to drawdown. While fishing activity is at its peak during August to September, it is adversely affected from April to July every year. About 50% of the members of Lachhipali society solely depend on reservoir fisheries for a living.

A variety of fishing gear is in used, including surface gill nets, drag nets, shore seines (operated by 10 to 15 people), hook and line, cast nets, and a variety of traps. The annual fish catch by the members of the society varied from 22.3 t in 1989-90 to 49.7 t in 1991-92. The society charges a commission of Rs. 2 per kg for big carp, small carp and big cat fish; Rs. 1.00 per kg of small cat fish and Rs. 0.50 per kg of miscellaneous fishes.

Table 8.10. Average annual catch of the cooperative societies of Hirakud

Sector	Area (ha)	Average catch (kg)*	Yield (Kg ha^{-1})
Govt. Sector	6 800	-	-
Mahamadpur	11 655	30 691	2.63
Thebra	11 137	122 682	11.02
Ib	9 842	30 201	3.07
Tamdei	6 734	29 510	4.38
Lachhipali	4 144	29 832	7.20

*Average of 1989-90 and 1990-91

Under a scheme sponsored by the National Bank for Agriculture and Rural Development (NABARD), the society received financial assistance of Rs. 608 400. The scheme, intended to extend assistance to the fishermen for procuring craft and gear, supported 36 fishing units in the reservoir at the rate of Rs. 2 500 for boat and Rs. 1 440 for nets each unit comprising three fishermen.

Situated at the Lakhimpur block of District Sambalpur and established in 1964, the Thebra primary fishermen cooperative society is the oldest society in the Hirakud reservoir and has a total membership of 657, which includes members belonging to the scheduled castes (233) and scheduled tribes (324). Enjoying the fishing rights over 11 137 ha of the reservoir area, it is the most prosperous among the cooperative societies and receives subsidy of Rs. 304 150. Most of the fishermen operating in the area are active and wholly dependent on the reservoir for their livelihood.

The Thebra Society collects a commission from the fishermen @ Rs. 3 for carps, Rs. 2 for catfish and Rs. 1 for small catfishes. In 1985, the society received financial assistance from the National Cooperative Development Corporation (NCDC) in the form of a loan and share capital worth Rs. 1 107 000. A total of 357 members benefitted from the NCDC loans. The package included a pick-up van and 2 motor launches for facilitating easy transport of catch by the society. Thebra Society is logistically well-equipped for undertaking the marketing of fish to the Howrah wholesalers. But, surprisingly, it leaves the catch at the disposal of private merchants who take away large chunks of the profit. The diesel pick-up van and the motor launch are kept idle, to the detriment of fishermen's interests.

The total fish landings from the area under the control of Tamdei PFPS varied from 28 570 kg in 1989-90 to 30 455 kg in 1990-91 and yield ranged from 3.35 to 5.27 kg ha^{-1}. The society normally deducts commission at a flat rate of Rs. 1 per kg of fish.

Mahamadpur is the largest Primary Fishermen's Cooperative Society in Hirakud, both in terms of membership and the area of water under its control. The society's fishing territory is confined to sector 2 and handles the lowest catch among all the cooperatives, varying from 14 132 kg in 1988-89 to 32 076 kg in 1989-90, with the corresponding yields of 1.75 kg ha^{-1} to 3.85 kg ha^{-1}.

More than 73% of the members of the Ib IPFCS belong to the scheduled castes and 17.34% to the scheduled tribes. Annual fish catch by the members varied from 15 525 kg in 1988-89 to 39 978 kg in 1989-90, equivalent to 2.04 and 3.59 kg ha^{-1}. The reservoir area under the control of Ib society covers one of the important sectors of the reservoir from biological and economic point of view. This stretch provides breeding grounds for major carps and it acquires significance while formulating conservation norms. The Ib sector also receives the maximum pollution load. A paper mill in Brajraj Nagar discharges effluents into the Ib river which find their way into the reservoir. Fish kills have been reported on account of pollution. Broodstock ascending the Ib river for breeding is susceptible to over-exploitation and strict vigil is needed to protect the fishes on their breeding run.

Development strategies

Hirakud, despite being the largest man-made watersheet in the country, remains one of the least researched in terms of its fishery potential. The fish production from this large water body is also one of the lowest. This reservoir has been in existence for the last 35 years and having passed through the initial phases of trophic burst and trophic depression, has now stabilised, but at a very low level of fish productivity. Fish species subsisting on higher links in the food chain and uneconomic varieties now dominate the catch, which results in the poor conversion of the primary energy into fish flesh.

The main accent on development of fisheries in the Indian reservoirs has been on optimum utilisation of the inherent production potential through stock manipulation. Since the lake has already crossed its formative years, any attempt towards substantial change in fish stock is difficult. Nevertheless, a holistic may the fish production. The key parameters identified for sustainable development of Hirakud reservoir fisheries include:

i) a better appreciation of the scientific attributes influencing productivity at different trophic levels
ii) appropriate stocking and harvesting policies
iii) adequate post-harvest infrastructure
iv) strong cooperative base and
v) sound conservation measures.

As the limno-chemical propensities of any water body largely govern its productivity rhythm, a deep insight into the production functions is essential before management measures are formulated. Hirakud with a poor database in terms of soil and water quality, needs a detailed scientific study to assess the flux of nutrients and related production dynamics.

More accurate and reliable data on total catch, effort, catch per unit of effort and size composition of the catches are essential for a proper understanding of the available resources, their extent of exploitation and the effect on the abundance on fish stock. This, in turn, will enable estimation of the level at which the stock has to be maintained for obtaining a sustained yield. A proper assessment of the existing populations of fishes, on the basis of a statistically designed, survey is needed. Such studies would form the basis of a long- term stocking and harvesting policy for the reservoir. This would, *inter alia*, include the stocking density, stocking schedule, species mix, gear selection and mesh size regulations.

Gear selection is an important component of fisheries management. Sufficient information exists on the efficacy of different fishing gear suitable for large water bodies, mainly through the experiments conducted by the Central Institute of Fisheries Technology. This expertise has to be utilised while formulating guidelines for gear selection.

Fisheries of Hirakud reservoir are totally under the cooperative sector as the societies enjoy the fishing rights of the entire reservoir. However, middlemen have an upper hand in the post-harvest operations, leading to erosion of profit earned by the societies. Market function has been performed by the private fish merchants without any valid reasons. There is a need for an apex body to coordinate the activities of the cooperative societies and to pool and share the resources available at the disposal of different societies. All the market functions including procurement of fish from the fishermen, their transport by water to the landing centres, and trade with the wholesale merchants at Howrah should be done under a properly organised cooperative network.

Since capture fisheries development depends heavily on the biotic communities present in the ecosystem, the management norms for such water bodies centre around principles of habitat conservation and optimum exploitation of the wild stock. Intervention in the ecosystem management should be restricted to augmentation of stock through artificial recruitment and fishing effort manipulations. Pollution from industries, agricultural wastes, thermal power plants and domestic effluents are the common causes of environmental degradation in reservoirs. Hirakud reservoir is already threatened with environmental hazards from:

i) industrial effluents from a paper mill, refractories and cement factories,
ii) wastes from collieries,
iii) thermal wastes from power plant,
iv) river bed modification through removal of gravel,
v) pesticide residues from agricultural operations in the peripheral areas of the reservoir.

The Ib river is a possible route to link carp breeding grounds with the reservoir. This and other areas having a bearing on the carp breeding run need to be studied to determine areas for sanctuaries for the conservation of desirable species.

8.3 RANGALI

Rangali reservoir, on the river Brahmani, is the second largest man-made lake in Orissa, covering 37 840 ha at full reservoir level and 28 000 ha at the mean level, storing 4 400 to 5 150 million m^3 of water. The reservoir has a catchment of 25 250 km^2, comprising mostly forests and wasteland and the nitrates during the rainy season are of 0.32 to 0.52 mg l^{-1}. The water temperature (20.25 to 29.0 °C) and pH (8.2) are conducive to reasonably high organic productivity (Pati and Sahu, 1992). This is further confirmed by the rich plankton community, dominated by Cyanophyceae (*Microcystis*, *Oscillatoria*, *Phormidium*, and *Lyngbya*), *Microcystis* being the most dominant form among them. The abundance of phytoplankton has a definite correlation with the inflow on account of the allochthonous input of nutrients (Pati and Sahu *op. cit.*).

The fish catch is estimated at 14.5 t, which is equivalent to 0.52 kg ha^{-1}, based on the average area (28 000) of the reservoir. No stocking has so far been done.

8.4 STATUS OF RESERVOIR FISHERIES IN ORISSA

Reservoirs of Orissa produce 4 192 t of fish. A district-wise break-up is given in Table 8.11. Yields vary depending on the size of the reservoir. Fifty-three small reservoirs in 10 districts range from 0.77 to 152.75 kg ha^{-1} (Table 8.12). Medium reservoirs produce fish at the rate of 0.51 to 53.13 kg ha^{-1} and the large reservoirs 0.52 to 15.74 kg ha^{-1} (Table 8.13). Yields of small, medium and large reservoirs in Orissa are estimated at 25.85, 12.76 and 7.62 kg ha^{-1}. This is just 4.5% of the total inland fish production. Considering that the 1 442 reservoirs (197 198 ha) in the State produce only 4 192 t of fish (21.26 kg ha^{-1}), there is much scope to increase the production through rational management.

The 65 047 ha of small reservoirs could produce 6 505 t of fish and the medium and large reservoirs could add an equal quantity with the final effect of 50 kg ha^{-1}. This would increase the share of reservoirs to inland fish production from the present 4.5% to 14%. If the total reservoir area of 256 000 ha is taken into account at the above mentioned yield rate, the production can go up to 18 991 t or 20% of the total catch. The reservoirs can play a vital role in increasing the fish availability in the State as their fisheries development is labour-intensive, environment-friendly and socially rewarding, compared to the aquaculture activities. The capital cost involved in producing a similar quantity of fish through aquaculture will be astronomical and the profit is siphoned off by a few investors, whereas the benefits accrued by the fisheries development percolate down to some of the weakest sections of the society. This is especially applicable to Orissa, considering the demographic and social conditions of the areas where these reservoirs are situated.

Table 8.11. Fish production in reservoirs of Orissa (annual)

District	Total fish production from reservoirs (t)	Total inland fish production of the state (t)
Cuttack	14.200	21 847
Puri	3.900	15 031
Balasore	-	7 217
Mayurbhanj	436.640	4 931
Ganjam	96.150	8 542
Gajapati	15.830	1 819
Phulbani	301.930	2 242
Kalahandi	654.580	1 585
Koraput	50.000	950
Nawarangpur	29.000	807
Malkangiri	153.000	757
Rayagada	3.000	241
Sambalpur	2 171.400	11 875
Bolangir	30.200	5 719
Sundargarh	120.000	3 132
Keonjhar	45.350	1 037
Dhenkanal	67.290	6 029
Total	**4 192.470**	**93 761**

Table 8.12. **Fish yield in small reservoirs of Orissa**

District	Number of reservoirs	Yield range (kg ha^{-1})	Average (kg ha^{-1})
Cuttack	1	270	0.77
Puri	2	200-717	5.61
Mayurbhanja	11	20-907	16.51
Sambalpur	7	56-427	7.05
Dhankanal	3	480-648	8.3
Sundargarh	9	44-456	15.76
Ganjam	7	70-700	14.08
Phulbani	1	120	152.75
Koraput	2	72-670	32.27
Kalahandi	10	39-500	5.39
Total	**53**		**25.85**

Table 8.13. **Fish yields in medium and large reservoirs of Orissa**

District	Reservoir	Yield
Medium		
Sambalpur	Gohira	0.51
Sundargarh	Mandira	53.13
Keonjar	Satandi	8.70
Koraput	Upper kolab	10.55
Sundar	Kalahandi	2.70
Kalahandi	Bhatrajore	0.97
Medium		*12.76*
Large		
Sambalpur	Hirakud	6.60
Dhenkanal	Rangali	0.52
Koraput	Balimela	15.74
Large		*7.62*

The Department of Fisheries and the Orissa Fish Seed Development Corporation (OFSDC) stock the reservoirs regularly. The district-wise details are given in Table 8.14. The OFSDC stocked 9 reservoirs in the year 1987-88 and by 1992-93, the number of reservoirs under its fold rose to 30. Total number of fry stocked in various reservoirs during the years 1985-86 to 1992-93 is given in Table 8.15.

Perusal of Tables 8.14 and 8.15 indicates that the stocking rate has been arrived at without any rationale. Perhaps, the easy availability and the nearness to fish seed farms were the criteria for stocking. Details of qualitative composition of stocking are also not known.

Table 8.14. Stocking of reservoirs in Orissa during 1985-86 to 1992-93

Year	No. of reservoir stocked	Area of reservoir stocked (ha)	Fry stocked in thousands	Stocking rate (No ha^{-1})
1985-86	37	63 159	8 437	133
1986-87	34	135 805	9 500	70
1987-88	38	62 726	20 437	236
1988-89	71	54 215	64 182	1183
1989-90	70	131 217	57 534	438
1990-91	63	125 776	64 490	513
1991-92	61	122 510	65 629	536
1992-93	39	46 557	36 360	781

Table 8.15. Details of stocking by Department of Fisheries in the reservoirs of Orissa

Districts	1985-86		1986-87		1987-88		1988-89	
	No. and area	Fry stocked	No. and area	Fry stocked	No. and area	Fry stocked	No. and area	Fry stocked
Cuttak	-	-	-	-	-	-	-	-
Puri	(1) 717	255	(1) 717	250	-	-	(5) 132	130
Balasore	-	-	-	-	-	-	-	-
Mayurbhanj	(4) 1 284	1384	(4) 1 284	1284	(4) 146	450	(9) 1 071	15 00
Ganjam	(7) 2 255	1555	(6) 1 555	1695	-	-	-	-
Phulbani	(1) 105	105	(1) 105	150	(2) 130	270	(1) 270	285
Koraput	(1) 19 440	726	(2) 20 075	524	(4) 24 283	1 602	(4) 4 880	881
Kalahandi	(9) 4 043	2118	(9) 4326	2118	(9) 407	12 64	(11) 2 027	774
Sambalpur	(1) 1 094	438	(2) 73 057	591	(2) 107	130	(5) 298	690
Bolangir	-	-	-	-	-	-	-	-
Sundargarh	(5) 3 686	252	(3) 3 566	1335	(3) 3 558	25 88	(3) 37 60	15 40
Keonjhar	(4) 2 110	768	(2) 1 990	408	(5) 2427	11 78	(9) 25 70	18 15
Dhenkaual	(4) 29 130	836	(4) 29 130	1145	-	-	-	
Total	(37) 63 864	8437	(34) 135 805	9500	(29) 31 058	7 484	(47 15 008	7 615

Districts	1989-90		1990-91		19891-92		1992-93	
	No. and area	Fry stocked	No. and area	Fry stocked	No. and area	Fry stocked	No. and area	Fry stocked
Cuttak	-	-	(1) 270	187	(1) 270	600	-	-
Puri	-	-	-	-	-	-	-	-
Balasore	-	-	-	-	-	-	-	-
Mayurbhanj	(13) 1 201	23.00	(11) 1 130	1 850	(9) 2 677	1 660	(2) 100	140
Ganjam	(2) 855	5.00	(1) 780	345	(1) 760	400	-	-
Phulbani	(1) 120	5.03	(1) 120	285				
Gagapati	-	-	-	-	-	-	-	-
Koraput	(4) 4 320	592	(3) 1 905	260	(4) 1 245	600	(2) 35 72	300
Nawarangpur	-	-	-	-	-	-	(2) 17 35	150
Malkangiri	-	-	-	-	-	-	(2) 11 63	420
Rayagada	-	-	-	-	-	-	-	-
Kalahandi	(8) 260	6.80	(5) 19[illegible]	400	(6) 299	740	-	
Sambalpur	(4) 745 453	26.72	(5) 71 393	3 102	(8) 72 576	4 416	-	-
Bolangir	-	-	-	-	-	-	-	-
Sundargarh	(2) 3 336	25.19	(2) 3 336	510	(11) 3 328	1404	(1) 3328	350
Keonjhar	(12) 3 071	18.51	(8) 1 037	7.45	(7) 643	912	-	-
Dhenkaual	-	-	-	-	-	-	-	-
Total	(46) 87 616	11 617	(37) 80 170	7 684	(47) 81 798	10732	(9) 9 898	1 360

Stocking figures in thousands

Figures in parantheses-number of reservoirs

9. GUJARAT

Situated in the northwest corner of India, Gujarat is the seventh largest State in India and comprises three distinct geographic regions, *viz.*, the undulating terrains of Saurashtra, which is a peninsula jutting into the Arabian sea, the Rann (desert) of Kachchh in the north, bordering Pakistan, and the mainland which is essentially the alluvial plains. The latter are drained by the big rivers Sabarmati, Mahi, Narmada and Tapti, and by the smaller rivers Banas, Saraswati and Damanganga. Northern districts are arid with scanty rainfall. In the rest of the State, rainfall varies between 65 to 127 cm. With the tropic of Cancer passing through it, Gujarat has a hot and cold climate. But, the nearness to Arabian sea and the Gulf of Cambay in the West and the forest covered hills in the east moderate the climatic extremes.

All major rivers in Gujarat are west-flowing. The Sabarmati has a catchment of 21 674 km^2, 81% of which lies in Gujarat. Similarly 34% of the Mahi basin and 11.5% of the Narmada basin lie in the State, mainstreams of all the three meeting the Arabian sea. A large number of minor streams radiate down from the hills of Saurashtra joining the Arabian sea, Gulf of Kachchh and the Gulf of Khambhat. This network of small rivers and undulating terrain creates ideal conditions for water resource development projects and thus the five districts of Saurashtra, *viz.*, Rajkot, Jamnagar, Junagarh, Amreli and Bhavanagar among themselves share nearly half of the reservoirs in the State (Fig. 9.1). Sabarkantha, Panchmahals and Surat also have large areas under man-made lakes.

9.1 RESERVOIRS OF GUJARAT

The State Fisheries Department of Gujarat has listed 50 reservoirs, covering an area of 242 205 ha. This is in addition to 561 small irrigation tanks, which are actually reservoirs, with a waterspread of 44 025 ha. Thus, the total area under reservoirs in the State is 286 230 ha. More than 95% (in number) of these man-made lakes belong to the small category, although they form only 29% of the total area (Table 9.1). The small reservoirs in the State have an average size of 349 ha. There is a large concentration of these small water bodies (< 1 000 ha) in the western half of the State, mainly the Saurashtra. The Kachchh district and the Saurashtra region together have 85 small reservoirs, out of the total 115 in the State. The medium reservoirs (1 000-5 000 ha) are spread over 12 districts (average size 2 000 ha). Rajkot district with 8 and Sabarkantha with 5 have the maximum concentration (Table 9.2). The seven large reservoirs (>5000 ha) in the State, *viz.*, Dharoi (Mehsana district), Kadana, Panam (Panchmahal district), Kakarapur, Vallabhsagar (Surat district), Damanganga (Valsad district) and Shetrunji (Bhavanagar district) have a total area of 144 358 ha and an average size of 20 622 ha. Among them, Dharoi, Kadana, Kakarapur and Ukai are in the above-10 000 ha class (Table 9.3).

Very few scientific investigations have been conducted on the reservoirs of Gujarat. The Central Inland Fisheries Research Institute conducted ecological studies on Ukai during the 1970s, as a part of the All India Coordinated Research Project on Reservoir Fisheries. Some information is also available on Sayaji Sarovar in Vadodara.

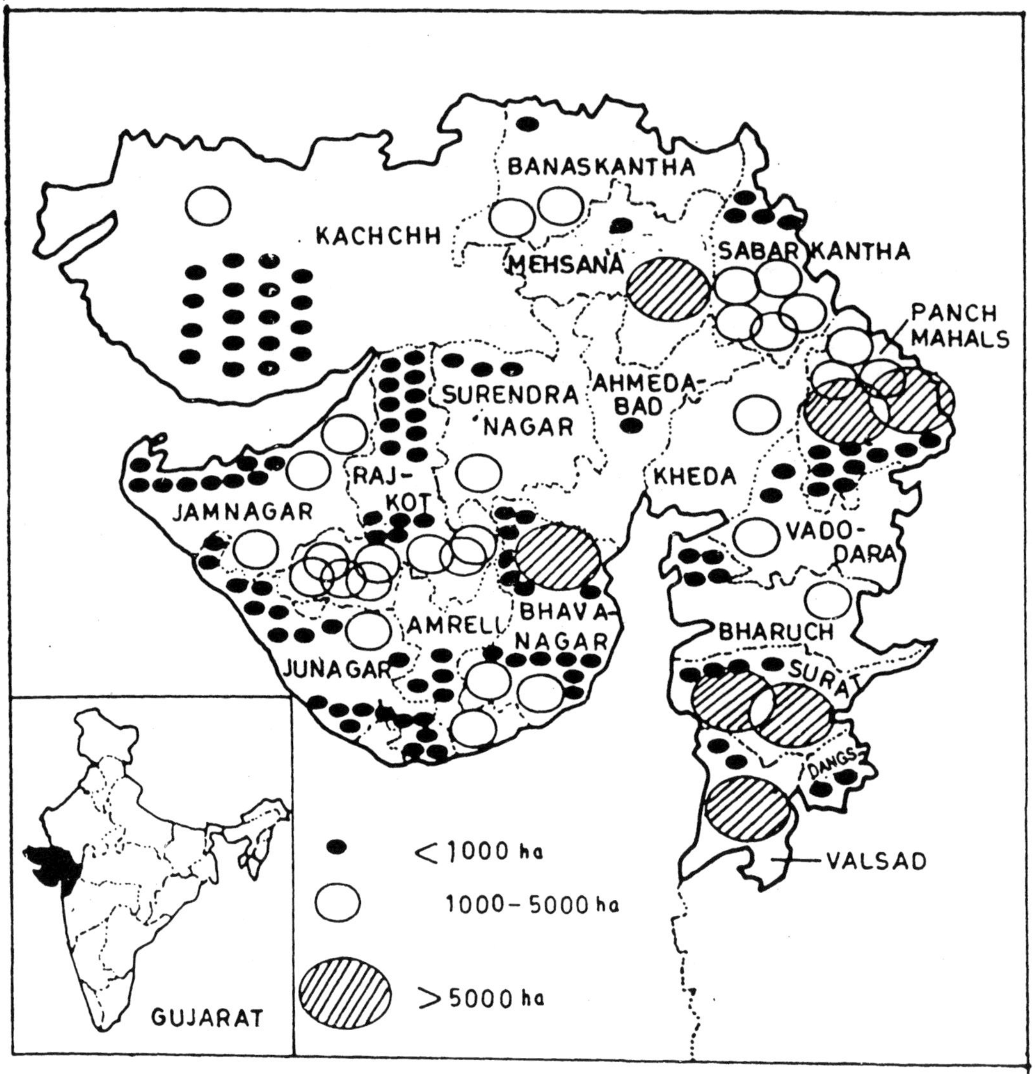

Figure 9.1. Distribution of reservoirs in Gujarat

Table 9.1. Reservoirs of Gujarat

District	Small (<1 000 ha) No.	Small Area (ha)	Medium (1 000-5 000 ha) No.	Medium Area (ha)	Large (>5 000 ha) No.	Large Area (ha)	Total No.	Total Area(ha)
Ahmedabad	1	296	-	-	-	-	1	296
Banaskantha	1	578	2	6 615	-	-	3	7 193
Bharuch	4	634	1	3 677	-	-	5	4 311
Kheda	-	-	1	2 086	-	-	1	2 086
Mehsana	1	277	-	-	1	10 700	2	10 977
Panchmahals	9	4 915	1	1 668	2	25 580	12	32 163
Sabarkantha	4	936	5	11 850	-	-	9	12 786
Surat	4	869	-	-	2	96 200	6	97 069
Vadodara	2	697	1	2 904	-	-	3	3 601
Valsad	2	674	-	-	1	5 144	3	5 818
Dang	2	17	-	-	-	-	2	17
Amreli	4	1 862	1	3 285	-	-	5	5 147
Bhavanagar	14	5 810	2	2 207	1	67 34	17	14 751
Jamnagar	10	2 989	3	5 018	-	-	13	8 007
Junagarh	20	5 802	1	1 581	-	-	21	7 383
Rajkot	16	5 166	8	13 798	-	-	24	18 964
Surendranagar	3	1 963	1	1 914	-	-	4	3 877
Kachchh	18	6 614	1	1 145	-	-	19	7 759
Total	115	40 099	28	57 748	7	144 358	150	242 205
Irrigation tanks	561	44 025	-	-	-	-	561	44 025
Grand Total	676	84 124	28	57 748	7	144 358	711	286 230

Compiled from the information provided by the Department of Fisheries, Gujarat

Table 9.2. Large reservoirs of Gujarat

Reservoir	District	Area at FRL (ha)
Dharoi	Mehsana	10 700
Kadana	Panchmahals	16 600
Panam	"	8 980
Kakarapur	Surat	44 200
Vallabhsagar	"	52 000
Damanganga	Valsad	5 144
Shetrunji	Bhavanagar	6 734

Table 9.3. Medium reservoirs of Gujarat

RESERVOIR	DISTRICT	AREA AT FRL (ha)
Datiwada	Banaskantha	4 047
Sipu	"	2 568
Karjan	"	3 677
Vanakobori	Kheda	2 086
Deo	Panchmahals	1 668
Hathmati	Sabarkantha	3 236
Meshwo	"	1 116
Watrak	"	4 475
Guhai	"	1 713
Mazam	"	1 310
Sukhi	Vadodara	2 904
Sankroli	Amreli	3 285
Ranghola	Bhavanagar	1 162
Sukhbhandar	"	1 045
Sasoi	Jamnagar	1 468
Uni	"	1 730
Sani	"	1820
Amipur	Junagarh	1 581
Bhandar	Rajkot	4 400
Moj	"	1 203
Phophal	"	1 207
Machhu I	"	1 200
Machhu II	"	1 396
Aji II	"	1 439
Demi II	"	1 023
Nyari I	"	1 930
Brahmi	Surendranagar	1 914
Tapar	Kachchh	1 145
TOTAL		**57 748**

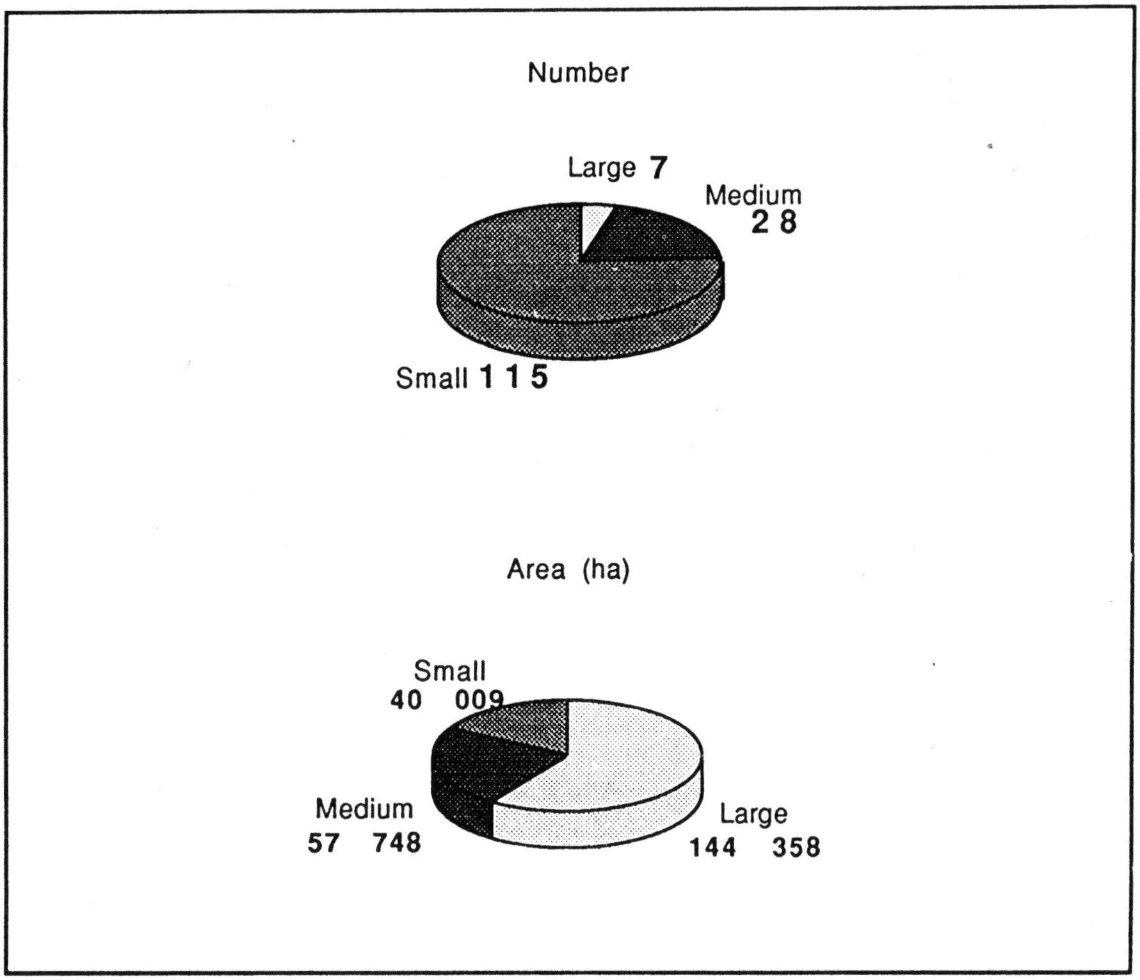

Figure 9.2. Areas under different categories of reservoirs in Gujarat

9.2 UKAI RESERVOIR

Ukai also known as Vallabhsagar, is the largest reservoir in Gujarat situated on the river Tapti at 21° 15' N and 73° 35' E. Completed in 1971, this multi-purpose reservoir has a catchment of 62 255 km^2 and a waterspread of 52 000 ha at the FRL of 105.16 m above MSL with a mean depth of 11.8 m. Vallabhsagar has a moderately irregular shoreline (shoreline development index 5.56). The water level fluctuates up to 23 m during different seasons. Information furnished in the following account has been obtained from various reports of the Central Inland Fisheries Research Institute, Barrackpore (Anon., 1980, 1983, 1984b).

The prevailing climatic conditions have a key role in determining the thermal and nutrient regime of the reservoir. The local precipitation is high, *e. g.*, during 1983-84 it was 156 cm. Ukai witnesses climatic extremes with the air temperature going above 45 °C (47.6 °C in May 1983) during summer and the minimum of 6.0 °C during winter months.

The water is rich in nutrients and dissolved salts, both from the allochthonous enrichments and due to the flooded fertile soil. The catchment area comprises forest land, which received precipitation of 79 cm. Apart from the cumulative addition from the inflowing waters, the *per se* nutrient status of the reservoir bottom is equally high, as 99% of the submerged area is either agricultural (58%) or forest (41%) land. The organic carbon in the flooded soil is high (0.31-3.6%), with matching available phosphorus (1.2 - 2.0 mg 100 g^{-1}) and nitrogen (12.00 - 32.48 mg 100 g^{-1}). The soil is alkaline in reaction (pH; 7.3-7.9).

Productive potential is also well-reflected in all measurable physico-chemical attributes of water quality (Table 9.4). The pH is uniformly high without deviating much from the mean of 8.3. Transparency, exhibiting a strong seasonality, ranges between 8.5 to 328 cm. The values of bicarbonates and the specific conductivity in consonance with the ions of magnesium and calcium indicate high production potential of the water. Nitrate nitrogen and phosphate phosphorus levels are very high, despite the absence of any anthropogenic sources of organic enrichment. Thus, the reservoir maintains its high productivity due to the allochthonous and intrinsic nutrient loading.

Table 9.4. Water quality of Vallabhsagar reservoir (surface values)

Parameters	Range in variation (n=24)	Mean±SD
Water temperature (^0C)	20.0-32.6	26.5±3.89
Transparency (cm)	8.5-328.0	102.4±83.98
pH	8.0-8.4	8.3±0.13
Dissolved oxygen (mg l^{-1})	3.7-14.0	8.3±2.97
Carbon dioxide (mg l^{-1})	nil-20.1	*9.5±5.81
CO_3 (mg l^{-1})	0.2-30.0	9.6±7.53
HCO_3 (mg l^{-1})	98.0-366.0	166.0±65.01
Sp. conductivity (μ mhos)	41.0-585.0	322.3±191.91
Ca^{++} (mg l^{-1})	12.0-58.0	29.6±10.64
Mg^{++} (mg l^{-1})	6.7-33.8	13.6±7.72
No_3 N (mg l^{-1})	0.02-0.09	0.050±0.02
PO_4P (mg l^{-1})	0.012-1.00	0.110±
Silicates (mg l^{-1})	0.06-0.82	0.093-
Dissolved organic matter (mg l^{-1})	0.02-28.0	11.85±6.60

* *Mean of eight values only* *(Anon., 1980)*

The sharp difference between the air temperatures of summer and winter seasons notwithstanding, thermal stratification does not occur during summer due to the continuous water uptake. Even during the peak of the dry season of April-May, when the inflow trickles down to 0.63 to 3.45 million m^3 (compared to the heavy influx of 5 320 million m^3 during September), a steady outflow of 142 to 176 million m^3 of water is maintained for various purposes. Thus, instable conditions prevail in the reservoir resulting in a vertical mixing of water.

The tropholytic zone is characterised by the decline in oxygen (7.2 mg l^{-1} at the surface to 5.6 mg l^{-1} at 16 m). Dissolution of carbonates as bicarbonates in the presence of CO_2 is evidenced by the increase in bicarbonates towards the bottom (150 mg l^{-1} at the surface to 168 mg l^{-1} at 12 m depth). This vertical stratification of chemical parameters indicates high rate of organic productivity.

The gross primary productivity rate during 1977-78 to 1982-83 ranged from 70.92 and 661.44 mg C m^{-3} day^{-1} (mean 218.42 ±122.5). The plankton community of the reservoir is exceptionally rich, the volumetric abundance ranging from 0.3 to 35.5 ml m^{-3} (Table 9.5).

Table 9.5. Abundance of plankton in Vallabhsagar

Year	Volumetric abundance of plankton (ml m^{-3})
1976-77	5.26 to 6.60
1977-78	12.5 to 22.5
1978-79	2.0 to 35.5
1979-80	1.0 to 6.3
1980-81	0.95 to 11.0
1981-82	0.3 to 6.7
1982-83	0.23 to 12.6

The organically rich bottom is favourable for the rapid growth of benthic macrofauna, although the sharp fluctuations in water level are an inhibiting factor. Among the benthic animals, gastropods, which have some ability to adjust with the receding and rising water levels, form the most dominant group. The lentic zone of the lake promotes a better growth of benthic community, than the lotic. Population density of gastropods in the lentic sector during the seven years from 1976-77 varied from 93 to 474 individuals m^{-2} with a mean of 225. Mean density of all benthic organism for the same period is 453, 300 and 178 individuals m^{-2} for the lentic, intermediate and lotic sectors (Table 9.6).

The indigenous ichthyofauna of the river Tapti before impoundment, comprised mainly the Indian hilsa, *Tenualosa ilisha, Tor tor, Aorichthys seenghala, A. aor, Labeo fimbriatus*, and *Wallago attu* (Menon, 1973). The hilsa has a smoth passage from the Arabian sea up to the Ukai dam through the Tapti estuary, though the migratory pathway is threatened to be obstructed by the proposed Navagam dam across the Tapti.

Soon after the impoundment, the fish stock in the reservoir became thinner and stocking was resorted to with Indian major carps for building commercial fish populations which proved to be very useful. The stocked major carps have established themselves firmly and have commenced breeding in the reservoir, contributing to autostocking. The young ones of Indian major carps which escape from the dam through the outlets have helped the stocking of downstream impoundments. The fish catch structure in the main river Tapti, 112 km downstream, has changed in favour of the Indian major carps, by virtue of the stock augmentation from the outflowing waters of Vallabhsagar reservoir.

Table 9.6. Sectoral variations in the benthic population density in Vallabhsagar reservoir

	1976-77	77-78	78-79	79-80	80-81	81-82	82-83	Mean
LENTIC SECTOR								
Gastropods	474	189	-	237	93	136	221	225
Bivalves	47	164	-	50	71	68	54	76
Oligochaetes	13	49	-	20	2	1	826	152
LENTIC TOTAL	**534**	**402**	-	**307**	**166**	**205**	**1 101**	**453**
INTERMEDIATE								
Gastropods	292	291	266	185	99	110	100	192
Bivalves	38	20	157	89	37	52	38	61
Oligochaetes	-	-	-	9	2	2	318	47
INTERMEDIATE TOTAL	**330**	**311**	**423**	**283**	**138**	**164**	**456**	**300**
LOTIC SECTOR								
Gastropods	199	92	180	133	77	68	99	120
Bivalves	38	48	50	41	32	-	19	33
Oligochaetes	158	20	-	-	-	1	-	25
LOTIC TOTAL	**395**	**160**	**230**	**174**	**109**	**69**	**118**	**178**
MEAN FOR THE YEAR	420	291	326	255	134	146	558	310

(Anon., 1984)

Stocking of the reservoir was initiated in 1972, when it was still in formative stage. After the initial stocking, fishing was not allowed for three years to facilitate building-up of a stock of carps. During the period between 1972 and 1977, more than 5 million fry/fingerlings (30-150 mm) of catla (40.9%), rohu (29.0%) and mrigal (30.1%) were stocked into the reservoir, along with 1 600 of their brooders. In the subsequent years, the stocking rate was erratic, varying between 65 000 to 3 million (Table 9.7). The species composition of 4:3:3 for catla, rohu and mrigal was more or less followed over the years.

Attempts were made to build up a stock of hilsa (*Tenualosa ilisha*) in the reservoir through stocking. In the year 1979, hilsa eggs fertilised through artificial fecundation were reared into fry in enclosures installed in the reservoir and 5 000 fry were released into the reservoir. Nevertheless, the fish failed to appear in the commercial catches.

Table 9.7. Stocking in Vallabhsagar reservoir

Year	No. stocked (in thousands)
1975-76	1 514
1976-77	494
1977-78	1 044
1978-79	65
1979-80	635
1980-81	329
1981-82	272
1982-83	-
1933-84	200
1984-85	439
1985-86	3 000
1986-87	1 836
1987-88	2 877
1988-89	3 000

The success of stocking carps in the reservoir is phenomenal. The bountiful fish food resources in the form of plankton and rich bottom detritus have proved to be a boon for the major carps, *Catla catla, Labeo rohita,* and *Cirrhinus mrigala* and the indigenous *Tor tor.* Apart from contributing 80 to 85% to the total catch, they represented higher year classes. Average size of catla was 5.7 kg, 5.1 kg, 3.0 kg, and 4.2 kg during 1977-78, 1978-79, 1979-80 and 1980-81. Similarly, the size range of *L. rohita* and *C. mrigala* were within 1.5 and 2.66 kg during the same period. The size of *T. tor* in the catches varied from 1.87 kg to 2.83 kg. This suggests that the fish stocks can sustain a good fishery even at a higher level of fishing effort, without risking over-exploitation.

Fisheries

Based on the figures available at the cooperative societies, the fish production from the reservoir ranged from 138 to 240 t during the period 1975-76 to 1979-80 (Table 9.8). This is only a fraction of the actual catch. Large quantity of fish is sent to various destinations bypassing the cooperative societies.

Fishing licences are issued to fishermen on a royalty basis. They pay to the Government Rs. 1 per kg of major carps and Rs. 0.50 per kg of other fishes they catch from the reservoir. A similar amount is charged by the cooperative society to cover their managerial expenses. The royalty charged by the Government is spent on the reservoir management. Banks provide financial assistance to the fishermen for procurement of nylon nets. Nevertheless, a large number of fishermen resort to illegal fishing and sell their catch to private merchants who abet this clandestine operation. Government often finds it difficult to curb this practice due to inadequate supervisor manpower and other logistic problems. At present, 4 069 tribal men are registered under the 12 cooperative societies as fishermen, while about 5 000 others are believed to be operating in the reservoir illegally.

Only a fraction of the catch is channelled through the cooperatives. According to one of the revealing estimates, the fish despatched to the Calcutta market from the railway station during 1975-76 to 1976-77 was 4 000 t, against the official reservoir catch statistics of 444 t. The State Fisheries Department has resorted to sample surveys for a more realistic estimation of catch since 1983-84. Information collected from the railway records, and other sources has also been utilised for this purpose. According to this estimate, the production ranged from 863 to 3 989 t (Table 9.9) and the yield rate touched 109 kg ha^{-1} during 1988-89. The steady increase in production since 1986-87 is attributed to the improved stocking (Tripathy, 1990). Vallabhsagar, is one of most productive reservoirs in the country.

Table 9.8. Fish catch sold through cooperative societies

	1975-76	1976-77	1977-78	1978-79	1979-80
Total catch (t)	240.0	204.3	149	204	138
Catla catla (%)	16.55	20.55	33.70	38.80	22.05
L. rohita (%)	51.85	36.00	30.60	27.20	24.05
C. mrigala (%)	13.25	14.85	14.67	13.50	17.20
Tor tor (%)	6.55	7.90	6.01	5.05	19.06
Misc. (%)	11.80	20.70	15.02	15.45	17.64

Table 9.9. Estimated fish production in Vallabhsagar

Year	Estimated production (t)	Yield rate (kg ha^{-1})
1983-84	863	23.6
1984-85	1 159	31.7
1985-86	1 610	44.1
1986-87	2 023	55.4
1987-88	2 869	78.5
1988-89	3 989	109.2

Displaced tribals

Vallabhsagar reservoir inundated 170 villages, affecting more than 16 000 families, 90% them belonging to the tribals. All the affected were rehabilitated in 17 clusters of villages on the periphery of the reservoir (Mod *et al.*, 1985). Fisheries has been considered as one of the means to rehabilitate the people who lost their source of livelihood on account of displacement. In the year 1975, when the commercial fishing was initiated in the reservoir, the tribal men were organised into fishermen's cooperative societies, under a tribal development scheme. Apart from distributing boats and nets to the members on a 50% subsidy basis, they were also trained in handling various fishing craft and gear. So far, 800 youths have been trained by the Government. At present, 1 200 fishing boats operate in the reservoir, employing thousands of units of gill nets.

The fact that there is so much interest in entering into fisheries is a testimony to the high remunerativeness of the fishing operations. The successful establishment of the carp populations facilitates the high fish production. The reservoir perhaps holds even a higher potential than what it produces today. The present annual fish landings of nearly 4000 t (1988-89) are only half the production potential of 8 000 t, as estimated by the Central Inland Fisheries Research Institute, based on the primary production rates.

Construction of roads for quick transportation of catch and effective measures to keep the middlemen and unscrupulous merchants at bay are prerequisites for bringing the fisheries operations into the legal fold.

9.3 SAYAJISAROVAR RESERVOIR

Located at 23° 30' N, 24 km distant from Vadodara, Sayajisarovar is a small reservoir covering 145 ha. Despite a low mean depth (3.4 m) and high shore development index (10.19) the reservoir does not show eutrophic tendencies (Ganapati and Pathak, 1969). Exposed to wind action, complete turnover takes place by 0600 hrs every day. Nitrogenous and phosphatic nutrients are in traces and silicate is in very low concentration (0.03 to 0.5 mg l^{-1}). Plankton population is scanty and photosynthesis is chiefly through the macrophytes, *Vallisneria spiralis and Potamogeton natans*, which absorb most of the nutrients. Phytoplankton, forming only 5% of the total plankton count, is represented chiefly by *Pediastrum* simple and, *Botryoccocus braunii*. Eutrophic species are conspicuous by their absence. Photosynthetic production of oxygen (gross) as estimated by Ganapati and Pathak (1969) was 0.42 to 7.64 g m^{-2} day^{-1} in 1963 and 0.45 to 1.57 g m^{-2} day^{-1} during 1964. The annual average photosynthetic efficiency during the two years was 0.63% and 1.18% respectively.

Gujarat is an important State from the fisheries point of view. At present, the main emphasis is on marine fisheries. With the longest coastline among the mainland States of India, Gujarat has a flourishing marine fisheries. Against the current (1992-93) marine fish production of 0.61 million t, the inland water bodies produce 51 154 t, or 7.7% of the total fish landings. The State has registered more than 3.5 times increase in inland fish production during the last two decades. During 1971-72, the total production was 14 167 t which increased to 19 323 t in 1981-82 and 51 154 t in 1992-93.

Per capita consumption of fish in Gujarat is one of the lowest in India. In rural and urban areas, the monthly consumption is 0.02 kg, against 1.06 kg (rural) and 1.59 kg (urban) of Kerala. The State needs quantum jumps in fish production in order to increase the availability of fish to its people. The reservoir fish production can assist in achieving this objective.

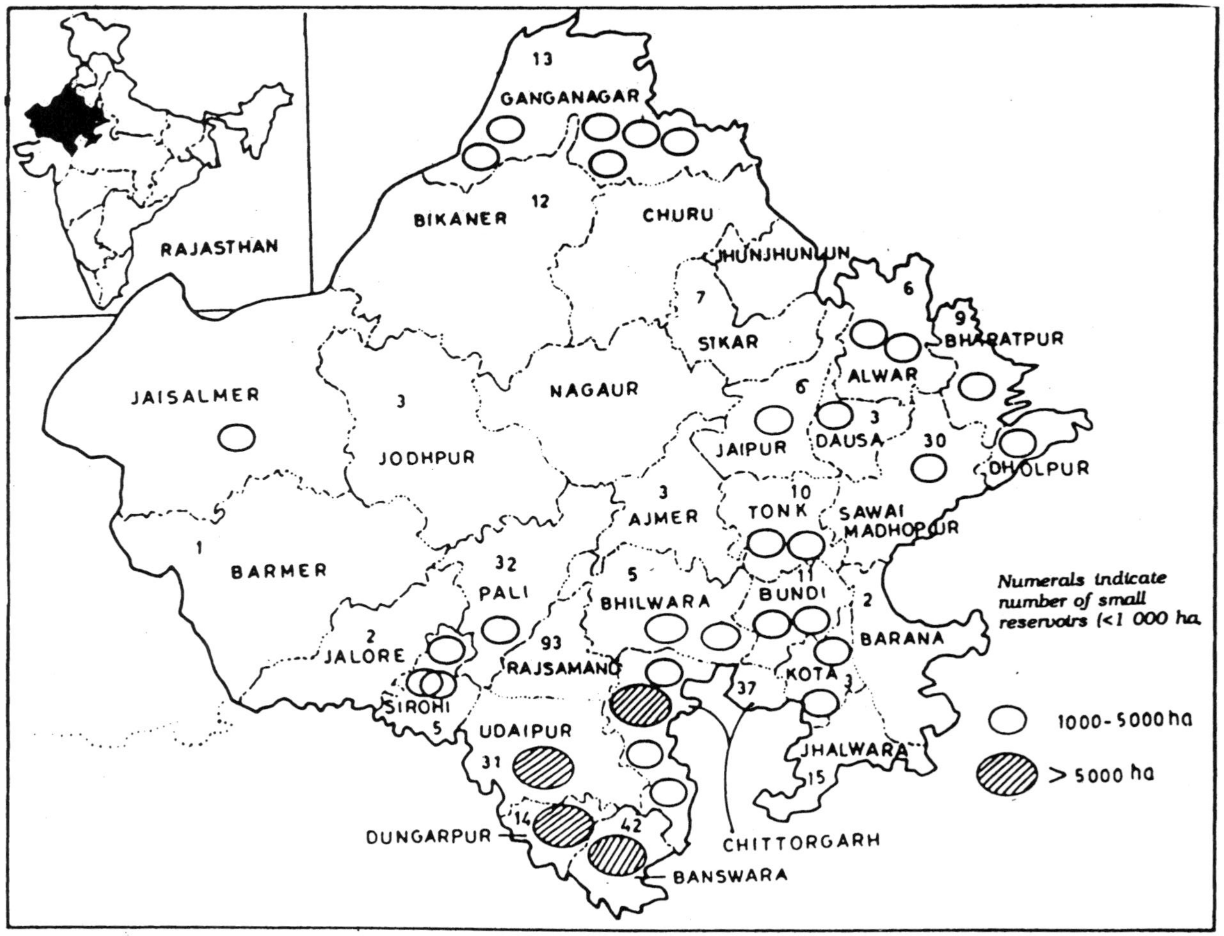

Figure 10.1. Distribution of reservoirs in Rajasthan

10. RAJASTHAN

Rajasthan is one of the border States of India, sharing the country's frontier with Pakistan in the west and northwest. The State encompassing an area of 342 239 km^2 shows great physiographic variations. Six distinct geographical divisions are clearly discernible. The sprawling *western region* is arid and virtually a desert land, where no reservoirs are located. The *semi-arid region* between the Aravalli ranges and the western deserts runs across the state from Jalor to Jhunjhunu. Southern part of this region is drained by the Luni river, while the northern part remains dry. The Indira Gandhi Canal, carrying water from Bhakra dam passes through this region in the Ganganagar district. The *Aravalli region* is undulating hill country dominated by mountains. Taking advantage of the uneven terrain, a large number of small impoundments have been created in the region, especially in the districts of Pali, Udaipur and Sirohi. The *eastern region* is extensively drained by the Banas River and its many tributaries. This region is rich in industries with a large number of reservoirs, especially in Sawai Madhopur, Bundi, Alwar and Bharatpur districts. The *southern region* of the State, embracing the districts of Banswara, Chittorgarh, Rajsamand, Jhalwara and Kota consists of stony uplands setting ideal sites for water resources development. The maximum number of man-made lakes are situated in this region. The *Chambal ravine region* lies along the river Chambal, forming the boundary between Rajasthan and Madhya Pradesh.

Alluvial soil occupies the northeastern parts of Rajasthan, comprising the districts of Alwar, Bharatpur, Jaipur, and Sawai of Madhopur. The State receives rain from both the Arabian Sea and the West Bengal branches of the southwest monsoon, 80 to 90% of the rainfall occurrs during June to October. On an average, the State receives 53.6 cm of rainfall per annum.

10.1 RESERVOIR FISHERIES RESOURCES

Rajasthan has about 153 444 ha of water area under reservoirs, more than a third of which fall under the small category. Of the 423 reservoirs listed, only four are in the large category, while the small and medium sized reservoir number 389 and 30 respectively. Rajsamand, Banswara and Chittorgarh have a large number of small reservoirs, though they do not contribute much to the total area. The large reservoirs, *viz.*, Rawatbhate (19 600 ha) in Chattisgarh district, Bajajsagar (13 500 ha) in Banswara district, Jaisamand (7 600 ha) in Udaipur district and Kadana (9 000 ha) in Banswara and Dungarpur districts, cover more than 32% of the total water area (Fig. 10.1; Table 10.1).

Fish production from reservoirs of Rajasthan

Among the 97 reservoirs for which details of current production (1993) are available, seventy-eight small reservoirs distributed in 18 districts produce fish to the tune of 0.1 to 467 kg ha^{-1} of fish (Table 10.2). Bankli reservoir in Jalore district is the most productive reservoir with a yield of 316 kg ha^{-1}. Average fish yield is high in the reservoirs of Ajmir (260 kg ha^{-1}), Pali (137 kg ha^{-1}), Sawai Madhopur (113 kg ha^{-1}) and Jodhpur districts (109 kg ha^{-1}). Bilwara district with a yield ranging from 0.1 to 7 kg ha^{-1} (average 2 kg ha^{-1}) and Bundi district with 1 kg ha^{-1} are among the poorest districts in reservoir yield.

Table 10.1. District-wise distribution of resevoirs in Rajasthan

District	Small (< 1 000 ha)		Medium (1 000-5 000 ha)		Large (>5 000 ha)		Total	
	No.	*Area*	*No.*	*Area*	*No.*	*Area*	*No.*	*Area*
Chittogarh	37	3 926	3	3 439	1	19 600	41	26 965
Ajmer	3	150					3	150
Banswara	42	4 910	-	-	1	13 500	43	18 410
Bundi	11	3 739	2	3 237			13	6 976
Jhalwara	15	1 392					15	1 392
Dungarpur	14	3 224	-	-	1	9 000	15	12 224
Sirohi	5	617	3	3189			8	3 806
Pali	32	3 822	1	2 000			33	5 822
Barana	2	800					2	800
Bikaner & Ganganagar			1	1 800			1	1 800
Sikar	7	357					7	357
Dausa	3	1 390	1	1 360			4	2 750
Bharatpur	9	1 061	1	1 160			10	2 221
Jodhpur	3	530					3	530
Jaisalmer			1	3 000			1	3 000
Jalore	2	280					2	280
Rajsamand	93	6 413					93	6 413
Sawai Madhopur	30	6 588	1	1 619			31	8 207
Alwar	6	1 763	2	3 000			8	4 763
Dholpur	6	1 282	1	2 402			7	3 684
Ganganagar	13	5 800	6	13 800			19	19 600
Udaipur	31	4 057			1	7 286	32	11 343
Kota	3	137	2	2261			5	2 398
Tonk	10	1 235	2	3 000			12	4 235
Jaipur	6	*200	1	*1 260			7	1 460
Barmer	1	100	-	-	-	-	1	100
Bhilwara	5	458	2	3 300	-	-	7	3 758
Total	389	54 231	30	49 827	4	49 386	423	153 444

*Area of only one reservoir *Compiled from the data provided by the State Govt. of Rajasthan*

Current fish production for17 medium reservoirs, covering a waterspread of 24 506 ha, is available. Medium reservoirs of the State have yields ranging from 0.2 to 100 kg ha^{-1}. The 2 000 ha Jawi reservoir in Jaisalmer district and 1 325 ha Gambhiri reservoir in Chittorgarh district are very productive with yield of 100 and 60 kg ha^{-1} respectively. On the whole, medium reservoirs of Rajasthan have a yield of 24.47 kg ha^{-1}.

Among the large reservoirs the Mahibajaj and Kadana produce 40 and 80 t of fish (3 and 9 kg ha^{-1} respectively) (Table 10.3). Fisheries management in some reservoirs in the State under the Rajasthan Tribal Areas Development Corporative Federation is good (Sreenivasan, 1993). These reservoirs receive financial support from NABARD in the form of material inputs and share capital.

The average fish yield of 24.89 kg ha^{-1}from the reservoirs of Rajasthan is one of the highest in India. This is the result of concerted efforts towards yield optimisation, already launched by the State (Jhingran, 1989b). However, detailed ecological investigations of reservoir ecosystems are still very few.

Table 10.2 Fish production in small reservoirs of Rajasthan

District	Number of reservoirs examined	Yield range ($kg\ ha^{-1}$)	Average yield ($kg\ ha^{-1}$)
Ajmer	3	171 to 467	260
Alwar	4	3.3 to 50	16
Banswara	10	4 to 58	25
Barana	2	5 to 20	9
Barmer	1	20	20
Bhilwara	5	0.1 to 7	2
Bundi	1	1	1
Chittorgarh	13	4 to 190	19
Dausa	3	50 to 100	64
Dungarpur	5	3 to 8	5
Jalore	1	316	316
Jodhpur	2	67 to 160	109
Pali	1	137	137
Rajsamand	2	7 to 105	76
Sawai Madhopur	11	29 to 377	113
Tonk	2	36 to 44	39
Udaipur	7	13 to 60	30
Bundi	5	2 t0 72	22
Pooled	78	-	46.43

*Based on the data provided by the State Fisheries Department.

Table 10.3. Fish production in medium and large reservoirs of Rajasthan

District	Reservoir	Area (FRL)	Yield ($kg\ ha^{-1}$)
<u>**Medium reservoirs**</u>			
Alwar	Jaisamand	1 200	10
-do-	Mangalsar	1 800	5
Banswara	Urvania	1 000	6
Bharatpur	Baretha	1 160	12
Bhilwara	Meja	1 500	1
Bikaner	Ghaghria	1 800	0.2
Chittorgarh	Gambhiri	1 325	60
-do-	Wagan	1 100	27
-do-	Meja	1 000	8
-do-	Morel	1 360	15
Ganganagar	Ghaggar	1 800	55
Jaisalmer	RD 1 356	3 000	0.2
Kota	Alania	1 200	13
-do-	Sawanbhadan	1 061	19
-do-	Jawi	2 000	100
Tonk	Mash	1 000	65
-do-	Tendisagar	1 200	15
Medium reservoir pooled		*24 506*	*24.47*
<u>**Large reservoirs**</u>			
Banswara	Mahibajaj	13 500	3
Dungarpur/ Banswara	Kadana	9 000	9
Large reservoirs pooled		*22 500*	*5.3*

10.2 RAMGARH RESERVOIR

The Ramgarh reservoir is situated at a distance of 72 km to the north of Jaipur at 27^0 12' N and 75^0 32' E at an elevation of 132.5 m above MSL. Construction of the dam across the river Banganga, a tributary of the Yamuna, commenced in 1897 and the reservoir filled during 1903. The area experiences an annual rainfall of about 53.6 cm which sometimes goes up to 75 cm. The catchment area of 769 km^2 drains mostly the Aravalli hill ranges. At full reservoir level, reservoir covers 1 260 ha and has a maximum length of 7 km. The capacity of the reservoir is 59 million m^3, the maximum and mean depths being 18 and 4.6 m respectively.

Ramgarh reservoir has a shoreline of 22.8 km and the shoreline development is 1.81. The volume development index is 0.61. Gopal *at el.* (1981) reported a heavy soil erosion and estimated 3.5 m of silt deposition on the reservoir bottom during the last 75 years, resulting in a reduction of its water volume. Jhingran (1989b) gives an account of ecology, production propensities and fisheries of Ramgarh. An evaluation of productivity and trophic status of the reservoir has been attempted to facilitate planning for fishery management.

Physico-chemical attributes

The basin soil is alkaline and contains moderate concentration of organic carbon (0.5-1.0%), available nitrogen (26.6-88.2 mg 100 g^{-1}) and low levels of available phosphorus (0.04-1.1 mg 100 g^{-1}). However, phosphate concentration of 0.11-0.16 mg 100 g^{-1} in the soil is relatively high. Calcium carbonate (12.0-14.5%) in soil is medium high. Thus, the edaphic features of the reservoir place it in the productive category (Table 10.4). The total alkalinity (129.33 to 198.66 mg l^{-1}), total dissolved solids (215.66 to 554.0 mg l^{-1}), and specific conductivity (307 to 791 µmhos) of the reservoir water indicate a high primary productivity. Transparency is variable spatially and seasonally. The inorganic turbidity results from influx of colloidal particles and suspended silt in flood water and the organic turbidity from phytoplankton blooms.

Water temperature ranges from 18 °C ((January) to 32 °C (June-July) with an annual fluctuation of 14 °C. The wide seasonal variation in water temperature has a bearing on the heat cycle of the reservoir. The annual heat budget is found to be 82.60 million cal m^{-2}. The heat content fluctuates between 64 (in winter) and 146.6 million cal m^{-2} (in summer). Thus, the amount of energy required to maintain the annual heat cycle of the reservoir appears to be of a high order. The nutrient status of the reservoir water with respect to phosphate, nitrate and silicate is optimal. On many occasions, phosphate is in the range of 0.11 to 0.16 mg l^{-1}. Nitrate nitrogen content of 0.22 to 2.13 mg l^{-1} (average 0.90 mg l^{-1}) is much higher than in many other reservoirs in India. Free ammonia is very low, the range of variation being traces to 0.1 mg l^{-1}. Silicate values vary from 5.6 to 18.46 mg l^{-1} (average11.53). Organic matter in the range of 4.40 to 8.05 mg l^{-1} (average 6.5 mg l^{-1}) indicates a high cocentration of oxidisable organic substances in water, yet another important parameter reflecting the productive nature of water. Chloride is fairly high in the reservoir (38.66-76.0 mg l^{-1}) but fluoride is low (traces to 0.7 mg l^{-1}).

Hydrological parameters clearly suggest the high productivity potential of Ramgarh reservoir. It is interesting to note that despite poor soil quality, especially with respect to phosphorus, the water remains rich in nutrients. Geochemical characteristics of the watershed and climatic factors like rainfall determine the productive potential of the reservoir to a considerable extent. The rich water quality reflects the transport of allochthonous dissolved nutrients. A high rate of runoff seems to favour the nutrient enrichment of the reservoir. There are very little spatial variations among different zones of the reservoir in respect to the limnochemical parameters.

Table 10.4. Important limno-chemical features of Ramgarh reservoir

Parameters	Range	Average
SOIL QUALITY		
pH	7.4-8.6	8.1
Organic carbon (%)	0.15-1.05	0.38
Available nitrogen (mg 100 g^{-1} soil)	26.6-88.2	52.5
Available phosphorus (mg 100 g^{-1} soil)	0.04-1.1	0.57
Free calcium carbonate (%)	15-22	16.8
Specific conductivity (μmhos)	200-1000	475.5
WATER QUALITY		
Water temperature (°C)	18-32	-
Transparency (cm)	80-108	-
pH	8.3-8.6	8.4
Alkalinity (mg l^{-1})	129.33-198.66	165.63
Calcium (mg l^{-1})	13.33-25.0	20.32
Magnesium (mg l^{-1})	9.33-23.0	15.52
Total dissolved solids (mg l^{-1})	215.66-554.0	359.17
Specific conductivity (μmhos)	307.0-791.0	513.0
Dissolved organic matter (mg l^{-1})	4.40-8.05	6.50
Dissolved oxygen (mg l^{-1})	6.3-9.2	8.06
Chlorides (mg l^{-1})	38.66-76.0	43.06
Fluorides (mg l^{-1})	Tr-0.7	0.28
Phosphates (mg l^{-1})	Tr-0.4	0.052
Nitrates (mg l^{-1})	0.22-2.13	0.90
Free ammonia (mg l^{-1})	Tr-0.1	0.025
Silicate (mg l^{-1})	5.6-18.46	11.53

(After Jhingran, 1989b)

Biotic communities

The reservoir with a favourable environment for organic .productivity harbours rich biotic communities at the primary and secondary producer levels, comprising plankton, macrophytes and benthos. The occasional blooms of phytoplankton and rapidly growing macrophytes and sedentary algae are the hallmarks of the ecosystem.

Plankton

Diatoms constitute the dominant group among phytoplankters and form on an average 19.08% (by number) of the total population. Two pulses, one in January (34.9%) and the other in March, though of a lesser magnitude, are observed. The main taxa in order of abundance are *Navicula radiosa*, *Nitzschia* spp., *Synedra ulna*, *Fragilaria* sp., *Cyclotella* sp., *Diatoma* sp., and *Rhopalodia* sp., *Gyrosigma* sp. and *Gomphonema* sp.

The green algae, as a group, form 15.0% of the total planktonic population and is mainly represented by *Chlorella vulgaris*, *Selenastrum gracile* and *Ankistrodesmus* sp. in the order of abundance. Other species, though insignificant in numbers, are *Crucigenia tetrapedia*, *Microspora* sp., *Ulothrix* sp., *Oedogonium* sp., *Pediastrum duplex* and *Scenedesmus* spp. Maximum density of this group is observed in March (23.0%) and minimum in July (7.2%).

The blue-green algae rank third in order of abundance and fluctuate from 19.1% in January to 5.7% in June. On an average, the group constitutes 10.89% of the total plankton population. Two pulses, one in January and the other in August are witnessed. The species in order of abundance are *Raphidiopsis indica*, *Phormidium* spp., *Merismopedia* spp., *Microcystis aeruginosa*, *Nostoc* sp., *Oscillatoria* spp. *and Aphanocapsa pulchera*. Lost concentration of 6.3% is recorded in August and, on an average, the blue-greens constitute 10.7% of the total planktonic population.

Cladocera is the most abundant group of Crustacea and, on an average, forms 58.4% of the crustacean population and 17.9% of the total plankton. It is most abundant in February (31.8%) followed by July (29.3%) and October (29.6%). The main genera are *Ceriodaphnia*, *Simocephalus*, *Chydorus*, *Daphnia* and *Moina* in order of abundance. Copepods form 36.9% of the crustacean population and 12.5% of the total plankton. Their population remains low during winter months and increases in June (29.0%) and August (21.6%). *Diaptomus*., *Cyclops*, *Mesocyclops* and their larvae form the main constituents of this group. Ostracods are very poorly represented and form only 4.7% of the crustacean population and occur only in April, August, and December. *Cypris* is the only representative of this group.

Rotifers, forming 13.9% of the total plankton, are represented by *Brachionus*, *Keratella*, *Lecane*, *Asplanchna* and *Filinia* in the order of abundance. They are more abundant in July (24.60%), followed by October (24.2%), but low in numbers in January (5.3%). In general, two peaks, one each in pre- and post-monsoon months, are observed .

Benthos and aquatic macrovegetation

The benthic fauna fluctuates from 933 to 2 453 individuals m^{-2}. The three peaks in the benthic populations coincide with the winter, summer and post-monsoon seasons falling in the months of December, June and September respectively. The main constituents of benthos are insects, molluscs, oligochaetes, and miscellaneous fauna. These groups form, on an average, 59.0%, 26.4%, 12.7%, and 1.9% respectively of the total benthic population.

Density of aquatic macrovegetation ranges from 715 g m^{-2} in September to 2 193 g m^{-2} in April,(averageb 1 429 g m^{-2}). Maximum abundance is recorded from January through March and a peak of small magnitude in June and July. The abundance of the macrophytes is found to have some direct relation with water level of the reservoir and an inverse relationship with water temperature. *Hydrilla*, *Ceratophyllum*, *Potamogeton*, *Najas*, *Azolla* and *Ipomea* are the major macrophytes encountered in Ramgarh reservoir.

Fish fauna

The reservoir harbours rich fish fauna, represented by 74 species, belonging to 17 families and 9 orders (Mahajan *et al.*,1984). Cypriniformes alone has 54 species. Nine species contribute significantly to commercial fisheries of reservoir *viz.*, *Catla catla*, *Labeo rohita*, *L. calbasu*, *Cirrhinus mrigala*, *Wallago attu*, *Channa marulius*, *L. bata*, *Notopterus notopterus*, and *Puntius sarana*.

C. catla subsists mainly on zooplankton whereas *L. rohita*, *L. calbasu* and *C. mrigala* live predominantly on detritus and algae. The food spectrum of *P. sarana* is very wide. Besides detritus and algae, it consumes gastropods, bivalves, and insect larvae. *W. attu* and *marulius* are voracious predators, while *N. notopterus* is an omnivore, subsisting on plant detritus, insect larvae, prawns and weed fishes.

A high shore development, relatively low mean depth, optimum nutrient levels, high concentration of phytoplankton, increasing aquatic macrophytes, and well-established zooplankton and benthos exert augmentative effects on the productivity of the reservoir at various trophic levels and hence the fishery. The reductive factors that exert stress on the system are the low inorganic transparency during floods and relatively low species diversitv of plankton, benthos and fish.

Community metabolism

Light energy available at the water surface of the reservoir ranges from 115 x 10^4 cal m^{-2} day^{-1} (December) to 238 x 10^4 cal m^{-2} day^{-1} (May) with an average value of 183 x 10^4 cal m^{-2} day^{-1}. The energy f ixed by producers is in the range 3 969 m to12 274 cal m^{-2} day^{-1}. Only 0.17 to 1.01% of the light energy available on the surface of the reservoir is fixed through primary production (Table 10.5). The average rate of energy transformation by producers in the reservoir is 8 236 cal m^{-2} day^{-1} (0.49% of light).

The fish faunistic spectrum of the reservoir being biased in favour of detritus feeders, the main pathway of energy flow in Ramgarh is through the detritus chain. Of the 98 532 k cal ha^{-1} yr^{-1} of energy output from the reservoir, the detritivores and primary consumers contribute 50 652 (51.4%) and 14 184 k cal ha^{-1} yr^{-1} (14.4%) respectively (Table 10.6). The share of secondary consumers is 15.9% and the tertiary consumers account for 18.3% of the energy. Thus, almost 66% of the energy output from the reservoir is contributed by primary consumers (mainly detritivores), owing to their dominance. This points towards a better utilisation of energy, than in many other Indian reservoirs.

The energy conversion efficiency, from light to fish (0.00146%) or primary fixed energy to fish (0.33%), observed in Ramgarh is better than the values observed in many other Indian reservoirs. The exploitation efficiency which gives the extent of utilization of net energy fixed by producers as fish is 0.44%.

Table 10.5. Energy transformation through primary production in Ramgarh reservoir

Months	Incident visible solar energy (cal m^{-2} day^{-1})	Gross production		Net production		Photosynthetic efficiency (%)
		Carbon (mg Cm^{-2} day^{-1})	Energy (cal m^{-2} day^{-1})	Carbon (mg Cm^{-2} day^{-1})	Energy (cal m^{-2} day^{-1})	
January	1 220 000	1249.92	12 274	999.96	9 820	1.01
February	1 490 000	1218.72	11 968	968.64	9 512	0.80
March	1 860 000	1218.72	11 968	968.64	9 512	0.64
April	2 110 000	1156.20	11 354	937.32	9 204	0.54
May	2 380 000	874.92	8 592	750.00	7 365	0.36
June	2 260 000	750.00	7 365	349.96	3 437	0.33
July	2 330 000	404.16	3 969	266.32	2 615	0.17
August	2 210 000	487.20	4 784	270.72	2 658	0.22
September	1 980 000	666.00	6 540	308.80	3 032	0.33
October	1 630 000	819.96	8 052	437.40	4 295	0.49
November	1 330 000	645.72	6 310	364.44	3 579	0.47
December	1 150 000	572.76	5 624	322.92	3 170	0.49
Average	1 830 000	838.37	8 236	578.76	5 683	0.49

(After Jhingran, 1989b)

Table 10.6. Photosynthetic energy fixation and energy conversion in Ramgarh reservoir

Latitude	27° 12' N
Total visible solar energy (cal m^{-2} yr^{-1} x 10^5)	6 680
Photosynthetic production energy (cal m^{-2} yr^{-1} x 10^6)	3.007
Efficiency of transformation of light to chemical energy (%)	0.49
Fish production as energy (cal m^{-2} yr^{-1})	9 853
Conversion of energy (%) fish/photosynthesis	0.33
Fish/light	0.00146
Photosynthesis (g C m^{-2} yr^{-1})	306.00
Fish yield (g C m^{-2} yr^{-1})	0.821
Conversion (%)	0.24
Exploitation efficiency ($\frac{\text{Energy output}}{\text{Energy fixed}}$ X 100)	0.44%

(After Jhingran, 1989b)

Fish production potential

The morpho-edaphic index (MEI) of Ramgarh reservoir, based on the average rate of total dissolved solids and the mean depth, is 69.68. This value of MEI suggests a fish production of 200 kg ha^{-1}. An alternative estimate derived from trophodynamic and energy flow model indicates a yield potential of 202 kg ha^{-1} (Jhingran, 1989b). Past catch records reveal that 80% of the standing stock is harvestable through maximum efficiency of fishing units. A yield of 160 kg ha^{-1} is not difficult to achieve with the help of a rational stocking programme. Thus, the actual mean fish yield of 82.11 kg ha^{-1} is only 40% of the potential.

Stocking

The energy fixed at the primary producer level is not transferred directly to higher trophic levels for want of adequate populations of phytophagous fishes in the reservoir. This calls for stocking suitable species to utilize the vacant niches. Moreover, maintenance stocking should be the top priority, owing to the reported feeble natural recruitment in the reservoir. Stocking of major carps, on a continuous basis, would prove beneficial in balancing wide annual fluctuations in natural recruitment. It would also help maintain the desired density of the component species during the unfavourable years when they are unable to reproduce.

Stocking of Indian major carps, has been erratic. The impact of stocking on the fisheries of Ramgarh reservoir can not be evaluated in the absence of a regular stocking policy. According to the available records, a total of 435 681 fingerlings of rohu, mrigal, and catla were stocked in the reservoir during February 1983-April 1984 (Table 10.7). This stocking, which works out to be 350 fingerlings ha^{-1}, appears to be devoid of any rationale other than the availability of seed. No standards on the size of the stocked material and the reason of stocking have been maintained, thereby ignoring these vital considerations.

Table 10.7. Stocking in Ramgarh reservoir during 1983 to 1984

Species stocked	Number (%)	Size range (mm)
L. rohita	325 768 (74.75%)	60-220
C. mrigala	82 803 (19.00%)	60-220
C. catla	27 116 (6.25%)	100-140

(After Jhingran, 1989b)

Optimum number of fish to be stocked per unit area in Ramgarh reservoir has been worked out, based on the potential fish yied and the average individual growth rate of the species to be stocked (Jhingran, 1989b). Giving allowance for natural mortality and possible escapement of fingerlings through irrigation channel, 530 late fingerlings of catla, rohu, mrigal and calbasu (4:2:2:2) ha^{-1} are required to be stocked. Thus, the number required would be 371 000 fingerlings, on the basis of an average waterspread of 700 ha.

Fisheries

Fish catch from Ramgarh varied from 12.6 to 98.9 t during 1969-70 to 1984-85 (Table 10.8). The fish output during the fishing periods of 1983-84 and 1984-85 is 55 984 and 58 967 kg respectively, equivalent to 79.98 kg ha^{-1} and 84.23 kg ha^{-1} on the basis of a mean waterspread. With the average yield of 77.8 kg ha^{-1}, Ramgarh can be placed amongst the productive reservoirs of the country. *Cirrhinus mrigala* contributing 33% of the catch by weight is the dominant species of the commercial catches followed by *C. catla* (15%) and *Labeo rohita* (14%) (Fig. 10.2). However, *N. notopterus*, *L. bata* and *P. sarana* have an edge over others in numerical terms.

Artificial impoundments pose many obstructions to successful fishing operations. In Ramgarh, the tree stumps, boulders and other submerged obstacles limit the use of many a fishing gear. Fishing in the reservoir is carried out by the commercial fishing parties who deploy mainly gill nets and shore seines, with limited use of cast nets, traps, and hook and line.

Gill nets of mesh size 37.5, 50.0, 62.5, 75-100.0 and 112.5 to 150.0 mm account for most of the catch. Of the total of 73 068 fishes weighing 55 984 kg, caught from the reservoir during 1983-84, the share of gill nets was 58 444 (80%) by number and 35 770 kg (63.9%) by weight. Gill nets of smaller mesh bars such as 37.5 and 50 mm are effective in catching *W. attu*, *P. sarana*, *N. notopterus*, *L. bata* and *C. marulius*.

Table 10.8. Estimated catch for different years in Ramgarh

Year	Total catch (kg)	Yield (kg ha^{-1})
1969-70	57 669	82.38
1970-71	42 915	61.31
1971-72	12 625	18.03
1972-73	55 600	79.43
1973-74	49 468	70.67
1974-75	n.a.	
1975-76	72 760	103.94
1976-77	73 651	105.22
1977-78	98 943	141.35
1978-79	38 600	55.14
1979-80	60 000	85.71
1980-81	30 300	43.29
1981-82	n.a.	
1982-83	n.a.	
1983-84	55 984	79.98
1984-85	58 967	84.23
	Average	***77.80***

n. a. - Data not available *(After Jhingran, 1989b)*

The fishing effort in terms of gear and manpower varies from month to month and the same is the case with the net fishing days per month, which vary from 9 during 1983-84 to 7 in 1984-85.

Commercial fishing records indicate that each fisherman generally catches 1 t of fish per yr from Ramgarh, as opposed to the harvest level of 2 to 4 t fish per yr obtained from some other Indian and African reservoirs. Assuming the potential of the reservoir to be 200 kg ha^{-1}, 100 to 160 kg ha^{-1} can safely be extracted, depending on the efficiency of fishing units and desired density of stock maintained through stocking. Taking the average productivity of gill nets as 8 kg 1 000 m^{-2}, it has been estimated that 12 000 to 20 000 m^2 of gill net area, manned by 70 to 112 fishermen and 35 to 56 boats shall have to be deployed for obtaining the targetted yield.

The fishing rights of the reservoir are leased out to contractors against a royalty paid to the Government according to the species of fish caught. The fishermen get wages for their labour and all the inputs like boats, gill nets, drag nets, traps and hooks and lines are provided by the contractors. The restrictions imposed by the Fisheries Department are limited to the mesh size, by not allowing the operation of nets < 37.5 mm mesh bar and the observance of a closed season in July and August to prevent indiscriminate killing of young and brood fishes.

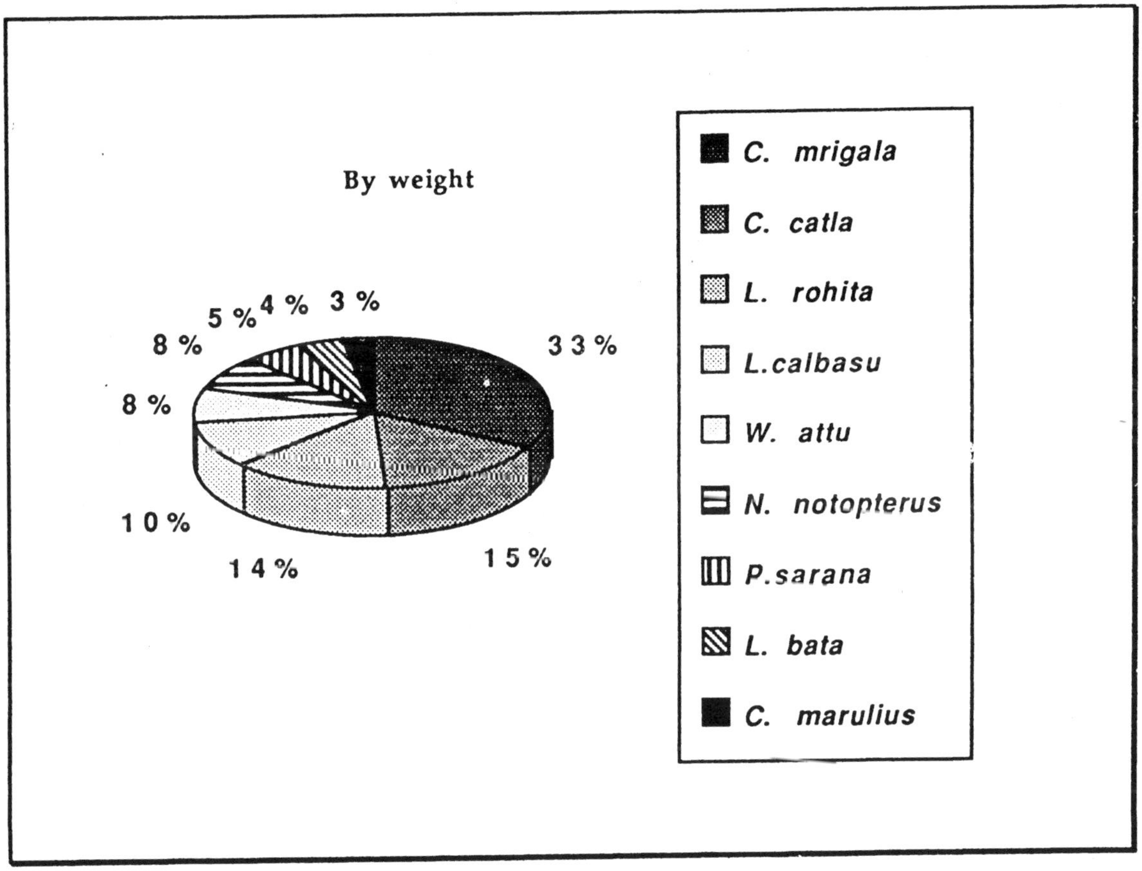

Figure 10.2. Percentage composition of fish catch in Ramgarh

During 1983-84 , a total of 48 fishermen worked for 145 days and earned Rs. 1 64 628. In 1984-85 fishing season, a work force of 61 fishermen toiled for 78 days for the contractor and earned Rs. 1 65 858. During the two fishing seasons, the net annual income for each fisherman was only Rs. 3 032.

A preliminary probe into notional cost and return functions reveals that fishermen, in the event of organising themselves into cooperatives, would have been better placed in terms of income. During the fishing season of October 1984 to June 1985, 48 fishermen would have incurred an expenditure of Rs. 69 000 on fishing crafts and gear, and the annual capital cost of Rs. 9 700. This, along with the opportunity cost, maintenance and repair, marketing cost and other miscellaneous cost would have added up to a total input cost of Rs. 162 500 to harvest fish worth 745 243. Thus, a net income of Rs. 5 82 743 could have accrued to 48 fishermen giving each fisherman an annual income of Rs. 12 140.48. Even after leaving a margin of 10% for royalty to the State Government as auction value, the fishermen would have earned more than three-fold of what they received as wages from the contractor (Table 10.9). Similar figures have been obtained for the fishing period from February to June 1984 when the net income worked out to be Rs. 10 435.

Ramgarh is a medium-size irrigation impoundment where fishing operations are for a limited period. The reservoir cannot offer employment for the whole year unlike the very large reservoirs. Ecological data indicate a potential for raising the yield from the current average of 77.4 kg ha^{-1} to 150 kg ha^{-1}. In that event, there is a fair possibility of creating employment avenues for an additional 48 fishermen families.

Table 10.9. Notional cost and return functions for fishing activities in Ramgarh reservoir

	Initial outlay cost (in Rs.)	Annual capital
A. Fixed cost (FC)		
1. Boats 10 small		
1 large	57 000=00	5 700=00
2. Nets	12 000=00	4 000=00
Total fixed cost (TFC)	69 000=00	9 700=00
B. Variable cost (VC)		
3. Opportunity cost @ Rs. 10 per day or Rs. 1.25 per hr (145 x 48 x 14 x 1.25)		1 21 80=00
4. Maintenance and repair		1 000=00
5. Marketing cost		24 000=00
6. Miscellaneous		6 000=00
Total variable cost (TVC)		1 52 800=00
Total cost (TC=TFC+TVC)		1 62 500=00
Total income (TI)		7 45 243=00
Net income (TI-FC)		5 82 743=00
Net income per fisherman*		12 140=00

*Without accounting for auction cost

(After Jhingran, 1989b)

10.3 CHHAPARWARA RESERVOIR

Chhaparwara is a small reservoir on the seasonal Mavshi river, the major portion of which dries up during the summer months. It was commissioned in 1894, mainly for irrigation purposes. Located at the geographical ordinates 26^0 37' N and 75^0 15' E, the lake extends to 200 ha at full level and can hold 35 million m^3 of water received from a catchment area of 515.2 km^2. The mean depth is 1.7 m.

Being a small, shallow, seasonal irrigation impoundment, high flushing rate. The determinants of primary production are allochthonous detritus and sessile algae. Limnological data available for this reservoir are scanty. Jhingran (1989b) made an assessment of the productivity based on total alkalinity (76 to 100 mg^{-1}), calcium (20 to 35 mg^{-1}) and phosphate (0.11 to 0.165 mg^{-1}) and he considered it to be fairly productive (Table 10.10).

The bottom soil is characterised by low organic carbon (0.15 to 0.34%), rich available nitrogen (36.6 to 61.6 mg 100 g^{-1}) and poor available phosphorus (1.1 to 1.6 mg 100 g^{-1}). The higher phosphate concentration in water emphasises the importance of allochthonous enrichment. The low phosphorus content of the soil perhaps indicates rapid assimilation of all available phosphorus in the biota.

Biotic communities

Fish food resources are reported to be uniformly distributed at all depths and zones of the reservoir (plankton 3 100 to 20 100 individuals l^{-1}; benthos 110 to 947 individuals m^{-2}). Loss of nutrients through irrigation canal seems to be amply replenished by allochthonous detritus entering the reservoir. Production rate is high on account of high water temperature and shallowness. High shore development and larger substratum for photosynthetic organisms are the other augmentative factors.

Community metabolism and fish yield potential

The net energy fixed by producers in Chhaparwara ranges between 2 380 and 6 137 ca m^{-1} day^{-1}. About 49 to 76% of the gross energy fixed by the producers is stored by them. The energy assimilation ratio is higher than that observed in some other Indian reservoirs (Bhavanisagar, Nagarjunasagar, Rihand, *etc.*). The rates of energy conversion potential both from light to fish (0.0014%) and from primary energy to fish (0.39%) are also higher. The fish yield potential of the reservoir is estimated at 158 kg ha^{-1}, the actual yield is 79 kg ha^{-1}. Thus, only 50% of the potential is actually being harvested, leaving enough room for enhancing fish yield.

Fisheries management

Chhaparwara reservoir, in stocked by the State government, with fingerlings of major carps on an arbitrary basis. Attention has not been given to suitable time, number and sizes of fingerlings to be stocked, perhaps owing to dearth of stocking material and insufficient knowledge of the production processes of the ecosystem. Elsewhere, intensive stocking has proved to be highly advantageous for small reservoirs (Keetham, Loni, Gulariya, Bachhra), which dry up completely during summer months or where the water level declines so much that complete harvesting is possible. In such reservoir, there exists a direct correlation between the stocking rate and catch, and a similar strategy could be applied for Chhaparwara reservoir.

The catch records of the reservoir do not show any relation between stocking intensity and catch. Harvesting is reported to have been done by gill nets and drag nets. The fish yield potential, based on very scanty limnological data, has been estimated at 158 kg ha^{-1}. The stocking rate for this reservoir has been arrived at (Jhingran,1989) aiming a yield of 200 kg ha^{-1}, which is the maximum taken out from the reservoir (1976-77). The absence of predator population, possibility of total harvesting, allochthonous enrichment, and high potential for stock maintenance above the natural carrying capacity of the system have been taken into consideration while fixing the target.

Table 10.10. Salient limno-chemical features of Chhaparwara reservoir

Parameters	RANGE	AVERAGE
SOIL QUALITY		
pH	7.80-8.65	8.40
Organic carbon (%)	0.15-0.34	0.23
Available nitrogen (mg 100 g^{-1} soil)	33.60-61.60	43.60
Available phosphorus (mg 100 g^{-1} soil)	1.10-1.60	13.40
Free carbon dioxide (%)	12.00-14.50	13.50
Spec. cond. (µmhos)	120.00-900.00	508.00
WATER QUALITY		
pH	8.00-8.40	8.30
Total alkalinity (mg l^{-1})	76.00-100.00	88.00
Calcium (mg l^{-1})	20.00-35.00	24.10
Magnesium (mg l^{-1})	8.00-12.00	10.58
Total dissolved solids (mg l^{-1})	426.00-997.00	535.00
Sp. conductivity (µmhos)	567.00-690.00	645.00
Dissolved organic matter (mg l^{-1})	7.57-13.27	9.10
Dissolved oxygen (mg l^{-1})	6.10-10.90	9.33
Chlorides (mg l^{-1})	70.00-136.00	106.00
Fluorides (mg l^{-1})	tr-1.20	0.46
Phosphates (mg l^{-1})	0.11-0.16	-
Nitrates (mg l^{-1})	0.40-1.10	0.84
Free ammonia (mg l^{-1})	0.09-0.11	0.10
Silicate (mg l^{-1})	1.92-8.00	4.17

tr=traces *(After Jhingran, 1989b)*

By adoption of annual cropping policy coupled with 8 to 9 months of growing period to enable stocked fish to attain average size of 500 g, it has been estimated that 600 fingerlings or 720 late fry ha^{-1} need to be stocked in the reservoir annually.

A total gill net area of 8 000 m^2 would be required to harvest the targeted 200 kg ha^{-1}. This fleet of net would need 15 to 40 fishermen and 7 to 20 boats, following the concept that 1 t catch per fisherman and 2 fishermen per boat would be most economic.

10.4 JAISAMAND RESERVOIR

Commissioned in 1730, Jaisamand is one of the oldest reservoirs in Rajasthan. The 7 286 ha reservoir has a catchment area of 1 813 km^2 and a water holding capacity of 56.6 million m^3. High values of specific conductivity (242 to 249 μmhos) indicate a high productivity. Nutrient enrichment in terms of phosphate (0.106 to 0.224 mg l^{-1}) and silicate (3.33 to 6.0 mg l^{-1}) is high and favourable for photosynthetic activities. Ranges in values of total alkalinity (54 to 91.5 mg l^{-1}) are also high (Sharma, 1980).

Despite an indiscriminate exploitation, with no closed season for a period of 240 years (till 1970), the reservoir continues to support a lucrative fishery (Durve and Kakkar, 1982). Durve (1976) reported 44 fish species, of which 23 contributed to commercial fisheries. Natural hybrids of catla x rohu and rohu x calbasu have been recorded.

Fishing is organised since1949. Presently, the major commercial fish are major carps and mahseer (*Catla catla, Labeo rohita, L. calbasu, L. fimbriatus, Cirrhinus mrigala*, and *Tor tor*) minor carps (*L. gonius, L. boggut, L. bata, Puntius sarana, Cirrhinus reba*) catfishes (*Wattago attu, Aorichthys seenghala, A. aor, Mystus cavasius, Heteropneustes fossilis* and *Ompok pabda*), the murrels (*Channa marulius, C. striatus*, and *C. punctatus*) and the miscellaneous fish (*Notopterus notopterus* and *Mastacembelus armatus*). In the absence of a closed season, fishes are also caught in the course of spawning migration, which leads to considerable decline in the fish stock and their size at capture.

Table 10.11. Changes in species composition in Jaisamand reservoir

	1962	1963	1964	1965	1969-70
Major carps (t)	382 (59.94)	394 (62.14)	368 (61.42)	221 (34.03)	60 (17.63)
Catfishes and murrels	64 (10.07)	61 (9.56)	62 10.38)	102 (15.81)	279 (82.37)
Others (t)	191 (29.99)	179 (28.30)	169 (28.20)	326 (50.16)	
Total (t)	637	634	599	649	339
Yield rate (kg ha^{-1})	88.8	88.5	83.6	90.5	47.2
Fishing days	298	283	258	243	223

*Fig. in parenthesis indicate percentage (*After Durve and Kakkar, 1982*)

Fish catch during 1962 to 1965 was less than 600 t yr^{-1} (Table 10.11), when the average fishing days were about 270. In the early 1960s, *Labeo rohita* was dominant species, followed by *Cirrhinus mrigala, Aorichthys seenghala* and *L. fimbriatus*. Specimens of catla weighing as much as 39 kg were common till 1964, especially at the beginning of the fishing season in November and again at the onset of rains. Mahseer were few in number but the individual specimens weighed more than 3 kg. The major carps contributed up to 60% of the catch till 1964. Subsequently, the decline started as a consequence of continuous fishing of 223 days under contract system. The minor carps, chiefly *L. gonius*, came to dominate the catch. While catla was rarely encountered, the major carps which formed a minor component of the catch were represented by *L. rohita*, *L. fimbriatus*, *L. calbasu* and *C. mrigala*. All fishes including the catfishes were < 2 kg by weight. Fishing gear employed in the reservoir is mainly of three types, *viz.*, surface gill nets, hook and line and drag nets.

10.5 OTHER RESERVOIRS

Fatehsagar

Fatehsagar, the oldest reservoir of Rajasthan, was constructed in 1678. It was reconstructed during the ninteenth centuary by the monarch, Maharana Fateh Singh, after whom it was named. It is surrounded by hills except on the eastern side, where a straight masonry dam of about 800 m is located. The reservoir lies on the northwest of Udaipur city at a latitude of 24^0 35' N and longitude of 73^0 42' E.

The reservoir covers 400 ha at the full level of 587 m above MSL. The maximum length of the reservoir is 2.6 km and the maximum width 1.8 km. The length of shoreline is 8.5 km at FRL, the shoreline development index 1.2. Maximum and mean depths of the reservoir are 13.4 m and 3.0 m respectively. Catchment area of the river encompasses 53.66 km^2 and the volume of reservoir at full level is 12.10 million m^3. Sharma (1980) evaluated the limnochemical attributes of Fatehsagar. High ionic content, phosphorus loading and hardness indicate a high productivity.

Lakhotia lake

Situated near Pali City at 25^0 8' N, 73^0 3' E, Lakhotia is a rain-fed man-made lake, with a maximum length of 1 825 m and breadth of 950 m. Khatri (1984a) examined the limnological characteristics of water, rate of productivity and the biotic communities.

The reservoir has a maximum depth of 45 m and the water has a transparency from 51 to 105 cm. Bicarbonate alkalinity and the pH values register wide fluctuations (4 to 158 mg l^{-1} and 6.3 to 9.0 respectively), their higher values recorded in summer. A rich community of plankton develops during summer, taking advantage of the stable conditions and high concentrations of nutrients. The onset of monsoon disturbs the equilibrium and all the physico-chemical attributes fall short of optimum level, with a concomitant decrease in biotic communities. A recovery starts during winter as the water level stabilises.

Among micronutrients, Fe, Zn, and Mn are present in detectable concentrations throughout the year whereas Cu is present only during April to August (Khatri, 1984 b). A positive correlation between phytoplankton and phosphate and nitrate is reported by Khatri (1987b) along with an inverse relation between diatoms and silicates.

Outlines of physico-chemical and morphometric attributes in respect of Baghala, Namana, Anasagar, Mansarovar, Ranapratapsagar, Bandha Bharti, Gambhiri and Rajasamund are given by Sharma (1980), and reproduced in Table 10.12 together with data on other reservoirs of the State.

Table 10.12. Salient features of some reservoirs in Rajasthan

	Baghala	Namana	Anasagar	Fatehsagar	Chhaparwara
Date closure	-	-	-	1678	1894
Area (ha) at FRL	100	528	-	400	200
Maximum length (km)	-	-	-	2.6	-
Maximum width (km)	-	-	-	1.8	-
Max depth (m)	-	-	-	13.4	-
Mean depth (m)	-	-	-	3.0	1.70
Volume (million m^3)	686 Mft	1875 Mft	-	12.10	1236.3
Catchment area (km^2)	261	939.6	-	53.66	515.2
Elevation (m above MSL)	-	-	-	587	-
Shoreline (km)	-	-	-	-	8.5
Shore dev. index	-	-	-	1.20	-
Volume dev. index	-	-	-	0.402	
Inflowing rivers	-	-	-	Mahshi	-
Latitude (N)	-	-	-	24^0 35'	26^0 37'
Longitude (E)	-	-	-	73^0 42'	75^0 15'
Soil					
Organic carbon (%)	-	-	-	-	0.15-0.34
Average nitrogen (mg $100g^{-1}$)	-	-	-	-	36.6-61.6
Average phosphorus (mg $100g^{-1}$)	-	-	-	-	1.1-1.6
Water					
Water temperature (^0C)	21.2-25.5	20.5-26.2	21.8-24.2	15.9-30.1	11-31
pH	8.1-8.5	8.5	8.5	7.4-9.2	8.0-8.4
Transparency (cm)	78-141	106-107	51.5-74.0	49-115	-
DO (mg l^{-1})	6.8-7.3	8.0	3.5-8.5	4.7-14.8	6.1-10.9
CO_2 (mg l^{-1})	-	2.0	-	0-8.8	0
Total alkalinity (mg l^{-1})	112.5-134	16.0-58.0	40.5-82.0	88.8-240	76.0-100.0
Spec. cond. (μmhos)	422.9-547.3	379.3-472.7	410.5-475.0	255-541.1	567-690
Total hardness (mg l^{-1})	-	-	-	31.6-157.6	-
Calcium (mg l^{-1})	12.4-15.2	10.7-18.0	7.2-11.2	8.8-29	20-35
Nitrate (mg l^{-1})	-	-	-	0.009-0.5	0.4-1.10
Phosphate (mg l^{-1})	0.34-0.392	0.154-0.163	0.192-0.262	tr.-0.929	0.11-0.16
Silicate (mg l^{-1})	6.0-8.76	4.33-17.04	4.33-7.28	1.12-21.30	1.92-8.0
Chlorides (mg l^{-1})	15-20	8-16	14.0-15.0	7.3-38.7	70-136
Organic carbon (%)	-	-	-	-	7.57-13.27

Table 10.12 (cont.). Salient features of some reservoirs in Rajasthan

	Ranapratapsagar	Bandh Bharti	Gambhiri
Area at FRL (ha)	22 000	11 600	2 090
Volume (million m^3)	-	76.40	59.01
Mean depth (m)	-	-	-
Catchment area (km^2)	27 840	205	870
Water temperature (^0C)	22.5-26.7	21.7-26.6	23.6-26.2
pH	8.0-8.6	7.8-8.5	7.9-8.5
Transparency (cm)	239-347	118-164	132-152
Dissolved oxygen (mg l^{-1})	7.5-10.1	7.5-12.2	8.7-7.2
CO_2 (mg l^{-1})	0-1.2	Nil-0.5	0-2.5
Total alkalinity (mg l^{-1})	72-83	61.3-75	55-140
Spec. cond. (μmhos)	239.4-248.5	217.7-230.1	106.7-136.8
Total hardness (mg l^{-1})	-	-	-
Calcium (mg l^{-1})	14.2-20.0	10.2-13.3	9.5-13.4
Nitrate (mg l^{-1})	-	-	-
Phosphate (mg l^{-1})	0.134-0.154	0.106-0.126	0.126-0.267
Silicate (mg l^{-1})	6.0-10.0	3.33-4.33	4.56-7.33
Chlorides (mg l^{-1})	5.8-6.0	3.0-13.0	7.0-8.5

Table 10.12 (cont.). Salient features of some reservoirs in Rajasthan

	Ramgarh	Jaisamand	Rajasamund
Date of closure	1903	1730	-
Area at FRL (ha)	1 260	7 286	950
Maximum length (km)	7.0	-	-
Maximum depth (m)	18	-	-
Mean depth (m)	4.6	-	-
Volume (million m^3)	59.0	56.6	-
Catchment area (km^2)	769	1813	
Elevation (m above MSL)	132.54	-	-
Length of shoreline (km)	22.80	-	-
Shore dev. index	1.81	-	-
Inflowing rivers	Banganga	Gomti, Jhameri Rupar Begar	
Outflowing rivers		Teri, Ghati	
Latitude (N)	27^0 10'	24^0 14'	-
Longitude (E)	75^0 32'	73^0 57'	-
Water temperature (^0C)	18-32	22.0-23.9	21.5-26.0
pH	8.3-8.6	8.5-8.9	7.8-8.3
Transparency (cm)	80-108	65-112	108-189
DO (mg l^{-1})	6.3-9.2	10.7	6.0-9.7
CO_2 (mg l^{-1})	0-1.5	0	-
Total alkalinity(mg l^{-1})	129-198	54-91.5	67.5-111
Spec. cond. (μ mhos)	307-791	242-249	427.3-897.6
Calcium (mg l^{-1})	13.33-25.0	14.0-16.7	11.6-13.2
Nitrate (mg l^{-1})	0.22-2.13	-	-
Phosphates (mg l^{-1})	tr.-0.4	0.106-0.224	0.154-0.163
Silicate (mg l^{-1})	5.6-18.46	3.33-6.0	3.300-13.33
Chlorides (mg l^{-1})	38.66-76.0	6.0-8.0	10-12

11. HIMACHAL PRADESH

Himachal Pradesh is a small State of 55 673 km^2, away in the the Himalayas. It is drained by a number of rivers, the most important among them being Chanab, Ravi, Beas, Sutlej and Yamuna. All the rivers are snow-fed and hence perennial. Topographically, the State has considerable potential for hydel power generation.

11.1 RESERVOIRS OF HIMACHAL PRADESH

The two major projects situated in the State, *viz.*, Bhakra Nangal (Gobindsagar reservoir) and Beas (Pong reservoir) are primarily meant for hydel power generation, apart from serving other needs such as irrigation and flood control. Pandoh, with an area of about 200 ha, is a small impoundment created across the river Beas in the Kullu valley and Chamera is a small reservoir (100 ha) across the river Ravi in Chamba district (Fig. 11.1). The four reservoirs in the State, have a total waterspread of 41 696 ha (Table 11.1).

Table 11.1. Reservoirs of Himachal Pradesh

Name of the reservoir	District	Area (ha)		Fish production (t)
		at FRL	*Average*	
Gobindsagar	Bilaspur	16 867	10 000	964.0
Pong	Kangra	24 629	15 000	468.0
Pandoh	Mandi	200	-	-
Chamera	Champa	100	-	-
Total		***41 696***		

11.2 GOBINDSAGAR

Gobindsagar is an important reservoir for more than one reason. The Bhakra-Nangal project was launched at the dawn of independence, signifying the beginning of an era of large developmental projects in India and the project was considered as an icon of the young nation's pride, aptly described by Jawaharlal Nehru as one of the temples of modern India. Secondly, the reservoir, by virtue of its location at a high latitude, transcending the limits of the tropics and cold water influx from the Beas system, simulates some sub-temperate environment that interests the limnologists. Thirdly, it provides a unique opportunity for the fishery biologists to study the behaviour of the exotic silver carp in an Indian reservoir. Gobindsagar has attracted the attention of a number of scientists including limnologists, fishery biologists, gear technologists and general fishery managers. The reservoir was studied by All India Coordinated Project on Reservoir Fisheries from 1971 to 1985.

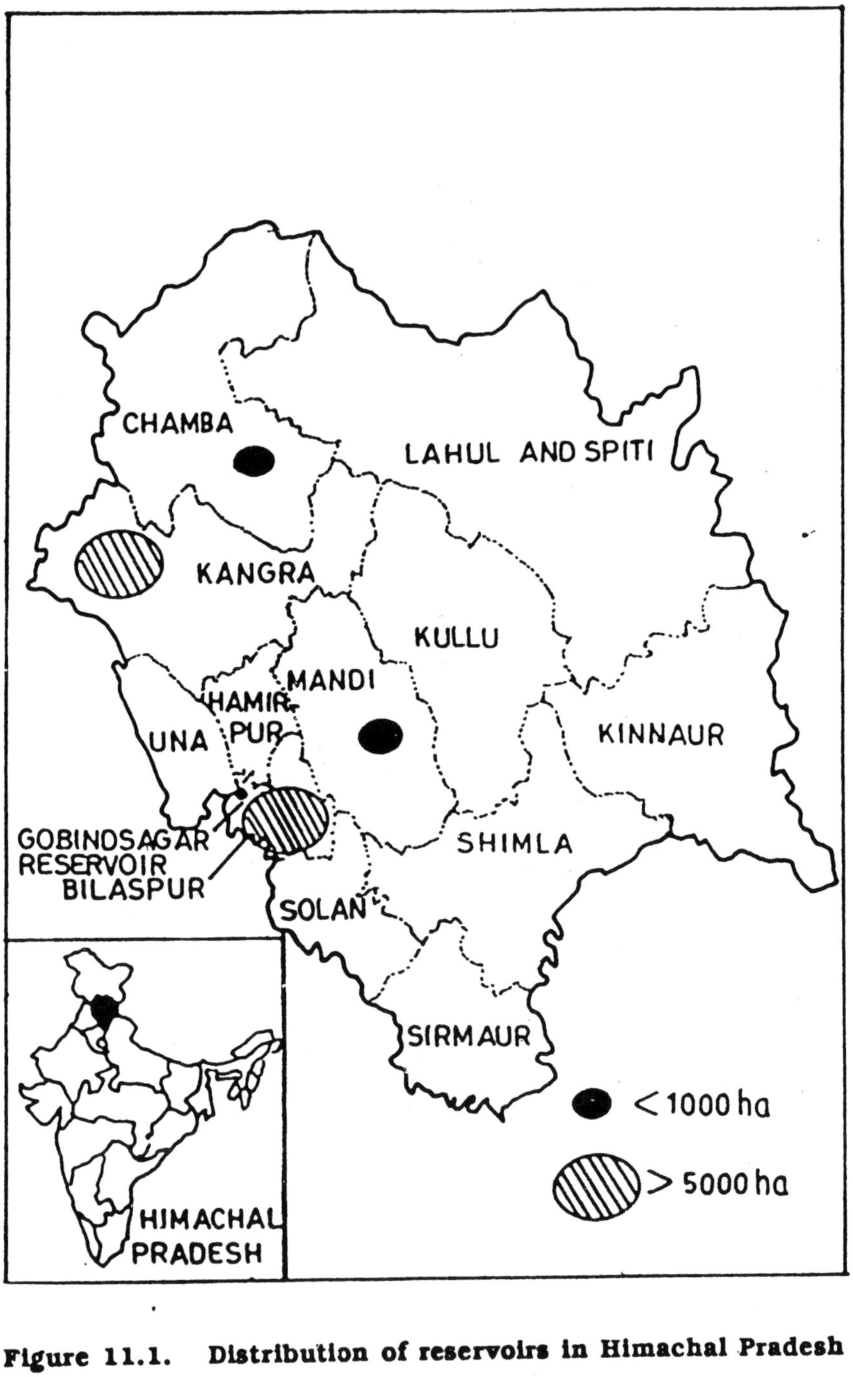

Figure 11.1. Distribution of reservoirs in Himachal Pradesh

The reservoir

Gobindsagar reservoir, one of the largest man-made lakes of India, was created by a 255 m high, straight gravity dam at Bhakra (31^0 25'N and 76^0 25'E) on the river Sutlej. Construction of Bhakra Nangal dam was initiated in 1947 and the reservoir was formed in 1963. The river Sutlej, originating from Mansarovar lake on the Tibetan plateau at an elevation of 4 572 m, flows in a southwest direction, cutting 915 m deep through the plateau and later entering India after traversing a distance of 320 km in Tibet. River Spiti, the largest tributary of the Sutlej, joins it near Pooh. Important tributaries (*khuds*), such as Lunkharkhud, Sirkhud, Gambharkhud and Alikhud enter the reservoir directly. Downstream of the reservoir, the Sutlej joins the river Beas at Harike and enters Pakistan. Located in the Bilaspur and Una districts, Gobindsagar reservoir covers 16 867 ha at the FRL of 514 m above MSL and 5 063 ha at the minimum water level. The reservoir has been designed to hold 9 621 million m^3 of water, when full, and 2 431 million m^3 at DSL. However, due to silt accumulation, the capacity has declined to 7 118 and 1 973 million m^3 respectively (Table 11.2).

The Sutlej river receives cool, snow-melt water from the upper catchment during the spring and summer months and from monsoon precipitation during July-September in its lower catchment. A number of seasonal rivers, *viz.*, Gambroula, Gamloha, Sir, and Lunkhar join the reservoir. The total catchment area of the reservoir is 56 980 km^2. The annual runoff is estimated at 16 775 million m^3, brought in from the drainage area receiving a precipitation of 100 to150 cm per year, apart from the snowmelt. Commissioning of the Beas Sutlej Link (BSL) scheme in 1978 added a new dimension to the hydrographic cycle of the water body. Under the BSL scheme, Beas water is diverted to Gobindsagar for augmenting the power generation and irrigation capacity of the reservoir. The Beas water joins the lotic sector at flow rates ranging from 57 to 212 m^3 sec^{-1} during diferent seasons, contributing 4 477 million m^3 annually. Blending of the cool Beas water and the warmer Sutlej water in the reservoir leads to some unique pattern in the thermal and oxygen regime and other processes which influence the life and production cycle of organisms.

The reservoir provides warm, mixed and cool lacustrine conditions for its biota. It has 3 distinct zones, *viz.*, Lunkharkhad zone, lentic zone and lotic zone, covering 26.0, 42.0 and 32.0% of the area respectively. At Dehar, the edge of the lotic zone, the water from Beas enters Sutlej. The Lunkharkhad joins the reservoir near the dam (Fig. 11.2).

Gobindsagar is a 96.56 km long, 6 km wide reservoir with an irregular shoreline, as indicated by the high shoreline development index of 12.86. The maximum and mean depths of the reservoir are 163.07 m and 55 m respectively, making it one of the deepest man-made lakes in the world. A volume development index of 1.01 indicates a concave bed level. Discharge through the outflowing channels varies from 1.21 million ha m in the year 1974-75 to 2.35 million ha m during 1978-79, causing a level fluctuations of 67.96 m. The two power plants at the base of the Bhakra dam and the Ganguwal and Kotla power houses on the Nangal hydel channel downstream, have a total installed capacity of 1 204 MW of power. The irrigation canals originating from Gobindsagar support a command area of >4 million ha of agricultural land making it one of the largest multi-purpose river valley projects in the world.

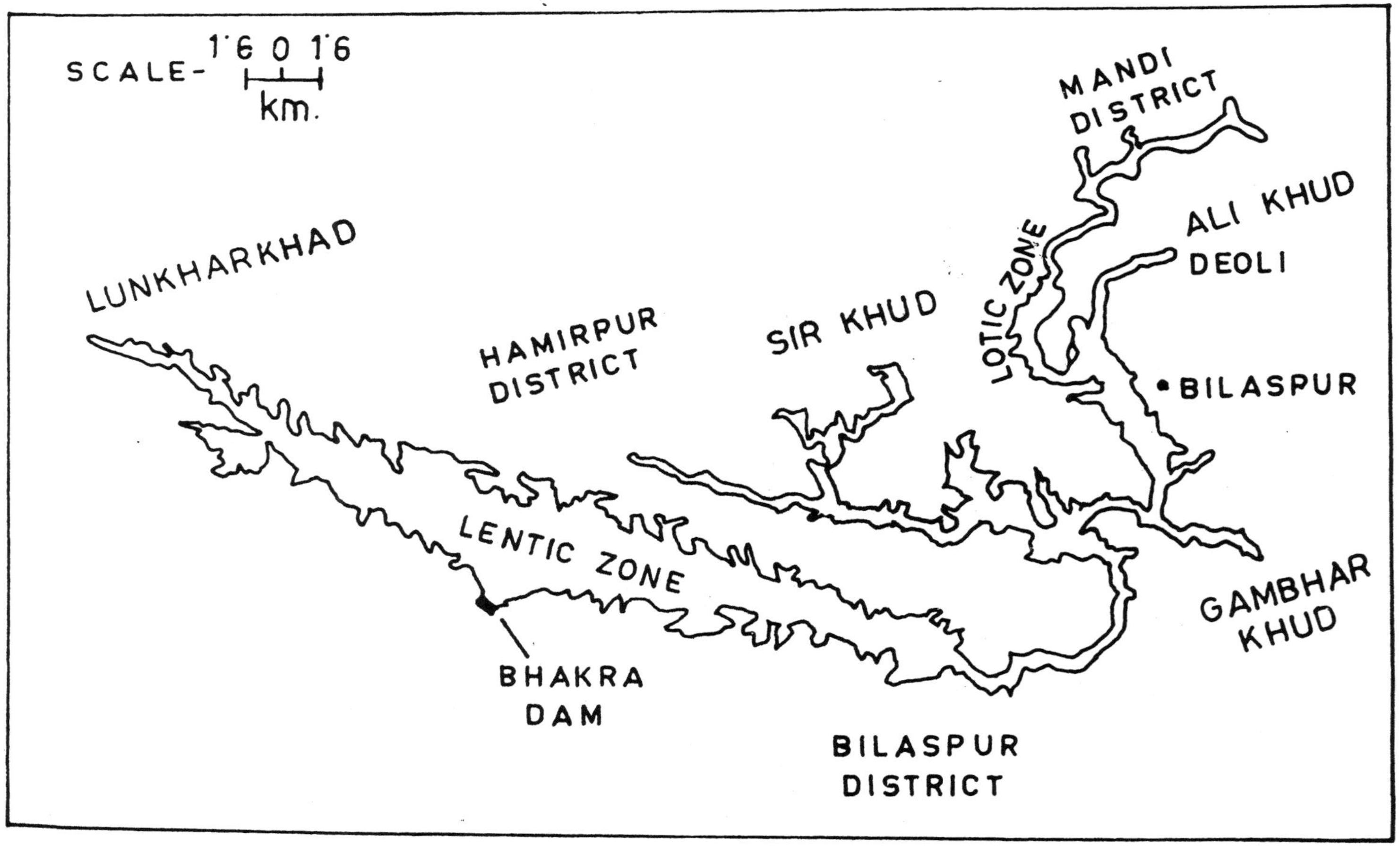

Figure 11.2. **Gobindsagar reservoir, Himachal Pradesh**

Table 11.2. Morphometric Features of Gobindsagar and Pong Reservoirs

The Dam		
1.	Location	Bilaspur
2.	Name of the river	Sutlej
3.	Water source	Snow melt & monsoon runoff
4.	Dam	Concrete straight gravity type
5.	Height of the dam (m)	226.0
6.	Elevation of the dam (m above MSL)	560.0
7.	Year of commissioning	1959
8.	Geographical ordinates	31^0 25' N; 76^0 25' E
9.	Purpose	Multi-purpose
10.	Catchment area (km^2)	56 980
The reservoir		
11.	Surface area at FRL (ha)	16 867
12.	Surface area at DSL (ha)	5 063
13.	Average surface area (ha)	10 000
14.	Mean depth (m)	55.0
15.	Total length (km)	96.56
16.	Widest stretch (km)	6.0
17.	Shoreline development index	12.26
18.	Volume development index	4.04
19.	Annual water level fluctuation (m)	450-507
20.	Max. water level fluctuation (m)	70.0
21.	Gross storage capacity (million m^3)	9 868
22.	Live storage capacity (million m^3)	7 771
23.	Inflow (million cusecs)	4.4-8.0
24.	Outflow (million cusecs)	4.9-7.0

Physico-chemical limnology

Sarkar *et al.* (1977) and Anon (1989b) describe the soil and water quality of Gobindsagar reservoir. The soil flooded by the reservoir is alkaline (pH 8.2 to 8.8) and silty, with less than 40% clay content in all sectors (24.2 to 36.6%). It is poor in nutrients. The available nitrogen ranges from 13.0 to 20.5 mg 100 g^{-1} and phosphorus 0.47 to 0.53 mg 100 g^{-1}. The organic carbon in the reservoir varies from 1.03 to 3.1%, calcium carbonate from 4.2 to 16.1% (Table 11.3). The soil differs considerably in the different zones. In the intermediate sector, the percentage of silt is high, with equal proportions of sand and clay, while in lotic and lentic sectors the clay portion is high. Chemical features also show seasonal differences. The flooded soil seems to have very little impact on the quality of water as the latter mainly derives nutrients from the catchment area.

Water quality

Water temperature at Slapper (confluence point of Beas water) used to range from 17.0 to 27.0 °C before commissioning of BSL in 1978. But after the Beas water started entering the reservoir, the range has come down to 9.0 to 11.0 °C in 1983-84, 10.5 to 17.5 °C in 1984-85, 9.0 to 18.0 °C in 1986-87 and 15.0 to 16.5 °C in 1987-88 with a concomittant rise in dissolved oxygen level which increased from 8.0 to 12.48 mg l^{-1}. The incoming cold water remains as a distinct layer at the bottom, further strengthening the thermal stratification. pH of water varies within a range of 7.3 to 8.4, suggesting a strong buffering capacity. The water is generally clear during post-monsoon (winter) months and the transparency is reduced from March to April due to turbidity caused by run-off from snowmelt and monsoons. The Secchi disc visibility depth varies between 135 and 690 cm.

Table 11.3. Physico-chemical parameters of the bottom soil of Govindsagar

Parameter	Lotic sector	Lentic sector	Lunkhar-khad
Clay (%)	24.2-29.6	32-36.6	34.7-35.6
Silt (%)	43.6-48.7	39.4-41.5	34.2-38.1
Sand (%)	27.1-26.8	24.0-26.5	27.2-30.2
pH	8.6-88	8.2-8.4	-8.2
Calcium carbonate (%)	12.2-16.1	5.6-9.9	4.2-4.7
Organic carbon (%)	1.03-1.20	1.92-2.12	2.57-3.16
Total nitrogen (%)	0.140-0.160	0.120-0.135	0.180-0.190
Available nitrogen (mg 100 g^{-1})	15.1-15.7	13.0-13.3	17.0-20.5
Total phosphorus (mg 100 g^{-1})	0.041-0.047	0.032-0.036	0.043-0.048
Available phosphorus (mg 100 g^{-1})	0.47-0.50	-0.52	0.50-0.53

Total alkalinity and specific conductivity are within the productive ranges of 50 to 88 mg l^{-1} and 169 to 331 µmhos repectively. Values of phosphate (0.05 to 0.08 mg l^{-1}) and calcium (22.0 to 24.6 mg l^{-1}) are high. The limnological characteristics of different zones are given in Table 11.4. The lotic sector is characterised by low temperature, high oxygen, and ion-rich water, more akin to a temperate regime, compared to Lunkharkhad, a warm, lentic bay of tropical nature with high carbon dioxide, CO_3 and HCO_3, moderately high values of phosphate, specific conductivity, and high silicate content. These also show that the cool Beas water exerts its influence up to Bilaspur in lotic zone. The features distinguishing Beas water are low temperature, low specific conductivity, low bicarbonate, phosphate, silicate and organic carbon, but high dissolved oxygen content.

Table 11.4. Physico-chemical characteristics of Gobindsagar reservoir

Parameter	Slapper	Lotic sector	Lentic ssector	Lumnhar-khad
Water temperature (^{0}C)	15.0-16.5	13.0-29.0	16.0-29.0	16.0-30.0
Transparency (cm)		25.5-320	86.0-456.0	29.0-382
pH	7.9-8.0	7.8-8.4	7.7-8.2	7.3-8.2
Dissolved oxygen	9.6-11.2	2.4-9.6	6.72-9.6	4.8-9.3
Free carbon dioxide	6.0-	4.0-6.0	nil-6.0	nil-6.0
Total alkalinity (mg l^{-1})	76.0-82.0	68.0-86.0	68.0-82.0	72.0-88
Calcium (mg l^{-1})	23.6-24.6	22.0-24.0	22.2-24.2	22.0-2.8
Magnesium (mg l^{-1})	2.5-2.7	2.4-2.6	2.3-2.7	2.3-2.8
Nitrite nitrogen (mg l^{-1})	0.065-0.075	0.06-0.09	0.06-0.10	0.06-0.09
Phosphate (mg l^{-1})	0.055-0.065	0.05-0.08	0.05-0.085	0.05-0.08
Silicate (mg l^{-1})	2.4-2.5	2.2-2.5	2.3-2.6	2.2-2.8
Specific conductivity (µmhos)	231.4-331.0	169.4-277.7	147.9-239.8	165.5-257.7

Thermal stratification

Thermocline, observed in Gobindsagar is sharper and longer in duration, compared to the reservoirs in the lower latitude. In the lotic sector and Lunkharkhad, thermal stratification is observed until September or October. The difference in temperature between the surface and bottom layers is 5 to 8 °C in the lotic sector and 6.5 to 9.5 °C in Lunkharkhad. A fall in values of dissolved oxygen and pH is generally associated with thermocline. CO_2, although generally absent at the top layer, is present at the bottom layer, which, in fact, triggers the release of H^+ ions and hence the fall in pH. All these suggest tropholytic activities at the bottom. The breakdown of thermal and chemical stratification during the monsoon turbulence releases nutrients to the euphotic zone, where it leads to a higher primary production. The limno-chemical indicators strongly point towards productive nature of water despite a poor soil quality, the nutrient and salts being received from the inflowing waters and the biodegradation processes within the lake.

Primary productivity and plankton

Gobindsagar being located at the highest latitude among the reservoirs studied in India, receives the least amount of incident solar radiation (172×10^4 cal m^{-2} yr^{-1}). The rate of energy fixation by primary producers is, however, higher, obtaining better energy conversion (0.68%) rate than in the reservoirs closer to the equator (0.20% to 0.49%). The gross primary production observed during seven years has an average range of 316.86 to 1 088 mg C m^{-3} day^{-1}. The net production for the same period is 186.6 to 670.28 mg C m^{-3} day^{-1}.

The annual average plankton biomass which ranges from 2.1 to 16 ml m^{-2}, increases from September onwards, reaching the maximum in December and February. There are two pulses of plankton in a year with spring and autumn blooms. The ratio between zooplankton and phytoplankton, as estimated by various workers, is 1:6.5. Zooplankton dominates only in September and October, while in other months the phytoplankton is dominant. Lentic sector is the most productive sector of the reservoir with planktonic biomass ranging from 0.-75 to 6.32 ml m^{-3} (1975-83). *Ceratium* is the dominant planktonic alga, the abundance of which varies in different zones.

Benthic invertebrates

The benthic macrofauna of the reservoir is not very rich both in qualitative and quantitative terms, the density varying from 345 individuals m^{-2} in the lotic to 677 in the Lunkharkhad sector. Oligochaetes (58.8%), culicids (22.4%), chironomids (12.6%) and molluscs (3.0%) form the benthic community. Benthos shows a bimodal pattern of fluctuation with two peaks, one in winter (January) and the other in summer (May). The overall abundance of benthos is high in the intermediate sector, whereas moderate in the lotic and poor in the lentic sectors due to their gravelly and rocky beds.

Fish fauna and fisheries

Ichthyofauna of the reservoir comprises 51 species belonging to 9 families, including the brown trout *Salmo truta* snow trouts, *Schizothorax* spp., and several species of hillstream fishes (Anon., 1989b). The species spectrum of the reservoir is distinct due to the sub-temperate climate and the zoogeographic affiliations with the Himalayan region. Prior to the construction of the dam, the cold water of the upper reaches of the river Sutlej used to sustain 30 species of fish, of which *Tor putitora*, *Labeo dero*, *L. dyocheilus*, *Schizothorax* sp. and *Aorichthys seenghala* were the dominant ones. A decline in the number of species has been reported by Dua (1993) on account of the changed ecological conditions.

especially the silt deposition at the bottom. Apart from the minnows, coldwater species such as *L. dero, Schizothorax plagiostomus* and *T. putitora* have also been reported to have been affected. Presently, besides the two exotic species *viz., Hypophthalmichthys molitrix and Cyprinus carpio var. specularis, Catla catla, T. putitora, L. rohita, L. dero, L. calbasu, L. bata, L. dyocheilus, Cirrhinus mrigala, Wallago attu and A. seenghala* constitute the commercial fisheries in order of their abundance. Common predatory catfishes of the tropical region such as *Wallago attu, Aorichthys aor, Bagarius bagarius* and *Silondia* spp. are absent. *Channa* spp. and *A. seenghala* are two predators known in Gobindsagar.

In 1961-62, *Catla catla, Labeo rohita* and *Cirrhinus mrigala* were stocked, followed by a regular stocking of *Cyprinus carpio.* Until 1974, the gid (*L. dero*), mahseer and common carp were the major components of catch and later, catla and rohu became dominant. Breeding of Indian major carps in the reservoir has been reported by Kaushal and Rao (1990). Silty, muddy and grassy areas at the upper Lunkharkhad are the spawning grounds of *Catla catla* and *Labeo rohita.* The emergence of silver carp in 1979 marked the beginning of a radical change in the catch structure. Having struck a favourable note vis-a-vis breeding and feeding in the reservoir, the fish has established an overriding dominance over all other species.

Fish yield

The current (1992-93) fish production from Gobindsagar is 964 t, which is 96.4 kg ha^{-1}, based on an average reservoir area of 10 000 ha, one of the highest yields from a large reservoir in India. The increase in fish yield of Gobindsagar over the last two decades has been remarkable. From a modest 160 t in 1970-71, the production rose steadily to 700 t in 1977-78. It remained above 700 t for the next four years with a peak of 754 t in 1978-79. After remaining steady (377 to 653 t) during 1981-82 to 1986-87, the yield started increasing rapidly every year, reaching an all time high of 964 t in 1992-93 (Fig. 11.3).

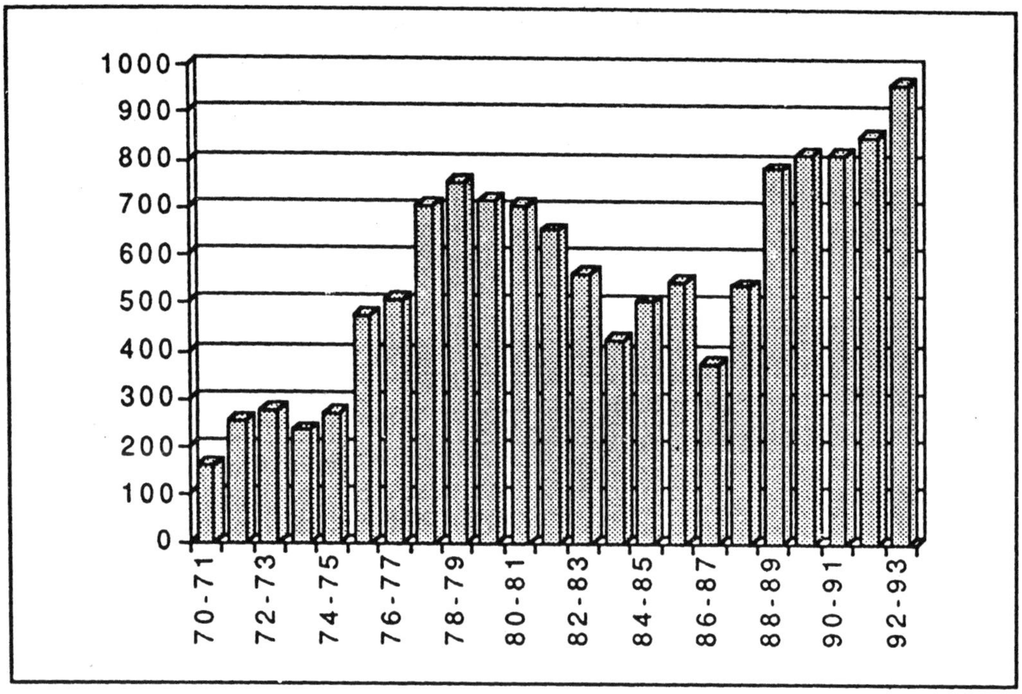

Figure 11.3 Fish production trends (t) in Gobindsagar 1971-72 to 1992-93

A six-fold rise in fish production has taken place during the 20 years in the wake of steeper increase in the fishing units. Number of licensed units in the reservoir registered a ten-fold increase from 100 in 1970-71 to 963 in 1992-93. A perusal of data presented in Table 11.5 clearly indicates that the catch per unit did not show wide variations, despite a noticeable hike in both fish yield and number of fishing units.

Major fisheries

Among the four Indian major carps, *L. rohita* has undergone a marked decline since 1979 due to inadequate recruitment and wanton killing of brooders in their breeding season during the period 1977 to 1982. The minor carps are represented mainly by *L. dero, L. dyocheilus, L. bata, C. reba* and *P. sarana* and the hillstream species. The share of these fishes in the catch started declining after 1982, reaching an all time low of 1.2 % in 1989-90 (Table 11.6). During 1990-91 to 92-93 they ranged from 3.5 to 4.5%.

The major carnivorous fish of the reservoir are mahseer (*T. putitora*) and *A. seenghala*. Their contribution has declined over the last two decades from 18.6 to 5%. *A. seenghala*, over the last 19 years, fluctuated between 0.4 and 4.0% of the total catch. The low percentage of the catfish has helped in fast build-up of Indian major carp during the initial years and silver carp in the post-1978 period. *T. putitora*, the prized game fish, constituted a major fisheries in the river Beas before the reservoir was filled. Although its percentage contribution has dropped from 16.7 to 5.8% (1974 to 1987), the actual production in terms of landings has not appreciably changed, when compared to the catch from the initial years of impoundment.

Exotic carps:

Silver carp, grass carp and common carp represent the exotic component in the fish landings. In recent years, the exotic carps, by virtue of their prolific breeding, have virtually dominated the whole reservoir. Their contribution to the total catch increased from 14.2% in 1974-75 to 87.4% in 1989. Despite a downward trend noticed in the subsequent years, the exotic carps especially the silver carp, are still dominating (Fig. 11.4).

Mirror carp (*Cyprinus carpio* var. *specularis*) constitutes an important fishery of Gobindsagar. The fish is well-entrenched in the lake as indicated by its percentage composition in the total landings, which ranged from 13.8 to 32.8 during the period 1974-87. Regular stocking has been carried out by the Department since 1965. This species attains an average size of 0.8, 1.4, 2.2 and 3.1 kg in its first, second, third and fourth year respectively. The dearth of aquatic plants in the reservoir prevents spawning of the mirror carp, as it has no suitable substrata to which to attach its eggs. Specimens of grass carp have rarely been encountered in the reservoir during the last 19 years. Kaushal and Rao (1982) observed a few large specimens of 542 to 1 225 mm (2 to 20 kg), including an oozing female among the commercial fish landings at Lathiani. The non-establishment of grass carp in the reservoir is attributable to the virtual absence of aquatic weeds.

On the basis of experimental gill netting, George *et al.* (1977) found a strong seasonality and zonal variations in fish landings. The catch per 100 m^2 of gill net was 6.34 kg in upper reaches of Lunkharkhad, as opposed to 2.37 kg in the lower reaches. The values in respect of the main river course and Sir arm were 0.758 and 2.758 respectively. While *Labeo diplostoma* was well represented in catch throughout the year with its peak in December, most of the Indian major carps were caught during the monsoon months.

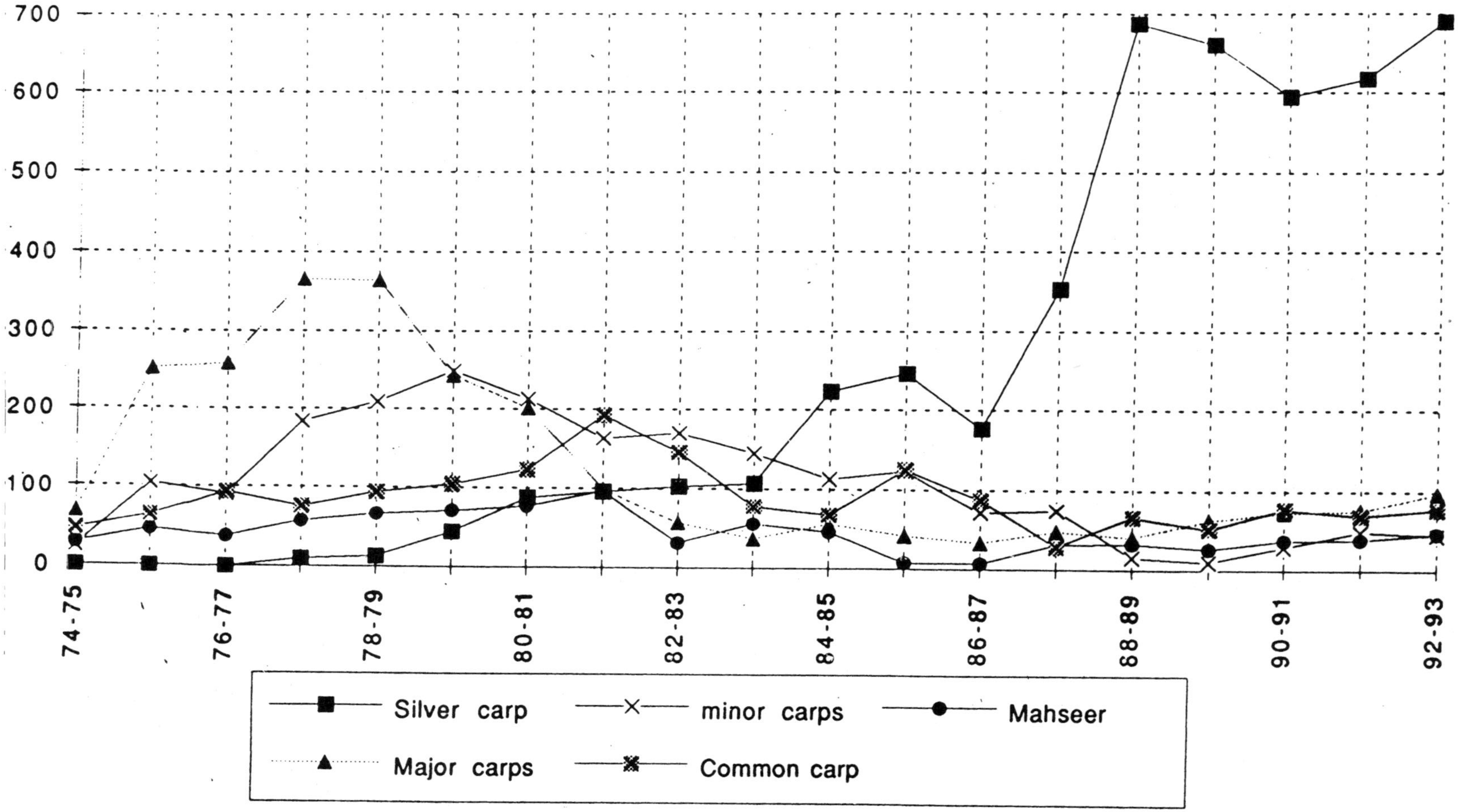

Figure 11.4. Changes in fish catch composition (t) in Gobindsagar

Table 11.5. Catch, yield and fishing effort in Gobindsagar reservoir
(1970-71 to 1992-93)

Year	Catch (t)	Yield (kg ha^{-1})	No. of licensed nets	Catch index (1970=100)	Fishing effort index (1970=100)	Catch per net
1970-71	160	16.0	100	1.00	1.00	1.60
1971-72	255	25.5	200	1.59	2.00	1.27
1972-73	277	27.7	280	1.73	2.80	0.99
1973-74	236	23.6	257	1.47	2.57	0.92
1974-75	171	17.1	321	1.06	3.21	1.24
1975-76	476	47.6	383	2.97	3.83	0.71
1976-77	509	50.9	710	3.18	7.10	0.71
1977-78	707	70.7	627	4.41	6.27	1.12
1978-79	754	75.4	617	4.71	6.17	1.22
1979-80	716	71.6	718	4.47	7.18	0.99
1980-81	707	70.7	770	4.41	7.70	0.92
1981-82	653	65.3	762	7.08	7.62	0.85
1982-83	562	56.2	748	3.51	7.48	0.75
1983-84	425	42.5	734	2.65	7.34	0.58
1984-85	505	50.5	663	3.15	6.63	0.76
1985-86	547	54.7	657	3.41	6.57	0.83
1986-87	377	37.7	671	2.35	6.71	0.56
1987-88	538	53.8	695	3.36	6.95	0.77
1988-89	784	78.4	703	4.90	7.03	1.11
1989-90	815	81.5	816	5.09	8.16	0.99
1990-91	816	81.6	995	5.10	9.95	0.82
1991-92	855	85.5	961	5.34	9.61	0.89
1992-93	964	96.4	963	6.02	9.63	1.00
Total	12809	1280.89	12351	79.96	143.51	21.42
Average	556.91	55.69	537	3.47	623	0.93
Range	160-964	16-96.4	100-995	10-6.02	1.0-9.95	0.53-1.60

Table 11.6. Fish landing from Gobindsagar reservoir in 1974-92 (t)

Year	Silver Carp	Minor Carp	Mahseer	*A. seenghala*	IMC	Mirror Carp	Miscellaneous	Total
1974-75	-	24.0	28.7	3.1	67.7	46.4	0.8	171
1975-76	-	104.0	46.3	9.2	252.3	64.2	-	476
1976-77	-	91.0	38.4	8.0	258.8	92.7	19.6	509
1977-78	10.0	184.0	58.0	16.0	366.0	76.0	-	707
1978-79	13.0	209.0	67.0	7.0	364.0	94.0	-	754
1979-80	44.0	249.0	71.0	5.0	243.0	104.0	-	716
1980-81	88.0	213.0	77.0	5.0	202.0	122.0	-	707
1981-82	96.0	163.0	97.0	5.0	99.0	193.0	-	653
1982-83	103.0	170.0	31.0	7.0	56.0	145.0	-	562
1983-84	107.0	144.0	55.0	5.0	36.0	78.0	-	425
1984-85	223.0	113.0	45.0	2.0	55.0	67.0	-	505
1985-86	246.0	123.0	8.5	2.4	40.8	124.6	0.7	547
1986-87	176.0	71.0	7.5	1.7	33.2	87.2	-	377
1987-88	354.0	74.0	31.0	2.6	46.6	27.8	0.6	538
1988-89	687.0	15.0	32.0	2.0	40.0	66.0	2.0	784
1989-90	661.0	10.0	26.0	4.0	62.0	51.0	1.0	815
1990-91	595.0	29.0	37.0	4.0	73.0	77.0	1.0	816
1991-92	617.0	48.0	38.0	4.0	76.0	69.0	3.0	855
1992-93	690.0	44.0	46.0	4.0	98.0	78.0	4.0	964

Biology of fishes

Feeding habits and reproductive biology of some of the fishes from the reservoir are known. Kaushal *et al.* (1980a) reported that the silver carp in Gobindsagar lived mainly on a phytoplankton diet comprising green algae, diatoms, dinoflagellates and blue-green algae in the order of importance. *T. putitora* subsists mainly on a fish diet, supplemented by zooplankton (Kaushal *et al.*, 1980b). Bhatnagar (1964) studied the spawning periodicity of some of the commercially important fishes of Gobindsagar. T. putitora spawns intermittantly throughout the greater part of the year. *Cyprinus carpio*, and *A. seenghala* spawn once a year with a prolonged spawning duration. *Labeo dero, L. bata, Mystus bleekeri, Crossocheilus latius latius* and *Garra gotyla* also spawn once, but with a short spawning season. The snow trout, *Schizothorax plagiostomus, Clupisoma montana* and *Mastacembelus armatus* are reported to have two periods of spawning in a year.

Silver carps in Gobindsagar

The silver carp, *Hypophthalmichthys molitrix* got an accidental entry in Gobindsagar in 1971, when 47 specimens in the size range of 290 to 530 mm (0.5 to 2.5 kg) were washed out into the reservoir from the nearby Deoli fish farm. Although stray individuals (2.5 to 10 kg) and a few juveniles were reported in the landings in 1974, a significant appearance of the species in the fishery was noticed only since October 1977, when a number of juveniles below 300 mm (< 400 g) appeared in the commercial landings. During 1977-78, 10 t of silver carp, were captured, which accounted for 1.4% of the total landings. In the subsequent years, the fish contributed up to 72% to the total catch. Its population increase had also aided in an unprecedented rise in the fish yield from the reservoir.

In view of the wide feeding spectrum and high fecundity (average of 167 000) of the silver carp, the species appears to have ecological advantage over the other species. The fish has extended through all zones of reservoir with stable winter and spring fisheries, whereas the fishing grounds of Indian major carps are location-specific. While 90% of the catla catch is harvested during the monsoon season, silver carp supports a steady fishery throughout the year, thereby apparently making it more lucrative from economic angle. However, the role of silver carp in the fisheries development of Indian reservoirs in general and the Gobindsagar in particular is a subject of animated controversy.

Major disadvantages of the fish are its poor shelf life and low market preference. Although the low price fetched by the fish is compensated to some extent by the higher production obtained by the fishermen, there is a widespread discontent among them against the change in catch structure over the years. Ecologists fear that the exotic carp, by virtue of its common food niche threatens the indigenous catla. Despite its gaining entry in a number of Indian reservoirs located at varying geoclimatic regions of the upper peninsula, Deccan plateau and the southern tip of the sub-continent, silver carp is not naturalised in any of them. Only in the cold water regime of Gobindsagar, where the fish was exposed to an environment akin to its original habitat, it commenced successful breeding and proliferation. Thus, the fear of silver carp obliterating the indigenous catla populations in warm water reservoirs seems to be unfounded.

Fisheries management

Stocking:

The exotic mirror carp, *Cyprinus carpio* var. *specularis*, introduced into Indian waters in 1939, entered the region in 1955 when it was cultured in a tank at Nahan. Because it thrived in cold waters and at high altitude, this species was preferred and introduced into the newly created Gobindsagar reservoir in 1962. Stocking of the mirror carp became regular after the fish farm at Deoli was built in 1964. Till 1979, on an average, 0.55 million fingerlings (size 50 to 80 mm) were stocked annually. In fact, mirror carp was the only fish stocked regularly, as the seed of Indian major carps was still hard to produce. Further, it was scarce and costly.

Based on a general consensus among the leading biologists that none of the local fish would be able to offer viable commercial fishery, it was recommended that the reservoir should be stocked with Indian and exotic carps. Taking this into consideration, during 1967-71, the State Fisheries Department stocked 3 500 gravid spawners and 0.5 million of fingerlings of Indian major carps ranging in size from 100 to 150 mm. By 1979, major carps had established themselves in the reservoir and started breeding.

Stocking the reservoir with spawners and seed of Indian major carps as early as 1960s was a positive management decision that led to the establishment of a lucrative fisheries. As per various estimates, the reservoir is to be stocked annually with 2.0 million fingerlings, but the average stocking rate was only 0.3 million fingerlings per year which was 16% of the required number. The stocked seed was of small size, *i.e.* 20 to 30 mm against the recommended 125 to 150 mm. No study was undertaken on the survival of the stocked seed, but in all probability it was very low.

Fishing:

Before the reservoir was created in 1959, fishing activities in the river were limited to angling and cast nets targetted against mahseers, *L. dero* and *Schizothorax* spp. Even after the impoundment, there were no organised fishing activities or management till the1960s, when new fish species were introduced into the lake. Fishing operations began with 16 nets in 1964-65 which yielded 29 t of fishes. By 1974-75, the fishing was extended to all zones by deploying 230 nets with the fish production going up to 250 t. Later on, management measures, including increase in fishing units and mesh sizes and sustained stocking, led to a remarkable yield of 96.4 kg ha^{-1}.

The subsistence fishery that existed in the rivers and streams before impoundment was of inconsequential nature. With the formation of the reservoir the lucrative fishery started attracting a large number of fishermen and the evacuees who lost their property due to submergence. At present the rehabilitated families settled near the reservoir account for about 20% of the active fishermen population.

The fishermen in Gobindsagar are full-time licencees and members of cooperative societies who own their boats. On an average, 40% of the fishemen have some education, their monthly income ranging from Rs. 800 to 3 000. Each fisherman has, on an average, 3 to 4 gill nets of 100 to 140 mm mesh size, which last for 1-2 years. The common craft is the wooden boat measuring 5m x 1m x 0.5m, costing approximately Rs. 2 000. Till 1974-75, the Fisheries Department issued licences at the rate of Rs. 10 per gill net and fishermen were free to dispose their catches without any restriction. Later, a new fishing policy was introduced, under which the fishermen were organised into cooperative societies and an apex body was entrusted with the responsibility of selling fish outside the state. A federation of local societies was made responsible for marketing of fish within the state.

Presently, there are 11 Fishermen Cooperative Societies functioning in the reservoir with a total membership of 1 479 active fishermen. Different areas of the reservoir have been earmarked for these societies for fishing and fish marketing

Marketing

The catch is brought to landing centres where the representatives of the Fisheries Federation receive the fish and keep daily records on the category and species of fishes in the presence of a Fisheries Officer. The Federation hands over the catch to the contractor appointed through general tenders at the beginning of the year. The contractor makes weekly payment to the Federation, besides keeping a fixed deposit of Rs. 500 000 to 600 000 with the Federation. The Federation makes the payment once in 15 days to the Societies after deduction of a royalty of 15% for the Department and a commission 5% of its own. The fishermen get their remuneration from the Societies after the deduction of a nominal commission, which is fixed by the General Body.

Rates for various categories of fish have been fixed on a year to year basis since 1988-89 (Table 11.7).

Table 11.7. Grade-wise rates of commercial fish species in Gobindsagar

Grade	Species	Price (Rs kg^{-1})				
		1988-89	1989-90	1990-91	1991-92	1992-93
A	*L. rohita, C. catla, M. seenghala*	13.90	16.70	20.58	20.52	20.68
B	*C. carpio, T. putitora, C. mrigala, C. idella*	9.45	10.25	11.85	11.80	12.02
C	*H. molitrix*	8.15	8.90	11.85	11.80	12.02
D	All fishes < 1 kg than	7.45	8.55	11.85	11.80	12.02
E	Minor carps < 1 kg	6.40	7.45	11.85	11.80	12.02

11.3 PONG RESERVOIR (Fig. 11.5)

Pong reservoir on the river Beas, commissioned in 1974, is primarily meant for power generation, irrigation and flood control. It has a catchment area of 12 561 km^2 and a surface area of 24 529 ha at FRL and 14 312 ha at DSL. Originating in the southern face of Rohtang Pass at an elevation of 4 062 m above MSL, the river Beas has a steep gradient in its headwater zone with a fall of 24.0 m km^{-1}. On its way, the river receives a number of tributaries, the principal ones being the Manalsu, the Parbati and the Sainj on the left bank and the Uhi on the right. Further downstream, the river takes a sharp turn towards the west and cascading down, receives the whole drainage of southern slopes of Dhauladhar chain of mountains. The Beas and its principal tributaries being snowmelt or glacier fed, are perennial, but the flow rate in the river fluctuates widely during the course of the year. The discharge is at its minimum during mid-February while floods occur from July to September. The inflow rate increases with the influx of snowmelt water and continues till July-August due to the monsoon runoff. The reservoir discharge is the highest during July and lowest in February (Table 11.8).

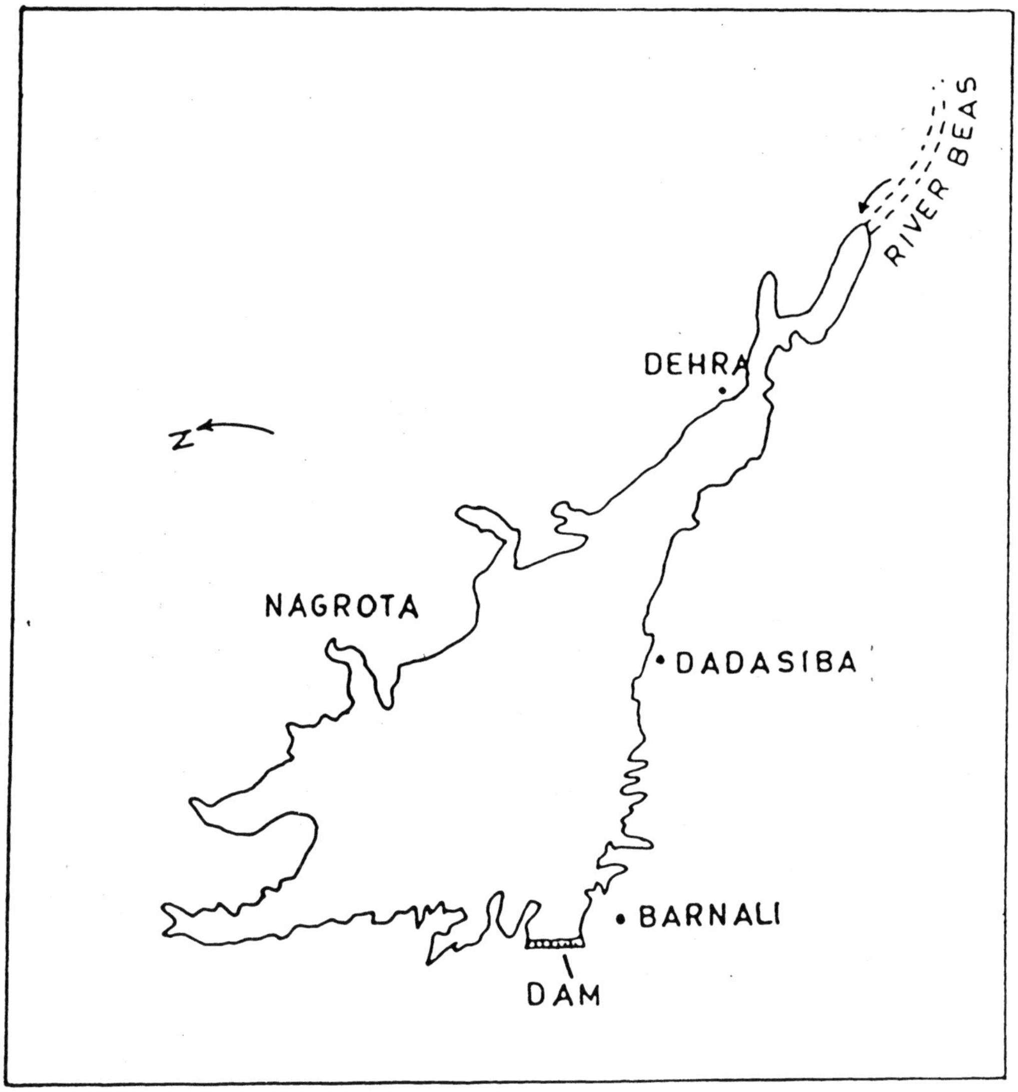

Figure 11.5. Pong reservoir, Himachal Pradesh

Table 11.8. Morphometric features of Pong reservoir

	General	
1.	Location (district)	Kangra
2.	Name of the river	Beas
3.	Water source	Snowmelt & monsoon run-off
4.	Dam	Earth core-cum gravel shell type
5.	Height of the dam (m)	115.8
6.	Elevation of the dam (m above MSL)	435.86
7.	Year of commissioning	1975
8.	Geographical ordinates	32 ° 25'N
9.	Purpose	Multi-purpose
10.	Catchment area (km^2)	12 561
	The reservoir	
11.	Surface area at FRL (ha)	24 000
12.	Surface area at DSL (ha)	6 000
13.	Average surface area (ha)	15 000
14.	Mean depth (m)	35.7
15.	Total length (km)	41.8
16.	Widest stretch (km)	19.0
17.	Shoreline development index	2.48
18.	Volume development index	-
19.	Annual water level fluctuation (m)	384-433
20.	Max. water level fluctuation (m)	49.0
21.	Gross storage capacity (million m^3)	8 570
22.	Live storage capacity (million m^3)	7 290
23.	Inflow (million m^3)	6 855-13 641
24.	Outflow (million m^3)	8 215-15 334

The reservoir bottom soil during the pre- and post monsoon months is alkaline. The soil is deficient in nitrogen (14.4 to 17.2 mg 100 g^{-1}) and phosphorus (0.27 to 0.34 mg 100 g^{-1}), with a low organic carbon due to intensive siltation (Table 11.9). The soil exhibits considerable variations in different zones in respect of its physical and chemical characteristics. Nevertheless, the reservoir soil seems to have a limited role in determining the quality of water as the latter mainly derives from the nutrients from the catchment area.

Excepting the rainy season (April to August), the water of Pong reservoir remains generally clear. While the surface water temperature of the main reservoir varies between 22.2 and 25.1 °C, the incoming river water has a temperature range of 6 to 26 °C. Table 11.10 shows average values of selected physico-chemical parameters recorded at four different stations of Pong reservoir for the year 1987-88.

The reservoir is rich in organic nutrients and planktonic and periphytic assemblages. The plankton community of the reservoir is very rich qualitatively and quantitatively. Plankton density of 1 053 to 18 661 individuals l^{-1} is matched by a high rate of photosynthetic activities by the phytoplankton. Mean gross primary production is estimated at 507.77 mg C m^{-3} day^{-1} with high net-gross production ratio of 0.55. Diatoms represented by 16 taxa constitute the largest component of plankton (69.75%), followed by green algae (22.22%) blue-green algae (3.72%) and dinoflagellates (0.7%). Chlorophyceae, with 18 genera exhibit the maximum diversity. The average standing crop of benthic macrofauna varies from 496 to 513 individuals m^{-2}. Oligochaetes, dipterans, ephemeropterans and the molluscs represent the major benthic organisms in the reservoir.

Pong is a shallow reservoir with a lower organic production than Gobindsagar and the fish catch was originally dominated by carnivores. On account of systematic stocking over a number of years with seed of mirror carp and Indian major carps, the catch structure of the reservoir was completely altered and carps eventually accounted for 61.6% of the total landings (1987-88). The fish yield of the reservoir also increased significantly from 6.5 kg ha^{-1} during 1976-77 to 53.1 kg ha^{-1} during 1987-88. Catfishes and carps accounted for 70.1 and 29.9% of the total production of the reservoir during 1982-83 against 27.8 and 67.6% during 1989-90 respectively.

Table 11.9. Physico-chemical parameters of soil of Pong reservoir

Soil quality	Pre-monsoon	Post-monsoon
Physical characteristics		
Sand	19%	28%
Silt	43%	42%
Clay	38%	30%
Chemical characteristics		
pH	6.5-9.0	7.5-8.5
Organic carbon (%)	1.26	1.39
Available nitrogen (mg 100 g^{-1}; average)	14.4	17.2
Available phosphorus (mg 100 g^{-1}; average)	0.27	0.34

Table 11.10. Physico-chemical characteristics of water of Pong reservoir (by sector)

Stations / *Parameters*	*Dehra*	*Nagrota Surian*	*Dada Siba*	*Pong*
Water temperature (^{0}C)	22.3	25.1	25.1	24.0
Transparency (cm)	121.0	112.0	132.8	112.3
pH	8.1	8.2	8.1	8.1
Dissolved oxygen (mg l^{-1})	9.9	9.8	9.2	9.4
Free CO_2 (mg l^{-1})	6.0	7.2	8.0	7.6
Total alkalinity (mg l^{-1})	53.9	81.5	73.9	74.0
Total hardness (mg l^{-1})	18.8	67.8	64.5	65.5
Chlorides (mg l^{-1})	7.8	28.0	25.0	22.0
Spec. cond. (μmhos)	177.3	189.9	185.6	172.8

Commercial fishing was initiated in the reservoir soon after the impoundment and the important species in order of abundance were *Labeo rohita, Aorichthys seenghala, Labeo calbasu, Tor putitora, Cirrhina mrigala, Wallago attu, Cyprinus carpio, Labeo dero, Catla catla* and *Channa* sp. The total catch during the first year of fishing operation was 98.1 t which increased progressively, attaining a peak of 797.4 t during 1987-88. During 1977-78 to 1986-87, the landings fluctuated within a narrow range of 443 to 596 t. In the year 1988, due to heavy rains and floods, the water level of the reservoir reached an alarming level of 442 m and forced the dam authorities to open the flood gates. This caused heavy loss of fish from the reservoir, obviously affecting the catches in the following years which plummeted to 475.8 t during 1988-89 to increase marginally to 489.2 t during 1989-90. The quantum of fishing effort, however, continued to rise in spite of the falling yield. The maximum yield of 53.1 kg ha^{-1} was obtained in 1987-88. On an average, the yield was 32.2 kg ha^{-1} (range 6.5 to 53.1 kg ha^{-1}) during the l ast 14 years (Table 11.11).

Catch of *L. rohita* increased progressively from 0.4 to 42.5% from 1982-83 to 1987-88. *Catla catla* failed to get a foothold in the reservoir despite the stocking effort. Its contribution to the total catch remained only 0.04 to 1.6%. The percentage composition of *C. mrigala* increased from 5.3 to 9.7 from 1982-83 to 1987-88. *Cyprinus carpio specularis*, the only exotic carp of the reservoir, sustains a low population, despite heavy stocking, its percentage fluctuating between 2.6 and 5.9. The major carnivores of the reservoir are *Tor putitora, A. seenghala, W. attu* and *Channa* spp. (Table 11.12).

Table 11.11. Fishing effort and yield in Pong reservoir during 1976-77 to 1989-90

Years	Catch (t)	Yield (kg ha^{-1})	Fishing effort No. of nets	Catch Index	Fishing Index	Catch per effort
1976-77	98	6.5	105	1.0	1.0	0.9
1977-78	265	17.6	186	2.7	1.7	1.4
1978-79	537	35.8	460	3.7	3.1	1.1
1979-80	596	45.7	670	3.8	3.5	0.9
1980-81	569	37.9	637	3.8	3.5	0.9
1981-82	443	29.5	622	3.6	3.5	0.7
1982-83	498	33.2	681	3.7	3.6	0.7
1983-84	470	31.3	758	3.7	3.7	0.6
1984-85	479	31.9	833	3.7	3.8	0.5
1985-86	552	36.8	758	3.8	3.7	0.7
1986-87	519	34.6	947	3.8	3.9	0.5
1987-88	797	53.1	986	4.3	3.9	0.8
1988-89	475	31.6	1 277	3.9	4.2	0.3
1989-90	489	29.6	1 096	3.9	4.1	0.4
Total	6 787	452	10 016	49.4	47.2	10.4
Average	**484.8**	**32.3**	**715**	**3.5**	**3.3**	**0.7**
Range	**98-797**	**6.5-53.1**	**105-1277**	**1.0-4.3**	**1.0-4.2**	**0.3-1.4**

Table 11.12. Catch composition and catch per unit of gill net area in Pong reservoir

	1990-91 (%)	1991-92 (%)	Catch per unit gill net (g 50 m-1)
Tor putitora	64.7 (14.6)	59.8 (12.3)	98
C. carpio	4.5 (1.0)	3.5 (0.7)	6
W. attu	24.1 (5.5)	20.3 (4.2)	33
A. seenghala	121.2 (27.4)	163.5 (33.6)	267
L. calbasu	41.7 (9.4)	49.1 (10.1)	80
C. mrigala	17.0 (3.9)	14.2 (2.9)	23
Channa spp.,	0.2 (0.1)	0.2 (0.1)	0.4
C. catla	7.1 (1.6)	12.1 (2.5)	20
L. rohita	137.8 (31.2)	150.6 (31.0)	246
Miscellaneous	23.2 (5.2)	12.5 (2.6)	20
Total	441.5 (100)	485.8 (100)	

Initially, the fish fauna of the reservoir consisted mainly of catfishes, minor carps and few *coarse* fishes mainly residual and acclimatised from the river. Stocking of 130 000 fingerlings of mirror carp was done during June 1974, followed by 149 000 seed of Indian major carps in 1976-77. It has been estimated that 30 million fish seed comprising *L. rohita*, *C. catla*, *C. mrigala* and *C. carpio* have so far been stocked in the reservoir in the ratio of 2:2:1:1. The overall stocking deficit is estimated at 53.4 to 80% in case of mirror carp and 83.9% for Indian major carps. However, overstocking of mirror carp had beeen reported during 1988-89 and 1989-90 to the tune of 18.8% and 329.1% respectively. Fisheries development in Pong reservoir has helped in rehabilitation of the families displaced due to inundation.The Government's initiative to stock the reservoir with the Indian major carps has gone a long way in establishing the population of *L. rohita* which contributes as much as 42.5% to the fisheries.

11.4 PANDOH RESERVOIR

Pandoh reservoir, located at 31^0 50' N,, at an altitude of 987 m above MSL has a waterspread of 200 ha. It is a part of the Beas-Sutlej Link Project, diverting the Beas water into the Sutlej basin. In the process, the water flows down from a height of 320 m and hydel power is generated at the Dehar power station. The earth-cum-rock fill dam was constructed in1977. The reservoir has a gross capacity of 4.1 million m^3. The morphometric, hydrographic and hydrological characteristics of the reservoir are detailed in Table 11.13.

Kaushal *et al.* (MS.) provide a limnological profile of Pandoh reservoir. Thermal stratification between 13 and 14 m has been reported, the temperture dropping from 16.5 to 10.5 ^{0}C from surface to 19 m. The dissolved oxygen, following a similar trend, declines from 9.56 mg l^{-1} at the surface to 7.84 mg l^{-1} at 19 m. Although total alkalinity and specific conductivity increase towards the botttom, no such difference in pH is reported. The low temperature regimen and nutrient status of the reservoir point towards oligotrophic tendencies. This is strengthened by the poor plankton count of 100 organisms l^{-1} and primary production rate (484 mg C m^{-3} day^{-1}. The main component of phytoplankton is Bacillariophyceae represented by *Rhizosolenia*, *Diatoma*, *Cocconeis*, *Frustulia*, *Mastogloia*, *Amphora*, *Tabellaria*, *Navicula*, and *Stauroneis*. The benthic community is represented by *Branchiura*, *Limnodrilus*, *Tubifex*, *Nais* and *Chironomus*.

Table 11.13. Morphometric and physico-chemical characteristics of Pandoh reservoir

Parameter	Mean value
Maximum reservoir level (m above MSL)	896.42
Minimum reservoir level	883. 92
Live storage capacity (million m^3)	4.1
Maximum length of the reservoir (m)	2134
Maximum width of the reservoir (m)	457
Area at FRL (ha)	200
Water temperature ^0C	14.00
pH	7.52
Total alkalinity (mg l^{-1})	42
Dissolved oxygen (mg l^{-1})	11.2
Free carbon dioxide (mg l^{-1})	2.0
Phosphate (mg l^{-1})	0.08
Organic carbon (mg l^{-1})	2.4
Total hardness (mg l^{-1})	58
Specific conductivity (µmhos)	90.9
Total dissolved solids (mg l^{-1})	45.5

(After Kaushal et al., MS.)

Common carp was stocked in 1978, but discontinued afterwards. There is no organised fishing in the reservoir, except occasional sport and recreational fishermen. *Salmo trutto fario, Schizothorax richardsonii, Labeo dero, L. dyocheilus, Tor putitora* and some hillstream fish are present in the reservoir.

11.5 RESERVOIR FISHERIES IN HIMACHAL PRADESH

The reservoirs in Himachal Pradesh serve primarily as flood control and irrigation systems and next as hydro-power generating units. Fish stocking and production in the reservoirs were initiated with the aim of rehabilitating the displaced families and ensuring gainful employment to people affected by the inundation. However, the State has made great strides in reservoir fisheries development. During 1992-93, the production from Pong and Gobindsagar reservoirs was 448 (30 kg ha^{-1}.) and 964 tonnes (96.4 kg ha^{-1}) respectively based on the average reservoir area. Compared to the national average, the two reservoirs in Himachal Pradesh are well-managed and the yield of Gobindsagar is one of the highest from a large reservoir in the country. The amendment of the State Fisheries Act 1976, the enforcement of mesh size regulations, the organization of fishermen in a cooperative, establishment of the Fisheries Federation, settling of fishermen during the initial stage from outside the State, establishment of fishermen welfare schemes, etc., are measures which have assisted in the optimal development of the reservoir fisheries and provided vocation to the oustees of the reservoirs.

Table 11.14. Salient features of some reservoirs in Himachal Pradesh

	Gobindsagar	Pong	Pandoh
Date of closure	1963	-	-
Area at FRL (ha)	16 835	24 000	200
Minimum area (ha)	5 063	6 000	-
Average area (ha)	10 000	15 000	-
Maximum length (km)	96.56	-	-
Maximum width (km)	6	-	-
Maximum depth (m)	163.07	-	-
Mean depth (m)	55	53 - 94	-
Volume (million m^3)	9 621	-	4.1
Catchment area (km^2)	56 980	-	-
Elevation (m above MSL)	514	440.6	899.16
Length of shoreline (km)	564	-	-
Shore dev. index	12.86	-	-
Volume dev. index	1.01	-	-
Annual level fluctuations (m)	67.96	-	-
Inflowing rivers	Sutlej	-	Beas
Outflowing rivers	Sutlej	-	Beas
Maximum outflow (m^3 $sec.^{-1}$)	11 326	-	9 939
Latitude (N)	31^0 25'	30^0 25'	31^0 50'
Longitude (E)	76^0 25'	75^0 45'	--
Water temperature (^{0}C)	9-31	22.3 - 25.1	16.5
pH	7.3-8.4	6.5 - 9.0	-
Transparency (cm)	14.0-639.0	-	-
DO (mg l^{-1})	1.76-13.44	9.2 - 9.9	9.56 - 7.84
CO_2 (mg l^{-1})	0-38.0	6.0 - 8.0	-
Total alkalinity (mg l^{-1})	50.0-104.0	53.9 - 81.5	24 - 35
Spec cond. (µmhos)	119.0-291.6	177.3 - 185.9	-
Total hardness (mg l^{-1})	-	53.9 - 81.5	86.4 - 97.2
Calcium (mg l^{-1})	18.4-30.4	-	-
Nitrate (mg l^{-1})	tr.-0.15	-	-
Phosphate (mg l^{-1})	tr.-0.12	-	0.10 - 0.15
Silicate (mg l^{-1})	0.8-9.6	-	3.2 - 4.1
Chlorides (mg l^{-1})	-	7.8 - 28.0	-
Organic carbon (%)	-	-	0.7 - 1.6
GPP (mg C m^{-3} d^{-1})	135-1875	-	-
Soil			
pH	8.2-8.8	-	-

12. UTTAR PRADESH

Uttar Pradesh, the northern province, is the most populous State of India, occupying 9% of the total geographical area of the country. The State is surrounded by Nepal in the north, Himachal Pradesh in the northwest, Haryana in the west, Rajasthan in the southwest, Madhya Pradesh in the south and southwest and Bihar in the east. Uttar Pradesh has three distinct regions *viz.*, the Himalayan mountains of the north, hills and plateaus of the south and the Gangetic delta in between. The lofty Himalayas embrace Uttar Kashi, Chamoli, Pithorgarh, Tehri-Garhwal, Garhwal, Almora Nainital, and Dehra Dun, covering about one sixth of the total area of the State. The southern hilly tracts lie at an altitude of 300 m above MSL. Between the two uplands lies the vast homogenous alluvial plain, one of the largest in the world.

Main rivers of Uttar Pradesh are the Ganga and its tributaries, *viz.*, the Yamuna, Ramganga, Gomti, and Ghagra. All of them except Gomti emerge from the Himalayas. The Ganga being too large to be tamed, no major irrigation projects have been created on its mainstream. The irrigation needs are being met by canal projects constructed early this century. In recent years, some of the northern and southern tributaries of Ganga and their minor feeder streams have been tamed, creating a few reservoirs. Nevertheless, the fourth largest State in India has a sparesely dotted reservoir map (Fig. 12.1).

12.1 RESERVOIRS OF UTTAR PRADESH

Sixty-six reservoirs with an area of 137 034 ha are distributed among 17 districts and this list is not exhaustive. The four large reservoirs, *viz.*, Rihand, Malhatila, Kalagarh, and Saradasagar, occupy 71 196 ha. Sonbhadra district with 52 000 ha has the largest area under reservoirs on account of the imposing presence of Rihand reservoir. The medium impoundments, having a total waterspread of 44 993 ha, are distributed in the districts of Jhansi, (5), Varanasi, Sonbhadra, Banda (3 each), Mirjapur (4), Hamirpur (2), Bijnaur and Pilibhit (1 each). Forty small reservoirs have been documented with a total area of 20 845 ha, dispersed almost uniformly in the districts. Mirjapur (7) Hamirpur (6) and Gonda (6) districts have the maximum number of small reservoirs (Table 12.1).

The State Fisheries Deparment classifies the water bodies in the State based on the area. Accordingly, total areas under small (<1 000 ha), medium (1 000 to 5 000 ha), and large (> 5 000 ha) reservoirs are 162 000 ha, 31 840 ha and 271 000 ha, respectively. Thus, the State has 464 840 ha of water bodies (individual units above 10 ha in size). But, this also includes the natural water bodies such as oxbow and upland lakes. By deducting the estimated 130 000 ha of natural lakes (Saxena, 1986), the total water area under man-made impoundments becomes 334 840 ha.

Present fish yield from the reservoirs is estimated at 8 to 10 kg ha^{-1}. Only very few reservoirs of Uttar Pradesh have been studied. Rihand was a part of the All India Coordinated Project on Reservoir Fisheries. Between 1971 and 1981, fisheries ecology of the reservoir was investigated in detail (Natarajan *et al.*, 1982).

Figure 12.1. Distribution of reservoirs in Uttar Pradesh

Table 12.1. Distribution of reservoirs in Uttar Pradesh (by districts)

District	Small < 1 000 ha		Medium 1 000-5 000 ha		Large > 5 000 ha		Total	
	No.	Area (ha)	No.	Area (ha)	No.	Area (ha)	No.	Area (ha)
Varanasi	2	951	3	4 571	-	-	5	5,522
Sonebhadra	-	-	3	5 463	1	46539	4	52,002
Mirjapur	7	2 666	4	11 878	-	-	11	14,544
Jhansi	3	1 789	5	9 262	-	-	8	11 051
Hamirpur	6	1 347	2	2 960	-	-	8	4 307
Banda	2	1 672	3	5 556	-	-	5	7 228
Lalitpur	-	-	-	-	1	10360	1	10 360
Allahabad	2	374	-	-	-	-	2	374
Bijnaur	-	-	1	1 214	1	7417	2	8 631
Agra	1	259	-	-	-	-	1	259
Eta	1	518	-	-	-	-	1	518
Pilibhit	-	-	1	4 089	1	6 880	2	10 969
Bahraich	2	232	-	-	-	-	2	232
Gonda	6	2 494	-	-	-	-	6	2 494
Nainital	4	7 962	-	-	-	-	4	7 962
Siddharthnagar	2	381	-	-	-	-	2	381
Basti	2	200	-	-	-	-	2	200
Total (Listed)	40	20 845	22	44 993	4	71,196	66	**137 034**
Unlisted								**197 806**
Grand Total								**334 840**

Area based on mean reservoir level Source: Department of Fisheries Uttar Pradesh

12.2 RIHAND RESERVOIR (Fig. 12.2)

Also known as Gobind Ballabh Pant Sagar (GB Sagar), Rihand reservoir was created on the river Rihand, a tributary to Sone, which in turn joins the Ganga on its right flank. The dam is located at 24^0 1' N and 83^0 3' E. The river originates from the hills of Madhya Pradesh, draining a catchment of 13 344 km^2. Commissioned in 1962, the reservoir has a capacity of 10 625 million m^3 at the FRL of 268 m above MSL and a surface area of 46 538 ha, which shrinks to 13 759 ha at the dead storage (mean 30 149 ha). It has a maximum length of 48 km and the mean depth at FRL is 22.8 m. Annual rate of inflow is estimated at 6 301 million m^3.

Based on texture, the bottom soils of the reservoir are classified as clay to clay loam. The soil pH ranges from 6.0 to 8.0 with the maximum during post-monsoon months. The low values in pre-monsoon months denote the higher rate of organic decomposition producing more carbon dioxide, which in association with water, gives carbonic acid, lowering the pH values of soil. The available nitrogen ranges from 15.68 to 30.52 mg 100 g^{-1}, being higher during post-monsoon months. The level of available phosphorus (1.10 to 2.30 mg 100 g^{-1}) is more or less uniform throughout the year. The organic carbon is high (1.19%) in the deeper region, compared to the shallows (0.50%) due to the low rate of organic decomposition. $CaCO_3$ is high (2.307 to 4.377%) during the post-monsoon period.

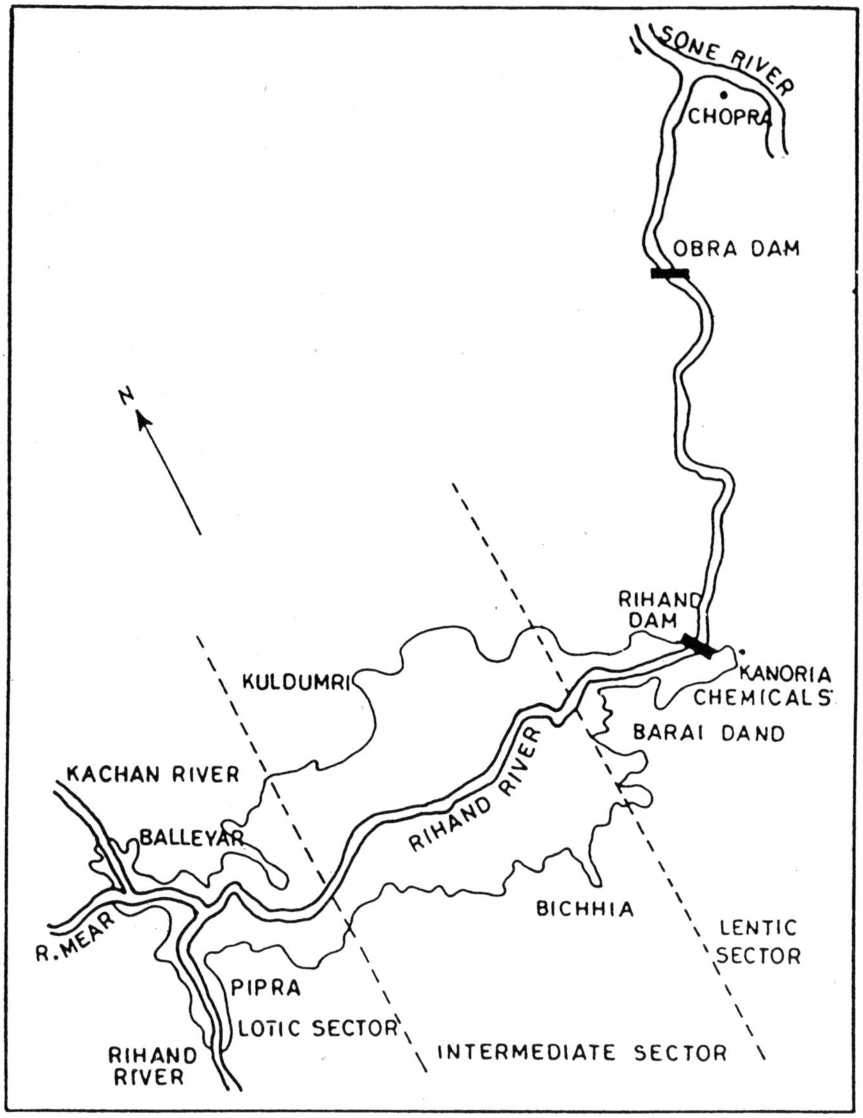

Figure 12.2. Rihand reservoir, Uttar Pradesh

The water remains moderately warm, the mean winter and summer temperatures being 19.9 and 31.1 °C respectively. Mean transparency of the reservoir water is low (14.58 cm), the range in values being 1.8 to 28.5 cm. It is neutral to alkaline in reaction (pH 7.0 to 8.8), and the lower values are recorded normally during monsoon months due to carbon dioxide-rich inflow. Range in dissolved oxygen is within 3 to 10 mg l^{-1}, with a saturation level of 50 to 107%. Free carbon dioxide values are from 3.5 to 34.0 mg l^{-1}.

Carbonates being absent in the presence of free CO_2, total alkalinity is in the form of bicarbonates with values ranging from 28.0 to 60 mg l^{-1} (average 43.87 mg l^{-1}). Total hardness of 8.4 to 21.6 mg l^{-1} confirms the low rate of dissolved salts. Phosphate (traces to 0.65 mg l^{-1}) and nitrate (0.15 to 0.70 mg l^{-1}) indicate a conducive nutrient status. Silicate is recorded in higher concentration (14.0 mg l^{-1}) during the post-monsoon months. The specific conductivity with an average of 92.24 μmhos suggests reasonably high productivity levels.

Despite the low total alkalinity and hardness, the total ionic concentration and the phosphate phosphorus create suitable conditions for primary productivity. A well-defined thermocline exists in May and June. The difference between surface and bottom temperatures is 8 to 10 °C. However, the thermal stratification is short-lived and broken by the influx of flood waters in July and August. No thermal stratification is observed in winter.

The weak chemical stratification indicates a low rate of tropholytic activities. While dissolved oxygen, pH, bicarbonates and specific conductivity decrease from surface to bottom, free carbon dioxide registers a reverse trend. Despite the presence of a strong thermocline, there is only a slight decrease in oxygen towards the bottom (8.3 to 5.1 mg l^{-1}), with enough oxygen dissolved in depth, indicating low oxygen demand for organic decomposition. This is further corroborated by the pH values which do not vary much in the tropholytic layer.

The mean primary productivity rates are 387.04 mg C m^{-3} day^{-1} (gross) and 161.44 mg C m^{-3} day^{-1} (net). The rate of carbon fixation was much higher during the initial years (1972-73), when *Microcystis* formed blooms (Singh and Desai, 1980).

Cyanophyceae, represented by *Microcystis* dominate the phytoplankton. Other forms, recorded occasionally, belong to Chlorophyceae, Bacillariophyceae and Dinophyceae (Singh *et al.* 1980). The mean annual plankton density is the highest (70 units l^{-1} in 1976) in lentic sector, followed by intermediate (63 unit l^{-1}), and lotic sectors (54 units l^{-1}). Plankton density in the reservoir, as a whole, was low, ranging from 4 to 41 units l^{-1} during the 6 year period from 1972-73 to 1977-78.

The benthic invertebrate community is poor, with densities of 32 to 70 individuals m^{-2}, mainly represented by the dipteran larvae, *Chaoborus* and *Chironomus*, caddis fly larvae, mayfly nymphs, dragonfly nymphs, and oligochaetes. Occasionally, the bivalve *Corbicula* sp. is encountered. The benthic macrofauna is richer in the lotic sector with an organically rich substratum, than in the intermediate and lentic sectors with gravelly and rocky bottom.

The original fish fauna of the river Rihand, before the impoundment, comprised 42 species (Hora, 1949). Motwani (1970) listed 44 species during the post-impoundment phase. Between 1971 and 1981, 44 species were present in the reservoir (Natarajan *et al.*, 1982). The total number of species recorded by various workers before and after impoundment is 74. Cyprinidae, with 35 species shows the maximum species diversity.

Fisheries

Commercial fishing started in 1963-64, by deploying fishermen from Mirzapur. These fishermen were organised into a cooperative society, under the State patronage and provided with nylon yarn and other fishing inputs. Essential marketing infrastructure was also created.

Craft and gear: The most common fishing gear in the reservoir is the surface gill net, operated from a boat. A normal fishing unit comprises 25 gill nets, each measuring 50 x 6m, a boat and 2 to 4 fishermen. The boat used in the reservoir is a flat-bottomed country craft, 7.3 m long and 1.2 m wide.

A single 400 m long, multi-meshed giant drag net was also introduced in the reservoir. The oval shaped net is 6 m wide at the middle and the tapering ends measure 2 m. The mesh sizes (bar) vary from 15 to 50 mm. The net is operated by 16 to 17 fishermen with the help of 2 boats and a single haul lasts more than two hours.

Desai (1993) reports in detail, the fish yield trends of Rihand reservoir. Total catch from gill nets and the catch per effort (effort = 50 m gill nets) during the last decade have been erratic (Table 12.2). The fish catch by gill nets increased progressively from 184.13 t in 1972-73 to 328.82 t in 1974-75 and declined thereafter. An unusually low catch was recorded during 1978-79 due to restricted fishing activities.

Table 12.2. Fish production and fishing effort in Rihand reservoir (1971-72 to 1980-81)

Years	Total catch (t)	Yield ha^{-1} (kg)	Number of nets	Number of 50 m gill net	Catch/ 50 m unit (kg)	Percentage increase Effort	Percentage increase Catch
1971-72	152.40	5.05	-	-	-	-	-
1972-73	184.13	6.11	126 597	75 285	2.44	-	-
1973-74	,442.62	7.85	105 029	62 656	3.87	-16.77	+31.77
1974-75	328.82	10.91	245 626	147 375	2.23	+95.76	+78.57
1975-76	160.75	5.33	1,50,644	90 386	1.78	+20.06	+12.67
1976-77	68.23	2.26	139 442	95 513	0.71	+26.87	-62.97
1977-78	83.49	2.76	144 095	86 457	0.96	+14.84	-54.65
1978-79	24.87	-	34 249	20 549	1.21	-72.71	-86.48
1979-80	148.15	4.91	410 106	246 063	0.60	+226.84	-19.54
1980-81	79.52	2.64	225 449	135 269	0.58	+78.08	-56.81

Effort = Number of 50 m gill net

Desai (*op. cit.*) describes an inverse relationship between the water level and fish yield in the reservoir. The high production of 242.62 t (1973-74) and 328.82 t (1974-75) were achieved at low water levels of 248.55 m and 248.22 m respectively. There was a similar correlation between the water level and catch for 1971-72, 1972-73 and 1975-76. Lower water level permits effective operation of gill nets, especially in areas of the reservoir, where column setting of gill nets becomes possible.

There is a progressive decline in catch per effort. During 1979-80, despite a catch level near to that of the first year (1971-72), the catch per effort was only 0.60 kg, compared to 2.44 kg for 1971-72. This is due to a three and half times increase in fishing effort.

Catch structure

The preponderance of *Catla catla* (73 to 99%) in the fisheries of Rihand is remarkable (Table 12.3). The other species are *Cirrhinus mrigala, Labeo calbasu, L. rohita, L. bata, Puntius sarana, Wallago attu, Silonia silondia, Mystus spp., Bagarius bagarius, Notopterus chitala, N. notopterus,* and *Channa marulius.*

Table 12.3. Fish catch structure of Rihand reservoir (1971-72 to 1980-81).

Year	Percentage	
	C. catla	Others
1971-72	99.2	0.8
1972-73	98.8	1.2
1973-74	95.6	4.4
1974-75	99.1	0.9
1975-76	90.2	9.8
1976-77	87.1	12.9
1977-78	90.5	9.5
1978-79	-	-
1979-80	84.2	15.8
1980-81	73.1	26.9

Catch per unit effort (when 50 m gill net is taken as a unit) remained unchanged despite a two-fold increase in the number of units during the period 1972-73 to 1974-75, indicating that the stock density of catla was not affected until mid 1970s. In 1975-76 and 1976-77, even with reduced fishing effort, the catch per unit effort declined (1.78 and 0.71 kg respectively) and the decline continued even when the fishing effort was raised. Thus, stock abundance in the reservoir is being affected.

A selective fishing practice is being followed by the fishermen who operate a large number of nets in the mesh bar range of 150 to 170 mm, targeted against larger size group of catla (900-1100 mm/14-25 kg). Percentage of catla in the large-meshed nets is high (97.2%), compared to the smaller nets (36.6%). This fact tempts fishermen to employ the former, as catla is at premium over the other fishes. Comparison of the catch structure of gill nets and drag nets is interesting (Table 12.4). The catch of carps in drag nets remains as low as 8.65%, the remainder being catfishes (*S. silondia, W. attu, Aorichthys* spp. *B. bagarius*) and other predators, *viz.*, *Notopterus* spp., and *Channa marulius*. This brings out the role of drag nets in controlling the predators in the reservoir, that are getting a foothold among the ichthyofauna.

Breeding

Breeding of major carps in Rihand, reported in July, is influenced mainly by the monsoon inflow during the month. During 1974 to 1977, the annual spawn collection (eggs) mainly yielded catla and mrigal (Table 12.5). However, the survival of the latter appears to have been poor.

Table 12.4. comparison of catches from gill nets and drag nets.

	Gill nets		Drag net	
	no (%)	wt (%)	no (%)	wt (%)
Carps	74.36	91.88	1.60	8.65
Catfish	21.45	7.41	88.46	49.87
Others predators	4.19	0.71	9.94	41.88

Table 12.5. Species composition of eggs collected during monsoon months

Year	No. of eggs collected	Percentage of				
		catla	mrigal	calbasu	rohu	others
1974	60 000	7.6	55.6	7.1	2.7	27.0
1975	81 000	43.6	38.5	5.1	5.1	7.7
1976	87 200	54.1	34.0	6.4	5.5	-
1977	4 500	57.1	26.6	6.5	5.4	4.4

Stocking

Stocking was initiated in Rihand reservoir during 1963-64. and by 1970-71 it was stocked with a total of 5.6 million fry/fingerlings at an annual average of 700 000 which works out to 23 ha^{-1}, on the basis of an average reservoir area of 30 148 ha. During the subsequent 11 years ending 1981-82, the stocking rate declined to 13 ha^{-1}.

Desai (*op. cit.*) reiterates the need for intensive stocking of Indian major carps, especially mrigal, rohu and *L. calbasu* to diversify the carp species spectrum. Despite its breeding success, catla is under threat as its spawn is vulnerable.

The limited availability of upstream spawning habitats leads to competition among carps for breeding grounds, resulting in the frequent occurrence of catla-rohu hybrids in the reservoir (Natarajan *et al.*, 1976).

Natarajan *et al.*, (1977) reports an interesting case of three ecological populations of *Catla catla* in the reservoir, distinguishable by the relative lengths of their pectoral fins and the body girth. Among the three sub-populations, the one with a medium pectoral fin merits special attention, as it is reported to feed mainly on phytoplankton, especially *Microcystis aeruginosa*, the ubiquitous blue-green alga of Indian waters.

Pollution

The effluents from the Kanoria Chemicals (P) Ltd., Renukoot, located on the northern bank of Rihand, are discharged into the reservoir. The factory manufactures caustic soda, bleaching powder and benzene hexachloride. The highly acidic effluents are characterised by high chloride and free chlorine content. Arora *et al.* (1970) recorded fish mortality due to chlorine-bearing wastes. Panwar *et al.* (1979) confirmed this by reporting mortality due to toxic effluents from the same industrial unit. The affected fishes included *Catla catla* (2.5 to 3.0 kg), *Aorichthys aor* (0.3 to 0.6 kg), *Cirrhinus mrigala* (0.35 to 1.5 kg) and *Clupisoma garua* (72 to 85 kg). Plankton and benthic communities are also affected.

There are five thermal power stations along the banks of Rihand reservoir, including the three super thermal power plants, which produce 10 600 MW of electricity (Fig 12.3). Adverse effects of the heated discharge and the deposition of fly ash into the water have been discussed by Chandra *et al.* (1985). The temperature of heated dischargre ranges from 46 to 52 ^{0}C and the deposition of ash extends up to 500 m downstream of the outfall region. All the power plants are located near the intermediate and lotic sectors which are known for concentrations of fish (Desai, 1993).

12.3 GULARIYA RESERVOIR

Gulariya is a typical small reservoir of the Gangetic plain, situated in the Allahabad district. It is created over a small rivulet, Gulariya. The reservoirs covers 300 ha at FRL and shrinks to 6.7 ha during summer. It dries up completely during extreme summers. The catchment area of 38 km^2 comprises rocky terrain with forest cover, receiving an average rainfall of 90 mm. The reservoir has a gross storage capacity of 9.3 million m^3 at FRL. The 3 km long earthen dam is provided with a 228 m long escape channel. Constructed for the primary purpose of minor irrigation, the reservoir releases water through a 24 km long channel, serving a gross command area of 3 395 ha.

Mehrotra and Jhingran (1986) described the soil and water quality parameters of the reservoir in detail. The soil of Gulariya reservoir is predominantly sandy (64-73%), with some clay (10-16%) and silt (15-20%). Available nitrogen is fairly rich (12-33 mg 100 g^{-1}), while values of available phosphorus are low (1.2-1.8 mg 100 g^{-1}). The soil is neutral, the $CaCO_3$ content is1 to 2%. The allochthonous dissolved nutrients entering the reservoir are responsible, to a great measure, for the high productivity. The values of total alkalinity (25 to 80 mg l^{-1}), hardness (13 to 40 mg l^{-1}), and calcium (12 to 26 mg l^{-1}) indicate a conducive chemical environment for organic productivity. Loss of nutrients through irrigation channels is replenished from allochthonous sources. The water is warm due to the shallowness, which allows rapid growth of organisms.

Annual average production of carbon, estimated on the basis of primary productivity rate, is 2 266 to 2 462 t yr^{-1}. The fish food organisms are abudant. Wishard and Mehrotra (1988) report total phytoplankton counts of 18 565, 13 766 and 12 084 l^{-1} during 1977, 1978 and 1979 respectively. Green algae dominate, followed by blue-greens. Occasional algal blooms and luxuriant growth of periphyton and macrophytes indicate organic enrichment.

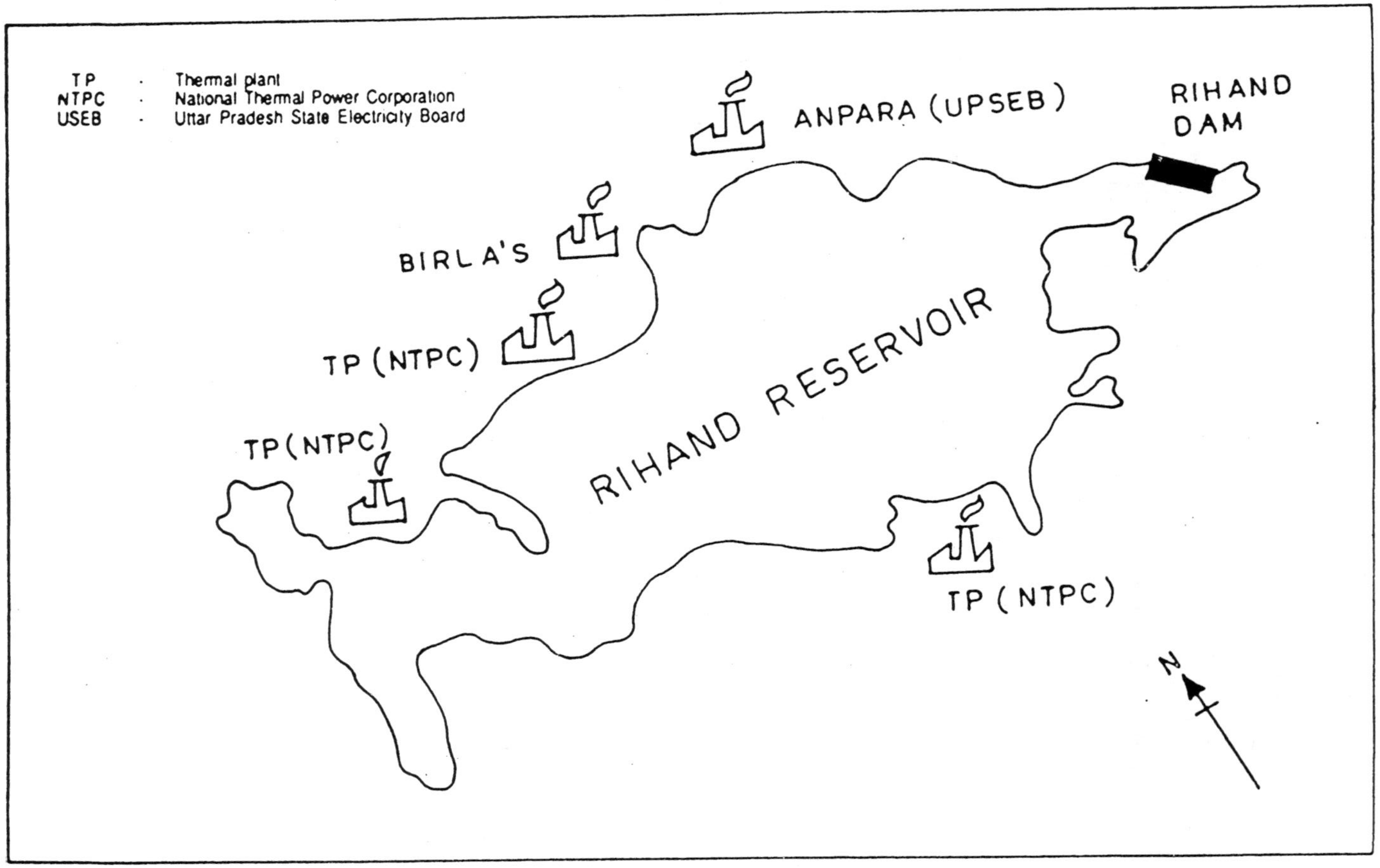

Figure 12.3. Location of thermal power stations along the Rihand reservoir

Fisheries

Gulariya represented one of the earliest testing grounds for CIFRI's efforts to develop management packages for small irrigation impoundments. During summer months, Gulariya either dries up completely or else the water level gets drastically reduced leaving scope for over-fishing. This results in the depletion of broodstock and affects recruitment in the succeeding years. Thus, despite the possibilities for breeding of major carps within the system, autostocking and self-sustained stocks are not within the realm of possibility. Fish populations in the reservoir have to be built up by compensating the loss of natural recruitment through the introduction of desirable species. There is a direct correlation between the stocking rates and the catch per unit of effort from such small impoundments, where stock can even be maintained above the natural stocking capacity of the system through supplemental fertilization.

Jhingran *et al.* (1981), after considering various methods of assessing the potential yield from the reservoir used a thermodynamic model to estimate the production potential, based on the energy projections. With much of its nutrients derived from allochthonous sources, Gulariya is placed in the category of *lotic* reservoirs, where the morpho-edaphic approach has limited application. Presuming a yield potential of 234 kg ha^{-1}, moderated to 200 kg ha^{-1}, the stocking rate was calculated at 500 fingerlings ha^{-1}, which included a 50% allowance for escapement through spillways. The annaul stocking requirement was worked out to be 75 000, considering 50% of the area at FRL as the mean waterspread.

An important consideration was to allow maximum growout period between the date of stocking and the final harvesting, *i.e.*, before the levels go below the critical mark. The possible loss due to the low size at harvest was to be made good by the number. During 1976-77, the total fish harvest from the reservoir was 5 t. In 1977-78, 75 800 fingerlings were stocked in the reservoir which resulted in a harvest of 15.15 t. This, along with the residual stock netted out during the next year (7t) amounted to 22 t, *i. e.*, 150 kg ha^{-1}. Considering that the total catch from the reservoir ranged from 0.076 to 8 t during 1971-72 to 1976-77, there has been a three-fold hike in the production rate after introducing the new management package.

12.4 BACHHRA RESERVOIR

Bachhra reservoir is created on a small rivulet of the same name in Allahabad district. The reservoir covers an area of 140 ha at the FRL of 111 m above MSL, and a mean depth of 5.22 m. At full storage, the reservoir holds 7.42 million m^3 of water which comes down to 0.03 million m^3 at the dead storage level. During the dry season, the waterspread becomes 4 ha in area. The dam is a 558 m earthen embankment, provided with a 98 m long waste weir. The catchment comprises predominantly rocky terrain with forest land and pastures receiving a rainfall of 90 cm. The reservoir bed is undulating and strewn with boulders which obstruct the use of many a fishing gear.

Khan *et al.* (1989, 1990) investigated the reservoir and placed it in the eutrophic category, based on a number of physico-chemical attributes. Due to the nutrient enrichment from the rain washings, the reservoir is rich in total dissolved solids, nutrients and mineral salts. With a favourable temperature regime, coupled with rich bicarbonate alkalinity and hardness, the water is very conducive for primary productivity (Table 12.6).

Table 12.6. Some important water quality parameters of Bachhra reservoir

Parameters	Range in values
Temperature °C	19-32
Total alkalinity (mg l^{-1})	72-167
Total hardness (mg l^{-1})	22-78
Ca^{++} (mg l^{-1})	18-25
Nitrates (mg l^{-1})	0.08-0.18
Phosphates (mg l^{-1})	0.06-0.25
Silicates (mg l^{-1})	0.73-15.0

(After Khan et al., 1989, '90)

The estimated annual primary production in terms of carbon varies from 1 250 t to 1 402 t. The plankton community of the reservoir is rich with the total count going up to 4 762 units l^{-1} during 1984. The major constituents of phytoplankton are *Microcystis aeruginosa*, *Aphanocapsa* sp., *Coelosphaerium* sp., *Anabaena* spp., *Merismopedia* sp;. (Cyanophyceae), *Pediastrum* spp., *Eudorina* sp., *Pandorina* sp., *Scenedesmus* sp., *Ankistrodesmus* sp., *Oedogonium* spp. *Spirogyra* sp., *Ulothrix* sp. (all Chlorophyceae), *Melosira* sp., *Navicula* spp., *Surirella* sp., *Fragilaria* spp. and *Synedra* sp. (all diatoms). The copepods (*Cyclops*, *Diaptomus* and their nauplii), rotifers (*Brachionus* spp., *Keratella* spp., *Filinia* sp., *Polyarthra* sp., and *Lecane* sp.), cladocerans (*Bosmina Ceriodaphnia* and *Diaphanosoma*) and Protozoa (*Arcella* and *Actinophrys*) represent the common zooplankton.

The rate of colonisation of periphyton and benthic macrofauna on the available substrata is reported to be very good. Ranges in density of periphyton on submerged objects had been reported as 30 to 2 500 units cm^{-2} during 1963 and 746 to 120 000 units cm^{-2} during 1987.

A progressive increase in the density of benthos is reported from 1982 to 1987 when the mean number of benthic invertebrates increased from 490 to 1 894 individuals m^{-2}. Insect larvae (52 to 71%) and annelids (15% to 36%) together form a substantial component of the benthic fauna, followed by molluscs. Chironomidae, *Phylopotamus*, *Culiocoides*, *Chaoborus* and coleopteran larvae constitute the insect fauna and the worms are accounted by *Aulodrilus plurisita* and *Branchiura sowerbyi*. *Pisidium*, *Gyraulus*, *Limnaea*, *Indoplanorbis*, and *Viviparus* are the common molluscs encountered in the reservoir.

Fish and fisheries

The indigenous ichthyofauna of the river consisted mainly of rheophilic species of a low population density. After impoundment, the lentic water body harbours 51 species of fish, mainly belonging to the cyprinide and catfish families, and the prawn, *Macrobrachium lamerrei*.

The Irrigation Department auctioned the fishing rights in 1981 and the contractor fished out all the fishes by gill netting which was followed by poisoning. Therefore, during the ensuing year, the entire stock had to be rebuilt by stocking. The reservoir was again auctioned in 1986 for a sum of Rs. 63 000/- when the contractor could catch 4.85 t of fish, corresponding to a yield of 67.4 kg ha^{-1}. The production increased to 10 t in 1988.

At the rate of 1.2% of the primary production, the fish yield potential of the lake was calculated as 240 kg ha^{-1}, against which, an actual yield of 139 kg (10 t) was obtained in 1988, at a conversion efficiency of 0.580%. The main pathway of energy flow is considered to be the detritus chain, the two detritophagus fishes, *viz.*, mrigal and calbasu forming 29.54% and 31.87% of the catch respectively.

The higher yield obtained in 1988 is the result of an optimum stocking density and maintenance of a rational stocking and harvesting schedule, as followed in the case of Gulariya reservoir. Economic evaluation of the fishing operations in Bachhra revealed that the fish business income was about four-fold of the fishermen's remuneration (Khan *et al., op. cit.*). Considering the remuneration of fishermen and period of their employment (active fishing period restricted to March to June), only 12 fishermen could be employed at a subsistence level of income, *i. e.*, Rs. 6 000 per annum. The seasonal nature of fishing in small reservoirs precludes the possibility of gainful employment for fishermen unlike the large reservoirs, where fishing activity is a year-round occupation.

12.5 BAIGUL RESERVOIR

Baigul (Sukhi) is a small tributary of the Ganga originating from the foothills of Kumaon Himalayas, which was harnessed in the year 1967 for irrigation purposes to form the Baigul reservoir. The dam is located at 28° 56' N and 79° 40' E, and provided with four gates and two sluices for outflow. The reservoir has an area of 2 995 ha at FRL of 211 m above MSL. The maximum spillway discharge amounts 566 m^3 sec^{-1}. The drainage area of 305 km^2 is fed by the southwest monsoon, comprising mainly the Baigul basin and local catchment of wooded forests. The water temperature ranges from 14 to 33 °C, the transparency from 24 to 230 mm and the pH values from 7.0 to 8.2 (Khan, 1986). Alkalinity values are high throughout the year (86 to 107 mg l^{-1}). Plankton community of Baigul reservoir is rich, comprising 17 genera of green algae, 4 genera of blue-greens, 10 genera of desmids and 14 genera of diatoms (Salim and Ahmed, 1985).

12.6 BAGHLA RESERVOIR

Baghla is a small irrigation impoundment, created on the rivulet Barica in 1952. Situated at a distance of 55 km southwest of Allahabad, the reservoir covers 250 ha at the FRL of 119 m above MSL. The water retention capacity is 9.58 million m^3 at FRL and 0.141 million m^3 at DSL. Catchment comprises 32 km^2 of hilly land, receiving a rainfall of 100 cm.

The Central Inland Capture Fisheries Research Institute conducted ecological studies on the lake from 1988 to 1991. The salient findings are presented below. The soil flooded by the reservoir is alkaline (pH 7.36 to 7.72) with a high specific conductivity of 322.5 µmhos, organic carbon (1.66%) and calcium carbonate (5.2%). Nutrients as well as the total ionic concentration indicate eutrophic conditons (Table 12.7). Thermal stratification in association with a drop in oxygen values in deeper layers was reported in 1990. Temperature dipped by 2 °C from surface (30.0 °C) to 1 m. Mean gross primary productivity was estimated at 98.58 mg C m^{-3} hr^{-1} in 1989-90 and 110.47 mg C m^{-3} hr^{-1} in 1990-91. The rich nutrients in water and soil enable luxuriant growth of plankton, benthos, periphyton and aquatic macrophytes. Average plankton density is 1 484, 4 363 and 3 075 units l^{-1} for the three years from 1988-89 to 1990-91, Cyanophyceae dominate with *Microcystis aeruginosa*, *Aphanocapsa* sp., and *Anabaena* spp., the major constituents. *Eudorina*, *Pediastrum*, *Pandorina*, *Scenedesmus*, *Ankistrodesmus* and *Spyrogyra* represent the Chlorophyceae. *Melosira*, *Synedra*, *Gyrosigma*, *Asterionella*, *Cymbella*, *Cyclotella* and *Navicula* are recorded among diatoms and *Ceratium* is the sole representative of Dinophyceae. Zooplankton forming 19 to 35% of the plankton includes copepods, rotifers and cladocerans.

Table 12.7. Water quality of Baghla reservoir

Parameter	Range in values	Mean
Water temperature (°C)	19.0-32.5	25.10
Transparency (cm)	9.0-204.0	96.36
Free Co_2 (mg l^{-1})	nil-3.0	0.59
Carbonate (mg l^{-1})	nil-19.0	6.97
Bicarbonate (mg l^{-1})	40.0-96.0	60.50
Dissolved oxygen (mg l^{-1})	2.4-12.8	8.34
Sp. conductivity (μmhos)	63.0-292.0	107.24
pH	7.1-8.8	7.83
Nitrate (mg l^{-1})	0.3-0.4	0.31
Phosphate (mg l^{-1})	0.2-0.4	0.30
Silicate (mg l^{-1})	2.4-4.9	3.27
Calcium (mg l^{-1})	14.0-22.0	18.24
Magnesium (mg l^{-1})	2.2-3.7	2.96

* Values pertain to 1988-89 to 1990-91 *(CICFRI Barrackpore)*

Benthic macrofauna comprises chironomid larvae, *Phylopotamus* sp., and *Chaoborus* sp., annelids *Aulodrilus plurisita, Branchiura sowerbyi*, and molluscs *Pisidium, Corbicula* sp., *Gyralus, Lymnaea, Indoplanorbis*, and *Viviparus* spp. The density of benthic organisms recorded during 1988-89 to 1990-91 ranged from 976 to 2 132 individuals m^{-2}. Periphyton density is estimated at 144 to 545 415 individuals cm^{-2}. Aquatic macrophytes are present during the months of May, November, December and January coinciding with the lesser rates of water level fluctuation. The population density (wet weight) ranges from 250 g to 2.2 kg m^{-2}. The macrophytes harbour a rich associated fauna. Molluscs, prawns, hemipterans (*Belostoma, Ranatra*), other insect larvae, and molluscs are often encountered among weeds. Annual carbon production of the reservoir varies from 1 150 to 2 180 t, based on which the fish yield potential is estimated at 11.5 to 13.1 t. This is equivalent to 184 and 210 kg ha^{-1}. Keeping in view the uneven, weed-infested basin, and the past experience, 60% of the potential yield is fixed as achievable target.

Commercial fishing

Fishing rights of the reservoir are auctioned by public bidding. The catch details are presented in Table 12.8.

Table 12.8. Fish catch from Baghla reservoir

Year	Total fish production (t)	Yield (kg ha^{-1})	Percentage: Major carps	catfish	Misc.
1988-89	6.5	106	33.80	31.30	34.90
1989-90	5.1	102	43.40	29.30	27.30
1990-91	5.1	102	41.92	34.06	24.02

(CIFRI Barrackpore)

12.7 KEETHAM RESERVOIR

Keetham reservoir in Agra city was created in 1925. The 306 ha lake has a capacity of 9 million m^3. The maximum depth is 7.2 m and the mean depth varies from 2.5 to 3.5 m. The climate of Agra is tropical, characterised by hot and dry summer. The catchment area of the reservoir receives a rainfall of 50 to 75 cm.

Dwivedi and Chondar (1980) report high total alkalinity (152 to 176 mg l^{-1}) in the reservoir. The other water quality parameters recorded are dissolved oxygen (3.8 to 17.6 mg l^{-1}), free carbon dioxide (0 to 11.28 mg l^{-1}), pH (7.2 to 9.2) and temperature (17.4 to 36.0 ^{0}C). The reservoir bottom soil is loamy sand with pH of 9.2 and organic carbon at the rate of 0.68%. Judging by the water and soil quality attributes, the reservoir is productive. The average primary productivity is 25 to 30.6 mg C m^{-3} hr^{-1}.

Keetham attracted national attention during 1969-70 when a bumper fish production of 148.6 t was obtained from the lake. This unprecedented yield of 485 kg ha^{-1} was, however, short-lived and the yield rate started declining soon, reaching 25.7 t (84 kg ha^{-1}) during 1974-75. This was attributed to the erratic stocking schedule. At present, major carps constitute 12% of the catch. *Labeo gonius*, contributes 39% of the catch, followed by *Gudusia chapra* (30%), catfishes (8%) and other miscellaneous groups (11%). Among major carps, *Cirrhinus mrigala* (by number) and *L. rohita* (weight) dominate.

Mahajal (a giant drag net), *bisar jal* (scoop net), drift gill net and hooks and lines are the common gear employed in fishing. Drag nets mostly bring in bigger fish including major carps, minor carps and catfishes, while the scoop net is mainly used for catching *Gudusia chapra*.

12.8 FISH PRODUCTION TRENDS IN RESERVOIRS OF UTTAR PRADESH

Reservoirs of Uttar Pradesh have a relatively low fish yield. Examination of 31 small reservoirs covering 11 475 ha, under 10 districts showed yields ranging from1.1 to 227 kg ha^{-1} (Table 12.9). The widest range of 3.2 to 227.3 kg ha^{-1} is observed in Hamirpur, which has a high district average of 19.8 kg ha^{-1}. Bahraich district with 28 kg ha^{-1}, Allahabad with 17.4 kg ha^{-1} (13.6 to 30.9 kg ha^{-1}) and Jhansi with 21.2 kg ha^{-1} (5.8 to 34.8 kg ha^{-1}) also have high yields .

The 44 ha Kabarai reservoir in Hamirpur district yielded 10 t (227.3 kg ha^{-1}) of fish during 1992-93. It is the most productive reservoir in Uttar Pradesh. However, this is more of an exception than rule, as the next highest yield obtained at Kohargaddi, Gonda district is only 59.67 kg ha^{-1}. On an average, the small reservoirs produce 14.6 kg ha^{-1}.

Medium reservoirs (based on 13 reservoirs, covering 21 733 ha, located in seven districts across the State) have a yield range of 1.2 to 15.04 kg ha^{-1}. Rihand is the only large reservoir, for which recent production trends are available. This 46 539 ha reservoir produced 50 t of fish during 1992-93, with the yield of 1.07 kg ha^{-1} (Table 12.10).

Table 12.9. Fish yields of small reservoirs in Uttar Pradesh (1992-93)

District	Number of reservoirs examined	Yield range (kg ha^{-1})	Average yield (kg ha^{-1})
Varanasi	2	8.9-9.5	9.1
Mirzapur	6	1.4-34.6	7.6
Jhansi	3	5.8-34.8	21.2
Hamirpur	6	3.2-227.3	19.8
Banda	2	3.9-6.4	5.1
Allahabad	2	13.6-30.9	17.4
Eta	1	11.6	11.6
Bahraich	2	17.6-44.4	28.0
Gonda	5	1.1-59.7	16.6
Siddhardhnagar	2	15.9-29.1	18.9
Total	**31**	-	-
Average	-	-	***14.60***

Table 12.10. Fish yield rates in medium and large reservoirs of Uttar Pradesh

District	Reservoir	Yield rate
Medium		
Varanasi	Bhusakhand	2.49
-do-	Chandraprabha	2.70
-do-	Nowgarh	1.20
Sonebhadra	Ghagherol	8.72
-do-	Nagwa	1.84
Mirzapur	Ahrora	2.96
-do-	Mera	4.70
Jhansi	Pahari	15.04
-do-	Parwar	13.80
Hamirpur	Arjun	12.35
Banda	Ragwa	10.42
-do-	Varua	8.67
Bijnaur	Pilli	11.53
Medium reservoirs average:		*7.17*
Large		
Sonebhadra	Rihand	1.07

Table 12.11. Salient features of some reservoirs in Uttar Pradesh

	Rihand	Baigul	Gulariya
Date of closure	1962	1967	1969
Area at FRL (ha)	46 538	2 995	300
Minimum area (ha)	13 759	-	64-73
Average area (ha)	30 149	-	15-20
Maximum length (km)	40.25	-	10-16
Maximum width (km)	-	-	0.91-1.1
Mean depth (m)	22.8	-	3
Volume (million m^3)	10 625	-	9.24
Catchment area (km^2)	13 344	305	39.5
Elevation (m above MSL)	268	-	-
Length of shoreline (km)	561.33	-	1.2-1.8
Shore dev. index	7.04	-	-
Volume dev. index	0.89	-	-
Annual level fluctuations (m)	32	-	13
Inflowing river	Rihand	-	Gulariya
Outflowing river	Rihand	-	Gulariya
Maximum outflow ($m^3 sec^{-1}$)	209 317	-	340
Latitude (N)	24^0 1'	28^0 56'	25^0 12'
Longitude (E)	83^0 3'	79^0 40'	82^0 10'
Soil			
pH	6.0-8.5	-	6.5-7.1
$CaCO_3$ (%)	0.45-4.8	-	-
Available N (mg 100 g^{-1})	15.5-30.5	-	1.2-1.8
Available P (mg 100 g^{-1})	1.1-2.3	-	-
Water			
Water temperature (^{0}C)	19.9 - 31.1	14-33.5	18^0-34^0
pH	7.0 - 8.8	7.0-8.2	7.0-8.4
Transparency (cm)	1.8 - 28.5	2.4-23.1	10-104
DO (mg l^{-1})	3.0 - 10.0	5.9-11.0	5.0-9.2
CO_2 (mg l^{-1})	2-34	-	0-5.2
Total alkalinity (mg l^{-1})	28- 60	-	25-80
Spec. cond. (μmhos)	92.24	-	-
Total hardness (mg l^{-1})	8.4 - 21.6	-	13-40
Calcium (mg l^{-1})	6.2-14.8	-	12-26
Nitrate (mg l^{-1})	0.15 - 0.70	-	0.07-0.16
Phosphate (mg l^{-1})	tr.- 0.65	-	0.04-0.13
Silicate (mg l^{-1})	14.0	-	5.0-14.0
Chlorides (mg l^{-1})	6-20	7.4-19.8	2.2-4.8
Organic carbon (%)	-	-	2.2-4.8
GPP (mg C $m^{-2} d^{-1}$)	37.17-1022	-	-

Table 12.11 (cont.). Salient features of some reservoirs in Uttar Pradesh

	Baghla	Bachhra	Nanaksagar	Keetham
Date of closure	-	1980	1957	1925
Area at FRL (ha)	250	140	4 662	306
Minimum area (ha)	-	4	-	-
Average area (ha)	-	72	-	-
Mean depth (m)	3.9	5.22	-	2.5-3.5
Volume (million m^3)	9.6	7.42	-	28 937
Catchment area (km^2)	32.4	-	64	-
Elevation (m above MSL)	119	-	210	-
Inflowing river	Baghla nullah	Bachhra	Chuka	-
Water temperature (^{0} C)	18.0-21.0	-	29.5-33.6	17.4-36.0
pH	7.76-8.68	-	7.1-7.8	7.2-9.2
Transparency (cm)	50-150	-	100-160	-
DO (mg l^{-1})	8.80-12.80	-	6.6-7.8	3.80-17.0
CO_2 (mg l^{-1})	0	-	-	0-11.28
Total alkalinity (mg l^{-1})	-	72-167	-	152.0-176.0
Spec. cond. (μmhos)	63-82.1	-	-	-
Total hardness (mg l^{-1})	-	22-78	-	-
Calcium (mg l^{-1})	-	18-24.5	-	-
Nitrate (mg l^{-1})	0.29-0.36	0.085-0.18	-	-
Phosphate (mg l^{-1})	0.27-0.33	0.06-0.25	-	-
Silicate (mg l^{-1})	2.4-2.6	7.3-15	-	--
Chlorides (mg l^{-1})	-	-	8.8-15.0	-
Organic carbon (%)	-	-	-	-
GPP (mg C $m^{-2} d^{-1}$)	75-125	50-112	-	25.0-30.5

13. BIHAR

Bihar is a land-locked State, stretching from the foothills of Himalayas in the north to Orissa in the south. The west is bordered by West Bengal and the east by Uttar Pradesh and Madhya Pradesh. The most striking geographical feature of Bihar is the sharp division between the north and south. North Bihar is an almost entirely level tract comprising the Ganga, its tributaries on either side and the maze of floodplains and wetlands. This alluvial land is extremely fertile, watered by the Sarayu, the Gandak and the mainstream Ganga. Southern Bihar, especially in and around the districts of Chhotanagpur and Santhal Parganas, is thickly wooded and consists of a succession of hills. The altitude varies from 300 to 1 300 m, the highest peak being 1 372 m above the sea level.

Palamu district is drained by the river North Koel, which joins the Son, a major tributary of the Ganga. The two main rivers originating in the Chhotanagpur hills and flowing down east are the Damodar and Subarnarekha. Both have been harnessed for irrigation and power generation. The Sankh, draining the Gumla district and the South Koel and Karo draining Gumla, Ranchi and Singhbhum districts are the tributaries of Brahmani, flowing to Orissa.

13.1 RESERVOIR FISHERIES RESOURCES OF BIHAR

Reservoirs, as a potential fishery resource, are yet to gain importance in Bihar. Documentation on reservoir fisheries resources in the State is not very comprehensive and exact, with the various estimates on the water area at variance with one another. Ahmed and Singh (1992) compiled a list of 105 reservoirs in the State, which is very close to the reality. Total area under reservoirs in the State is estimated at 96 695 ha (125 reservoirs), the share of small, medium and large reservoirs being 12 461 ha (112 reservoirs), 12 523 ha (5 reservoirs) and 71 711 ha (8 reservoirs) respectively. The landscape of Palamu district, with a large number of small streams radiating down to the North Koel, provides an ideal setting for small dams and 58 small reservoirs are situated in this district (Table 13.1). Ranchi, Monghyr, Bhagalpur and Gumla districts also have many small reservoirs. Large reservoirs (> 5 000 ha) are situated in Hazaribagh, Dhanbad (two each), Santhal Parganas, Gumla, Ranchi and Giridih (one each) districts (Fig. 13.1). Except the Latratu reservoir in Ranchi and Maithon reservoir in Dhanbad district, all the large reservoirs are in the <10 000 ha group.

Of the 125 reservoirs in the State, 112 (90%) are small and the water area occupied by them is only 13% of the total. The medium reservoirs also occupy similar percentage and thus 74% of the waterspread is under large reservoirs.

No major scientific studies on the reservoir ecosystems and their production functions have been done in Bihar, with the exception of the research conducted by the Central Inland Fisheries Research Institute in the Damodar Valley Corporation (DVC) reservoirs an Getalsud reservoir during the 1960s and 1970s respectively.

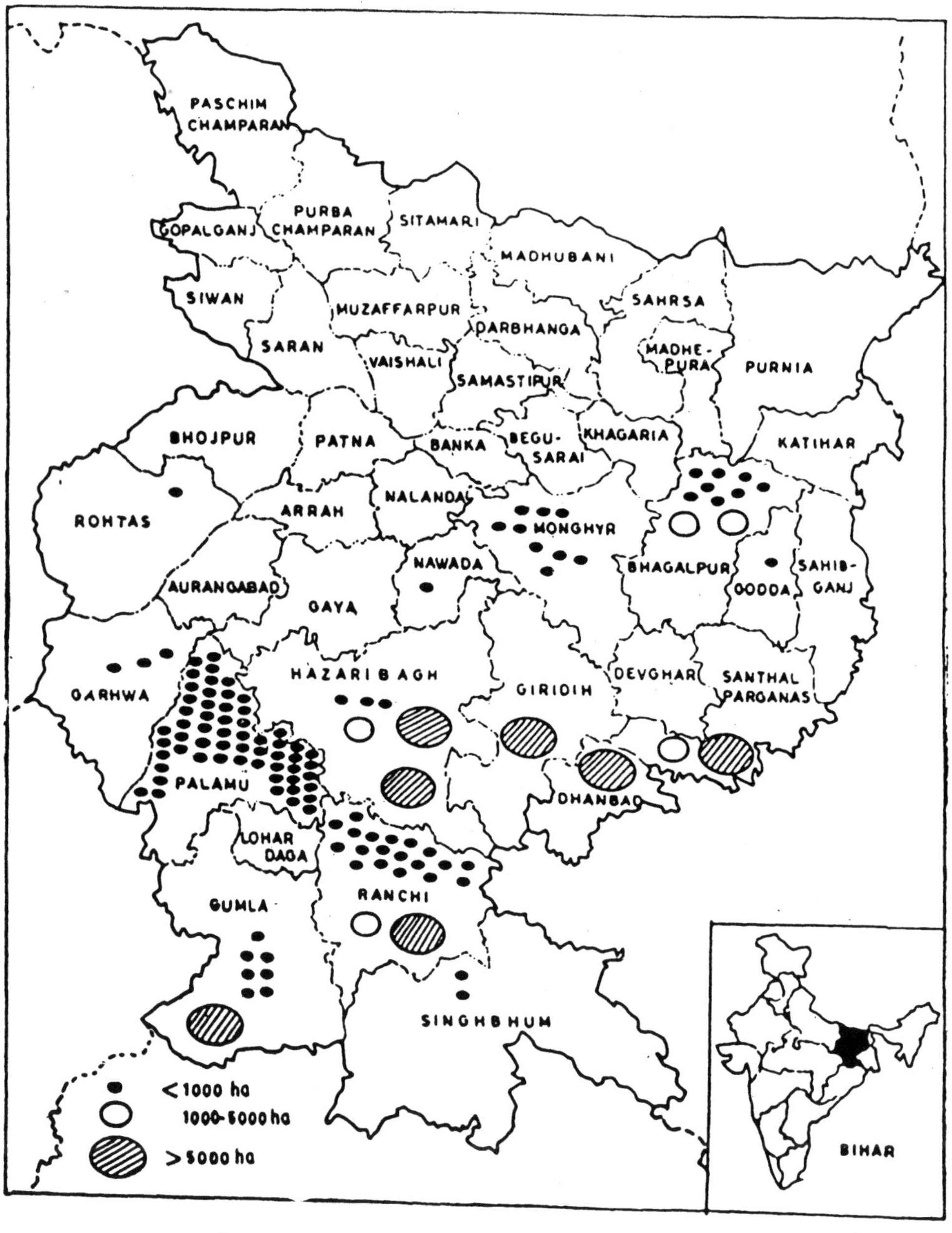

Figure 13.1. Distribution of reservoirs in Bihar

Table 13.1. Reservoirs of Bihar

District	Small < 1000 ha		Medium 1 000-5 000 ha		Large > 5000 ha		TOTAL	
	No. of units	Area (ha)	No. of units	Area (ha)	No. of units	Area (ha)	No. of units	Area (ha)
Santhal Parganas	-	-	1	3 846	1	10 000	2	13 846
Garhwa	3	340	-	-	-	-	3	340
Bhagalpur	8	1 164	2	2 385	-	-	10	3 549
Ranchi	19	3 363	1	3 500	1	16 000	21	22 863
Hazaribagh	3	243	1	2 792	2	13 209	6	16 244
Giridih	-	-	-	-	1	6 000	1	6 000
Monghyr	8	1 999	-	-	-	-	8	1 999
Godda	1	257	-	-	-	-	1	257
Dhanbad	-	-	-	-	2	19 002	2	19 002
Lohardaga	1	305	-	-	-	-	1	305
Rohtas	1	513	-	-	-	-	1	513
Nawada	1	11	-	-	-	-	1	11
Singhbhum	2	161	-	-	-	-	2	161
Palamu	58	1 994	-	-	-	-	58	1 994
Gumla	7	2 111	-	-	1	7500	8	9611
Total	*112*	*12 461*	*5*	*12 523*	*8*	*71 711*	*125*	*96 695*

Compiled from Ahmed and Singh (1992), supplemented by data provided by the Director of Fisheries,

13.2 GETALSUD RESERVOIR (Fig. 13.2)

Getalsud reservoir is located at 23^0 27' N and 85^0 33' E, across the river Subarnarekha, 40 km east of Ranchi. Completed in 1971, the multi-purpose reservoir supplies water to Ranchi town, and for the industrial needs of the Heavy Engineering Corporation and other factories. The reservoir has a maximum surface area of 3 459 ha and a capacity of 288.5 million m^3 at the FRL of 595.70 m above MSL. At the lowest level, the area is only 987 ha. River Subarnarekha, the main source of inflow, originates at Nagari, in the Chhotanagpur plateau of South Bihar, about 50 km upstream of Getlasud dam. Getalsud was studied by the CIFRI under the All India Coordinated Project on Reservoir Fisheries (Anon, 1984c).

The river has a total catchment area of 15 540 km^2. The local catchment receives rainfall of 100 to 1 198 cm. The reservoir has a shoreline of 80 km and the shoreline development index is 3.85. Maximum depth being 32.6 m, volume development index is less than unity (0.6839) and the mean depth is 8.46 m.

The air temperature at the reservoir drops down to 4.9 ^{0}C in winter and the summer temperature reaches 40.3 ^{0}C. These extreme variations in air temperature have an impact on the thermal regime of water. The winter water temperature is close to 16 ^{0}C and when the surface water starts heating up during summer, a thermocline is formed with a cool water layer in the depth. The temperature difference between the epilimnion and hypolimnion often reaches 7 ^{0}C. In 1980, the thermal discontinuity formed between 9 and 10 m with a temperature difference of 2 ^{0}C. The difference between the surface and bottom is 5.5 ^{0}C (Anon., 1984).

The bottom soil is acidic and poor in organic carbon (0.46 to 0.60 %). Available phosphorus and nitrogen are present in modest quantities (2.45 to 6.08 mg 100 g^{-1} and 24.38 to 35.06 mg 100 g^{-1}).

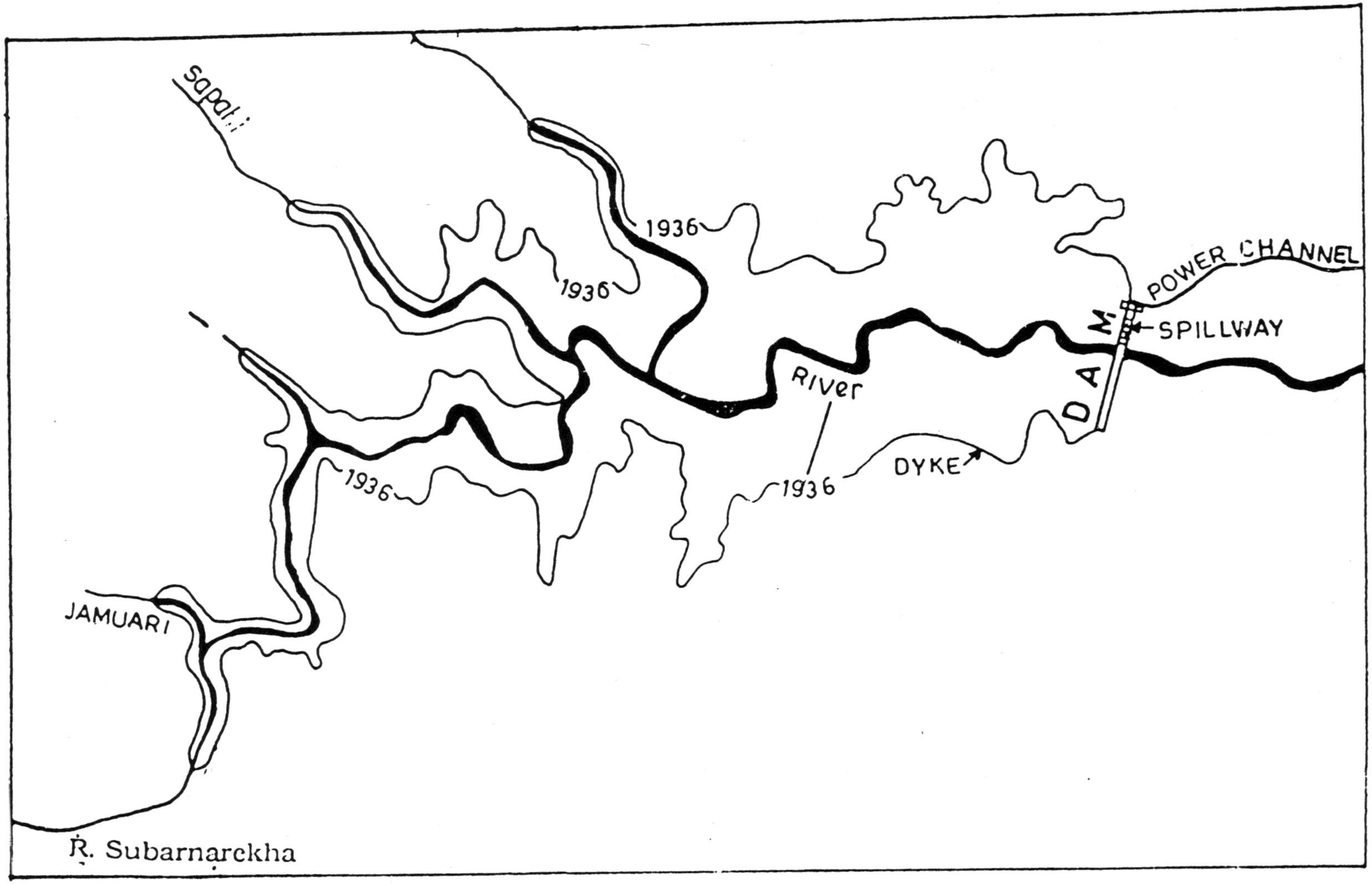

Figure 13.2. Getalsud reservoir, Bihar

The morphometric and edaphic indicators of productivity and the estimates on the standing crop send conflicting signals on the status of the water body. A low shoreline development index and nutrient status in the water and sediment phases suggest oligotrophy, while the klinograde distribution of oxygen, accompanied by an increase in carbon dioxide towards the deeper layers clearly suggest high photosynthetic and nutrient releasing activities.

The river Subarnarekha receives factory effluents from the Heavy Engineering Corporation and sewage wastes emanating from the industrial township. A drop in dissolved oxygen, accompanied by decrease in plankton, has been reported below the discharge point. The water recovers its quality further downstream.

An average gross primary production of 289.8 and net production of 154.4 mg C m^{-2} day^{-1} have been reported in the lentic sector of the reservoir, compared to 257.85 and 137.12 mg C m^{-2} day^{-1} in the intermediate sector and 295.5 and 150.64 mg C m^{-2} day^{-1} in the lotic sector. The mean gross and net production rates are 281.07 and 147.4 mg C m^{-2} day^{-1}.

Plankton

The moderately rich plankton community of the reservoir occasionally produces blooms. The mean density of plankton varies between 129 and 5 227 individuals l^{-1}. During the blooms, it gmay reach 55 777 units l^{-1} (Table 13.2). The total plankton count in the lotic sector is higher than that of the lentic sector, suggesting a subdued production process at the main reservoir, due to lack of autochthonus enrichment. Good standing crop of plankton in the lotic sector is attributed to the effects of sewage effluent discharge into the upstream stretches of the river.

Numerically, phytoplankton contributes 72.1 to 94.6% to the total plankton in the lotic sector, 70.44 to 94.27% in the intermediate sector and 54.56 to 87.79% in the lentic sector. Diatoms have the maximum species diversity, followed by Chlorophyceae and blue- greens. Desmids and Dinophyceae (*Ceratium* sp.) are also present. Zooplankton is represented by 12 genera of rotifers, 6 of protozoa, 5 cladocerans and 2 of copepods.

Benthic fauna

Bottom substrate in the lotic sector is soft and rich in organic matter, comprising decaying vegetation, but during monsoons, silt deposition is reported to blanket the organic matter. Density of benthic organisms is high in the lotic sector, their number attenuating towards the lentic sector. During 5 years of observations from 1975-76, abundance of benthic invertebrates varied from 353 to 1 693 units m^{-2} (average 978.4) in the lotic, 75 to 348 units m^{-2} (average 217.2) in the intermediate and 81 to 168 units m^{-2} (average 116.8) in the lentic sectors. Oligochaetes (*Tubifex, Nais*), Diptera (*Culiocoides, Dixa, Chironomus, Chaoborus*), Anisoptera (*Aphylla*), Trichoptera (*Philopotamus)* Coleoptera (*Promoresia*) and molluscs (*Corbicula, Pisidium, Goniobasis, Sphaerium* and *Gyralus*, constitute the benthic macrofauna. Rich deposition of periphyton in densities ranging from 690 to 1 745 units cm^{-2} has been reported on artificial substrata provided on experimental basis.

Forty-three species of fishes have been recorded from the reservoir. They include 24 species of Cyprinidae and 6 species of catfishes belonging to Siluridae, Bagridae, Clariidae and Heteropneustidae. Commercial fishing in Getalsud reservoir using gill nets was initiated by the Fisheries Department of the State Government in 1976-77 and continued up to June 1979. The reservoir was auctioned on 1st July 1979 for a period of 3 years.

Table 13.2. Plankton density (by number) in Getalsud reservoir

	1976-77	1977-78	1978-79	1979-80
Lotic sector	5 227 (55 777)	2 275 (18 876)	415 (1 236)	1 568 (11 957)
Intermediate sector	310 (728)	200 (701)	437 (1 171)	809 (4 375)
Lentic sector	337 (963)	129 (251)	180 (583)	446 (1 446)

*Figures within brackets indicate the maximum values recorded during the year

Fish and fisheries

The contractor deployed both gill and drag nets. Normally, four parties operate gill nets each having 4 men in one boat. The boats carry 40 nets, each measuring 30 m in length. A total of 4 800 m of gill nets in the mesh bar range of 70 to 150 mm are used. Major carps are mostly caught by the gill nets.

A normal unit of drag net, measuring 200 x 6 m, is operated by a group of fishermen with the help of a country boat. Drag netting, generally done during night between 2100 to 0400 hrs., is reported to be much more profitable in Getalsud reservoir, as large quantities of small fishes are caught.

Fishing is continued in the reservoir till the contractor is satisfied with the catch and thereafter moves his men and equipment to some other reservoir. The fish catch is brought to Ranchi and then handed over the retailers. As per information supplied by the State Fisheries Department, the commercial catch during 1976 to 1981 was in the range of 481 to 5 980 t (Fig. 13.3).

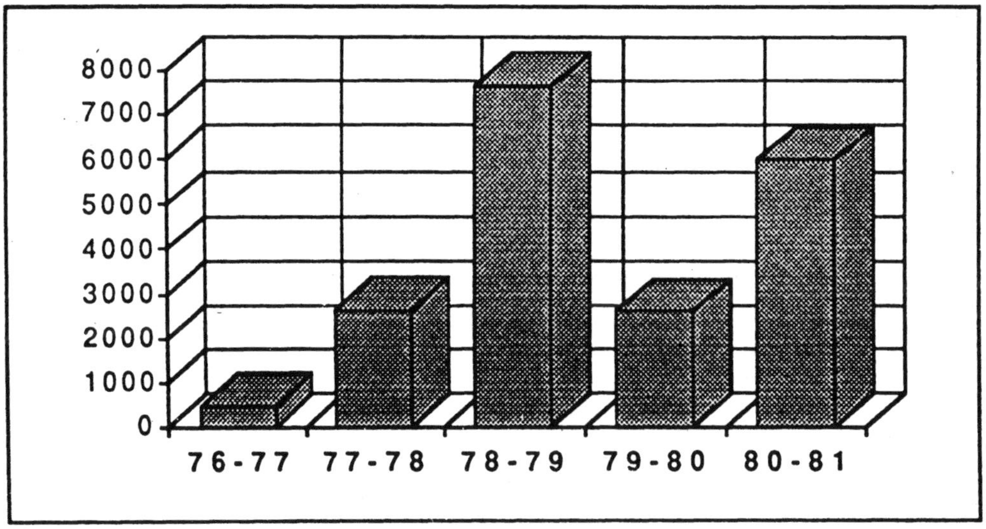

Figure 13.3. Fish catch (t) from Getalsud reservoir

Effect of impoundment on reproduction and recruitment

About 50 km upstream of Getalsud reservoir, the river Subarnarekha is dammed to form the Hatia reservoir. Jamuar, a tributary of the Subarnarekha joining the river between the dam and the confluence of Hatia, is also closed by a dam. When the southwest monsoon starts, the two upstream reservoirs discharge excess water towards Getalsud and this flowing water attracts brooders. However, they never reach breeding grounds and thus, the two dams above have an adverse effect on the spawning of fish in Getalsud reservoir.

Attempts to collect spawn from Jamuar were made in July and August during 1975 to 1979 and in 1981. The first indication of breeding of Gangetic major carps during the flood season was noticed during 1981, when 28.6% of the reared hatchlings were found to be *Labeo calbasu* and the rest minor carps. However, mostly, the spawn comprised fertilised eggs of minor carps and a few young ones of *Cirrhinus mrigala* . Bulk of the eggs, on rearing, turned out to be *Cirrhinus reba.* Operation of shore seines in various parts of the reservoir yielded hatchlings and fry of minor carps and trash fishes with no trace of major carps, indicating their poor recruitment.

Stocking

Stocking of Getalsud reservoir was initiated in 1974-75 and continued until 1980-81. During this period, a total of 3.35 million seed were stocked (Table 13.3). This included 120 992 minor carps stocked incidently. On the basis of average reservoir area, the stocking density was 203 ha^{-1}. Fingerlings numbering 2 680 of silver carp, *Hypophthalmichthys molitrix* were introduced into the reservoir in 1974. Based on the recovery of six specimens between 1974 to 1976, the growth of this fish was 715 mm and 4.5 kg in 866 days (Table 13.4).

Table 13.3. Stocking in Getalsud reservoir

Year	Major carps	Minor carp	Misc.	Total
1974-75	98 784	-	-	98 784
1975-76	159 700	19 832	248	179 780
1976-77	715 000	-	-	715 000
1977-78	849 975	-	86 775	915 000
1978-79	757 495	101 160	-	858 655
1979-80	281 112	-	-	281 112
1980-81	280 000	-	-	280 000
Total	***3 142 066***	***120 992***	***87 023***	***3 350 081***

The river Damodar used to be described as the sorrow of Bengal due to the frequent devastation it caused by flooding. Four impoundments were created on the river system under the Damodar Valley Corporation (DVC) for power generation during the 1950s, which moderated the flow regime. These four man-made lakes, *viz.*, Konar, Tilaiya, Maithon and Panchet are collectively referred to as DVC reservoirs (Fig. 13.4). While Panchet is situated on the main river Damodar, Tilaiya and Maithon are on the tributarly, Barakar. Konar reservoir is on the river Konar, another tributary of the Damodar. The four reservoirs cover an area of 20 116 ha.

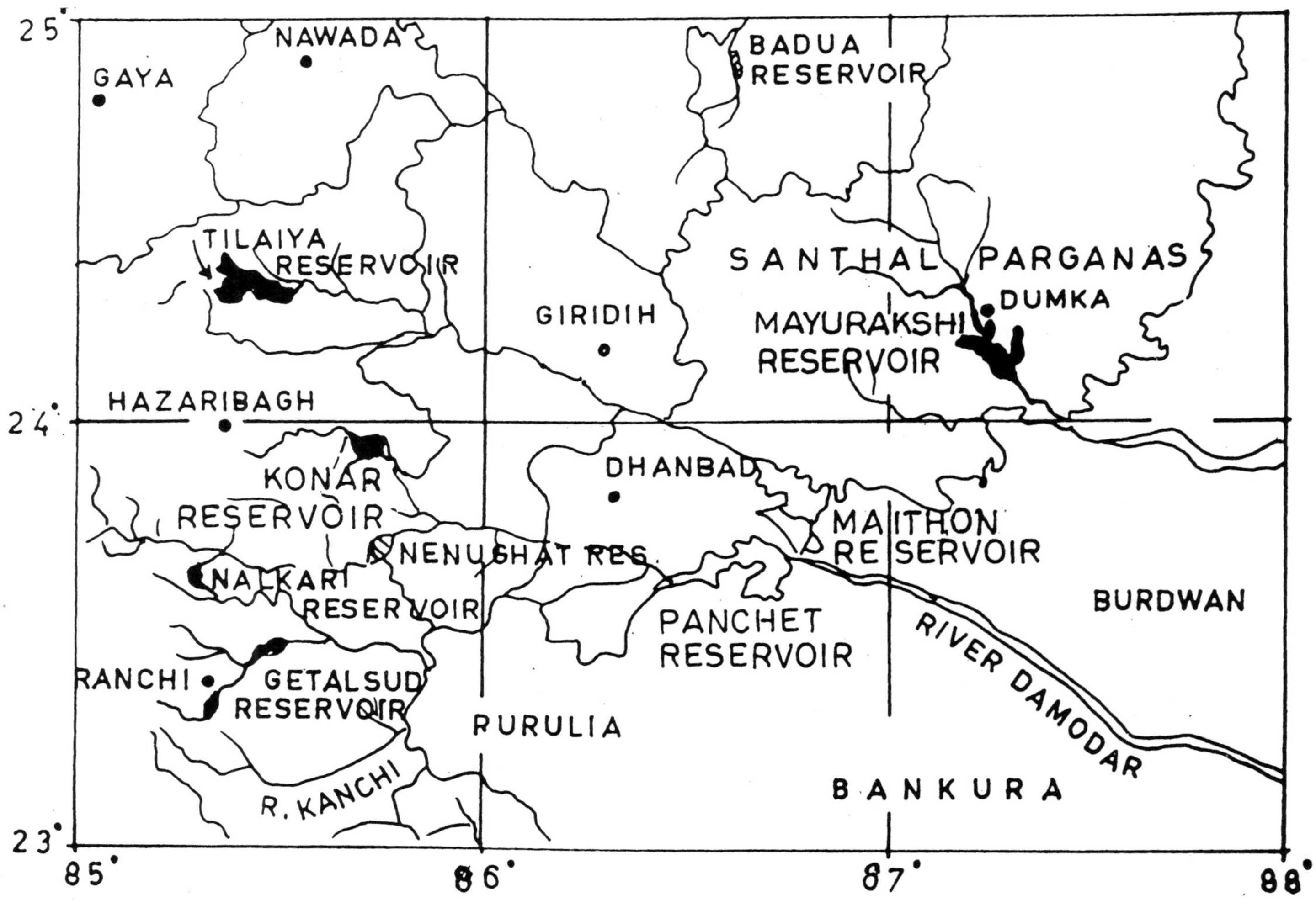

Figure 13.4. Damodar Valley Corporation (DVC) reservoirs, Bihar

Table 13.4. Growth of silver carp in Getalsud reservoir

Date of capture	Size on capture length (mm)	weight (g)	Increase in size length (mm)	weight (g)	Period of growth (days)	Growth rate day^{-1}) (mm)	(g)
Average initial size: 77 mm / 7 g.							
10.9.74	375	400	298	393.2	178	1.67	2.20
23.8.75	583	2 100	506	2 093.2	525	0.96	3.98
12.6.76	750	4 750	673	4 743.2	819	0.82	5.79
25.7.76	785	4 280	708	4 273.2	862	0.82	4.95
29.7.76	785	4 250	708	4 243.2	866	0.82	4.89
29.7.76	792	4 550	715	4 543.2	866	0.82	5.24

Date of release of silver carp in Getalsud reservoir - 15.3.1974

THE DVC RESERVOIRS

13.3 KONAR (Fig. 13.5)

Completed in 1955, Konar reservoir covers 2 590 ha at FRL. The inflowing river Konar is a seasonal stream joining the river Damodar. Konar, has a catchment area of 997 km^2, comprising thick jungles, wastelands and cultivated areas. At FRL, the reservoir has a capacity of 336.16 million m^3 and a mean depth of 12.97 m. The shoreline is 106 km long, the shoreline development index 8.78, volume development index 0.85. Basin soil of Konar reservoir is sandy loam to clay with 0.3 to 0.9% of organic carbon. The concentration of available phosphorus is a low 1.2 to 8.0 mg P_2O_4 100 g^{-1}. Organic nitrogen of 0.069 to 0.101% is medium to high and pH of soil is acidic (5.4 to 6.0).

Water temperature of Konar reservoir varies from 17.8 to 28.9 °C (Ramakrishniah and Sarkar, 1982). Thermal stratification develops during summer (Natarajan, 1976). Winds prevailing during summer (5 to 8 km hr^{-1}) can disturb only the epilimnetic layers which extend down to 9 m depth. A relatively stable warm hypolimnetic layer aids chemical stratification. While the inflowing floodwaters break the thermocline, the stronger monsoon winds of up to 11 km hr^{-1} help the water circulation (Ramakrishniah and Sarkar, 1982). Although depletion of oxygen is reported from the hypolimnion, stratification in terms of bicarbonates, free carbon dioxide and pH is not recorded.

Among the DVC reservoirs, Konar is the richest in ionic concentration. The total alkalinity varies in the range of 43.5 to 83.0 mg l^{-1} and specific conductivity within 81.20 to 147.6 µmhos during different seasons of the year. Nitrate (0.10 to 0.6 mg l^{-1}) and phosphate (traces to 0.047 mg l^{-1}) are high during the monsoon and post-monsoon months. Silicate (1.3 to 1.98 mg l^{-1}) is, however, poor.

The planktonic abundance peaks during August, in conformity with many other Indian reservoirs. The plankton in Konar is characterised by a poor species diversity and an overwhelming dominance of *Microcystis aeruginosa* and *Diaptomus* sp. among the phyto- and zooplankton respectively. During the peak periods of abundance, the phytoplankton count goes up to 5 256 units l^{-1} in the intermediate sector of the reservoir. The highest density was recorded (3 570 individuals l^{-1}) during the summer of 1973. *Anabaena, Merismopedia, Botryoccocus, Spirogyra, Oedogonium,*

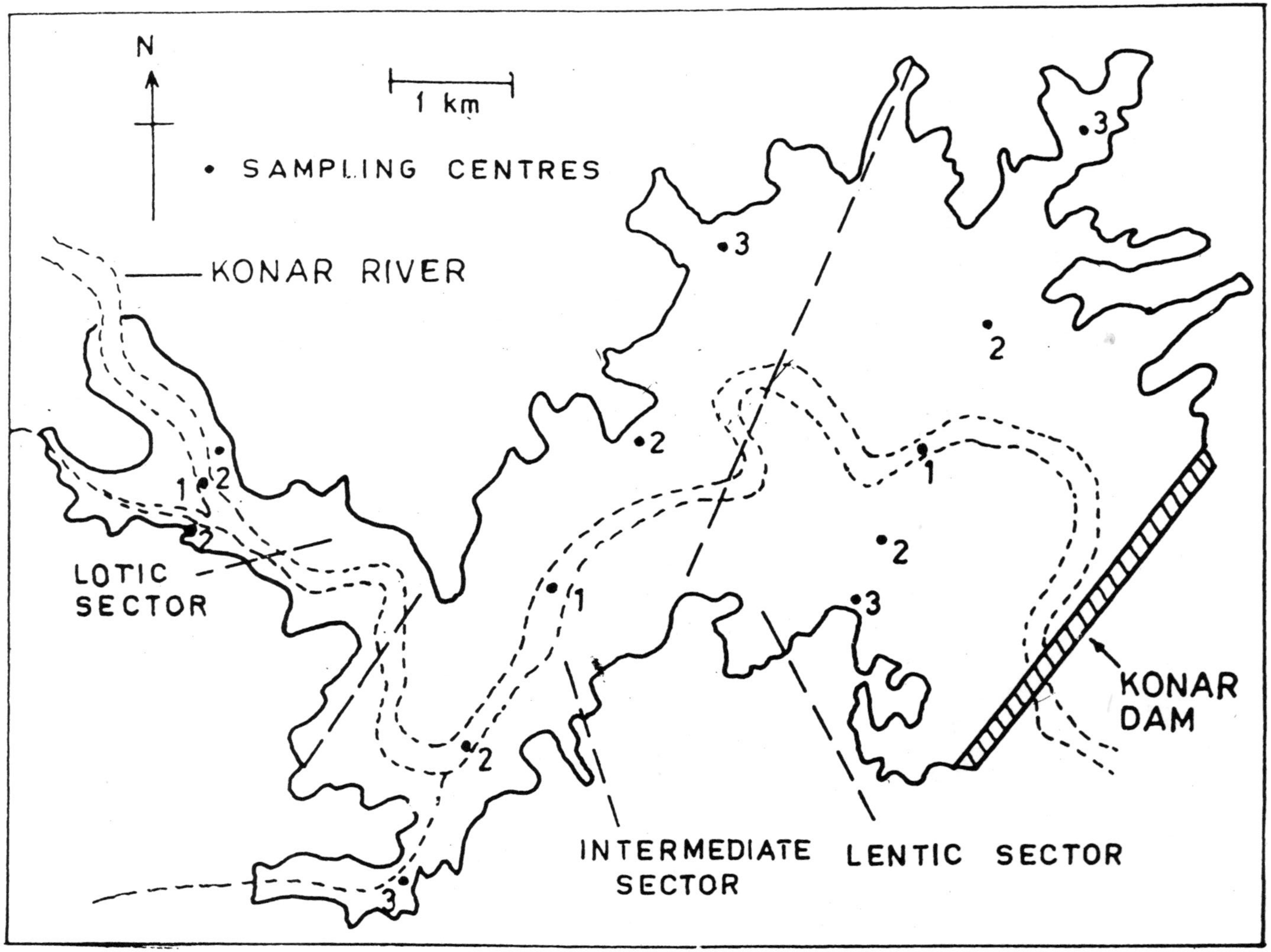

Figure 13.5. Konar reservoir, Bihar

Pleodorina, *Ceratium* and *Peridinium* are reported by Ramakrishniah and Sarkar (1982). Zooplankton comprises *Difflugia*, *Arcella*, *Actinosphaerium*, *Keratella*, *Brachionus*, *Filinia*, *Polyarthra*, *Diaphanosoma*, *Ceriodaphnia*, *Cyclops* and *Diaptomus*.

Construction of the dam and the consequent lake formation have brought radical changes in the fish composition. Prior to the damming, Konar was a torrential stream sustaining flood fisheries, such as minor carps and weed fishes. Job *et al.* (1952) reported 12 species of fishes before the impoundment. During the post-impoundment phase, 20 additional species of weed fishes have established themselves in the reservoir. Most of them were probably introduced accidently with the stocking material. The absence of large predatory catfishes has resulted in low mortalities of these uneconomic species chiefly, *Ambassis nama*, *Esomus dandrica*, *Oxygaster bacila*, *Amblypharyngodon mola*, *Osteobrama cotio*, *Puntius ticto*, *P. sophore* and *Barilius barna*.

Fisheries management

Stocking in Konar reservoir was initiated in 1957 and continued until 1959. During this period, a total of 0.55 million fingerlings of Gangetic major carps were stocked. Virtually, no stocking of carps was done during the 1960s and 1970s, except the release of 125 000 fingerlings of catla (in 1968) and 64 000 mrigal (in 1970).

The annual fish production from Konar was 5.7 t in 1968, 2.4 t in 1969, 1.7 t in 1970 and 2.9 t in 1971 with corresponding yields of 3.6, 1.5, 1.1 and 1.1 kg. The constituent species were *Catla catla*, *Cirrhinus mrigala*, *Labeo calbasu*, *L. rohita*, *L. bata*, *L. dyocheilus*, *Puntius sarana*, *Cyprinus carpio*, *Notopterus notopterus* and *Ompok bimaculatus*, major carps contributing 90.54 to 94.76% of the catch.

There has been a decline in fish yield since 1968. The yield of catla declined sharply from 1.6 t in 1968 to 0.49 t in 1969, but improved to 1.1 t in 1970 and 1.8 t in 1971 as a direct result of stocking done in 1968. Jhingran and Natarajan (1969) rated the recruitment of catla as very feeble and felt that stocking on a continuous basis was essential. Mrigal showed a steep decline in the reservoir and its recruitment was described as moderate to poor. A similar decline in the stock of *L. calbasu* was noted, when the catch dropped from the level of 1.3 t in 1968 to 0.717 t in 1969, 0.2 t in 1970 and 0.3 t in 1971, due to an unsteady recruitment.

Jhingran and Natarajan (1969) estimated a catch net^{-1} (15.25 m length, 4.25 m depth and 635 to1270 mm mesh bar) of 0.353 kg and a total catch of 5.6 t during 1962-63. During the next two years, due to increased fishing effort (which included the fishermen deployed from Mettur), the catch went up to 23 t with a matching increase in catch per effort. However, in 1967-68, with a nearly identical fishing effort, the production fell to 10 t and the catch per unit of effort decreased to 0.226 kg clearly showing overfishing. Catla was the dominant component of catch, especially in the nets laid by the Mettur fishermen. By 1968, the total catch and catch per net decreased to 1 t and 0.075 kg respectively in the nets of Mettur fishermen. Nevertheless, the average weight of catla went up to 12.66 kg with a modal value of length at 900 mm (7 to 8 years old). The above observations confirm the belief that the stock of catla was thinning and the average size and age increasing. Changes in species spectrum over the years (1963-64 to 1967-68) indicate a decline in the populations of catla and an increase in *C. mrigala* (Table 13.5).

Fish carrying capacity of Konar was estimated at 40 kg ha^{-1}, which at biomass/yield ratio of 1.0 : 0.6, can yield 25 kg ha^{-1} (Anon., 1968).

13.4 TILAIYA

Tilaiya dam, constructed across the river Barakar was sealed in 1952. The reservoir has an area of 5 921 ha (Fig. 13.6) at FRL with a volume and mean depth of 394 million m^3 and 6.65 m respectively. The shoreline is very irregular with a shoreline development index of 9.1. The volume development index is 0.766. The river rising from the hilly forests of Hazaribagh district, at an elevation of 610 m has a catchment area of 984 km^2, comprising mainly forests, pastures, cultivated land and wastelands. The annual rainfall in the area is 127 cm.

Table 13.5. Catch composition in Konar reservoir (%)

	1963-64	1964-65	1966-67	1967-68
C. catla	97.22	64.26	31.59	33.38
L. rohita	0.30	6.63	5.24	3.94
L. calbasu	1.07	17.85	26.24	16.31
L. bata			0.92	6.41
C. mrigala	1.30	7.23	32.92	38.78
C. carpio			1.90	1.03
P. sarana			0.46	0.17
Notopterus spp.			0.20	0.14
Channa spp.			0.52	0.52
Miscellaneous fishes	0.09	4.21		
Total landings (t)	**23.024**	**22.714**	**6.550**	**10.291**

The bottom soil is acidic (pH 5.6 to 5.9) and silty clay loam with low organic carbon (0.31 to 0.50%) and available phosphorus (3.0 mg P_2O_5 100 g^{-1}). Organic nitrogen is in medium productivity range (0.073 to 0.101%). Water is rich in ionic concentration. Alkalinity (53.5 mg l^{-1}) and specific conductivity (98.7 to 21: µmhos) are indicative of the the reservoir's proclivity to good organic productivity. Values of nitrate (0.25 mg l^{-1}) and phosphate (0.03 mg l^{-1}) are satisfactory. Gross primary productivity is estimated at 199 to 300 mg C m^{-3} day^{-1}.

Species composition of the fish catch during 1960 to 1968 is detailed in Table 13.6. Gopalakrishnan *et al.* (1966) observed large-scale breeding of Indian major carps in the reservoir. Ramakrishniah and Banerjee (1979) estimated 11.30% of the total spawn collected from the breeding grounds connected with the reservoir to be that of *L. rohita*, *C. mrigala* and *L. calbasu*. Job *et al.* (1952) recorded 40 species of fishes belonging to 7 families in the river stretch before the impoundment. Among them, *Labeo dero*, *Puntius chola*, *P. conchonius*, *Crossocheilus latius*, *Carra mullya*, *G. gotyla*, *Macrognathus aculiatus*, *Oxygaster clupeoides* and *Aspidoparia morar* have disappeared after the reservoir formation. The new species appearing after the reservoir formation include *Notopterus notopterus*, *N. chitala*, *Wallago attu*, *Ompok bimaculatus*, *Rhinomugil corsula*, *Ambassis nama*, *A. ranga*, *Eutropichthys vacha*, *Labeo bata*, *Ailia coila*, *Mastacembelus armatus* and *M. pancalus*.

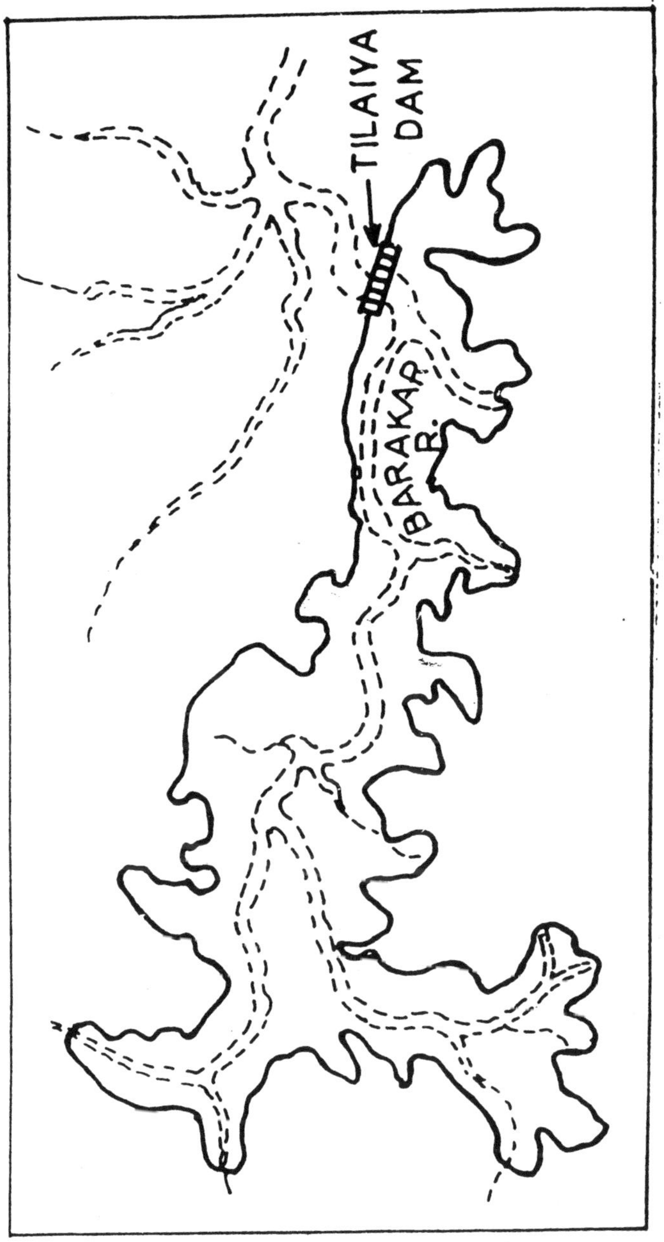

Figure 13.6. Tilaiya reservoir, Bihar

In Tilaiya reservoir, 181 449 fingerlings of mrigal, 103 480 fingerlings of rohu and 27 211 fingerlings of catla were stocked in the year 1968. This heavy stocking did not have any impact on the fishery till 1970 and 1971, largely due to the formidable presence of the predatory catfish *W. attu*. Other predators such as *Notopterus chitala* and *Barilius bola* also were common in the reservoir. Experimental fishing carried out in Tilaiya showed that mesh bars 45, 50, 55 and 60 mm were very effective for catching *W. attu* and 30 and 35 mm for *Barilius bola*.

The total fish landing from Tilaiya reservoir was estimated at 9 t in 1968 (six months only), 13.4 t in 1969, 12.2 t in 1970 and 13.5 t in 1971. Species that figured prominently in the catch were *C. catla, Cirrhinus mrigala, C. reba, Labeo calbasu, L. rohita, L. bata, L. boggut, L. dyocheilus, Puntius sarana, Barilius bola, Wallgo attu, Mystus cavasius, Aorichthys seenghala, Notopterus notopterus* and *Rhinomugil corsula*. Among them, the share of major carps ranged between 57.54 and 76.61%. The fish yield rate from the Tilaiya was 3.6, 3.3 and 3.6 kg ha^{-1} in 1969, 1970 and 1971 respectively.

The fishing effort in Tilaiya in terms of units of 15.25 m length of gill net was estimated at 55 477 units in 1969 and 69 660 units in 1970. The total catch per unit of effort was 241 g in 1969 and 170 g in 1970. The catla in Tilaiya shows signs of reduction in stocks as reflected by the catch trends (effort presumed nearly uniform) in 1969 (2.1 t), 1970 (1.3 t) and 1971 (2.8 t). Jhingran and Natarajan (1969) emphasised the need of stocking of catla for a better yield.

Table 13.6 Species composition of Tilaiya reservoir (%)

	1960	1961-62	1967-68
C. catla	20.40	45.28	32.75
C. mrigala	50.26	29.54	43.13
L. rohita	21.76	17.02	10.19
L. calbasu	7.58	8.16	3.13
L. bata	-	-	0.23
P. sarana	-	-	0.06
T. tor	-	-	0.04
W. attu	-	-	8.91
Notopterus sp.	-	-	0.43
Miscellaneous	-	-	1.13
Total landings (t)	4.604	3.992	22.473

There was no decline in the stock size of mrigal during 1969-71. The stocking of this species in 1968 helped maintain the stocks in 1970 and 1971. Spawning of mrigal was poor in 1968. There were indications of poor recruitment of calbasu and intensive stocking of this species was recommended. The stocks being low, intensive stocking of this fish was recommended. The low stock size of rohu in Tilaiya was attributed to spawning failure. Stocking of fingerlings of this fish in 1968 did not improve the fishery in 1970 and 1971. It seems that rohu was unable to adjust to the ecological conditions of Tilaiya.

Guidelines for fisheries development of Konar and Tilaiya reservoir

Cyanophyceae constitute the dominant abundant part of plankton in Konar and Tilaiya and these algae also dominate the guts of catla collected from Konar. However, the levels of utilization of the blue-green algae as food is not known. The indigenous fish *Thynnichthys sandhkol* and the exotic silver carp are known to feed on planktonic blue-green algae, although these fishes are not favoured by consumers. The Central Inland Fisheries Research Institute has formulated a set of guidelines for the fisheries development of Konar and Tilaiya reservoirs, which are summarised below:

1. Despite the recruitment failure of Indian major carps in Konar and Tilaiya reservoirs, the spawn of some of these carps occur in the form of eggs. These may be salvaged, reared in the nursery facilities available in Bacchai fish farm attached to Tilaiya and released into the reservoir. A similar procedure can be adopted in Konar after creation of nursery facilities near the reservoir.

2. Trash and uneconomic fishes form dense populations in both Konar and Tilaiya reservoirs and some of them compete with major carps for food. Elimination of these fishes by operation of drag nets has been suggested to improve the productivity of the reservoirs.

3. Dense population of predatory fishes like *W. attu, N. chitala* and *Barilius bola* occur in Tilaiya which should be controlled for greater survival of stocked fry and fingerlings of economic fishes. Experimental fishing with gill nets showed that *W. attu* of the sizes that occur in Tilaiya were vulnerable to mesh bars of 45, 50, 55 and 60 mm. Similarly, 30 and 35 mm mesh bars were found very effective for *Barilius bola*. Long line fishing may also be attempted for the control of *W. attu.*

4. Charting out of the optimum concentration of fish stocks in various zones in different months has been done in Konar and Tilaiya reservoirs. Fishing should be re-orientated in the light of the above findings to improve the fish productivity and catch per unit of effort.

5. Recruitment of catla, mrigal, calbasu and rohu being poor in Konar, stocking of these fishes should be done on a continuous basis. Similar stocking is needed in Tilaiya. There are indications that rohu is unable to adjust to the ecological conditions of Tilaiya. Therefore, this fish may be excluded from stocking.

6. Size limits (Table 13.7) have been suggested for the capture of major carps in Konar and Tilaiya and suitable mesh sizes prescribed for them.

Table 13.7. Mesh size prescribed for different species of fishes in Konar reservoir

Fish	Prescribed mesh bar *(mm)*	Size *(mm)*
C. catla	91	587
C. mrigal	46	460
L. calbasu	52	375
L. rohita	*	450

* Information not available

(After Natarajan, 1976)

7. Flash floods and unfavourable spawning habitat on the upstream stretches of Barakar and Konar do not favour normal breeding of major carps. This renders observance of closed season redundant in the two reservoirs.

8. Selective clearance of reservoir bed by removing tree stumps and other obstructions is sugggested at least upto minimum draw-down limit in Konar reservoir for facilitating better fishing operations.

13.5 MAITHON RESERVOIR

Created in the year 1957, the Maithon reservoir covers 11 490 ha. Stocking of the reservoir with the fingerlings of Indian major carps commenced in 1958 and continued in successive years. Jhingran and Natarajan (1969) observed during 1967-68 that the stocking did not have any impact on the catch even after a decade. The catch did not rise commensurate with the increase in efforts. During 1962-63 to 1967-68 the catch increased only by 4 t (9 to 13 t), despite two and half times increase in fishing effort. The reservoir has predatory catfishes *Wallago attu* and *Aorichthys* aor, which take a heavy toll of the stocked carps (Table 13.8).

Table 13.8. Species composition of catch (%) in Maithon reservoir

	1961	1962-63	1964-65	1966-67*	1967-68
C. catla	26.13	53.27	80.06	63.98	66.73
C. mrigala	52.06	18.61	5.58	20.36	21.67
L. rohita	1.11	0.66	1.74	6.01	2.28
L. calbasu	0.43	0.96	1.29	1.09	0.99
L. bata					0.15
C. carpio				0.02	0.02
P. sarana					0.09
M. aor		366	2.63	2.14	2.74
W. attu		2.43	1.96	4.38	5.08
B. bagarius				0.48	0.15
Notopterus spp.					0.12
Miscellaneous	20.27	21.00	5.54	0.81	
Total landings (t)	**3.498**	**9.134**	**15.551**	**13.422**	**13.425**

*Catch composition between July, 1966-March, 67. No fishing from April to June, 1966.

13.6 PANCHET RESERVOIR

Panchet dam was commissioned in 1958. The 7 511 ha reservoir (Fig. 13.7) is rather poor in terms of total dissolved salts as reflected by the specific conductivity values of 12.33 to 19 μmhos. Dissolved oxygen is low (0.66 to 2.8 mg l^{-1}) and the pH values range between 7.23 to 7.67. The reservoir was stocked only in 1958. The occurrence of catla, rohu, mrigal and *L. calbasu* in the fishery, without stocking indicates that the carps now reproduce in the reservoir. This was confirmed by the spawn collection survey, conducted by the Central Inland Fisheries. Research Institute during 1964, which noted the breeding of rohu and mrigal in the reservoir.

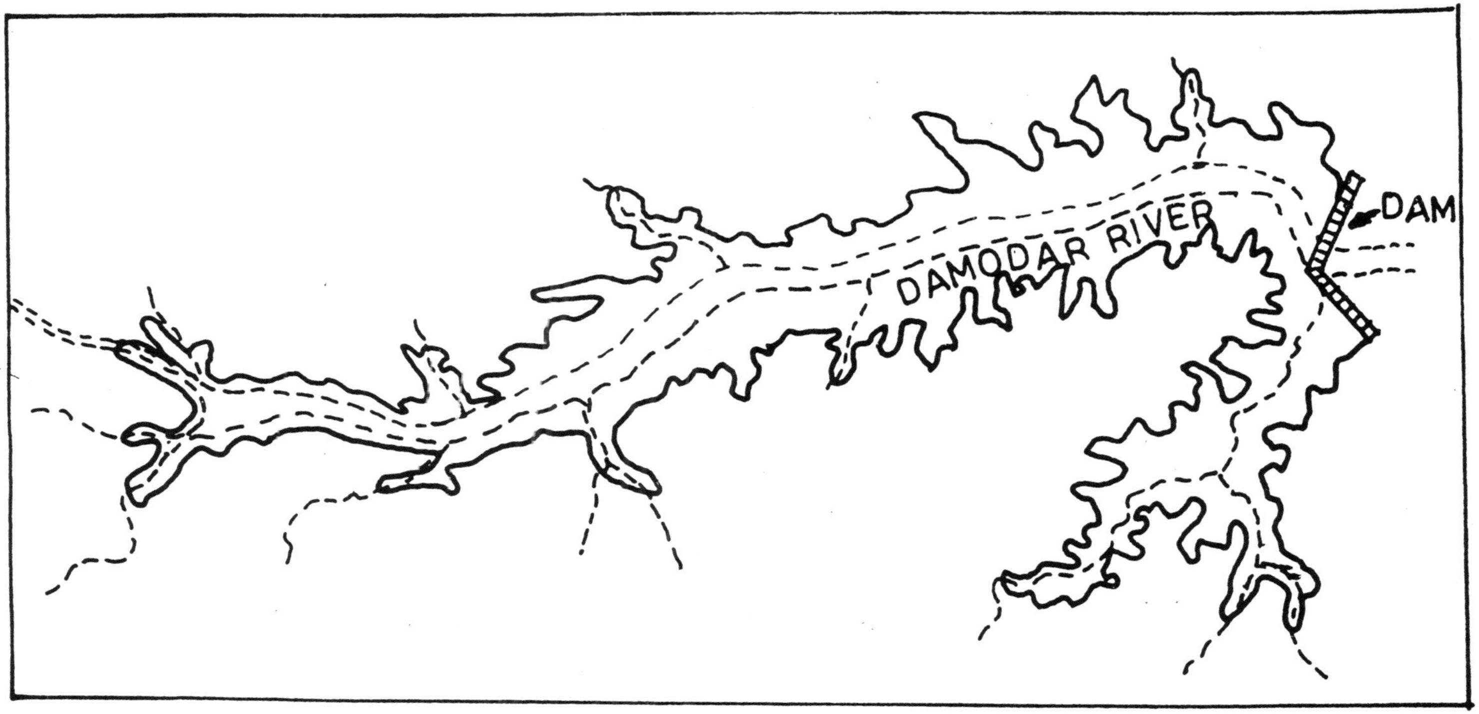

Figure 13.7. Panchet reservoir, Bihar

Fish catch structure of Panchet during 1960-61 was heavily biased in favour of major carps, mrigal and rohu. In 1967-68, mrigal was the dominant species and catla the second, closely followed by rohu. By 1967-68, the catfishes (*Aorichthys aor, Wallago attu* and others) formed the major component of the fish harvest (Table 13.9).

Table 13.9. Species composition of catch (%) in Panchet reservoir

	1960	1961	1966-67	1967-68
C. catla	6.60	4.00	1.97	9.08
C. mrigala	42.91	32.79	21.76	35.97
L. rohita	15.61	13.56	1.18	1.60
L. calbasu	6.06	16.72	0.71	0.43
L. bata			0.01	0.01
M. aor	16.48	5.68	69.19	46.19
W. attu	2.99	1.08	4.90	5.60
B. bagarius				0.15
Notopterus spp.			0.2666	0.92
Miscellaneous	9.34	26.15	0.03	0.05
Catch(t)	**4.645**	**6.617**	**30.755**	**10.575**

A relatively recent report on Panchet reservoir depicts deterioration in the environment due to pollution (Mukherjee *et al.*, 1986). Apart from a heavy sediment input of 0.67 ha m m^{-2} yr^{-1}, arising from the denuded catchment area of 6 120 km^2, the reservoir receives effluents from 31 different industrial establishments. The 21 coal industries alone add 0.9 million t of effluents per minute, bearing 4% fly ash. Four thermal power plants add huge quantities of effluents with pH above 10 and total solids above 5 000 mg l^{-1}. All the effluents, which include heavy metals, phenol and cyanide, exert stress on the ecosystem.

13.7 BADUA RESERVOIR

Badua reservoir (86° 30' E and 25° 30' N), 80 km south of Bhagalpur was commissioned in 1963. The dam comprises an earthen dyke storing 128.29 million m^3 of water for irrigation purposes. The reservoir has a surface area of 880 ha, the maximum and mean depths being 32.0 m and 14.57 m respectively. The drainage covers 297 km^2, receiving rainfall of 163 to 468 mm during July-October. The discharge from outlets is estimated at 1 000 m^3 $second^{-1}$, feeding the right and left canals irrigating 425 km^2 of agricultural land.

Verma and Munshi (1983), described Badua reservoir as monomictic, having thermal stratification during summer and isothermal conditions during monsoon and winter months. Alkaline in reaction (7.5 to 8.1), the water is saturated with oxygen (80-100%) throughout the year. Oxygen depletion in the bottom layer is reported (5.6 to 6.9 ml l^{-1}), accompanied by accumulation of nutrients (silicate, phosphate, and nitrate) in the deeper layers of water.

The maximum rate of primary productivity is at 0.5 m depth @ 98 mg C m^{-3} hr^{-1} (gross) and 70 mg C m^{-3} hr^{-1} (net). Phytoplankton constitute 65 to 67% of the plankton community with Cyanophyceae and Bacillariophyceae as the dominant groups. *Cymbella turgeda, Synedra ulna, Gomphonema sphaerophorum, G. lanceolatum*, and *Navicula cuspidata* are the common diatoms recorded from the reservoir, while *Microcystis aeruginosa* constitute the bulk of the Cyanophyceae.

Other blue-greens are *Merismopedia glauca*, *Anabaena unispora* and *Gloeocapsa punctata*. Members of Chlorophyceae present are *Chlorococcum* sp., *Closterium setaeum*, *Cosmarium puntulatum*, *Bulbochaete obliqua* and *Oedogonium oviforme*.

The State Fisheries Department has been stocking the reservoir with fingerlings of *Catla catla*, *Labeo rohita* and *Cirrhinus mrigala* in the ratio 4:3:3, since 1975. An estimated 1 128 500 fingerlings were stocked until 1979. Total fish landings during 1975-76, 1976-77, 1977-78 and 1978-79 were estimated at 4.03, 3.90, 4.70, and 5.88 t respectively.

13.8 NALKARI RESERVOIR

Nalkari is a tributary of the river Damodar, which is dammed to create the 991 ha reservoir. The lake was filled in the year 1967 to meet the water requirements of Patratu thermal power station and the associated township. Sarkar (1982) described the soil and water quality of the reservoir. The soil is acidic (5.4 to 6.2) with texture varying from sandy loam to sandy clay loam. Calcium carbonate content of soil varies from 0.5 to 1.0% and the organic carbon is in the range of 0.26 to 0.67%. Available nitrogen (47 to 150 mg 100 g^{-1}) shows medium to high productivity while phosphorus is in high productivity range (10 to16 mg P_2O_5 100 g^{-1}).

During the summer stratification, there is a fall of 9 ^{0}C in temperature in the hypolimnion, which is accompanied by a strong chemical stratification. Oxygen decreases from 7.6 mg l^{-1} at the top to 0.9 mg l^{-1} at 14 m depth. The tropholytic layer is characterised by accumulation of free carbon dioxide and bicarbonates, absence of carbonates and low pH. The surface water is above neutral in pH (7.8 to 8.3) and the total alkalinity ranges from 48 to 55 mg l^{-1}. The water is medium hard. Among nutrients, nitrate and phosphate are poor, while the silicate is moderately rich.

13.9 STATUS OF RESERVOIR FISHERIES IN BIHAR

Bihar is the second largest inland fish producing State in India, contributing 11% (0.185 million t) to the national production of of 1.7 million t (1991-92). The track record of the State in the aquaculture sector is impressive. Bihar's total nursery area of 350 ha is the largest in India and its annual fish seed production of 348 million fry is the third highest. However, the large water bodies do not get similar attention and consequently they do not contribute to the inland fish production to the extent they could. The per capita fish consumption in the State ranging from 0.08 to 0.13 kg is one of the lowest in India, thereby calling for an improvement in fish production.

Although Bihar is endowed with rich and diverse water resources in the form rivers, floodplain lakes, wetlands and man-made lakes, no serious developmental efforts have been directed towards increasing fish production from these large water bodies. Fisheries development from the open waters is considered to be environment-friendly and labour- productive. Considering that Bihar is also plagued with a very low per capita income and protein intake, fishery development in open waters assumes greater significance in the State, from social and economic perspectives. Reservoirs that spread over an area of nearly 100 000 ha can play a pivotal role in increasing the fish availability in the State and providing gainful employment to its rural poor. However, the State has a long way to go towards optimum utilisation of this important resource.

Factors that hamper the growth of reservoir sector are many. One of the main constraints that inhibits the development process is the infirm database, both on resources as well as the production processes. Documentation on reservoir fisheries resources in the State is grossly inadequate and even a reliable estimate of the total area under different categories of reservoirs is not available. The State Fisheries Department neither have the jurisdiction nor the mandate to initiate fisheries development in all the man-made lakes. Out of more than 125 reservoirs in the State, covering an area of 96 695 ha, only 14 with a total waterspread of 9 845 ha are under the control of Fisheries Department. resulting in insufficient information on their production dynamics. This serious lacuna has affected the development process. In the absence of a meaningful R&D back-up, almost all the reservoirs in the State are managed on ad-hoc basis, leading to poor realisation of their fish production potential.

Information on fisheries management and fish yields from the reservoirs of Bihar is equally scanty. Ahmed and Singh (1992) estimated the yield from reservoirs under the Department of Fisheries at 5 kg ha^{-1}, based on the fish production of 13 small and one medium reservoirs. As per a recent communication from the State Department of Fisheries, 24 small reservoirs in the size range from 21 to 469 ha, with a yield of 3.91 kg ha^{-1}, calculated on the basis of their mean area. If the area at FRL is considered, the yield rate will drop further. Similarly, three medium reservoirs in the size range 1 050 ha to 1 400 ha (at average level), covering a total area of 3 785 ha produce just 7 158 kg of fish (1.89 kg ha^{-1}). Under the large category, the 7 500 ha Kansjor reservoir produces 800 kg of fish with a yield of 0.107 kg ha^{-1}.

The production level and stocking rate for 37 reservoirs are available (Table 13.10). Covering an area (based on mean reservoir level) of 55 163 ha, they produce 29 942 kg *i. e.*, 0.54 kg ha^{-1} of fish. The fish yield would be much lower, if calculated on area at FRL. Thus, the reservoirs of Bihar have one of the lowest fish yields in the country.

Table 13.10. Stocking and fish yields of some reservoirs of Bihar

Name of the reservoir	Av. area *(ha)*	Rate of stocking *(ha^{-1})*	Av. annual catch *(kg)*	Major fisheries
Ranchi District				
1. Hatia	176	500	1200	Catla 80%, mrigal 10%
2. Getalsud	1 400	250	4000	Rohu 70%, calbasu 10%
3. Patratu	140	200	550	Rohu 60%, catla 30%, mrigal 5%
4. Doriya	110	500	250	Rohu 50%, Catla 4% mrigal 5%
5. Latratu	16 000	-	-	-

Cont.

Table 13.10 (cont.). Stocking and fish yields of some reservoirs of Bihar

Name of the reservoir	Av. area (*ha*)	Rate of stocking (*ha*$^{-1}$)	Av. annual catch (*kg*)	Major fisheries
6. Chirgal	150	1 000	300	Rohu 80%, catla 15%
7. Koel Sindri	175	-	300	Rohu 80%, mrigal 10%
8. Pandra	340	-	-	-
9. Biramkela	130	-	-	-
10. Chirgaldih	200	-	450	Rohu 70%, minor fishes 25%
11. Bhandra	250	500	250	Rohu 40%, catla 30%
Gumla District				
12. Rakamshera	16	5 000	300	Rohu 90%, catla 10%
13. Kansjor	7 500	-	800	Rohu 70%, catla 10%, mrigal 10%, misc. 10%
14. Tapkara	150	1 000	600	Rohu 60%, catla 30%, mrigal 5%, misc. 5%
15. Khajoori	480	200	400	Catla 80%, rohu 15% misc. 5%
16. Ram Rekha	400	2 000	700	Catla 60%, rohu 30%, misc. 10%
17. Parar Dam	375	2 000	300	Catla 30%, rohu 30%, mrigal 5%, misc. 35%
18. Ceergaon	615	-	250	Catla 30%, rohu 10%, mrigal 10%, misc. 50%
19. Masaria	75	-	-	-
Banka District				
20. Badua	1 335	1 500	2 000	Major carps 85%, *Notopterus* 15%
21. Chandan	1 050	1 500	11 55	Major carps 95%, minor carps 5%
22. Orahni	460	-	-	-
23. Madhyagiri	156	3 000	7 800	Major carps 90%, exotics 10%.

Cont.

Table 13.10 (cont.). Stocking and fish yields of some reservoirs of Bihar

Name of the reservoir	Av. area *(ha)*	Rate of stocking *(ha⁻¹)*	Av. annual catch *(kg)*	Major fisheries
24. Amahara	21	1 000	1 050	Major carps 95%, exotics 5%
25.Belharna	25	1 000	750	Major carps (rohu) 80%, catla 20%
Jamui/Monghyr				
26. Nagi	469	1 500	703	Major carps 60%, catfish 40%
27. Nakti	143	15 00	214	-do
28. Garhi	150	-	-	Major carps 60%, catfish 40%
29. Kharagpoor	202	500	1 515	Major carps 10%, catfish 50%, small fishes 40%
Daltongung District				
30. Batane	300	5 000	1 000	Catla 60%, rohu 10%, mrigal 10%, misc. 20%
31. Malayjalashaya	250	5 000	1 200	Catla 30%, rohu 50%, mrigala 20%
Garhwa District				
32. Balaha	120	-	105	Catla 30%, misc. 50%, rohu 20%
33. Panghatwa	200	-	700	Catla 40%, rohu 30%, misc. 30%
34. Annraj	200	-	800	Catla 60%, rohu 30%, misc. 10%
Lohardaga District				
35. Nandini	300	1000	300	Catla 70%, rohu 25%, mrigal 5%
Hazaribagh District				
36. Konar	2700	-	-	-
37. Tilaiya	6400	-	-	-
Dhanbad District				
38. Tenughat	12000	200	-	-

Misc. = Miscellaneous

*Source: State Dept. of fisheries

In the absence of data, it is difficult to estimate the fish production of the reservoirs. If the available literature on the environmental parameters and production dynamics from a few reservoirs is any guide, the water tends to be rich in terms of the ionic concentration and nutrients, mainly by virtue of rich drainage in case of the South Bihar reservoirs and due to the intrinsic productivity of the basin soil in the alluvial plains of the North. The abundance of fish is high. However, the conducive environment and the high primary and secondary productivity are not reflected in a correspondingly high fish yield. The very low fish yields reported from 6 districts of Bihar, ranging between 0.8 and 7.5 kg ha^{-1} (Table 13.11), are the poorest in India.

Table 13.11.Fish yields in Bihar reservoirs

District	Number of reservoirs	Yield range (*kg ha⁻¹*)		
Ranchi	8	1.0	to	6.8
Gumla	7	0.8	to	0.4
Monghyr	3	1.5	to	7.5
Daltongung (Palamu)	2	3.3	to	4.8
Garhwa	3	0.9	to	4.0
Lohardaga	1	1.0	-	-

Table 13.12. Salient features of some reservoir in Bihar

	Panchet	Tilaiya	Getalsud	Konar	Nalkari	Badua
Date of closure	1958	1952	1971	1955	1967	1963
Area at FRL (ha)	7 511	5 921	3 459	2 590	991	880
Minimum area (ha)	-	1545	987	528	-	-
Average area (ha)	-	3733	2 223	1 559	-	-
Maximum length (km)	-	-	11.2	-	-	-
Maximum depth	-	-	32.6	-	-	32
Mean depth (m)	5.2	6.65	8.46	12.97	-	14.57
Volume (million m^3)	391.11	394	288.5	336.16	-	128.29
Catchment Area (km^2)	6 120	984	716.80	997	-	29
Elevation (m above MSL)	-	-	595.7	-	-	-
Length of shoreline (km)	-	277.2	11.07	106	-	
Shore dev. index	-	9.1	0.68	8.78	-	-
Volume dev. index	-	0.766	0.6839	0.85	-	1.36
Inflowing rivers	Damodar	Barakar	Subarnarekha	Konar	Nalkari	Badua
Outflowing rivers	Damodar	Barakar	Subarnarekha	Konar	-	-
Maximum outflowq (m^3 $sec.^{-1}$)	1 000					
Latitude (N)	-	24^0 33'	23^0 27'	23^0 53'	23^0 40'	25^0 30'
Longtitude (E)	-	-	85^0 33'	-	85^0 19'	86^0 30
Soil						
Organic carbon (%)	-	0.31-0.50	0.46-0.60	0.3-0.9	0.26-0.67	-
Available P (mg 100 g^{-1})	-	3.0	2.45-6.08	1.2-8.0	10-16	-
Available N(mg 100 g^{-1})	-	-	24.38-35.06	-	47.15	-
pH	-	5.6-5.9	-	5.4-6.0	5.4-6.2	-
	-	-	85^0 33'	-	85^0 19'	86^0 30'
Water						
Water temperature (^0C)	24.9-40.2	24.5-33.0	16.0-28.5	17.6-28.9	22.0-31.0	29.0-30.5
pH	7.2-7.6	6.7-7.0	7.3-8.2	-	7.0-8.3	7.5-8.1
Transparency (cm)	-	31.5	10-176	-	109	-
DO (mg l^{-1})	0.66-2.80	4.7-8.3	5.8-9.6	-	0.9-7 6	5.6-8.3
CO_2 (mg l^{-1})	-	-	0.6-15.1	-	nil-9.0	0-2.0
Total alkalinity (mg l^{-1})	-	53.5	-	43.5-83.0	48.0-55.0	-
Spec. cond. (μmhos)	12.33-19.00	98.7-211.4	163.7-306.0	81.20-147.6	-	-
Total hardness(mg i^{-1})	-	-	-	-	-	-
Calcium (mg l^{-1})	-	16.2	5.3-18.8	13.2-16.2	-	-
Nitrate (mg l^{-1})	-	0.25	tr.-0.36	0.10-0.60	0.07	0.24-0.85
Phosphate (mg l^{-1})	-	0.03	0.001-0.12	tr.-0.047	tr.-0.06	0.067-0.321
Silicate (mg l^{-1})	-	-	2.8-18.2	1.3-1.98	9.0	23.0-33.0
Chlorides (mg l^{-1})	11.66-18.00	-	-	-	-	-
Organic carbon (%)	-	-	1.0-4.3	1.2-6.15	-	-
GPP (mg C m^{-2} d^{-1})	-	199-300	289.85	199-300	-	-

14. THE NORTHEAST

Situated within the geographical ordinates, 21^{0} to 29^{0} 30' N and 89^{0} 46' to 97^{0} 30' E, the Northeastern Region consists of seven States, *viz.*, Arunachal Pradesh, Assam, Manipur, Meghalaya, Mizoram, Nagaland and Tripura, popularly known as the *seven sisters*. The region, surrounded almost on all sides by Bhutan, China, Myanmar and Bangladesh, is divided into five physiographic units, the Assam Himalayas, the Brahmaputra valley, the Shillong plateau, the Barak valley and the Southeastern hill region. Except the Brahmaputra and Barak valleys in Assam and the Imphal valley in Manipur, the terrain is hilly and mountainous, only making 65% of the northeast an undulating land with elevations ranging from 200 to 900 m above MSL.

14.1 RESERVOIR FISHERIES RESOURCES OF THE NORTHEAST

The northeast has a copious rainfall during March to May, when precipitation is minimal in the rest of the country. Cherrapunji and Mawsyngram, recording highest rainfall in the world, are situated in the northeastern India, the annual rainfall of the region ranging from 100 to 1142 cm. The tall hills, punctuated by deep valleys receiving abundant drainage make the region ideally suited for tapping the hydro-electric power potential of water. It is estimated that Brahmaputra and Barak basins together constitute 50% of the total hydel power potential of the country (Sinha, 1990). However, very little of the potential is presently being utilised. Only eight reservoirs have so far been created, six of them for power generation under four hydro-electric projects. This, together with small irrigation impoundments, makes the total area under man-made lakes 8 206 ha (Table 14.1; Fig. 14.1). No organised fishing is practised in any of them except on the Gumti reservoir in Tripura.

Table 14.1. Reservoirs of the northeastern region

Name of the reservoir	State	Area *(ha)*
Barapani (Umiam)	Meghalaya	500
Kyrdemkulai	-do-	90
Nongmahir	-do-	70
Khamdong	-do- & Assam	1 335
Umrong	Assam	979
Gumti	Tripura	4 500
Khoupum	Manipur	100
Palak	Mizoram	32
Others	-	600
Total		***8 206***

(Anon., 1991 and Sinha, 1990)

Figure 14.1. Reservoirs of the North Eastern States

14.2 GUMTI RESERVOIR

Gumti reservoir is located at 23^0 25' 45" N and 91^0 49' 20" E, across the river Gumti in the South Tripura district. The 4 500 ha watersheet draws water from Barak, Raima and Sarma river basins. Completed in the year 1977, the 103 m long, 30 m high, straight gravity dam is made of brick and stone concrete, holding a storage of 235.62 million m^3 at the FRL. The river at the dam site has a catchment area of 547 km^2 receiving rainfall of 173 cm. The maximum capacity of spillways is 381.38 $m^3 sec^{-1}$ and the power intake tunnel has a maximum capacity of 31.15 $m^3 sec^{-1}$. Very little is known about the water and soil quality of the reservoir. Raha and Sarkar (1980) describe the soil as red to light yellowish brown and dark brown with a medium water holding capacity of fine loamy texture. Turbidity of water is very high due to soil erosion on account of denudation of forests in catchment area by shifting cultivators (*jhoom*). The soil is acidic (pH 5 to 6) but the water is alkaline within the pH range of 7.5 and 9.

Chaudhari (1992) reports a rich macrophytic community in the reservoir, comprising *Eichhornia crassipes*, *Nymphaea* sp., *Lymnanthemum* sp., *Trapa natans*, *Vallisneria spiralis*, *Hydrilla verticillata* and *Potamogeton nodosus*. The phyto- and zooplankton exhibit strong seasonal and spatial cycles with a domination of blue-green algae and copepods.

Details on the indigenous ichthyofauna of the river before the impoundment are not available. More than 47 species have been reported from the reservoir, commercially important among them being *Aorichthys aor*, *Channa marulius*, *C. punctatus*, *Mastacembelus pancalus*, *M. guentheri*, *Puntius stigma*, *P. ticto*, *P. sarana*, *Xenentodon cancila*, *Amblypharyngodon mola*, *Chanda ranga*, *Heteropneustes fossilis*, *Clarias batrachus*, *Notopterus notopterus*, *Tor tor and Labeo gonius*. Besides, the Indian and exotic carps, *viz.*, *Catla catla*, *Labeo rohita*, *Cirrhinus mrigala*, *Hypophthalmichthys molitrix*, *Ctenopharyngodon idella* and *Cyprinus carpio*, have been introduced successfully in the reservoir. Catla-rohu and rohu-catla hybrids are also reported from the reservoir (Raha and Sarkar, 1980). Indian major carps breed in the upper reaches of the river at Ultacherra, Chitrajhari, Gandacherra, Rangajhari, Sarmacherra, Thakurcherra and Raimacherra. However, this isolated breeding success does not seem to lead to recruitment to the populations (Chaudhari, 1992).

Fish production

The highest fish production of 246.6 t was reported in 1981-82, with a steep decline thereafter (Table 14.2). The average production for the last 15 years was 141.74 t (31.5 kg ha^{-1}). Variations in fishing effort, in terms of the number of licenses issued, have been erratic for the last 15 years with an increasing trend after the mid- eighties. The mean catch per licence is 379 kg.

Fishes caught from Gumti reservoir are graded into five categories (Table 14.3) for pricing and marketing. *Notopterus* spp., non-airbreathing catfishes and minnows dominated the catch till 1987-88. The share of Group IV, comprising the Indian and exotic carps, has been increasing since 1988-89.

Despite a species spectrum dominated by minnows and catfishes, the fish yield as per the official records is much higher than the national average. Although sustained stocking is often attributed to the high rate of productivity, it is clear from the catch data that the stocking has made some impact on fisheries only after 1988-89 and the yield was high even during the earlier periods. The growth performance of introduced species in the reservoir is reported to be very impressive with silver carp growing to 12.9 kg within a period of 3 years. Very high growth rates have also been recorded for catla (10.5 kg), catla-rohu and rohu-catla hybrids (6.7 kg), and for the common carp (6.5 kg) for the same time span.

Table 14.2. Fish production from Gumti reservoir

Year	Fishing intensity *(No. of licences issued)*	Stocking of fish seed *(x 10^5)*	Fish harvest *(t)*
1978-79	285	1.64	169.000
1979-80	293	6.89	191.708
1980-81	430	0.45	193.398
1981-82	379	2.50	246.611
1982-83	383	0.15	90.891
1983-84	352	5.00	117.552
1984-85	210	7.04	41.124
1985-86	198	3.05	105.064
1986-87	197	7.70	133.986
1987-88	588	13.45	322.043
1988-89	442	12.03	138.693
1989-90	495	18.21	137.393
1990-91	300	22.22	121.331
1991-92	433	36.66	101.630
1992-93	620	28.48	15.731
Total	**5 605**	**165.47**	**2 126.155**
Average	***374***	***11.03***	***141.740***
Average yield (kg ha^{-1}) : 31.50			
Average catch $licence^{-1}$ $year^{-1}$: 379			

(Anon., 1994b)

Table 14.3. Grade-wise fish catch composition in Gumti reservoir

Year	Total catch (t)	Percentage composition of grades I	II	III	IV	V
1978-79	169.0	-	-	-	-	-
1979-80	191.7	26.2	-	56.5	0.6	0.2
1980-81	193.4	15.4	12.1	67.0	5.3	0.2
1981-82	246.6	35.0	22.8	37.8	3.9	0.5
1982-83	90.9	46.3	18.6	31.7	1.2	1.1
1983-84	117.5	25.1	23.2	34.4	13.9	3.3
1984-85	41.1	48.6	11.2	31.1	7.3	0.8
1985-86	105.1	35.0	14.0	41.9	8.2	0.9
1986-87	134.0	37.1	7.6	45.7	9.1	0.5
1987-88	322.0	50.6	4.6	28.9	15.3	0.5
1988-89	138.7	34.3	10.5	20.7	32.3	2.1
1989-90	137.4	38.4	2.4	42.1	16.5	0.5
1990-91	121.3	28.6	3.4	43.5	23.9	0.6
1991-92	101.6	4.6	3.2	26.6	65.5	0.1
1992-93	15.7	4.0	5.1	62.1	28.7	0.1

Grade I - All minnows and weed fishes except *Puntius* spp.
II - *Puntius* spp., eels and murrels (*Channa* spp.)
III - *Notopterus* spp. and non-airbreathing catfishes
IV - Indian and exotic major carps
V - *Clarias batrachus* and *Heteropneustes fossilis*

Stocking

In the absence of any recruitment of major economic species in the reservoir, sustained stocking assumes greater importance as a management measure to improve the fisheries, both qualitatively and quantitatively. Although stocking was initiated as early as 1978-79, it picked up momentum only after 1987-88, when the fish seed farm was commissioned adjacent to the reservoir. Detailed species-wise break-up of seed stocked is not available. So far, 16.5 million seed have been stocked in the reservoir over a period of 15 years at an annual average of 1.1 million and a stocking rate of 245 ha^{-1}. During 1992-93, the stocking rate stood at 632 ha^{-1}. The stocking did make an impact on the catch structure by increasing the share of Indian major carps (Fig. 14.2), but it did not increase the yield.

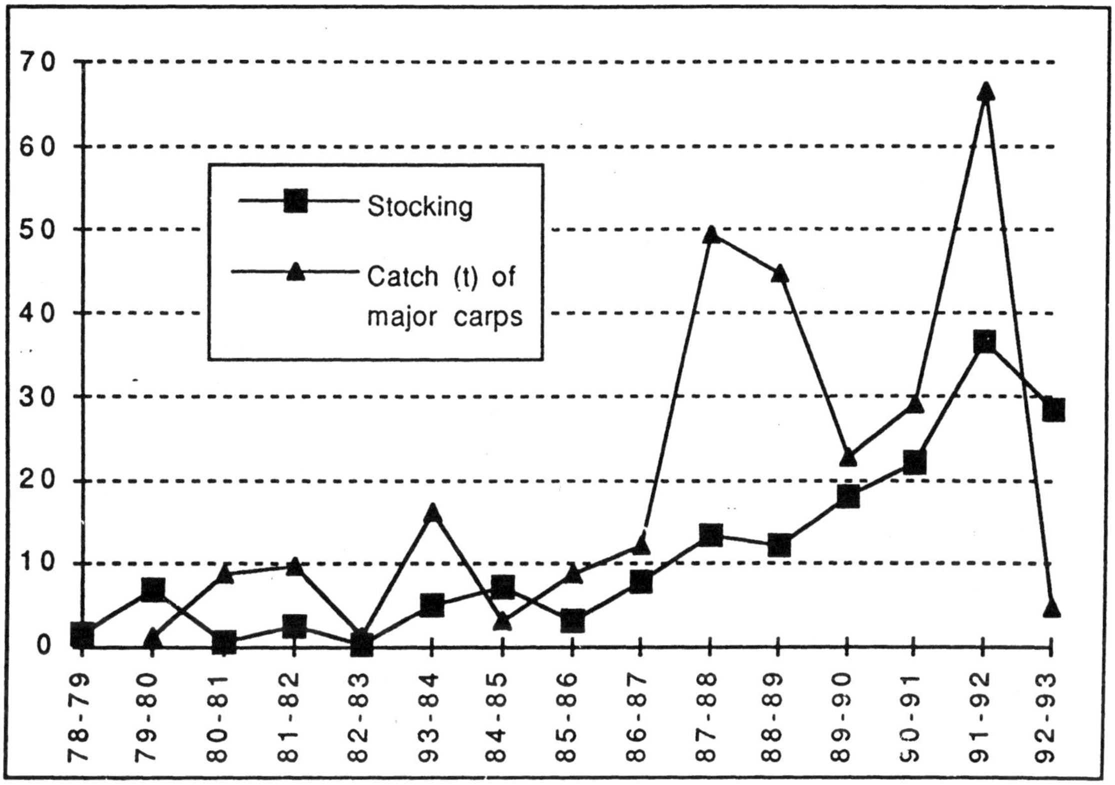

Figure 14.2. Stocking (in numbers x10^5) and major carp catches from Gumti reservoir

Craft and Gear

Gill nets of three different mesh sizes, ranging from 5 to 20 cm, are the most commonly used gear, followed by long lines, hook and line, traps, drag nets and cast nets. Hook and line and traps are targetted at murrels and *Mastacembelus* spp., while drag nets with 0.8 cm mesh are commonly employed for catching small fishes. A locally fabricated fishing boat usually 6 m x 1.25 m, with a V shaped hull, is used by the fishermen. One motor launch and two speed boats, owned by the Fisheries Department, are operated for transport of fish from landing to assembly centres.

Organization of fisheries

Organised fishing was initiated in Gumti in 1978-79. At present, more than 1 000 families, belonging to the Scheduled Caste and Scheduled Tribe communities, are engaged in fishing under seven cooperative societies. Fishing inputs are supplied to the fishermen under various welfare schemes. Boats, nets and nylon twines are distributed either free of cost or at a subsidy upto 50%. In 1990-91, 580 fishermen have benefitted from the welfare schemes (Table 14.4).

Table 14.4. Free distribution of fishing inputs to fishermen in Gumti reservoir

Year	Boats (number)	Nets (number)	Nylon twine (kg)	Total number of beneficiaries
1986-87	10	85	75	140
1987-88	7	105	100	212
1988-89	27	213	100	340
1989-90	50	333	125	508
1990-91	20	40	520	580

(After Chaudhari, 1992)

Post-harvest arrangements

The licensed fishermen are required to hand over the catch by 0800 hrs to the officials of the Fisheries Department at any of the 5 assembly centres along the reservoir. The Department pays remuneration to the fishermen at the rates fixed on the basis of categories of fish. During 1991-92 the fishermen got Rs. 3.00 to 20.00 per kg of fish belonging to groups I to V. After paying the fishermen, the Department hands over the catch to the respective cooperative societies which sell it to retail outlets through an Apex Cooperative. The Fisheries Department levies a royalty ranging from Re. 1.00 to Rs. 3.00 per kg of fish sold by the society. The small fish, especially *Puntius* spp. harvested from the reservoir, support a flourishing dry-fish trade.

14.3 KYRDEMKULAI

Kyrdemkulai and Nongmahir are two pick-up reservoirs created below the Sumer power station of the Umtru-Umiam stage I hydroelectric project. They are the only two reservoirs in the region subjected to some investigations. The Central Inland Capture Fisheries Research Institute, under a consultancy assignment with the North Eastern Council, conducted feasibility studies for fisheries development in the two reservoirs (Sugunan and Yadava, 1991a). The salient findings of the study follows.

Kyrdemkulai reservoir, formed in 1979, is a part of the Umtru-Umiam Hydro-electric Project Stage III. River Umiam has a catchment of about 150 km^2. The immediate catchment of the reservoir comprises forested hills, rarely exceeding the elevation of 700 m above MSL. The 106.7 m concrete gravity type diversion dam with three drainer gates has a height of 27.5 m and is provided with a 42.7 m long spillway having a capacity of 1 700 m^3 sec^{-1}. The river downstream is known as Umtru. The 167.7 m long diversion tunnel carries water from Kyrdemkulai to Nongmahir, another downstream reservoir.

The reservoir is elongated, with 10 major and minor pockets (bays). Besides the river Umiam, the reservoir receives 10 to 15 creeks, draining the forested catchment. With a total surface area of 80 ha, the reservoir maintains a full storage level (FSL) of 3.7 million m^3 and a minimum storage level (DSL) of 2.78 million m^3 It has a shoreline of 19.6 km. The water level in the reservoir fluctuates between 675.0 and 683.4 m above MSL. The level fluctuation is primarily determined by the requirements of the Nongmahir forebay. The water regime is also governed by incursion of Umiam waters, precipitation and runoff from the catchment. The mean depth of the reservoir at FRL is estimated at 4.63m.

Kyrdemkulai reservoir is situated in the low slopes of the East Khasi Hills, where climate varies from warm sub-tropical to cool and bracing. The region has a high rainfall, the average daily rainfall (calculated on monthly basis) varying from insignificant to 35.43 mm during 1987 and 1988.

The bottom sediments of the reservoir are fed by sand and clay fractions with average organic carbon content of 1.13%. Soil pH is on the alkaline side. Average values of organic matter (3.91%), available phosphorus (0.83 mg $100g^{-1}$), available nitrogen (21.56 mg 100 g^{-1}) and specific conductivity (105.55 µmhos) reflect a low soil productivity.

Kyrdemkulai has a deep profile and its pockets with large littoral zones add considerably to the shoreline (Fig. 14.3). Limnological investigations do not portray significant sectoral differences, although there is a well expressed vertical profile of some physico-chemical parameters.

Water temperature, ranging from 19.5 °C in November -December to 27.0 °C in July, depicts moderate seasonal variations. Water pH, in consonance with the underlying soil, is alkaline. Dissolved oxygen and free carbon dioxide in the sub-surface have optimum values with a normal trend in their seasonality. Low alkalinity characteristic of the north-eastern waters is observed in Kyrdemkulai. Low hardness (18.56 to 27.84 mg l^{-1}) places the reservoir in the soft water category. In contrast to the low nutrient status of the sediments, optimum availability of nutrients in the water phase is a significant indicator of the reservoir's productive nature. However, total dissolved solids and specific conductivity are low (Table 14.5).

Physico-chemical characters governing the productivity of Kyrdemkulai reservoir, indicate favourable conditions for fish yield optimisation.Vertical profiles of temperature, pH, dissolved oxygen, free carbon dioxide and total alkalinity reflect the formation of thermocline and chemical stratification during the summer months. During winter (December-January) and monsoon months, especially August, this thermal stratification is disturbed, resulting in a turnover and subsequent water mixing. River Umiam, draining a sizeable catchment, contributes richly to the allochthonous nutrient enrichment. The nutrient flux and its optimum availability in the water phase can play a pivotal role in sustaining optimum levels of biotic communities.

Biotic communities

The low standing crop of net plankton is an indicator of a poor community, both in numerical and gravimetric terms. The dry weight of the total plankton estimated from different sectors covering three seasons varies from 0.21 to 0.40 mg l^{-1} with an average of 0.31 mg l^{-1}. The average plankton density in numerical terms is 8 420 units l^{-1}. Despite a low plankton density, the primary carbon synthesis rate is high, suggesting a high nannoplankton concentration in the reservoir and a possible channelisation of production process through the nannoplankton chain. There is no significant difference in the plankton concentration and composition among various sectors. Chlorophyceae account for

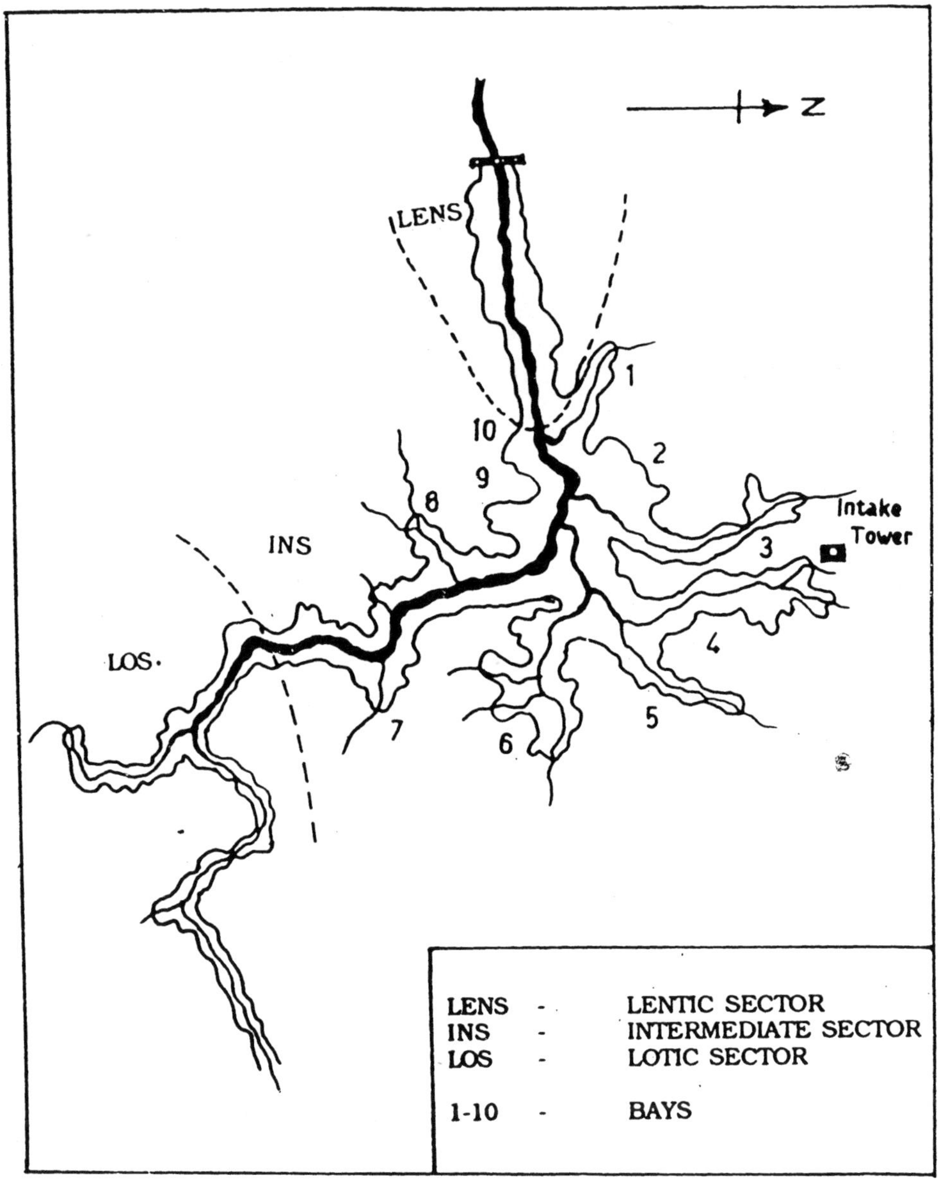

Figure 14.3. Kyrdemkulai reservoir, Meghalaya

the bulk of the plankton (5 220 units l^{-1}), represented mainly by *Staurastrum*, *Chaetophora* and *Dinobryon*. Cyanophyceae are well represented in Kyrdemkulai, strongly indicating that the reservoir is at a more advanced phase of eutrophication. The blue-greens form nearly 36% (3 030 units l^{-1}) of the total plankton. The plankton species diversity is high.

The reservoir is marked by the absence of aquatic macrovegetation, except for some isolated patches. The benthic macrofauna is rather poor, oligochaetes forming a sizeable component, followed by insect larvae. Gastropods are present in the shallows, especially the protected bays. Those parts of the bays receiving forest creeks, characterised by accumulation of detritus, harbour oligochaetes, larvae of *Chironomus* and *Chaoborus*, other insect larvae and nymphs. The density of benthos in the reservoir is 134 individuals m^{-2}. The acidic nature of the soil might inhibit the colonisation of benthic communities.

Primary productivity

The reservoir has moderate rate of photosynthesis with the gross carbon production ranging from 39.37 to 56 mg C m^3 hr^{-1}. The net productivity is 18.75 to 20 mg C m^3 hr^{-1}. These productivity figures are equivalent to an average of 1 464.58 mg C m^2 day^{-1} (gross) and 583.55 mg C m^2 day^{-1} (net).

Table 14.5. Soil and water quality of Kyrdemkulai reservoir

	WINTER	SUMMER	MONSOON	AV.
Soil				
pH	5.90	7.60	6.20	6.57
Organic carbon (%)	1.63	0.29	1.46	1.13
Organic matter (%)	2.81	6.49	2.42	3.91
Available phosphorus (mg 100 g^{-1})	0.79	1.12	0.59	0.83
Available nitrogen (mg 100 g^{-1})	24.64	7.70	32.33	21.56
Spec. cond. (µ mhos cm^{-1})	121.64	87.70	107.30	105.55
Water				
Temperature (°C)	19.50	23.50	27.00	23.30
Transparency (m)	2.84	2.44	2.20	2.49
pH	6.70	7.00	6.70	6.80
Dissolved oxygen (mg l^{-1})	7.10	7.10	6.70	6.80
Free carbon dioxide (mg l^{-1})	2.00	2.00	2.60	2.20
Total alkalinity (mg l^{-1})	28.00	22.00	32.00	27.33
Total hardness (mg l^{-1})	18.56	27.84	23.20	23.20
Chlorides (mg l^{-1})	18.80	12.72	13.16	14.89
Iron (mg l^{-1})	0.16	*	tr.	0.08
Ammoniacal nitrogen (mg l^{-1})	0.16	3.8	0.90	1.62
Nitrate nitrogen (mg l^{-1})	0.18	3.61	0.02	1.27
Silicates (mg l^{-1})	1.00	3.00	2.80	10.67
Phosphate (mg l^{-1})	tr.	0.02	0.02	0.01
Calcium (mg l^{-1})	14.84	*	9.28	12.06
Magnesium (mg l^{-1})	3.72	*	13.92	8.62
Total dissolved solids (mg l^{-1})	32.00	*	62.00	47.00
Spec. cond. (µmhos)	*	*	102.00	102.00

* Values not recorded (Sugunan and Yadava, 1991b)

Fish fauna

In the absence of regular fishing, it was not possible to prepare a complete list of fish species of the reservoir. The following fish are reported. *Cyprinus carpio, Acrossocheilus hexagonolepis, Tor putitora, Puntius sophore, Clarias batrachus, Dania rerio, Dania acquipinnatus, Dania dangila, Heteropneustes fossilis, Aplocheilus* sp., *Noemacheilus multifasciatus, Tor tor, Catla catla, Labeo rohita* and *Cirrhinus mrigala*. Large murrels and *Wallagu attu* are either absent or their number is too low to create any serious problems. Similarly, there seems to be a poor population of weed fishes in the reservoir, which would a compete with the commercial carps, if introduced into the system.

Present status of fisheries

There is no organised fishing in the reservoir. Tribals living in the neighbouring villages and the employees of the Meghalaya State Electricity Board (MeSEB) are engaged in subsistance or recreational fishing. Gill nets and long lines are the popular fishing tackle. The lake also attracts anglers who catch common carp, *A. hexagonolepis* and small fishes. Five to ten people angle in a day, each having 4 rods, on an average. No catch statistics are available. The villagers make improvised pen enclosures on the periphery of the reservoir to entrap fishes. A trap, locally called *roh*, is set in the pens which attracts common carp, *H. fossilis*, and small prawns. Six to seven pen enclosures are erected in the reservoir. There is only one fisherman who is engaged in full time fishing. He has gill nets, long lines and cast nets, his catch varying from 2 to 20 kg per day during the monsoon season.

There are five to six locally made boats in the reservoir, in the size range of 3 to 3.5 m, costing Rs. 500 to Rs. 1 000 apiece. Excepting the one used by the lone full time fisherman, these are primarily used for collecting forest wood and for carrying anglers. Experimental gill netting yielded chocolate mahseer and common carp, while cast nets brought in a good number of mahseer juveniles which were released back into the reservoir.

Productivity status

The temporal and spatial variations in the limno-chemical parameters and the biotic communities offer vital clues to the productivity status of water bodies. A probe into the above parameters in respect of Kyrdemkulai reservoir suggests a combination of positive and negative characters. The temperature regime of the reservoir is relatively conducive to the production processes and the ambient dissolved oxygen in water is high in all seasons, which points towards high photosynthetic activity. The high shoreline development index of 11.65 denotes a highly irregular shoreline, a positive indicator to productivity. Whereas dissolved oxygen, water temperature and shoreline index suggest productive nature of the water body, the average pH, total alkalinity and specific conductivity are not in the productive range. However, the moderately high rate of carbon synthesis is a clear signal of productivity potential at the basic level. The other supporting limnological evidence for the high productivity of the water body are the oxycline and chemocline observed during the warmer months. Low pH of water and soil is an indicator of low productive nature, but the small size of the reservoir offers opportunities to adopt corrective measures to overcome this constraint. Bottom soil is rich in organic matter and this clearly suggests scope for detritivorous fishes in the reservoir. The poor benthic communities despite rich soil quality may be the result of the low soil pH. Notwithstanding a few discouraging limnological features, there is enough scope for fish yield optimisation in the reservoir through efficient management.

Fish yield potential

Based on the quantity of carbon produced, the reservoir can sustain a fish yield of 534 kg ha^{-1}. In terms of calories, the average energy produced at primary level is 14 382 cal m^2 day^{-1}. At a conversion rate of 1% from carbon to fish; this primary productivity has the potential to yield 570.6 kg of fish (fish weight) ha-1 year-1. Considering the scope for increasing the fertility status of the reservoir through manuring, the target of production can be initially fixed at 600 kg ha^{-1}, with considerable scope for gradual increase up to 1 000 kg ha^{-1} in due course.

Reservoir management

With the main accent of fisheries management in Kyrdemkulai on the yield optimisation, emphasis should be also placed on the conservation of the natural populations of chocolate mahseer in the river Umtru. The high primary production rate in Kyrdemkulai can be channelled to fish production. Apart from the usual reservoir management measures like regular stocking, population management through gear selection and maintaining a stocking and the harvesting schedule, there is enough room for increasing the productivity of the reservoir through artificial eutrophication. The small size of the reservoir and the fact that it does not form a source of drinking water make the option of fertilization easy. The fertilization can push up the fish production potential to much higher levels than what is calculated on the basis of primary productivity.

The management strategy formulated for the Kyrdemkulai reservoir (Sugunan and Yadava, 1991) falls broadly under the following lines :

Conservation and management of mahseer fisheries

1. Establishment of a sanctuary for mahseers
2. Banning the capture of mahseer juveniles
3. Artificial breeding of chocolate mahseer for stocking the reservoir

Management of carp fisheries

1. Fertilization
2. Staggered stocking of carps
3. Harvesting schedule
4. Monitoring of fishing effort
5. Post-harvest management.

Conservation and management of mahseer fisheries

The chocolate mahseer, *Acrossocheilus hexagonolepis*, the golden mahseer, *Tor putitora* and *T. tor* are the important mahseer species of the region. These fish are highly preferred by the sport enthusiasts as well as commercial fishermen. Among them, the chocolate mahseer is specially liked as food fish by the local people. Mahseers, in general, are enlisted as endangered and threatened species in the country. Unlike their counterparts in the Himalayas, the North Eastern mahseers are least subjected to the anthropogenic environmental pressures. This makes the job conservation of these species easier. Mahseers in the North East command a premium price and attract anglers from far and wide. The fisheries development plan for Kyrdemkulai, therefore, should have an in-built component of conserving the indigenous ichthyofauna, especially the chocolate mahseer.

Sanctuaries for mahseers

The lotic sector of the reservoir merges into the sandy river course of Umiam which traverses thick virgin forests of East Khasi Hill district of Meghalaya. This river stretch and its further upper reaches are ideal habitats for mahseer breeding and recruitment. One of the main factors that contributed to the decline in the mahseer populations elsewere in the country is the destruction of such habitats. Therefore, this habitat needs to be protected by declaring an 8 km stretch upstream the reservoir as mahseer sanctuary. In the sanctaury, all human activities leading to environmental degradation like deforestation and agricultural operations need to be prohibited. Special care should be exercised to preserve the sandy shores which provide clear water rich in dissolved oxygen for the breeding of mahseers. Removal of sand from the area should be totally prohibited. A complete ban is suggested on catching any fish from the sanctuary. The State Departments of Environment and Forests have a pivotal role to play in managing the sanctuary by implementing the regulations.

Artificial recruitment of mahseers

During the post-monsoon months, the juvenile mahseers of 5 to 10 cm size abound in the reservoir. They become easy prey to the cast net and small meshed gill nets. They can also be caught easily by operating drag nets in the shallow bays where they congregate for feeding. The investigating team, while operating cast nets in the vicinity of the tunnel intake tower and shallow bays of the reservoir in the month of July, captured 10 juveniles of chocolate mahseer per cast net. *Tor putitora* and *T. tor* were also caught occasionally in the cast nets. Fortunately, there is only one fisherman in the reservoir who operates cast nets. In future, if more fishermen employ cast nets and drag nets to catch the juveniles of mahseers, it may have disastrous effects on the mahseer stocks. A strict ban on the capture of juvenile fish should be a part of the management programme. Conservation measures as suggested above are expected to result in substantial recruitment to the mahseer populations. However, as an additional precautionary and supplementing measure, breeding and stocking of chocolate mahseer are recommended. The State Fisheries Department has a mahseer breeding farm at Barapani, where attempts to breed and rear the chocolate mahseer are going on.

Management of carp fisheries

Since the environment of Kyrdemkulai is suitable for Indian major carps, they should be introduced in the reservoir to constitute the main fishery. The carp fisheries management strategy for the reservoir should include fertilization, regular stocking and harvesting, monitoring the number of fishing units and post-harvest management.

Fertilization

Since the energy produced at the primary level in Kyrdemkulai is transformed to higher trophic levels through nannoplankton chain, there is an urgent need to enhance the microplankton community through artificial eutrophication. This will enable diversification of the pathway of energy tranfer through plankton chain. Basic objectives of fertilizing the reservoir are to increase the plankton density and to enhance primary productivity. The plankton biomass of Kyrdemkulai, based on the dry weight, is estimated at 0.31 mg l^{-1}, far below the ideal range for planktophagous fishes. A plankton density of 7 to 10 mg l^{-1} is considered to be ideal for a reservoir, where the major fisheries depend on filter feeders. A higher plankton biomass of 30 mg l^{-1}, if achieved, can substantially increase the fish yield.

With a view to increasing the standing crop of plankton, a fertilization programme has been prepared for the reservoir, which envisages application of poultry manure at the rate of 10 t ha^{-1} yr^{-1}, along with urea and superphosphate at the rates of 40 and 20 kg ha^{-1} yr^{-1}. After a high standing crop of plankton is established, there is a fair possibility that the community may sustain itself without artificial eutrophication. In that event, it may even be possible to scale down and ultimately discontinue the application of fertilizers.

There is a need to lime the reservoir at the bays in order to increased the pH water and soil. The lime application also acts as a prophylactic cover against the outbreak of fish diseases like epizootic ulcerative syndrome.

Stocking

The basic purpose of stocking in Kyrdemkulai is to establish the quick growing carp populations into the ecosystem. The other important fish species, the mahseers are mainly carnivorous and do not directly compete with the carps for food. A breeding population of common carp is already available in the reservoir. Common carp is a well known detritivore and it has good market value in the region (unlike in many other parts of the country, common carp is not at discount in Meghalaya). Stocking of other bottom feeding detritivore is not recommended.

Since the fertilization is expected to increase the phytoplankon production, the stocking emphasis should be on filter feeders and the natural choice falls on the indigenous species, *Catla catla*, which has a proven track record. The other species suggested for stocking is rohu which is a versatile feeder. The reservoir needs to be stocked annually at the rate of 150 000 advanced fingerlings (above 100 mm) of catla and rohu in the ratio 3:2.

A circular hatchery (concrete, synthetic or metallic), a nursery farm for transit rearing and pen enclosures are the infrastructure needed for seed raising. Bay sectors of the reservoir provide ideal location for installing pens. The site lies between 675 to 679.7 m above MSL, which dries up during extreme summer, when the pens can be erected with ease. They would gradually fill up with the onset of monsoon and retain at least 1 to 2 m of water for sufficient duration to facilitate seed rearing (Fig. 14.4).

Harvesting schedule

Continuous stocking and harvesting system is suggested for Kyrdemkulai. In the beginning, the stocked carps need to be allowed to complete two years without fishing. Commercial exploitation may start from the third year onwards, but catla and rohu less than 400 g in size are required to be protected from fishing and if capture, to be released. On an average, 65 fishermen together can operate 390 units of gill nets (a unit being 50 m hung length). Average daily carp harvest is expected to be around 167 kg (calculated on the basis of 240 fishing days).

June 15 to August 15 may be declared as closed season and a complete ban on fishing strictly enforced during this period. A fishing party, comprising two fishermen, can share a boat and operate 12 units of gill nets of assorted mesh size but not less than 35 mm bar. Gill nets of less than 35 mm bar should be strictly prohibited, except for weed/trash fish population control (if necessary). Use of such nets could, however, be allowed if:

i) Nets are operated in the presence of State officials

ii) All the juveniles of commercially important fishes, specially the stocked ones, are released back into the reservoir.

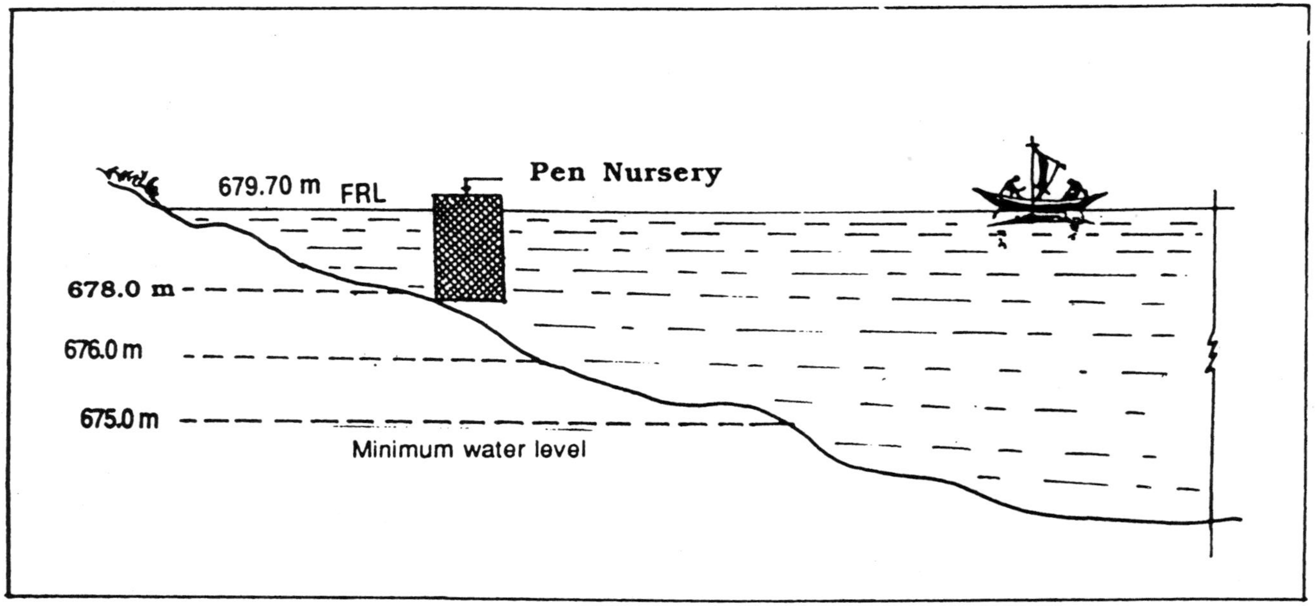

Figure 14.4. Site for pen nursery (diagrammatic sketch) in Kyrdemkulai

After a couple of years of such management, larger mesh size nets should be introduced in order to exploit fishes of 2 year plus age groups. The nets with head rope, breast line and foot rope with floats only should be used with a recommended hanging coefficent 0.5 horizontal and 0.87 vertical.

It is important to monitor the fishing effort both qualitatively and quantitatively. A low fishing effort may result in under-exploitation. Similarly, lower mesh size and excessive effort may have harmful effect through catching fish of an uneconomic size, thereby rendering the system unremunerative. In order to keep track of the fishing effort in terms of nets, boats and manpower, it is essential to register the units with State Fisheries Department. It is also suggested that a licence fee be levied. This will partially defray the cost incurred by the State for development and act as a check to keep the fishing units within the desired level. Experience shows that free fishing often leads to licentiousness and irrational proliferation of units which is detrimental for a sustainable capture fisheries management.

Post-harvest and other management

As the gill nets are usually lifted in the morning, the fishermen should get facilities to dispose off their catch instantly, at a reasonable price. State Government may have to look after this aspect. Preferably, a cooperative society may be formed to undertake marketing functions so that fishermen do not have to waste time and resources on post-harvest operations. Such facilities should be common for Nongmahir and Kyrdemkulai. Such arrangements, however, need to be worked out in close association with the fishermen, to prevent middlemen siphon away a good part of profits. The State Government may provide a boat which will serve to transport fish from fishing areas the landing centre. The same boat can be used for patrolling the reservoir to check unauthorised fishing activities.

Provision of fishery guards is very essential to ensure that all regulations are strictly implemented. They should act in close liaison with the security staff of MeSEB and the law enforcing authorities to implement the ban during the breeding season and to curb dynamiting, poisoning *etc.*, and to ensure that mesh regulations are adhered to. They will also guard the mahseer sanctuary. A total of 10 fishing guards have been suggested for Kyrdemkulai, in view of the need to patrol the sanctuaries. Provision for two additional country boats for the guards' movement in the riverine stretch is also made. The State Departments of Environment and Forests have pivotal role to play in managing the sanctuary and therefore their active cooperation may be sought to implement the regulations.

Mahseer fisheries

The expected mahseer fishery is 8 000 kg a year. Assuming that half of this will come from gill nets, the 65 member gillnet fishermen may remove 4 000 kg mahseers from the system, leaving the rest as the anglers' quota. Angling may be regulated by issuing a licence fee. Assuming 50 angling days per year, a total of 1 200 angling mandays can be allowed *i.e.*, at the rate 24 anglers a day on an average. An angling fee of Rs. 20/- can be realised from each angler per day. The bag limit may be at 6-8 fishes per angler, with the minimum size of fish of 250g.

Total Catch

Taking both carps and mahseers into consideration, Kyrdemkulai may yield an annual catch of 48 t, comprising 8 t of mahseer and 40 t of carps. The gill net catch will comprise 40 t of carps and 4 t of mahseers (total 44t). The anglers' share of mahseers will be 4 t per year.

Benefits

The two major objectives of fisheries development programme of Kyrdemkulai are fish yield optimisation and conservation of mahseer fisheries. The increased fish productivity from the lake is expected to improve the availability of fish in the state. Fish is a very costly commodity in Meghalaya. The scientific management of Kyrdemkulai reservoir can produce at least 48 t of fish per year which will considerably ease the availability position of fish in the local markets as well as in Shillong. It is estimated that fishing activities take place for 240 days a year with an average daily catch of about 184 kg. Apart from this, the project envisages to provide sport fishery for 50 days in a year with 24 anglers per day. Anglers' catch is expected to be 80 kg per day.

Employment

The reservoir can provide direct employment for 65 fishermen who will share a total annual fish harvest of 44 t of fish comprising 40 t of carps and 4 t of mahseers. Each fisherman will catch 615 kg of carps worth Rs. 18 450 at the rate of Rs. 30/- per kg and 61.5 kg of mahseers worth Rs. 2 460, at the rate of Rs. 40/- per kg. These sixty-five fishermen represent sixty-five families and, at an average of four members per family, the scheme offers direct benefits to 260 people. The mahseers in the reservoir are expected to provide sporting delight and additional income to 1 200 anglers. Many more will benefit indirectly. Boat building, net making and their maintenance are the main ancilliary activities. The reservoir is directly connected to the main centres of fish consumption, like Zero Point, Shillong and Guwahati. This eliminates the chances of exploitation by traders.

Remunerativeness of fishing operations

Fishermen of Kyrdemkulai will require assistance from financial institutions to meet their capital requirements. The overall remunerativeness of the fishing activities has been projected (Sugunan and Yadava, 1991a), based on a fishing unit comprising two fishermen, who share a boat and 12 units of gill nets (50 m hung length each). The initial outlay for boat and net is Rs. 10 500 involving an annual capital cost of Rs. 2 250. The variable cost, including the imputed value of wages that the two fishermen would have earned, is Rs. 11 600. The total annual cost including interest on capital, annual capital cost and all variable costs is estimated at Rs. 19 625. The two fishermen net out 1 230 kg of carps and 123 kg of mahseers which would earn them Rs. 41 820, at the rate of Rs. 30 per kg for carps and Rs. 40/- for mahseer (the sale price is kept at the lower side. Rs. 40 to 45 is the current market price for carp and Rs. 50 to 60, for mahseers).

Even after meeting the repayment obligations, and paying interest on capital and licence fee, the two fishermen between themselves can earn a net profit of Rs. 27 585, *i.e.*, each fisherman gets Rs. 13 792/- per year, a comfortable income indeed. A fisherman of Kyrdemkulai will spend Rs. 7 117.50 for a catch worth Rs. 20 910 in a year. In other words, he gets a daily catch of 2.8 kg worth Rs. 87.00 from which he earns nearly Rs. 57.50 as net profit. This clearly shows remunerativeness of fishing in Kyrdemkulai, provided proper management measures are taken and the fishing units are regulated as per scientific norms.

14.4 NONGMAHIR

Nongmahir is a pickup reservoir (forebay), formed in the year 1979 by conduiting the waters of the Kyrdemkulai reservoir into a steep gorge through a 2 839 m long tunnel. The main earthen dam is 150 m long with a height of 50 m above the bed level of 638.72 m above MSL. The reservoir level is maintained by a series of five saddle dams, varying from 33 to 68.5 m in length. The Ogive spillway has a clear waterway of 12.2 m with the designed flood water discharge of 158.4 m^3 sec^{-1}. The reservoir is more or less circular in shape with a number of islands. It has a total surface water area of 70 ha with a storage capacity of 5.8 million m^3 at FRL and 2.16 million m^3 at DSL. The small local catchment comprises hillocks bearing loose laterite soil. A conspicuous feature of the deep and steep basin is the limited littoral zone restricted to the embayments. The shoreline is 13.25 km long. The water level fluctuates between 669.20 and 672.07 m above MSL. After the power house became functional, there is a regular outflow of water from the reservoir. The level fluctuation is thus determined by the outflow, inflow from Kyrdemkulai, evaporation, and the runoff from the local catchment. The mean depth of the reservoir at FSL is estimated at 5.28 m.

One of the important morphometric characteristics of the reservoir is the deep drawdown. The elevation at outflow point is just 3 m above the bed level *i.e.*, at a depth of 10.4 m (at FRL). This prevents stagnation and keeps the water in slow circulation. The conventional demarcation of lotic, intermediate and lentic sectors is not of much relevance in case of Nongmahir, since it is a pickup reservoir storing the Kyrdemkulai waters. The only sectoral distinction having any limnological significance is between the central sector and the bays (Fig. 14.5). Sugunan and Yadava (1991b) have described the reservoir and discussed the feasibility for fisheries development.

Habitat variables

The sediments in the reservoir are dominated sand and clay fraction with organic carbon percentage varying from 1.74 to 2.13. The soil, is acidic (pH 4.6 to 6.0), contains low to moderate concentrations of available nitrogen (29.91 to 44.59 mg 100 g^{-1}) and phosphorus (0.38 to 1.12 mg 100 g^{-1}). The specific conductivity is of low order (107.3 to 367.3 μmhos) and organic matter ranges within 2.3 to 3.61%.

Water temperature is marked by seasonal variation from 25.0°C in November-December to 31.0°C in April. The water pH is largely governed by the underlying sediments, *i. e.*, it remains mostly acidic, except during the summer when pH is slightly higher than neutral. Oxygen saturation in the sub-surface columns of the reservoir water remains fairly high with minor drop during monsoon months, coinciding with the increase in turbidity. The water has low total alkalinity, ranging between 11.20 and 25.52 mg l^{-1}. Based on hardness values (16.84 - 25.52 mg l^{-1}), Nongmahir is a soft water body. Specific conductivity and total dissolved solids, notwithstanding seasonality, are low (Table 14.6). However, the concentrations of phosphate, nitrate and silicate indicate the productive potential of the reservoir.

There is a distinct vertical thermal gradient in summer, when the temperture decreases from 23.5 °C at the surface to 22.5 °C at 3 m and 21.0 °C at 6 m respectively. The dissolved oxygen declines from 7.4 mg l^{-1} at the surface to 5.8 mg l^{-1} at 9 m with concomitant increase in free carbon dioxide values, suggesting klinograde tendency in the reservoir. The prevailing duration of warm and cold phases appears conducive for the annual turnover and subsequent mixing of nutrients in the lake. This is amply reflected by the availability of phosphate,

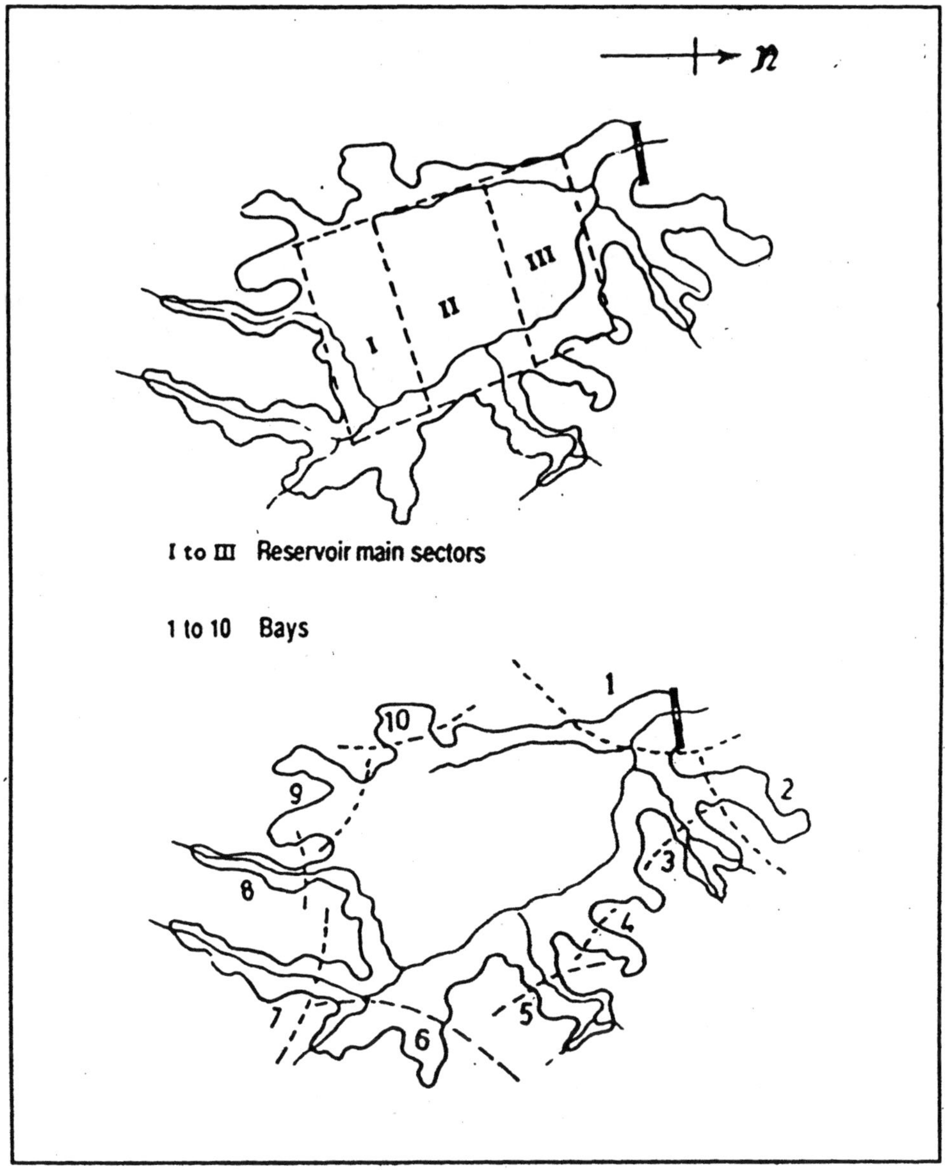

Figure 14.5. Nongmahir reservoir, Meghalaya

nitrate and silicate in water. Although the total alkalinity is low, the omnipresence of free carbon dioxide contributes to the photosynthetic activity. However, low concentrations of ions and total dissolved solids prevent a higher productivity of Nongmahir reservoir.

Table 14.6. Soil and water quality of Nongmahir reservoir

	WINTER	SUMMER	MONSOON	AV.
Soil				
pH	4.60	5.2	6.00	5.30
Organic carbon (%)	1.74	2.13	1.92	1.93
Organic matter (%)	2.30	3.61	3.27	3.06
Available phosphorus (mg 100 g^{-1})	1.11	1.12	0.38	0.87
Available nitrogen (mg 100 g^{-1})	33.60	44.59	29.91	36.03
Spec. cond. (µmhos)	133.40	367.30	107.30	202.67
Water				
Water temperature (°C)	25.00	31.0	29.00	28.33
Transparency (m)	2.60	2.40	1.00	2.00
pH	6.60	7.40	6.70	6.90
Dissolved oxygen (mg l^{-1})	6.90	7.28	6.90	7.03
Free carbondioxide (mg l^{-1})	2.00	0.16	5.00	2.38
Total alkalinity (mg l^{-1})	28.00	20.00	12.00	20.00
Total hardness (mg l^{-1})	16.84	25.52	23.00	19.90
Chlorides ((mg l^{-1})	6.66	12.74	22.56	13.98
Iron (mg l^{-1})	0.09	*	0.30	0.19
Ammonical nitrogen (mg l^{-1})	0.14	0.24	*	0.19
Nitrate nitrogen (mg l^{-1})	0.20	0.82	0.02	0.35
Silicates ((mg l^{-1})	1.00	4.00	6.60	3.86
Phosphate (mg l^{-1})	tr	0.02	0.08	0.03
Calcium (mg l^{-1})	14.84	20.88	13.90	14.37
Magnesium (mg l^{-1})	2.56	4.64	9.10	3.60
Total dissolved solids (mg l^{-1})	26.00	22.10	14.00	20.70
Spec. cond. (µmhos)	150.21	104.40	69.60	108.07

* *Values not recorded* *(Sugunan and Yadava, 1991b)*

Biotic communities

The dry weight of total plankton, estimated from various sectors covering the four seasons, varies from 0.20 mg l^{-1} to 0.45 mg l^{-1}, with an average of 0.32 mg l^{-1}, is the result of low densities of plankton. The average plankton density by number is 5 440 units l^{-1}. Despite a low plankton population, the primary productivity of the reservoir is high, suggesting high nannoplankton density in the reservoir.

The phytoplankton constitutes more than 96% (by number) of the total plankton in the main reservoir. It is dominated by Chlorophyceae (70.86%), represented mainly by *Staurastrum* sp; *Chaetophora* sp. and *Dinobryon* sp. The bay plankton is characterised by a higher density of the blue-greens, indicating eutrophic character of the shallows. The bays represent the trophogenic zone of the reservoir which has a rich and diverse plankton community, represented by 24 genera.

Macrovegetation is absent. The main reservoir and the areas with high percentage of sand yield little benthic fauna. The bays receiving forest creeks and characterised by accumulation of detritus, harbour oligochaetes, *Chironomus* spp., *Chaoborus*, other insect larvae and nymphs are also observed in the bay areas. Molluscs are conspicuous by their absence in the reservoir. The density of benthic fauna in the reservoir is 38 individuals m^{-2}.

Primary productivity

Estimates of the gross carbon production range from 34.38 to 68.0 mg C m^3 hr^{-1}. Corresponding figures for net productivity fluctuate from 25.0 to 43.62 mg C m^3 hr^{-1}. These figures are equivalent to an average of 1 187 mg C m^2 day^{-1} (gross) and 817.44 mg C m^2 day^{-1} (net).

Fish fauna

Cyprinus carpio, Acrossocheilus hexagonolepis, Tor putitora, Puntius sophore, Dania rerio, D. acquipinnatus, D. dangila, Clarias batrachus and *Heteropneustes fossilis* are the fish species encountered in experimental fishing. Besides, Indian major carps (*Catla catla, Labeo rohita* and *Cirrhinus mrigala*) and the exotic silver carp are reported to be occasionally caught by the local fishermen. There is a total absence of large murrels and predatory catfishes (*Wallago attu, etc.*) and poor population of weed fishes. The fact that weed/trash fishes have not been able to gain foothold in the reservoir can be ascribed to the absence of macrophytes and an insignificant littoral zone. Their small population favours the stocking of Indian major carps, since the commercial carps would face less competition food.

Present status of fisheries

There is no commercial fisheries and the pressure is restricted to angling and seasonal use of bamboo traps. Angling is a favourite past time in Meghalaya. Apart from the local inhabitants, domestic toursists, from as far as Shillong, frequent the reservoir for angling. It is estimated that two to six people angle every day. The catch is very low and mostly comprises small specimens of common carp.

Traps, lures and snares form the basic capture gear. The most common is the use of bamboo split mat pens, erected in the marginal areas of the reservoir. The pen forming a semi-circle, extends its arms beyond the water phase, completely barricading the littoral zone. The pen facilitates entry of fish from the reservoir through a trap door, but prevents their return. Normally, the fish trapped in the pen during the night are harvested in the morning, using hand nets, cast nets, etc. Weed/trash fishes and *H. fossilis* are commonly caught in these pens. Common carp is caught in the pens during monsoon months.

Productivity status of the water body

The water temperature favours all biotic communities, except during the peak of winter (December to February). Of the morphometric attributes, the shoreline development index of 8.93 indicates productive nature of the lake. High values of ambient dissolved oxygen during all seasons suggest prolific photosynthetic activity and an unstressed aquatic environment. The average pH, total alkalinity and specific conductivity values, lying in the unproductive range, do depict deficiency in the aquatic phase. However, the optimum rates of carbon synthesis denominate productive potential at the primary level. The other supporting evidence for this potential is the presence of oxycline and chemocline during warmer months. Notwithstanding the few discouraging limnological features, the lake leaves enough room for yield optimisation through fisheries management both at the environmental and biotic communities levels.

Fish yield potential

At a conversion rate of 1%, primary productivity could support a yield of 433 kg of fish per hectare per year. Considering the scope for increasing the fertility status of the reservoir through manuring, the target of production has been fixed at 500 kg ha^{-1}, with a scope for steady increase up to 1 000 kg ha^{-1} in due course.

Nongmahir being a forbay of Kyrdemkulai, does not receive runoff from any watershed. This deprives the reservoir of allochthonous nutrient loading which might have acted as a catalyst for trophogenic activities at the lower trophic levels. However, the small size of the reservoir and the fact that it does not form a source of drinking water, enables the fisheries manager to adopt the option of fertilization. A fertilizer and liming schedule similar to Kyrdemkulai has been recommended for Nongmahir.

Stocking

Stocking is the most important management tool for Nongmahir reservoir fisheries for the following reasons :

1 There is a negligible native fish stock (since the reservoir is a forebay of Kyrdemkulai).

2 The common carp is the only major commercial species present.

3 The *'log guard'* provided at the intake tunnel restrics the Kyrdemkulai fish fauna to gain entry into Nongmahir.

4 In the absence of any streams and rivulets joining the reservoir and coupled with a deep basin without littoral formations, chances of natural recruitment of stocked fish (except free spawners) are very remote.

Common carp is already present in the reservoir. It is reported to breed there. Hence, no stocking of bottom feeding detritivore is recommended. Since the fertilization promises a good standing crop of plankton, emphasis is on filter feeders, *Catla catla* and rohu. Annual stocking of 110 000 advanced fingerlings (>100 mm) of catla and rohu in 3:2 ratio is recommended for the reservoir. Experience in other small reservoirs indicates the desirability of phased stocking and multiple harvesting. Harvesting schedule and other management measures similar to those for Kyrdemkulai are recommended for Nongmahir (Sugunan and Yadava, 1991b).

14.5 OTHER RESERVOIRS

Barapani (Umiam), the oldest reservoir in the region was commissioned in 1965 by damming the river Umiam as stage I of the Umtru-Umiam Hydro-electric Project. Located within 25° 39' 30" N and 91° 43' 51" E, at an elevation of 900 m above MSL, the reservoir has a waterspread of 500 ha. It stores 182.72 million m^3 of water to feed the turbines of hydro-electric power station at Sumer. The river Umiam eventually joins a tributary of Brahmaputra. No fisheries development activities have been reported from the reservoir, except for stocking *Cyprinus carpio* in early 1970s. The reservoir is under the jurisdiction of East Khasi Hill District Council, an autonomous local self government. Presently the fishing is limited to angling, which is not organised. It is reported that the common carp in the lake has a slow growth due to overpopulation (Sinha, 1990). The city wastes emanating from Shillong find their way into the reservoir through various streams, causing eutrophication. Denudation of forest in the catchment, mainly due to the practice of shifting cultivation(*jhoom*) is accelerating the siltation rate in the reservoir.

Khandong and **Umrong** are two reservoirs created under the Kopili Hydro-electric Project. Presently, the lakes are under the jurisdiction of the North Eastern Electric Power Corporation (NEEPCO) and fishing has not been initiated in either of them. The NEEPCO and CICFRI have started feasibility studies for fisheries development in the reservoirs. Khandong is a 1 335 ha reservoir across the river Kopili situated at an elevation of 719.0 m above MSL, receiving drainage from an extensive catchment of 1 256 km^2, comprising thickly wooded forests. The reservoir has a maximum storage of 16.6 million m^3 at full level. The water is reported to be productive. Umrong is a pick-up reservoir, receiving and storing the water coming from the Khandong Power House. Situated at an elevation of 602 m above MSL, the 90 ha impoundment also receives drainage from the local catchment of 62 km^2. The reservoir has a capacity of 78.5 million m^3 with a mean depth ranging from 2.58 to 8.0 m. The reconnaissance survey conducted by CICFRI suggests that the water body is reported to be productive.

14.6 STATUS OF RESERVOIR FISHERIES IN THE NORTHEAST

Among the existing six reservoirs in the region, only Gumti has an organised fishing activity, the remaining ones are either under-exploited or remain unexplored. Even at a modest yield of 150 kg ha^{-1}, the reservoirs can contribute 1 211 t to the total fish production. The present level of yield, however, is not known.

Keeping in view the long-term solutions to the recurring flood problems in the region and tapping the enormous hydel power potential, a large number of projects are on the anvil. A total of 19 projects are envisaged in the Brahmaputra basin alone, with a colossal storage capacity of 9.33 million ha m. Of these, three are under construction and are expected to be completed during the VIII Five Year Plan (1996 to 1997). In the Barak basin, the proposed Taipaimukh dam at 24° 14' N, 93° 1' 36" E, is expected to create a waterspread of 6 455 ha with a gross storage of 0.18 million ha m. Five more projects in the basin are under various stages of planning. According to available projections, 68 760 ha of man-made lakes will soon appear in four States (Table 14.7). The region, poised for a quantum jump in the resource size, needs a well-conceived fisheries development strategy to tap the fish production potential from the proposed reservoirs.

Table 14.7. Projected increase in reservoir area during the VIII Plan

State	Existing area (ha)	Projected additional area (ha)
Arunachal Pradesh	-	160
Assam	2 314	-
Manipur	100	40 000
Meghalaya	660	-
Mizoram	32	-
Nagaland	-	27 100
Tripura	4 500	1 500
Others	600	-
Total	***8 206***	***68 760***

(*Anon., 1991; Sinha, 1990*)

Table 14.8. Salient features of Kyrdemkulai and Nongmahir reservoirs of the northeast

	Kyrdemkulai	Nongmahir
Date of closure	1979	1979
Area at FRL (ha)	80	70
Mean depth (m)	4.63	5.28
Volume (million m^3)	3.7	5.8
Catchment area (km^2)	150	-
Elevation (m above MSL)	683	672
Length of shoreline (km)	19.6	13.25
Shore dev. index	11.65	8.93
Annual level fluctuations (m)	8.4	2.9
Inflowing river	Kyrdemkulai	Tunnel (Kyrdem-kulai)
Outflowing river	Umtru	-
Maximum outflow (m^3 $sec.^{-1}$)	1 700	-
Latitude (N)	25^0 43'	-
Longitude (E)	91^0 47'	-
Soil		
Sand (%)	49.19	55.10
Clay (%)	39.29	33.65
Silt (%)	11.52	11.25
Organic carbon (%)	0.3-1.6	1.74-2.13
Available P (mg 100 g^{-1})	0.6-1.1	0.38-1.12
Available N (mg 100 g^{-1})	7.7-32.3	29.9-44.6
Spec. cond. (μmhos)	87.7-121.6	107-367
pH	5.9-7.6	4.6-6
Water		
Water temperature (^{0}C)	19.5-27	25-31
pH	6.7-70	6.6-7.4
Transparency (cm)	2.2-2.8	1.0-2.6
DO (mg l^{-1})	6.7-7.1	6.9-7.3
CO_2 (mg l^{-1})	2.0-2.6	0.16-5.0
Total alkalinity (mg l^{-1})	22.0-32.0	12-28
Spec. cond. (μ mhos)	102	69.6-150.2
Total hardness(mg l^{-1})	18.6-27.8	16.8-25.5
Calcium (mg l^{-1})	9.3-14.8	13.9-20.8
Nitrate (mg l^{-1})	3.7-13.9	0.02-0.82
Phosphate (mg l^{-1})	tr.-0.02	tr-0.02
Silicate (mg l^{-1})	1.0-3.0	-
Chlorides (mg l^{-1})	12.7-18.8	6.6-22.6
GPP (mg C m^{-3} hr_{-1})	39.4-56.0	34.4-68.0

REFERENCES

Abbott, W., 1967. Microcosm studies on estuarine waters 11. The effect of single dose of nitrate and phosphate. *J. Water Poll. Control Fed.*, **39** (1): 113-122.

Abraham, M., 1979. Studies on bottom macrofauna of Bhavanisagar reservoir (Tamil Nadu). *J. Inland Fish. Soc. India,* **11** (2): 41-48.

Abraham, M., 1980a. Rearing of spawn in pen. *J. Inland Fish. Soc. India*, **12** (2): 98-100.

Abraham, M., 1980b. Studies on plankton of Bhavanisagar reservoir with special reference to limnological conditions, inflow and water level. *J. Inland Fish. Soc. India,* **12**(2): 25-37.

Adholia, U. N., 1991. Primary productivity of Mansarovar reservoir, Bhopal. *Him. J. Env. Zool.,* **5**: 79-81.

Adholia, U. N., 1992. Bioenergetics of Mansarovar reservoir with reference to fish production. *Technical Report of the ICAR Ad-hoc Scheme,* Dept. of Zoology, Motilal Vigyan Mahavidyalaya, Bhopal, India, pp 121.

Agarwal, M. and H. D. Kumar, 1978. Physico-chemical and physiological assessment of two mercury polluted effluents. *Indian J. Environ. Hlth.*, **20**: 141-155.

Ahmed, S. M. J., 1993. A case for stocking our reservoirs with *Macrobrachium malcolmsonii. Fishing Chimes,* 12(11) 43-44.

Ahmad, S. H. and A. K. Singh, 1992. Present status potentialities and strategies for development of reservoir fisheries in Bihar. *Fishing Chimes,* **12** (8): 49-57.

Anon., 1968. Annual Report 1968, Central Inland Fisheries Research Institute, Barrackpore, India, pp. 93.

Anon., 1976. Report of the National Commission on Agriculture, 1976, Pt. VIII. Fisheries. Government of India, Ministry of Agriculture, New Delhi, pp 207.

Anon., 1977. Proceedings of the Fifth Workshop of All India Coordinated Project on Ecology and Fisheries of Freshwater Reservoirs, Ukai, Gujarat, India, 7- 8 Sept. 1977 pp 39.

Anon., 1979. Annual Report, 1979, Central Inland Fisheries Research Institute, Barrackpore, India, pp 132.

Anon., 1980. Proceedings of the Sixth Workshop of All India Coordinated Project on Ecology and Fisheries of Freshwater Reservoirs, Simla 25-26 November 1980, Central Inland Fisheries Research Institute, Barrackpore, West Bengal, India, pp 121.

Anon., 1982. Final Report All India of Coordinated Project on Ecology and Fisheries of Freshwater Reservoirs, Nagarjunasagar. *Research Information Series,* 3, March 1983 Central Inland Capture Fisheries Research Institute, Barrackpore, India, pp 66.

Anon., 1983. Proceedings of the Seventh Workshop of All India Coordinated Project on Ecology and Fisheries of Freshwater Reservoirs, Barrackpore, 9-10 March 1983, Central Inland Fisheries Research Institute, Barrackpore, West Bengal, India, pp 148.

Anon., 1984a. Proceedings of the Eighth Workshop of All India Coordinated Project on Ecology and Fisheries of Freshwater Reservoirs, Barrackpore, Patna, 11-12 July, 1984, Central Inland Fisheries Research Institute, Barrackpore, West Bengal, India, pp. 117.

Anon., 1984b. New low cost coracle for fishermen. CIFRI Newsletter, 7(1 & 2): 3-4.

Anon., 1987. Irrigation reservoirs in Tamil Nadu. *Publication No. 1*, Irrigation Management and Training Institute, Taramani, Tamil Nadu, pp 58.

Anon., 1989a. Statistical bulletin of fisheries- 1988-89. *Published by the Directorate of Fisheries,* Government of Karnataka, Bangalore, India, pp 43.

Anon., 1989b. Final Report of Coordinated Project on Ecology and Fisheries of Freshwater Reservoirs, Gobindsagar, Central Inland Capture Fisheries Research Institute, Barrackpore, India (MS.).

Anon., 1991. Project Report on Integrated Inland Fisheries Development Project in North Eastern States. Agricultural Finance Consultants Ltd., pp 498.

Anon., 1994a. Handbook on fisheries statistics of Andhra Pradesh 1992-93. *Published by the Commissioner of Fisheries* Andhra Pradesh, Hyderbad, pp 74.

Anon., 1994b. Detailed information on Gumti reservoir, Tripura. Communication from the Adviser (Fisheries) North Eastern Council, Central Inland Capture Fisheries Research Institute, Barrackpore, India, pp. 6.

Anon., 1984c. Final Report of All India Co-ordinted Research Project Ecology and Fisheries of Freshwater Reservoirs-Getalsud. *Inland Fisheries Research Information Series 4,* CIFRI, Barrackpore, pp 54.

Arora, H. C., S. N. Chattopadhyaya and U. P. Sharma, 1970. A probable occurrence of fish mortality in Renusagar, Renukoot due to chlorine bearing wastes. *Environ. Hlth.*, 12: 260-272.

Bandopadhyay, M. K., S. K. Singh and B. Sarkar, 1991. Growth and production of *Ctenopharyngodon idella* reared in plastic cages. *Proc. Nat. Symp. Freshwater Aquaculture,* CIFA, Bhubaneswar, 35-37.

Banerjee, R. K. and P. Ray, 1979. Soil and water quality of Tungabhadra reservoir as indices of biological productivity. *Lecture* delivered at the Summer Institute on Culture and Capture Fisheries of Man-made Lakes in India, Central Inland Fisheries Research Institute, Barrackpore, India, July-August 1979, pp 46-53.

Bennet, G. W., 1970. Management of lakes and ponds. *New York Van Nostrand Reinhold Co.*, pp 375.

Bhatnagar, G. K., 1963. On some aspects of biology of *Puntius kolus* of the Tungabhadra reservoir. *Indian J. Fish.*, 10(2): 500-520.

Bhatnagar, G. K., 1964. Observations on the spawning frequency and fecundity of certain Bhakra Reservoir fishes. *Indian J. Fish.*, 11 (1): 485-502.

Bhatnagar, G. K., 1979. Observations on production and recruitment of economic fishes in Tungabhadra reservoir with an account on the need for stocking for improving productivity. *Lecture* delivered at the Summer Institute on Culture and Capture Fisheries of Man-made lakes in India, Central Inland Fisheries Research Institute, Barrackpore, India, July-August 1979, pp 390-396.

Bhatnagar, G. K. and V. V. Sugunan, 1978. New records of fish from Nagarjunasagar reservoir on Krishna river in Andhra Pradesh. *J. Inland Fish. Soc. India,* 10: 146-149.

Bhatnagar, G. K., S. J. Karamchandani and M. D. Pisolkar, 1977. Observations on the spawning of major carps in Govindgarh Resevoir, M. P. *J. Inland Fish. Soc. India,* 9 : 193-195.

Bhukaswan, T., 1980. Management of Asian reservoir fisheries. *FAO Fisheries Technical Paper No. 207,* FAO, Rome, pp 69.

Birasal, N. R., V. B. Nadkarni and B. Y. M. Gouder, 1989. Ecology of zooplankton of Supa reservoir of Kali river during the filling phase (Western Ghats, Karnataka). *J. Freshwater Biol.*, **1**(1): 15-23.

Birasal, M. R., V. B. Nadkarni, S. R. Krishnamurthy and S. G. Bharati, 1991. The physico-chemical status of the Kali River after the closure of the Supa dam - A preliminary report. *J. Freshwater Biol.*, **3**(4): 251-257.

Brungs, W. A., 1972. Effects of pesticides and industrial wastes on surface water use. *In* Oglesby, R. T. and C. A. Carlson (Eds.). River Ecology and man, *Academic Press*, New York, San Fransisco London pp. 465

Chacko, P. I. and G. K. Kuriyan, 1948. Survey of the fisheries of Tungabhadra river. *Proc. Indian Acad. Sci.*, **28**(5): 165-76.

Chacko, P. I. and Dinamani P., 1949. Preliminary report on the fishery resources of the Bhavani river in relation to the lower Bhavani Project. *Proc. Indian Sci. Congr.*, **36**(3): 166.

Chacko, P. I., G. K. Kuriyan and Thyagarajan, 1955. A survey of fisheries of the Cauvery river. *Cont. Freshw. Fish. Biol. Sta. Madras*, **12** (1954): 1-19.

Chandra, K., R. S. Panwar and D. N. Singh, 1985. Possible pollution problems from wastes of thermal power plants in India-A case study of Rihand Reservoir, Mirzapur (U. P.). *Proceedings of the Symposium on Environmental Pollution*, pp 283-294. *The Academy of Environmental Biology Publications*, Muzaffarnagar, U. P.

Chaudhari Swapan, 1992. Role of Gumti reservoir of Tripura. *Fishing Chimes*, **12** (5): 23-26 .

Das, V. M., S. Balakrishnan and N. D. Hartman, 1993. Reservoir fisheries in Kerala: Study on marketing of fish from five reservoirs in Palghat and Trichur Districts. *Indo-German Reservoir Fisheries Development Project*. PN: 88. 2134 0-01. 200 Malampuzha, Palakkad district, Kerala.

Dash, M. C., P. C. Mishra, G. K. Kar and R. C. Das, 1983. Hydrobiology of Hirakud dam reservoir. *In:* Mishra, P. C. and R. K. Trivedy (Eds), Ecology and pollution of inland lakes and reservoirs, Ashish Publishing House, New Delhi, 317-338.

David, A. and K. V. Rajagopal, 1969. On some aspects of the fish populations of the Tungabhadra reservoir. *In* Proceedings of the seminar on Ecology and Fisheries of Freshwater Reservoirs, 27-29 Nov. 1969, Central Inland Fisheries Research Institute, Barrackpore: 457-474.

David, A., P. Ray, V. B. Govind, K. V. Rajagopal and R. K. Banerjee, 1969a. Limnology and fisheries of the Tungabhadra reservoir, *Bulletin* **13**. *Central Inland Fisheries Research Institute*, Barrackpore, pp 188.

David, A., N. G. S. Rao and M. F. Rahman,1969b. Preliminary Survey report on hydrobiology and fisheries at Nagarjunasagar, Andhra Pradesh. *Survey Report*, 5, CIFRI, Barrackpore, pp 36.

David, A., N. G. S. Rao and P. Ray, 1974. Tank fishery resources of Karnataka. *Bulletin, Central Inland Fisheries Research Institute*, Barrackpore : p 87.

David, A., N. G. S. Rao and S. L. Raghavan, 1975. Preliminary observation on the periphyton production in a lentic water mass. *Indian J. Exp. Biol.*, 13(2): 205-208.

Davies, I., 1966. Chemical Changes in cooling water towers. *Int. J. Air Water Pollut.*, **10**: 853.

David, A., P. Ray, B. V. Govind and Rajagopal, 1969 a. Limnology and fisheries of the Tungabhadra reservoir, Bull. No. 13. *Central Inland Fisheries Research Institute*, Barrackpore, Nov. 1969, pp 188.

David, A., N. G. S. Rao and M. F. Rahman, 1969 b. Preliminary survey report on hydrobiology and fisheries of Nagarjunasagar, Andhra Pradesh, Survey Report, 5, CIFRI, Barrackpore, pp 36.

Desai, S. S., 1980. Fisheries of Nathsagar Reservoir. *India Today & Tomorrow,* **8** (4): 181-161.

Desai, V. R., 1992. Fishery of Sukta reservoir (Dist. Khandwa, Madhya Pradesh) in the context of Mahseer fish. *Punjab Fish. Bull.*, **14** (1): 37-43.

Desai, V. R., 1993. Fisheries of Rihand dam- progression and retrogression. *In* (Ed. R. K. Jain) Advances in fisheries and Fish Production, *Hindustan Publishing Co.*, Delhi (In press).

Devaraj, M., 1973. Biology of the large snake head *Ophiocephalus marulius* (Ham) in Bhavanisagar reservoir. *Indian J. Fish.*, **20**(2): 280-307.

Devaraj, K. V., P. Gopalagowda, N. M. Krishnamurthy, A. R. Bala Ramagowda, H. S. Mahadeva and A. A. Fazal, 1987. Fish and fisheries of the Hemavathy reservoir. *University of Agricultural Sciences*, Bangalore, pp 83.

Dorairaj, K. H. and N. Pankajam, 1956. Limnology of Bhavanisagar. *Fish. Sta. Report, 1954-55,* Department of Fisheries, Madras, pp 323-345.

Drewes E., 1994. Reservoir fisheries management by cooperatives- Issues and scope. *Indo-German Reservoir Fisheries Development Project* PN: 88.2134 .0-01.200 Malampuzha, Palakkad district, Kerala.

Dua, A., 1993. Fisheries in Gobindsagar. *Fishing Chimes,* **13** (9): 53-54.

Dubey, G. P. and R. K. Mehra, 1959. Fish and fisheries of Chambal river. *Proc. All India Congr. Zoo.*, **1**(2): 647-665.

Dubey, G. P. and S. N. Chatterjee, 1976. Case study of Gandhisagar Reservoir, Madhya Pradesh India. *Indo- Pacific Fisheries Council/76/Sym/2/Food and Agriculture Organisation, Rome,* pp. 232-245.

Dubey, G. P., 1994. Current status and scope of fisheries management of major reservoirs in Narmada basin in Madhya Pradesh. *J. Inland Fish. Soc. India.* (In press).

Durve, V. S., 1976. The fisheries of the lake Jaisamund, Rajasthan. *J. Inland Fish. Soc. India,* **8**: 19-23.

Durve, V. S. and Virendra Kakkar, 1982. Catch analysis in 1974 and fishing trends in the Jaisamund Lake (Rajasthan). *Indian J. Fish.*, **29** (1 & 2) : 177-184.

Dwivedi, S. N. and S. L. Chonder, 1980. Fisheries of Keetham Lake. *India Today & Tomorrow,* **8** (4) : 182.

Dwivedi, R. K., S. J. Karamchandani and H. C. Joshi, 1986. Limnology and productivity of Kulgarhi Reservoir, M. P. *J. Inland Fish. Soc. India,* **18** (2) : 65-70.

Ehrlich, S., 1960. The production aspect of waster body typology. *Hydrobiologia,* **17**(4): 326-332.

Franklin, T., 1969. Phytoplankton of Bhavanisagar. *Madras. J. Fish.*, **5**: 112-118.

Ganapati, S. V., 1940. Geographical aspects of Red hill Lake. *J. Madras Geogr. Assoc.*, **25**(2): 152-254.

Ganapati, S. V., 1955. Diurnal variations in dissolved gases, hydrogen ion concentration and some of the important dissolved substances of biological significance in three temporary rock pools in the stream bed at Mattur Dam. *Hydrobiologia,* **7**: 285-303.

Ganapati, S. V., 1956a. Hydrobiological investigations of the Hope reservoir and of the Thambaraparani river at Papanasam, Tirunelveli district, Madras State. *Indian Geogr. J.*, **31**: 1-2.

Ganapati, S. V., 1956b. The limnology of two minor irrigation reservoirs near Madras 1. Earrakuppam reservoir. *Hydrobiologia*, 8 (3 & 4): 365-380.

Ganapati, S. V., 1957. Limnological studies of the upland waters in Madras State. *Arch Hydrobiol.*, 53: 30-61.

Ganapati, S. V. and K. H. Alikunhi, 1949. Factory effluents from the Mettur Chemical and Industrial Corporation Ltd., Mettur Dam, and their effects on the fisheries in the river Cauvey. *Proceedings of the Indian Science Congress,* Abstract No. 26, p. 26.

Ganapati, S. V. and C. H. Pathak, 1969. Primary productivity in the Sayaji Sarovar (A man-made lake) at Baroda. *Proceedings of the Seminar on the Ecology and Fisheries of Freshwater Reservoirs.* 27-29 Nov, 1969, Central Inland Fisheries Research Institute, Barrackpore, pp 37-45.

George, V. C., 1979. Fishing gear experimentation in Gobindsagar and Hirakud for improved catchability. *Lecture* delivered at the Summer Institute on Culture and Capture Fisheries of man-made lakes in India, CentralInland Fisheries Research Institute, Barrackpore, 7 July to 6 August 1979, pp 534-536.

George, N. A., A. A. Khan, O. P. Pandey and T. Joseph Mathai, 1977. Productivity and seasonal abundance of commercially important fishes of Gobindsagar Reservoir. *Fish. Technol.*, 14 (1): 7-12.

Ghosh, A. and J. P. George,1989. Studies on the abiotic factors and zooplankton in a polluted urban reservoir-Hussain Sagar, Hyderabad: Impact on water quality and embryonic development of fishes. *Indian J. Environ. Hlth.*, 31(1): 49-59.

Gopal, B., P. K. Goel, K. P. Sharma and R. K. Trivedy, 1981. Limnological study of a freshwater reservoir, Jamwa Ramgarh (Jaipur). *Hydrobiologia,* 283-294.

Gopalakrishnan, V., R. N. Pal and P. K. Chakraborti, 1966. Observations on the breeding of major carps in Tilaiya and Panchet Reservoir. *Bulletin* 9, Central Inland Fisheries Research Institute, Barrackpore, India, pp 24.

Gopinath, P. and T. N. Jayakrishnan, 1984. A study on the piscifauna of the Idukki reservoir and catchment area. *Fish. Technol.*, 21(2): 131-136.

Govind, B. V., 1963. Preliminary studies on plankton of Tungabhadra reservoir. *Indian J. Fish.,* 10(1) A: 148-58.

Govind, B. V., 1969. Planktological studies in the Tungabhadra reservoir and its comparison with other storage reservoirs in India. *Proceedings of the Seminar on the Ecology and Fisheries of Freshwater Reservoirs,* 27-29 Nov, 1969, Central Inland Fisheries Research Institute, Barrackpore, pp 72 - 98.

Govind, B. V. (MS). Cage culture of common carp and silver carp in Sankery tank in Bangalore (Karnataka), India, unpublished manuscript.

Govind, B. V., 1988. Culture of catla (*Catla catla*) in floating net cages. *Mysore J. Agric. Sci.,* 22: 515-522.

Gulbadamov, 1962. Improvement of fishing techniques in Inland reservoirs in India. *Expanded Programme of Technical Assistance, Report to the Govt. of India,* FAO Report No. 1499, pp 37.

Gupta, S. D., 1976. Macrobenthic fauna of Loni Reservoir. *J. Inland Fish. Soc. India,* 8: 49-59.

Gupta, B. P., 1979. Studies on soil characteristics of Bhavanisagar. *Lecture* delivered at the Summer Institute on Culture and Capture Fisheries of Man-made Lakes in India, Central Inland Fisheries Research Institute, Barrackpore, India, July-August 1979, pp 82 - 91.

Harikrishnan, K. & P. K. A. Azis, 1989. Ecology of the Neyyar reservoir-a Preliminary report. *Proceedings of the Kerala Science Congress,* February 1989, Cochin, Kerala State, India, pp140-150.

Hayes, F. R., 1957. On variation in bottom fauna and fish yield in relation to trophic level and lake dimensions. *J. Fish. Res. Bd. Canada,* **14**: 1-32.

Hayes, F. R. and J. E. Phillips, 1958. Lake water and sediments IV. Radio phosphorus equilibrium with mud, plants and bacteria under oxidised and reuced conditions. *Limnol. Oceanogr.,* 3: 459-475.

Hingorani, H. G., 1980. Case history of Hussainsagar Reservoir. *India Today & Tomorrow,* 8(4): 168-169.

Hingorani, H. G., K. Madhusudhana Rao and R. Ramachandran, 1977. An attempt to study pollution and fish mortality in Hussain Sagar Reservoir in Hyderabad. *Journal of the Indian Fisheries Association,* **3 & 4** (1 & 2).

Hora, S. L., 1949. Fish fauna of Rihand river. *J. Zool. Soc. India,* **1**: 1-17.

Inaba, D. and M. Nomura, 1956. On the digestive system and feeding habits of young Chinese carps collected in the Tone river. *J. Tokyo Univ. Fish.,* **42**(1): 17-24.

Iyer, R., L. Gulati, R. Nair and W. D. Hartmann, 1993. Socio-economic baseline survey of five reservoir fishing communities. *Indo-German Reservoir Fisheries Development Project* Malampuzha, Palakkad District, Kerala, PN 88.2134 .0-01.200.

Jayachandran, P. and S. Paul Raj, 1975. Mercury in fish from Mettur reservoir. *Curr. Sci.,* **44**(22): 828-829.

Jha, B. C., 1979. Qualitative composition and seasonal abundance of periphyton in Getalsud reservoir. *In* Proceeding of the Summer Institute on capture and culture fisheries of man made lakes in India, 7 July-6 August, 1979. CIFRI, Barrackpore, India, 120-125.

Jhingran, A. G., 1988. Reservoir fisheries management in India. *Bulletin* **45**, Central Inland Capture Fisheries Research Institute, Barrackpore, India, pp 68.

Jhingran, A. G., 1989a. Role of exotic fishes in capture fishery waters of India. *In* (eds. Jhingran, A. G. and V. V. Sugunan) Conservation and management of inland capture fisheries resources of India. *Inland Fisheries Society of India,* CIFRI, Barrackpore, 743101, pp 275.

Jhingran, A. G., 1989b. Limnology and production biology of two man-made lakes in Rajasthan (India) with management strategies for their fish yield optimisation. *Final Report, IDA Fisheries Management in Rajasthan,* Central Inland Capture Fisheries Research Institute, Barrackpore, India, 1-63.

Jhingran, A. G.,1990. Recent advances in the reservoir fisheries management in India. *In* Sena De Silva (Ed.), Reservoir fisheries of Asia, Proceedings of the 2nd Asian reservoir fisheries workshop held in Hangzhou Peoples Republic of China 15-19 Oct., 1990.

Jhingran, A. G., 1991. Performance of tilapia in Indian waters and its possible impact on the native ichthyofauna. IPFC Proceedings, Bogor, Indonesia, 24-29 June, 1991. *FAO Fisheries Report No. 458,* FAO Rome, 143-161.

Jhingran, V. G. and A. V. Natarajan, 1969. An assessment of fisheries of the DVC reservoirs in relation to stocking. *In* Proceeding of the Seminar on the Ecology and Fisheries of freshwater reservoirs 27-29 November, 1969, Central Inland Fisheries Research Institute, Barrackpore, India : 429-456.

Jhingran, V. G. and A. V. Natarajan, 1978. Recommendations for stocking silver carp in Gobindsagar (H. P.) and Nagarjunasagar (A. P.) together with an account of scope and limitations of silver carp stocking in rivers and reservoirs of India. *Bulletin No. 28,* Central Inland Fisheries Research Institute, Barrackpore, India, pp 8.

Jhingran, A. G., R. K. Dwivedi, K. P. Srivastava and D. N. Singh, 1981. An ecological approach towards stocking policy formulation in Gulariya, a small irrigation impoundment. *Silver Jubilee Symp., Tropical Ecology*, Bhopal, India, 5-10 October, 1981, Abstract No. 88: p 98.

Job, T. J., A. David and M. P. Motwani, 1952. Observations on the fish and fisheries of the Damodar basin with reference to the multipurpose projects of the Valley. *J. Asiat. Soc. (Sci).*, 18 (2) : 147-172.

Job, T. J., A. David and K. N. Das, 1955. Fish and fisheries of the Mahanadi in relation to the Hirakud dam. *Indian J. Fish.*, 2 (1): 1-36.

Job, S. V. and V. Kannan, 1980. The detritus limnology of Sathiar Reservoir. *Hydrobiologia,* 72(1-2): 81-84.

Joshi, H. C., 1990. Water Pollution problems in Indian reservoirs. *In* Jhingran, Arun G. and V. K. Unnithan (eds), Reservoir fisheries in India, Proceedings of the national Workshop on Reservoir Fisheries, 3-4 January, 1990. *Special Publication 3,* Asian Fisheries Society Indian Branch, Mangalore, India, p 21-28.

Joshi, H. C. and P. K. Sukumaran 1987. The impact of effluents from Harihar Polyfibre (HPF) and Gwalior Rayon (GR) factories on the aquatic life in the river Tungabhadra near Harihar in Karnataka- A report *Bull. No. 50,* CICFRI, pp 18.

Karamchandani, S. J. and D. N. Mishra, 1980. Preliminary observations on the status of silver carp in relation to catla in the culture fishery of Kulgarhi reservoir. *J. Bombay Nat. Hist. Soc.*, 77: 261-269.

Kartha, K. N. and K. S. Rao, 1990. Experimental trawling in Gandhisagar Reservoir. *In* Jhingran, Arun G. and V. K. Unnithan (eds), Reservoir Fisheries in India. Proceedings of the National Workshop Reservoir Fisheries, 3-4 January 1990. *Spl. Publication, 3, Asian Fisheries Society, Indian Branch, Mangalore, India,* pp 144-149.

Kartha, K. N. and K. S. Rao, 1993. A study on the piscifauna of Gandhisagar Reservoir with particular reference to the abundance of commercial varieties of fish. *Fish. Technol.,* 30: 11-14.

Kaushal, D. K. and Y. Rama Rao, 1982. A note on the occurrence and biological parameters of exotic grass carp, (*Ctenopharyngodon idella*) in Gobindsagar Reservoir. *J. Inland Fish. Soc. India,* 14(1): 98-100.

Kaushal, D. K. and Y. Rama Rao, 1990. Observations on the breeding of major carps in Gobindsagar Reservoir H. P. *J. Inland Fish. Soc. India,* 22(1 & 2): 96-97.

Kaushal, D. K., M. D. Pisolkar and Y. Rama Rao, 1980a. A note on the food habits of silver carp, *Hypophthalmichthys molitrix* (Val.) from Gobindsagar Reservoir (H. P.). *J. Inland Fish Soc. India,* 12(1): 129-130.

Kaushal, D. K., M. D. Pisolkar and Y. Rama Rao, 1980b. Observations on the food habits of *Tor putitora* (Hamilton) from Gobindsagar Reservoir, H. P. *J. Inland Fish. Soc. India,* 12(1): 138-139.

Kaushal, D. K., H. C. Joshi and S. Kumar, (MS.). A note on limnological features of Pandoh reservoir, Himachal Pradesh (Unpublished manuscript).

Khan, A. I., 1986. A note on the limnology of the Baigul Reservoir U. P. *Indian J. Fish.,* 33 (1): 117-123.

Khan, M. A., K. P. Srivastava, D. N. Singh, R. K. Tyagi, R. K. Dwivedi and S. N. Mehrotra, 1989. Role of ecological parameters in fisheries management of a newly impounded reservoir-Bachhra reservoir. *In* Jhingran, A. G., V. K. Unnithan and Amitabh Ghosh (Eds), Contribution to the management of fisheries in inland open water systems of India, *Inland Fisheries Society of India*, Barrackpore, pp 100-108.

Khan, M. A., R. K. Dwivedi, S. N. Mehrotra, K. P. Srivastava, R. K. Tyagi and P. K. Katiha, 1990. Stocking as a management tool in optimising fish yield from a small reservoir (Bachhra) in Ganga basin. *In* Jhingran, A. G., V. K. Unnithan and Amitabh Ghosh (Eds), Contribution to the management of fisheries in inland open water systems of India, *Inland Fisheries Society of India*, Barrackpore, 71-76.

Khan, A. A., Percy Dawson, J. Sita Rama Rao and M. D. Varghese, 1992. Fishing in impoundment waters- A case study of Hirakud Reservoir-Orissa, *CIFT, Matsyapuri* P. O. Cochin-682029, pp 20.

Khatri, T. C., 1984a. Seasonal variation in the ecosystem of the Lakhotia lake in Rajasthan. *Indian J. Fish.,* **31**(1): 122-129.

Khatri, T. C., 1984b. Distribution of some micronutrients in the Lakhotia lake, Pali (Rajasthan). *Indian J. Fish.,* **31**(3): 400-403.

Khatri, T. C., 1985. A note on the limnological characters of the Idukki reservoir. *Indian J. Fish.,* **32** (2): 267-269.

Khatri, T. C., 1987a. Seasonal distribution of phytoplankton in Idukki Reservoir of Kerala, India. *Environment & Ecology,* **5**(1): 71-74.

Khatri, T. C., 1987b. Nutrient level and its influence on phytoplankton population in Lakhotia lake of Pali (Rajasthan). *J. Inland Fish. Soc. India,* **19**(2): 37-46.

Khatri, T. C., 1988. Seasonal distribution of zooplankton in the Idukki Reservoir of Kerala, India. *Environment & Ecology,* **6**(1): 241-243.

Krishnamoorthy, K. N., 1966. Preliminary studies on the bottom macrofacuna of Tungabhadra reservoir. *Proc. Indian. Acad. Sci.,* **2**(B): 96-103.

Krishnamoorthy, K. N., 1979. Fish production trends in certain catfish dominated reservoirs- A case study Tungabhadra reservoir. *Lecture* delivered in the Summer Institute on culture and capture fisheries of the man-made lakes in India, Central Inland Fisheries Research Institute, Barrackpore, India, pp. 366-373.

Kulshrestha, S. K., M. P. George, Rashmi Saxena, Malini Johri and Manish Shrivastava, 1992. Seasonal variations in the limnochemical characteristics of Mansarovar reservoir of Bhopal. *In* S. R. Mishra and D. N. Saksena (Eds), Aquatic Ecology Ashish Publishing House, New Delhi, pp 275-292.

Kumariah, P. and S. Parameswaran 1985. Design of a circular cage for fish culture. *J. Inland Fish. Soc. India*, **17** (1 & 2) : 78-80.

Kumaraiah, P., S. Parameswaran and P. K. Sukumaran, 1986. Culture of tilapia *Oreochromis mossambicus* (Peters) in cages. *Proc. Nat. Symp. Fish and Env.*, 145-147 (Eds), B. D. Joshi, A. K. Chopra and Dinesh Bhat. Dept. of Zoology, Gurukula Kangri Viswavidyalay, Hardwar, India.

Kumaraiah, P., S. Parameswaran and N. M. Chakraborty, 1991. New perceptions in cage fish culture with reference to the growth and production of silver carp (*Hypophthalmichthys molitrix* Val.) in cages. *Proc. Nat. Symp. Freshwater Aquaculture,* CIFA, Bhubaneswar, 246-249.

Kunhi, K. V. M., 1990. Limnological studies on man-made lake-(Kuttiadi lake, North Kerala). Dissertation for *M. Phil.* Submitted to the Aligarh Muslim University Aligarh, Uttar Pradesh, India.

Lamarque, P., Y. Therezien and N. Charlon, 1975. Etude des conditions de pieche a lelectricite dans les eaux tropicales. 1 Estudes conduites a Madagascar et ev. Zambie (1972). *Bull. Cent. Etud. Rech. Sci. Biarritz* **10**(3): 403-554, **10**(4):475-665.

Lonkar, R. L., 1992. Case studies on selected reservoirs of Godavari basin with emphasis to reinforce their fishery. Dissertation submitted to the Central Institute of Fisheries Education Extension Training Centre, Kakinada 533007, A. P. India, pp 75.

Lu Xiangke, 1986. A review of reservoir fisheries in China. *FAO Fisheries Circular*, No. 803, 37 p.

Madeley, J. W., 1914. The Madras city water supply., published by the *Madras Corporation.*

Mahajan, C. L., H. S. Narayana, B. V. Ratnam and B. L. Jain, 1984. Final report on pilot project on reservoir fisheries management in Rajasthan, 112 p.

Mathew, P. M., 1975. Limnology and productivity of Govindgarh Lake, Rewa, Madhya Pradesh. *J. Inland Fish. Soc. India,* 7: 16-24.

Mathew, P. M., 1977. Primary productivity of Govindgarh lake.. *Indian J. Anim. Sci.*, 47(6): 349-356.

Mathew, P. M., 1985. Seasonal trends in the fluctuations of plankton and physico-chemical factors in a tropical lake (Govindgarh lake, M. P.) and their interrelationships. *J. Inland Fish. Soc. India,* 17(1&2): 11-24.

Mathew, P. M. and M. V. Mohan, 1990. Reservoir fisheries in Kerala- Status and scope for development. *In* Jhingran, Arun G. and V. K. Unnithan (Eds). Reservoir Fisheries in India. Proceedings of the National Workshop Reservoir Fisheries, 3-4 January, 1990. spl. publication 3, Asian Fisheries Society, Indian Branch, Mangalore, India, p 111-117.

Mehrotra, S. N. and A. G. Jhingran, 1986. Evaluation of limnological characteristics and trophic status of Gulariya Reservoir. *Proc. Nat. Acad. Sci. India,* LVI (B) part III: 204-209.

Menon, K. N., 1973. Development of fisheries in Ukai Reservoir, Gujarat. *Seafood Export J.*, 5(1): 57-61.

Menon, M. D. and S. T. Chari, 1959. Annual Report of the Freshwater Biological Station Bhavanisagar for the year 1955-56. *Government of Madras, Fisheries Station Reports and Year Book,* 20-44.

Mod, R. G. and P. K. R. Panicker, 1985. Fisheries of Vallabhasagar (Ukai Reservoir) in Gujarat: A boon to the displaced tribals. *Fishing chimes,* 4(12) : 37-41.

Motwani, M., 1970. Report on fish and fisheries of Rihand reservoir. Dept. of Fisheries, Uttar Pradesh, Lucknow, India.

Moyle, P. B. 1976. Fish Introductions in California: History and impact on native fishes. *Biol. Conserv.*, 9(2): 101-118.

Mukherjee, A. B., 1990. Wave dissipating structures for protection of floating net cages in reservoirs. *In* Jhingran, Arun G. and V. K. Unnithan (eds), Reservoir fisheries in India, Proceedings of the national Workshop on Reservoir Fisheries, 3-4 January, 1990. *Special Publication 3,* Asian Fisheries Society Indian Branch, Mangalore, India, pp 33-38.

Mukherjee, Ashok, Gautam Mukherjee and R. K. Banerjee, 1986. Physico-chemical character of Panchet reservoir in the context of industrial discharge-A preliminary study. *J. Inland Fish. Soc. India,* 18(1): 45-51.

Murugesan, V. K. and P. Kumaraiah 1978. Fourth Workshop Report of All India Coordinated Project on Airbreathing Fish Culture, CIFRI, Barrackpore 12-13 December, 1978: 1-16.

Murthy, H. S., S. S. Gonda and V. Ayyar, 1986. Fisheries of Kabini Reservoir in Karnataka. *Fishing Chimes,* 6(7): 36-38.

Nair Achuthan, V., 1973. Fish culture in the reservoirs of Kerala. *Seafood Export J.*, 5(5): 19-22.

Nair, N. B., 1986. River ecology in relation to man-made changes in the Western Ghats. *Final Technical Report* submitted to the Department of Environment and Forests, Govt. of India, New Delhi.

Nair, P. G., 1988. Length-weight relationship of *Tilapia mossambicus* of Idukki reservoir. *Fish. Technol.,* **25**: 18-20.

Nair, A. Kesavan, V. C. George, A. A. Khan and M. D. Varghese, 1981. Estimation of fish production from Hirakud reservoir. *Fish. Techno.* **18**: pp 17-23.

Natarajan, V., 1971a. Biology and fishery of *Labeo calbasu* (H) in Bhavanisagar reservoir. *Madras J. Fish.,* **6**:14-56.

Natarajan, V., 1971b. Report on spawn prospecting investigations in the Cauvery river. Breeding of fish upstream of Bhavanisagar in Mettur reservoir. *Madras J. Fish.*, **6**: 57-76.

Natarajan, A. V. 1976. Ecology and state of fishery development in some of the man-made reservoirs in India. *Proc. Indo-Pacif. Fish. Counc., Colombo.,* 7th Session 27-29 Oct. 1976: 258-267.

Natarajan, A. V., 1979a. Ecosystem oriented approach for reservoir fisheries development in India. *In* Golden Jubilee Souvenir, in commemoration of ICAR Golden Jubilee Year, 1979, Central Inland Fisheries Research Institute, Barrackpore, part I, p 107-111.

Natarajan, A. V., 1979b. Planning strategies for development of reservoir fisheries in India. *Madras J. Fish.,* **8**: 69-77.

Natarajan, A. V., 1988. Ecological and aquacultural roles of exotic fishes in aquatic productivity in India-a risk benefit analysis. *In* Mohan Joseph, M (Ed.). Exotic aquatic species in India. Proceeding of the workshop on Exotic Aquatic species in India 25-26 April, 1988. Special publication No. 1, Asian Fisheries Society, Indian Branch, Mangalore, India, p 11-17.

Natarajan, A. V. and Pathak V., 1983. Pattern of energy flow in freshwater tropical and subtropical impoundments. *Bull. No. 36 Central Inland Fisheries Research Institute,* Barradkpore, India : pp 37.

Natarajan, A. V., V. R. Desai and D. N. Mishra, 1976. On the natural occurrence of intergeneric catla-rohu hybrid in Rihand (U. P.) with an account of its potential role in reservoir fisheries development in India. *J. Inland Fish. Soc. India,* **8**: 83-90.

Natarajan, A. V., D. N. Mishra and N. P. Srivastava, 1977. Some new light on population ecology of the Gangetic major carp. *Matsya,* **3**: 46-59.

Natarajan, A. V., Ch. Gopalakrishnayya, G. K. Bhatnagar, M. Ramakrishnaiah, V. V. Sugunan and V Pathak, 1979. Fish yield in relation to primary production in Nagarjunasagar reservoir, Andhra Pradesh, with suggestion to enhance fish yield. *In* Proceedings of 66th Session of Indian Science Congress, Hyderabad, 3-7 January, 1979. Abstract No. 222, p 93.

Natarajan, A. V., S. D. Tripathi, Ch. Gopalakrishnayya and Mathew Abraham, 1981. Final Report- All India Coordinated Research Project on Ecology and Fisheries of Freshwater Reservoirs-Bhavanisagar. *Research Information Series (1).*, Central Inland Fisheries Research Institute, Barrackpore, 1981.

Natarajan, A. V. et al., 1982. Final Report- All India Coordinated Research Project on Ecology and Fisheries of Freshwater reservoirs- Rihand. *Research Information Series, 2,* pp 68.

Nayar Gopalan, S., 1979. Improved gear for reservoir fishing. *Madras J. Fish.,* **8**: 188-192.

Northcote, T. G. and P. A. Larkin, 1956. Indices of productivity in British Columbia lakes. *J. Fish. Res. Bd. Canada,* **13**: 515-540.

Pandian, T. J., 1987. Energy flow and present status of small irrigation reservoirs in southern Tamil Nadu. *In* S. S. De Silva (Ed.) Reservoir fishery management and development in Asia. Proceedings of Workshop held in Kathmandu 23-28 Nov., 1987. International Development Research Centre, Ottawa, Canada, P 69-73.

Pankajam, N., 1956. Limnology of Bhavanisagar reservoir part B. Phytoplankton. *Fisheries station reports and year book* 1954-55. *Madras Govt. Press,* pp. 333-345.

Parameswaran, S., 1993. Unconventional aquacultural systems- cages and pens. *In* Proceedings of the National Meet on Aqua farming systems-Practices and potentials. Central Institute of Freshwater Aquaculture, Bhubaneswar, India, 16 p.

Parameswaran, S. and V. K. Murugesan 1984. A note on an inexpensive coracele. *Indian J. Fish.,* 31 (3): 387-389.

Parameswaran, S., M. A. Khan and M. Ramakrishniah, 1969 (Published in 1978). Pre-recruitment studies in Konar reservoir, Bihar Proceedings of the Seminar on Ecology and Fisheries of Freshwater Reservoirs. Central Inland Fisheries Research Institute, Barrackpore

Parasuraman, M., 1979. Organisation and functioning of inland fisheries cooperatives in Tamil Nadu. *Madras J. Fish.,* **8**: 185-187.

Panwar, R. S., K. Chandra, D. N. Singh, R. N. Seth and D. Kapoor, 1979. Studies on pollutional effects of industrial wastes on Rihand reservoir ecosystem. *Proceedings of the symposium on environmental biology Muzaffarnagar,* 1978 (Technical Papers/Eds). S. R. Verma *et al.* Muzafarnagar, Academy of Environmental Biology: 465-479.

Pathak, S. C., 1975. Length-weight relationship condition factor and food study of *L. calbasu* (Hamilton) from Loni Reservoir (M. P.). *J. Inland Fish. Soc. India,* **7**: 58-64.

Pathak, V., 1979. Evaluation of productivity in Nagarjunasagar reservoir (Andhra Pradesh) as a function of hydrological and limnochemical parameters. *J. Inand. Fish. Soc. India,* **11**(2): 49-68.

Pathak, S. C. and A. G. Jhingran, 1977. Maturity and fecundity of *Labeo calbasu* of Loni Reservoir, Madhya Pradesh. *J. Inland Fish. Soc. India,* **9**: 72-83.

Pati, S. and B. K. Sahu, 1992. Studies on the correlation between rainfall and phytoplankton in a tropical reservoir. *J. Zool. Res.,* **5** (1 & 2) : 73-79.

Paul, S., 1990. Inland Fisheries marketing & marketing intervention. *In* Sugunan, V. V. and U. Bhawmik (eds.) Technologies for inland fisheries development. Central Inland Capture Fisheries Research Institute, Barrackpore, India, pp 219.

Paul, S. and V. V. Sugunan, 1983. Fishing In reservoirs-certain economic considerations. *In* Proceedings of the seventh workshop on Ecology and Fisheries of Freshwater reservoirs, Central Inland Fisheries Research Institute, Barrackpore, India, 9-10 March, 1983, p 148.

Paul, S. and V. V. Sugunan, 1990. Development of reservoir fisheries, areas of hopes and despair. *In* Jhingran, Arun G. and V. K. Unnithan (eds), Reservoir fisheries in India, Proceedings of the national Workshop on Reservoir Fisheries, 3-4 January, 1990. *Special Publication 3,* Asian Fisheries Society Indian Branch, Mangalore, India, p 29-32.

Prabhavathy, G. and A. Sreenivasan, 1979. Fishery biology studies of a hard water impoundment, Sathanur reservoir Tamil Nadu. *Madras J. Fish.,* **8**: 78-139.

Prahalad, A. K. and G. Seenayya, 1988. *In Situ* partitioning and biomagnification of mercury in industrially polluted Hussainsagar lake hyderbad India. *Water,* Air and Soil Pollution, **39**: 81-87.

Prasad, M. N. V., 1993. Metal pollution in Hussainsagar and Banjara lakes of Hyderabad, evolution of possible metal detoxification and tolerence mechanisms by aquatic algae an overview. *In* Mishra, P. C. and R. K. Trivedi (Eds.), *Ecology and Pollution of Indian lakes and reservoirs. Ashish Publishing House, New Delhi* pp. 347.

Raghavan, S. L., P. K. Sukumaran, B. V. Govind and I. Naija, 1977. A case of fish mortality in Byramangala reservoir. *J. Inland Fish. Soc. India,* **9**: 203-204.

Raha, K., and M. K. Sarkar, 1980. Salient features of the Gomti reservoir in Tripura. *India Today & Tomorrow,* **8**(4): 178-180.

Rajts Francois, 1993. Technical report on fry and fingerlings production for reservoir stocking. *Indo-German Reservoir Fisheries Development Project.* PN: 88.2134 .0-01.200 Malampuzha, Palakkad District, Kerala.

Rahman, M. F., 1993. Development of fisheries in Kali Nadi hydro-electric project reservoirs. *Fishing Chimes,* **13** (6): 5-6.

Rahimullah, M., 1943. Fish survey of Hyderabad State. Part II. Fishes of Hyderabad city and its suburbs. *J. Bombay Nat. Hist. Soc.,* **XLIV** (1): 88-91.

Ramakrishniah, M., 1984. Studies on the fishery and fish productivity and the biology of *Mystus aor* of Nagarjunasagar reservoir in Andhra Pradesh. *Ph. D. thesis* Andhra University Waltair, India, PP. 185.

Ramkrishniah, M., 1994. Role of reservoirs as sanctuaries for rare and endangered fishes. *Fishing Chimes,* **13**(11): 12-13.

Ramakrishniah, M. and B. K. Banerjee, 1979. A note on the breeding of carps in Tilaiya reservoir, Bihar, in the context of stocking. *J. Inland Fish. Soc. India,* **11**(1): 128-131.

Ramkakrishniah, M. and S. K. Sarkar, 1982. Plankton productivity in relation to certain hydrological factors in Konar reservoir (Bihar). *J. Inland Fish. Soc. India,* **14**(1): 58-68.

Ranganathan, V. and N. V. Radha, 1966. Biology and fisheries of *Mystus (Osteobagrus) aor* (Sykes) in the Bhavanisagar reservoir. *Madras J. Fish.,* **2**: 11-14.

Ranganathan, V. and E. R. Venkataswamy, 1966. On the efficiency of the Mettur Rangoon nets in the Bhavanisagar reservoir. *Madras J. Fish.,* **3**: 46-60.

Ranganathan, V., V. R. Menon and N. V. Radha, 1962. Biology and fisheries of *Barbus dubius* in Bhavanisagar reservoir. *Madras J. Fish.,* **I** : 1-24.

Rao, K. L., 1979. India's water wealth. Orient Longman Ltd., New Delhi, pp 267.

Rao, M. B., 1990. Effect of eutrophication on the biology of fishes in reservoir-A case study at Hussainsagar. *In* Jhingran, Arun G. and V. K. Unnithan (eds.), Reservoir Fisheries in India. Proceedings of the National Workshop on Reserver Fisheries, 3-4 January 1990, *Special Publication 3, Asian Fisheries Society,* Indian Branch, Mangalore, India.

Rao, J. B. and R. K. Dwivedi, 1972. On the recovery and growth of silver carp (*Hypophthalmichthys molitrix*) from Kulgarhi reservoir (M. P.). *J. Inland Fish. Soc. India,* **4**: 214-215.

Rao, N. G. S. and K. Gopinathan, 1972. Observation on the spawning of *Cirrhina reba* in the Cauvery and Bhavani rivers. *J. Inland Fish. Soc.,* **4**: 1972. 69-73.

Rao, K. S. and U. Choube, 1990. Studies on phytoplankton dynamics and productivity fluctuations in Gandhisagar reservoir *In* Jhingran Arun G. and V. K. Unnithan (eds.). Reservoir Fisheies in India. Proceedings of the National Workshop on reservoir fisheries, 3-4 January 1990, *Special publication 3, Asian Fisheries Society, Indian Branch,* Manglore India, p 106-112.

Rao, K. S., K. N. Kartha, D. R. Gupta, S. S. Pandya and H. K. Iyer, 1988. Studies on morphometry and hydrology of Gandhisagar reservoir with special reference to its fisheries. *Fish. Technol,* **25**(1): 21-28.

Rao, K. S., K. N. Kartha, D. R. Gupta, S. Shrivastava, Usha Chaubey and S. S. Pandya,1990. Studies on morphometric, hydrological and sedimental characteristics of Gandhisagar lake in relation to its fisheries. *J. Inland Fish. Soc. India,* **22**(1): 55-65.

Rawson, D. S., 1952. Mean depth and the fish production of large lakes. *Ecology,* **33**: 513-521.

Ray, P., 1969. Hydrology of Vanivalas Sagar. *Proceedings of the Seminar on the Ecology and Fisheries of Freshwater Reservoirs,* 27-29 Nov, 1969, Central Inland Fisheries Research Institute, Barrackpore: pp. 93-98.

Sahib, S. S., P. K. A. Azis, 1989. Post impoundment water quality of the Kallada river-a preliminary report. *Proceedings of the Kerala Science Congress,* February 1989, Cochin, 153-160.

Salim, M. and Z. Ahmed, 1985. Environmental factors and planktonic communities of Baigul and Nanaksagar Reservoirs, Nainital. *J. Bombay Nat. Hist. Soc.,* **82**: 13-23.

Sarkar, S. K., 1982. Observations on some limnological features of Nalkari reservoir in Bihar. *Proc. Nat. Acad. Sci. India,* **52**(B), III: 259-262.

Sarkar, S. K., B. V. Govind and A. V. Natarajan, 1977. A note on some distinctive limnological features of Gobindsagar reservoir, *Indian J. Anim. Sci.,* **47**(7): 435-439.

Sarma, B. C., 1990. Keynote address. National Workshop on Reservoir Fisheries, 3-4 January, 1990. Central Inland Capture Fisheries Research Institute, Barrackpore, India.

Saxena, R., 1986. Strategy for development of fisheries in U. P. *Fishing Chimes,* **6**(9): 16-17.

Schack, U., 1993. Economic evaluation of fingerling production experiments. *Indo-German Reservoir Fisheries Development Project.* PN: 88.2134 .0-01.200 Malampuzha, Palakkad District, Kerala.

Seenayya, G. and A. K. Prahalad, 1987. *In situ* compartmentalisation and biomagnification of chromium and manganese in industrially polluted Hussainsagar lake, Hyderabad India. *Water Soil and Air Pollution,* **35**: 233-239.

Selvaraj, C. and V. K. Murugesan, 1990. Management techniques adopted for achieving a record fish yield from Aliyar reservoir Tamil Nadu. *In* Jhingran, Arun G. and V. K. Unnithan (eds.), Reservoir Fisheries in India. Proceedings of the National Workshop on Reserver Fisheries, 3-4 January 1990, *Special Publication 3, Asian Fisheries Society,* Indian Branch, Mangalore, India. : 86-96.

Sharma, L. L., 1980. Some limnological aspects of Udaipur waters in comparison to selected waters of Rajasthan. *Ph. D. thesis* submitted to the University of Udaipur, New Campus Udaipur-313001, India., pp 251.

Sharma, R. and A. P. Diwan, 1989. Limnological studies on Yeshwantnagar reservoir: an assessment of its potential for fish culture. *J. Hydrobiol,* **5** (1): 67-69.

Sidharthan, T. and C. Mohana Kumaran Nair,1988. Fish seed production in Kerala-Demand and supply. *In* Proceedings Seminar on Problems and prospects of freshwater aquaculture in Kerala, Kerala Agricultural University College of Fisheries, Cochin.

Singh, R. K., 1986. Fluctuations in the composition of zooplankton population in relation to hydrobiological conditions of a reservoir. *J. Hydrobiol*, **2**(3): 37-42.

Singh, R. K. and V. R. Desai, 1980. Limnological observation on Rihand reservoir III-Primary productivity. *J. Inland Fish. Soc. India*, **12**(2): 63-68.

Singh, R. K., N. P. Shrivastava and V. R. Desai, 1980. Seasonal and diurnal variations in physico-chemical conditions of water and plankton in lotic sector of Rihand reservoir (U. P.). *J. Inland Fish. Soc. India*, **12**(1): 100-111.

Singit, G. S., 1987. Alvinet fishing unique to Tungabhadra Reservoir. *Fishing Chimes*, **7**(4): 49-55.

Singit, G. S., P. Ramamohan Reddy and H. V. Krishnamaraju, 1985. Carp fry rearing in cages. *Fishing Chimes*, **5** (5): 50-55.

Singit, G. S., P. Ramamohan Reddy and H. V. Krishnamaraju, 1987. Post impoundment changes and effects of conflicting uses on the fisheries of Tungabhadra reservoir, India. *In* S. S. De silva (Ed.) Reservoir fishery management and development in Asia, Proceedings of Workshop held in Kathmandu., 23-28 No. 1987. International Development Research Centre, Ottawa Canada, P. 136-144.

Sinha, M., 1972. Observations on the biology *Puntius sarana* of Loni reservoir (M. P.)-1. Length-weight relationship, food and condition factor. *J. Inland Fish. Soc. India*, **4**: 122-131.

Sinha, M., 1976. Observations on the biology of *Puntius sarana* (Hamilton) of Loni Reservoir (M. P.)- 2. Maturity and fecundity. *J. Inland Fish. Soc. India*, **7**: 49-57.

Sinha, M., 1990. Reservoir fisheries- its present status and future potentials in North Eastern region. *In* Jhingran, A. G. and V. K. Unnithan (eds) Reservoir fisheries in India. Proceedings of the National workshop on reservoir fisheries 3-4 January 1990, *Special publication 3, Asian Fisheries Society Indian Branch*, Mangalore, India, p. 57-64.

Sinha, M. P., 1986. Limnobiotic study on trophic status of polluted freshwater reservoir of coalfield area. *Poll. Res.* **15**: 13-17.

Sounder Raj, R., T. Franklin and A. Sreenivasan, 1971. Limnology and productivity of Poondi reservoir. *Madras J. Fish.*, **6**: 101-112.

Spencer, P. H. N., 1964. The macrophytic vegetation on freshwater lakes, swamps and associated form. *In* The vegetation of Scotland J. H. Burnall (Ed.) *Oliver and Boyd.*, Edinburgh and London.

Sreenivasan, A., 1962. A note on the hydrology of Bhavanisagar for 1956-61. *Madras J. Fish.*, **1**: 70-71.

Sreenivasan, A., 1964a. A hydrological study of a tropical impoundment, Bhavanisagar 1956-61. *Hydrobiologia*, **24**(4): 514-539.

Sreenivasan, A., 1964b. Limnological studies and fish yield in three upland lakes of Madras State. *Limnol. Oceanogr.*, **8**(4): 501-516.

Sreenivasan, A., 1965a. Limnology of tropical impoundments III Limnology and productivity of Amaravathy reservoir. *Hydrobiologia*, **26**(3): 501-516.

Sreenivasan, A., 1965b. An instance of unusual oxygen production in a tropical impoundment. *J. Mar. Biol. Assoc. India*, **7**(2): 469-71.

Sreenivasan, A., 1966. Limnology of tropical impoundments I Hydrobiological and fish production in Stanley reservoir, Mettur Dam. *Inst. Rev. Ges. Hydrobiol.*, **51**(2): 295-306.

Sreenivasan, A., 1967. *Tilapia mossambica*: its ecology and status in Madras State, India. *Madras J. Fish.*, **3** : 33-39.

Sreenivasan, A., 1968. Limnology of tropical impoundments V. Studies of two upland impoundments in Nilgiris, Madras State, *Phykos.*, **7**(1 & 2): 144-160.

Sreenivasan, A., 1969. The hydrobiological features, primary production and fisheries of Stanley reservoir, Mettur-dam for the years 1958-1965. *Madras J. Fish.*, **5**: 1-18.

Sreenivasan, A., 1970a. Limnology of tropical impoundments. A comparitive study of major reservoirs in Madras state. *Hydrobiologia,* **36**(3 & 4): 443-460.

Sreenivasan, A., 1970b. Eutrophication trends in a chain of artificial lakes in Madras. *Env. Health,* **2**: 392-401.

Sreenivasan, A., 1970c. Limnological studies on the Parambikulam-Aliyar Project 1. Aliyar reservoir (Madras State) India. Schweiz. Zcitach. *Hydrol.*, **32**: 405-417.

Sreenivasan, A., 1971. Liming of an upland lake. *Madras J. Fish.*, **6**:9-13.

Sreenivasan, A., 1976. Fish production and fish population changes in some South Indian reservoirs. *Indian J. Fish.*, **23** (1 & 2): 133-152.

Sreenivasan, A., 1979a. Status of reservoir fisheries in Tamil Nadu. *Madras J. Fish.*, **8**: 60-68.

Sreenivasan, A., 1979b. Problem of pollution of natural waters in Tamil Nadu. *Madras J. Fish.*, **8**: 172-175.

Sreenivasan, A., 1979c. Limnology and fisheries of Amaravathy reservoir. *Lecture* delivered in the Summer Institute on culture and capture fisheries of the man-made lakes in India, Central Inland Fisheries Research Institute, Barrackpore, India, 315-322.

Sreenivasan, A., 1984. Influence of stocking on fish production in reservoirs in India. *Fishing Chimes*, **4** (1): 54-70.

Sreenivasan, A., 1993. Reservoir fisheries of India-Present status. *Fishing Chimes*, **13** (1): 18-21.

Sreenivasan, A. and K. Venkatanarasimha Pillai 1979. Fertilization of fishery waters - Experimental fertilization of a small reservoir. *Madras J. Fish.*, **8**: 143-145.

Sreenivasan, A., R. Soundar Raj and Felicy Antony, 1964. Limnological studies of tropical impoundments II Hydrobiological features and plankton of Bhavanisagar reservoir for 1961-62. Proc. *Indian Acad. Sci.*, **59**(2): 51-71.

Srinivasan, T. K., K. Srinivasan, G. K. Seth, P. Murahari Rao and D. Venkata Rao, 1965. Studies on raw water characteristics of the lakes in and around Hyderabad. *Environ. Hlth.*, 7(3): 177-187.

Srivastava, U. K., D. K. Desai, V. K. Gupta, S. S. Rao, G. S. Gupta, M. Raghavachari and S. Vatsala, 1985. Inland fish marketing in India-Reservoir fisheries, vol. 4(A&B), *Concept Publishing Co.*, New Delhi, pp (A) 403 & (B) 1184.

Subba Rao, D. and B. V. Govind, 1964. Hydrology of Tungabhadra reservoir. *Indian J. Fish.*, **11**(1): 321-344.

Sugunan, V. V., 1980. Seasonal fluctuations of plankton of Nagarjunasagar reservoir, A. P., India. *J. Inland Fish. Soc. India*, **12**(1): 79-91.

Sugunan, V. V., 1991. Changes in the phytoplankton species diversity indices due to artificial impoundment in river Krishna at Nagarjunasagar. *J. Inland Fish. Soc. India*, **23**(1): 64-74.

Sugunan, V. V. and R. K. Das, 1983. Studies on the bottom macro-fauna of Nagarjunasagar reservoir, A. P., India. *J. Inland Fish. Soc. India,* 15(1&2): 1-12.

Sugunan, V. V. and V. Pathak, 1986. Temporal and spatial distribution of periphyton in Nagarjunasagar reservoir, Andhra Pradesh (India) and a new method for periphyton collection from Indian reservoirs. *J. Inland Fish. Soc. India,* 18(2): 20-33.

Sugunan, V. V. and G. K. Vinci, 1981. Length-weight relationship and food study of *Rhinomugil corsula* (Hamilton) with a note on its spawning and fecundity from Nagarjunasagar reservoir, A. P., India. *J. Inland Fish. Soc. India,* 13(1): 26-35.

Sugunan, V. V. and Y. S. Yadava, 1991a. Feasibility studies for inland fisheries development of Kyrdemkulai reservoir. Central Inland Capture Fisheries Research Institute, Barrackpore, pp 34.

Sugunan, V. V., Y. S. Yadava, 1991b. Feasibility studies for fisheries development of Nongmahir reservoir. Central Inland Capture Fisheries Research Institute, Barrackpore, pp 30.

Sulochanan, P., George, V. C. and Naidu R. M., 1968. Experimental fishing in Hirakud reservoir. *Fish. Technol.,* 5(2): 81-96.

Sukumaran, K. K., S. B. Singh, D. S. Murty and P. C. Chakraborty, 1968. Studies on compatibility and competition between silver carp, *Hypophthalmichthys molitrix* (val.) and catla, *Catla catla*.(Ham.). *Proc. Indo-Pacif. Fish. Coun.,* 13(2): 185-194.

Taege, M., D. M. Peters and K. V. M. Kunhi, 1993. Fish stocks and fish yields of the Malampuzha reservoir, Kerala, India. Malampuzha, *Indo-German Reservoir Fisheries Development Project,* Malampuzha, Palakkad District, Kerala, pp 45.

Tripathy, R. K. (MS), 1990. A case study-Fishery in Vallabhasagar reservoir, Ukai. *National Workshop on Reservoir Fisheries,* 3-4 January, 1990, Central Inland Capture fisheries Research Institute, Barrackpore, West Bengal, India, pp 8.

Trivedy, R. K., 1993. Water quality of Dhom reservoir Maharashtra, India. *In* Mishra P. C. and R. K. Trivedi (Eds.), Ecology and pollution of Indian lakes and reservoirs, *Ashish Publishing House,* New Delhi, pp. 347.

Unni, K. S., 1982. Limnological studies and Sampma reservoir, Betul, M. P. Proc. *Nat. Acad. Sci. India,* 52(B) 4: 365-372.

Unni, K. S. 1993. Limnology of central Indian lakes, reservoirs and ponds-a review. *In* Mishra, P. C. and R. K. Trivedi (Eds.). Ecology and pollution of Indian lakes and reservoirs. *Ashish Publishing House,* New Delhi, pp. 237-267.

Valsangkar, S. V., 1980. Economic rehabilitation of fishermen in Yeldari reservoir. *India Today & Tomorrow,* 8(4): 162-163.

Valsangkar, S. V., 1987. Common carp-an asset to Girna reservoir. *Fishing Chimes,* 7(5):: 29-39.

Valsangkar, S. V., 1993. Mahseer in Bhandardara Reservoir resilient. *Fishing Chimes,* 13(6): 43-46.

Varghese, M. D., A. A. Khan and V. C. George, 1993. Activity periods of certain fishes in Hirakud Reservoir. *Fish. Technol.,* 30: 9-10.

Venugopal, R., 1993. Status and development of reservoir fisheries in Kerala. Project Report Submitted to the University of Kerala for the Degree of MBA T. K. M. College of Engineering, Kollam-5, Kerala, pp. 87.

Verma P. K. and J. Datta Munshi, 1983. Limnology of Badua reservoir of Bhagalpur, Bihar. *Proc. Indian Nat. Sci. Acad.*, B **49**(6): 598-609.

Vinci, G. K., 1984. Some aspects of the biology of *Silonia childreni* (Sykes) from Nagarjunasagar reservoir, Andhra Pradesh, India. *J. Inland Fish. Soc. India*, **16**(1&2): 25-31.

Vinci, G. K., 1986. Biology of *Mystus seenghala* (Sykes) from Nagarjunasagar reservoir, Andhra Pradesh. *Indian J. Anim. Sci.*, **56**(7): 814-821.

Vinci, G. K. and V. V. Sugunan, 1981. Biology of *Labeo calbasu* (Hamilton) of Nagarjunasagar reservoir, A. P., India. *J. Inland Fish. Soc. India*, **13**(2): 22-39.

Wilson, 1920. On the methods of capture and supply of fish in the rivers of Nilgiri district. *Report No. 4 Bulletin 12.* Madras Fisheries Department (As quoted by Ranganathan *et al.*,1962).

Wishard, S. K. and S. N. Mehrotra, 1988. Periodicity and abundance of plankton in Gulariya reservoir in relation to certain physico-chemical conditions. *J. Inland Fish. Soc. India*, **20**(1): 42-47.

Yadava, Y. S. and Sugunan, V. V., 1992. Hirakud reservoir - Strategies for fisheries development. *Bulletin* **66**. Central Inland Capture Fisheries Research Institute, Barrackpore, India pp. 18.

Yang, Q., Z. Zhao, H. Guo and T. Yu, 1990. Study of comprehensive techniques of fish culture in Shishantou reservoir. Water Resources Department of Jiangsu Province, Nanjing Agriculture University, pp 11.

Zafar, A. R., 1966. Limnology of the Hussainsagar Lake, Hyderabad, India. *Phykos*, 5(1 & 2): 115-126.

Zafar, A. R., 1986. Seasonality of phytoplankton in some South Indian lakes. Hydrobiologia, 138: 177-188.

****

GLOSSARY

@	At the rate of
Alivi, Mahajal	A giant shore seine
Catla	*Catla catla*
Chocolate mahseer	*Acrossocheilus hexagonolepis*
CICFRI	Central Inland Capture Fisheries Research Institute
CIFRI	Central Inland Fisheries Research Institute
Coracle	A saucer shaped fishing craft
Common carp	*Cyprinus carpio*
CIFT	Central Institute of Fisheries Technology
FRL	Full reservoir level
DSL	Dead storage level
MSL	Mean sea level
Grass carp	*Ctenopharyngodon idella*
Mrigal	*Cirrhinus mrigala*
Mahseer	*Tor tor, T. putitora, Tor* spp.
Murrels	*Channa* spp.
NABARD	National Bank for Agriculture and Rural Development
NCDC	National Cooperative Development Corporation
Rohu	*Labeo rohita*
Scheduled castes	Socially backward castes, whose rights are protected constitutionally
Scheduled tribes	Socially backward tribes, whose rights are protected constitutionally
Tilapia	*Oreochromis mossambicus*
Silver carp	*Hypophthalmichthys molitrix*

****